PROCEEDINGS OF THE
AMERICAN SOCIETY FOR COMPOSITES

PROCEEDINGS OF THE AMERICAN SOCIETY FOR COMPOSITES

THIRD TECHNICAL CONFERENCE

INTEGRATED COMPOSITES TECHNOLOGY

Co-Sponsored by
POLYMERIC COMPOSITES LABORATORY
UNIVERSITY OF WASHINGTON
and
THE BOEING COMPANY

September 25–29, 1988
Stouffer-Madison Hotel
Seattle, Washington

TECHNOMIC
PUBLISHING CO., INC.
LANCASTER · BASEL

Published in the Western Hemisphere by
Technomic Publishing Company, Inc.
851 New Holland Avenue
Box 3535
Lancaster, Pennsylvania 17604 U.S.A.

Distributed in the Rest of the World by
Technomic Publishing AG

Printed in the United States of America
10 9 8 7 6 5 4 3 2 1

Main entry under title:
  Proceedings of the American Society for Composites—Third
  Technical Conference

A Technomic Publishing Company book
Bibliography: p.

Library of Congress Card No. 88-71942
ISBN No. 87762-638-3

Table of Contents

SYMPOSIUM VII—Session A
Advances in Materials and Process Engineering

SYMPOSIUM VII—Session B
Advances in Materials and Process Engineering

SYMPOSIUM VIII
Impact and Damage Tolerance

*Paper received at press time . . . included at end of text.

*Paper received at press time . . . included at end of text.

SYMPOSIUM XIV
Fatigue and Fracture

Preface

Composite material production and utilization are reaching a crucial point in their development cycle where performance of a composite cannot be simply evaluated with a one-sided perspective of a user or a manufacturer. Real integration of disciplines as well as of the individual players traditionally viewed as distinct entities in composites production must be accomplished in a relatively short time if the true potential of these material systems is to be realized without loss of resources currently committed for their development.

Appropriately, the theme of the 3rd Technical Conference of the American Society for Composites, Integrated Composites Technology, reflects this belief. Furthermore, the commitment of our respective institutions, the University of Washington's Polymeric Composites Laboratory and The Boeing Company, who are jointly co-sponsoring the conference, is an excellent example of industry/university integration of meeting future challenges. Indeed, a major challenge in this conference was the organization of the individual contributions received which ranged from biocomposites to structural design in a coherent integrated program. Accordingly, the papers were organized into fourteen different symposia which conveniently provide the subject headings for these proceedings.

The overwhelming response from prospective authors and presenters at the conference (over 100 papers submitted) demonstrates that, indeed, ASC although a young society is providing a much needed forum for integrating composite developments and applications. Furthermore, this response allowed us to select for inclusion in these proceedings from some excellent and original papers. Although this selection will become more difficult in future conferences, it is the intention of the society that its proceedings will reflect quality work that the participants want to present and publish rather than a compilation of all papers presented regardless of content, originality, and usefulness of the work. By appointing a scientific committee for the conference which consisted of the individual symposia chairmen and invited speakers, we were able to establish a peer review body that assisted in screening the different contributions. Accordingly, the collection of papers in these proceedings should be a true representation of the work that was presented at the Third Technical Conference of the American Society for Composites without including every paper that was presented. It should be emphasized, however, that of the papers presented at the conference, some are not part of these proceedings, the majority of which were because either the authors chose not to submit a manuscript or the paper was not available at press time.

In closing, we would like to express our appreciation to the organizing and scientific committees of the conference as well as to the officers and staff of the society for all their work and support.

We hope that some of the organizational arrangements, both technically and not, will provide some precedent on which future conferences can build upon. We are particularly grateful, however, to all speakers and attendees who made this conference possible.

Conference Co-Chairmen

J. C. SEFERIS
Polymeric Composites Laboratory
University of Washington
Seattle, Washington 98195

J. T. QUINLIVAN
The Boeing Company
Seattle, Washington 98195

Scientific Program

The Third Technical Conference on Composite Materials of the American Society for Composites has been specifically structured to provide an interdisciplinary forum for technical presentations in the critical engineering, scientific, and economic disciplines that are required to advance composite technology. Appropriately, the theme of the conference, Integrated Composites Technology, as well as the individual papers that have been organized into fourteen different symposia, should attract participation by an unusually broad cross-section of the different disciplines involved in composite research and development.

The program begins with a single morning symposium of invited presentations covering the spectrum of topics in the conference theme. Three parallel sessions with invited and contributed presentations from the different symposia are held each morning and afternoon of the conference. In the following detailed program, the symposia, parallel sessions, times of presentations, and author information are provided. From the time schedule, it should be noted that in the parallel sessions participants have an opportunity to follow presentations in the different symposia since session chairmen have been instructed to adhere strictly to the times allotted. Finally, it should be noted that the scheduled luncheon and banquet events are an integral part of the scientific program since prominent invited speakers at these events are expected to place composites within a national and international framework of materials and technology development.

OFFICERS OF THE SOCIETY
James M. Whitney, President
Keith T. Kedward, Vice-President
David Bonner, Recording Secretary
C. T. Sun, Membership Secretary
Dina N. Kapur, Treasurer
Som R. Soni, Editor
Charles E. Browning, Executive Committee
Norman J. Johnston, Executive Committee

ORGANIZING COMMITTEE
L. B. Ilcewicz (Boeing Commercial Airplane Co.)
J.-A. Manson (Polymeric Composites Laboratory)
T. Masuda (Polymeric Composites Laboratory)
J. O'Brien (University of Washington)
J. T. Quinlivan (Boeing Commercial Airplane Co.)
J. C. Seferis (Polymeric Composites Laboratory)
Som R. Soni (AdTech Systems Research)
J. Whitney (American Society for Composites)
E. M. Woo (Polymeric Composites Laboratory)
S. Yuh (Polymeric Composites Laboratory)

CONFERENCE CHAIRMEN
J. C. Seferis
Polymeric Composites Laboratory
University of Washington
Seattle, WASHINGTON 98195 USA

J. T. Quinlivan
Boeing Commercial Airplane Co.
Seattle, WASHINGTON 98195 USA

SCIENTIFIC COMMITTEE
Symposium Chairmen and Invited Speakers

Professor I. Aksay
University of Washington
Seattle, Washington

Dr. L. D. Bravenec
Shell Development Co.
Houston, Texas

Dr. C. Browning
Wright Patterson AFB
Dayton, Ohio

Dr. A. R. Bunsell
Ecole National Superieure
 des Mines de Paris
FRANCE

Dr. M. Carrega
Rhone-Poulenc
FRANCE

Dr. T. Cervenka
Shell Research, Ltd.
ENGLAND

Dr. C. C. Chamis
NASA Lewis Research Center
Cleveland, Ohio

Dr. F. Crossman
Lockheed Missiles and Space Co.
Palo Alto, California

Professor L. Drzal
Michigan State University
East Lansing, Michigan

Professor G. Dvorak
Rensselaer Polytechnic Institute
Troy, New York

Professor T. G. Gutowski
Massachusetts Institute of
 Technology
Cambridge, Massachusetts

Professor T. Hahn
Pennsylvania State University
University Park, Pennsylvania

Dr. J. Halpin
Wright Patterson AFB
Dayton, Ohio

Dr. L. B. Ilcewicz
Boeing Commercial Airplane Co.
Seattle, Washington

Professor J. L. Kardos
Washington University
St. Louis, Missouri

Dr. J. K. Lees
E. I. du Pont de Nemours & Co.
Wilmington, Delaware

Dr. J.-A. Manson
University of Washington
Seattle, Washington

Dr. A. G. Miller
Boeing Commercial Airplane Co.
Seattle, Washington

Dr. D. R. Moore
ICI Petrochemicals & Plastics Division
ENGLAND

Professor L. Nicolais
University of Naples
Naples, ITALY

Professor R. B. Pipes
University of Delaware
Wilmington, Delaware

Dr. J. T. Quinlivan
Boeing Commercial Airplane Co.
Seattle, Washington

Professor K. Reifsnider
Virginia Polytechnic Institute
Blacksburg, Virginia

Dr. D. Scola
United Technologies
 Research Center
East Hartford, Connecticut

Dr. R. C. Schiavone
Wright Patterson AFB
Dayton, Ohio

Professor J. C. Seferis
University of Washington
Seattle, Washington

Professor S. S. Sternstein
Rensselaer Polytechnic Institute
Troy, New York

Professor P. S. Theocaris
Athens National Technical University
Athens, GREECE

Professor S. S. Wang
University of Illinois
Urbana, Illinois

Dr. A. R. Wedgewood
E. I. du Pont de Nemours & Co.
Wilmington, Delaware

Dr. J. Whitney
Wright Patterson AFB
Dayton, Ohio

Professor R. Wilkins
University of Delaware
Wilmington, Delaware

Professor P. Zoller
University of Colorado
Boulder, Colorado

SYMPOSIUM II

Processing and Manufacturing Science

The Permeability of Aligned and Cross-Plied Fiber Beds During Processing of Continuous Fiber Composites

R. C. LAM AND J. L. KARDOS

ABSTRACT

Two of the most important input parameters needed to simulate the processing of continuous fiber laminated composites are the fiber bed permeability and the portion of the autoclave load born by the consolidating fiber network (compressibility). In this study we have experimentally examined how both these parameters change with resin volume fraction as pressure is applied and consolidation proceeds. For a unidirectional fiber bed, the Kozeny-Carman equation can be used to predict both the transverse (perpendicular to the laminate plies) permeability (Kozeny constant, $K_z' = 11$) and the axial (parallel to the fibers) permeability (Kozeny constant, $K_x' = 0.57$). The axial permeability was found to be dependent on the surface tension of the permeant. For a unidirectionally aligned fiber, the measured transverse permeabilities varied from $1.1 \times 10^{-10} \text{cm}^2$ to $1.2 \times 10^{-9} \text{ cm}^2$ while the axial values varied from $2.1 \times 10^{-9} \text{ cm}^2$ to $4.4 \times 10^{-8} \text{ cm}^2$ for a liquid volume fraction range of 0.25 to 0.5. Transverse and axial permeability measurements indicate that the permeability decreases with increasing off-axis angle α (measured from the laminate axial direction). The off-axis permeability behavior can be described by a modified Kozeny-Carman equation. The fiber network compressibility can be described with a logarithmic relation which has been found valid for a large number of consolidated soils.

INTRODUCTION

The development of a master process model to simulate the autoclave processing of continuous fiber reinforced epoxy composites requires an accurate description of resin flow through the fiber bed as the laminate consolidates. The fiber bed, through which the resin flows, is anisotropic in general because the fiber orientations may be different in each ply and the sequence in which the plies are stacked may change. The effect of fiber bed anisotropy on the bed permeability during consolidation of the bed has not yet been established.

R. C. Lam, Raybestos Products Co., 1204 Darlington Ave., Crawfordsville, IN 47933.
J. L. Kardos, Materials Research Laboratory, Box 1087, Washington University, St. Louis, MO 63130.

Models have been proposed to predict laminate consolidation during autoclave processing. Three main approaches have been reported in the literature, namely a Darcy's Law approach [1,2,3], a "squeezing" flow approach [4] and a "viscoelastic" approach [2,5-7]. Both the Darcy's Law and "viscoelastic" approaches assume that resin flow is generated by applying a pressure gradient to the liquid which flows through the fiber bed as if the bed were a porous medium [8]. Darcy's Law can be written as:

$$\underline{V} = \frac{-\underline{\underline{K}}}{\eta} \, \underline{\nabla}P \tag{1}$$

where $\underline{V}$ is the velocity of flow, $\underline{\nabla}P$ is the applied pressure gradient, η is the viscosity of the fluid, and $\underline{\underline{K}}$ is the permeability of the medium. $\underline{\underline{K}}$ is in general a symmetric tensor.

Employment of Darcy's Law requires an experimental knowledge of the effective permeabilities, K_i, as a function of the volume fraction of the resin, V_R, in the laminate [9]. In this paper, we present both the fiber bed permeability and fiber bed mechanical response as consolidation of the beds proceeds. The permeability and mechanical load response were measured experimentally for unidirectional and for alternating 0° and α° ply beds, where α is 30°, 45°, and 90°.

EXPERIMENTAL MEASUREMENTS

The permeability and the mechanical response were measured experimentally on a modified "flexible" wall permeameter as described in detail in a recent paper [9]. A permeability test as well as an incremental loading consolidation test can be run simultaneously for the same fiber specimen on this equipment. In general, the permeability test was run on the fiber sample at the end of each consolidation pressure increment by applying a fluid pressure differential across the fiber bed. The permeameter readings were recorded at suitable time intervals until consistent values were obtained for the permeability constant of the fiber bed, K'' [9]. The permeability K is related to the permeability constant K'' by:

$$K = \frac{K'' \, \eta}{\rho \, g} \tag{2}$$

where η is the viscosity of the fluid, g is the gravitational constant and ρ is the density of the fluid. The value of K depends only on the properties of the porous medium and has the dimension of length squared.

Samples were made by filament-winding tows of graphite fibers onto a square mandrel. In the case of off-axis orientation, plies of fibers were stacked up in the sequence $[0°,\alpha]_n$ where α was 30°, 45°, and 90° and n was 10 and 15.

Flow Perpendicular to the Plane of the Fibers (Vertical Flow)

Each of the aligned, off-axis laminates was aligned so that flow was perpendicular to the fibers, as shown in Figure 1. The fiber bed was then encased laterally in silicone rubber to permit only vertical motion of the fluid. Pressure was applied normal to the fibers. In order to prevent pressure from being applied to the periphery of the sample (parallel to the fibers), a metal rigid hollow block was placed around the sample and the porous disc, as shown in Figure 1.

Flow Along the Fibers (Horizontal Flow)

Each of the aligned off-axis laminates was aligned in such a way that flow was parallel to the laminate plies (perpendicular to the plane of the porous discs), as shown in Figure 2. The sample was encased in silicone rubber again to permit only vertical motion of the fluid. In this case, however, an additional membrane was placed around the platens, porous discs and silicone-encased sample in order to assure only vertical flow and at the same time allow lateral consolidation perpendicular to the laminate plies. A metal rod was fastened to the top of the upper platen and upper porous disc to prevent the top of the sample from being deformed vertically, as shown in Figure 2.

RESULTS AND DISCUSSION

The permeability-liquid fraction and effective pressure-liquid fraction data were obtained for different fiber sample thicknesses (numbers of plies) using water and silicone oil (Dow Corning 200) as permeants. These liquids were selected to give different viscosities and surface tensions; typical values for water were 1 centipose and 75 dynes/cm, and for silicone oil 5 centiposes and 21 dynes/cm at 20°C. All the experiments utilized Hercules AS4 graphite fibers.

Consolidation

The experimental data for the effective pressure on the fibers, P_f, and the volume fraction of the liquid, V_R, were fitted to a logarithmic relation which has been found valid for a large number of normally consolidated soils [10],

$$(\frac{V_R}{1-V_R}) = -C_c \; \log \frac{P_f}{P_{f1}} + (\frac{V_R}{1-V_{R1}}) \tag{3}$$

where C_c is the compression index of the porous medium, V_{R1} is the volume fraction of the liquid when the effective pressure on the fibers is P_{f1}. Small consolidation pressure increments of 5 psi were used.

Figure 3 shows the compressibility plots obtained from tests on 20 and 30 plies of uniaxially aligned graphite fiber tows permeated with water and silicone oil. It was found that there are two regions for each plot. The first region covers a liquid fraction range of 0.5 to 0.44 and is non-linear. The second region is reasonably linear and covers a liquid volume fraction range of 0.44 to 0.25 which is the practical region for fiber-reinforced composites processing. For flow of water perpendicular to the planc of the fibers, the compression indices, $C_c = 1.25$ and 1.18, were determined from the linear region for 30- and 20-ply samples, respectively. A compression index of 1.01 was determined for the 20-ply sample from the linear region using silicone oil as the permeant.

For flow along the fiber direction in aligned fiber specimens, similar compressibility behavior was found from tests on 20- and 30-ply aligned graphite fiber tows permeated with water and silicone oil. The linear region covers a liquid fraction range of 0.25 to 0.43. The compression indices, C_C = 1.18 and 1.20, were determined from the linear region from the 20- and 30-ply samples respectively.

Unidirectional Fiber Bed Permeability

A) Flow Perpendicular to the Plane of the Fibers

Fiber bed permeabilities were determined for flow perpendicular to unidirectionally aligned fibers and are summarized in Figure 4 for various fiber bed thicknesses. The experimental permeabilities were fitted to the Kozeny-Carman relation [11] for permeability, which is given below:

$$K_z = \frac{V_R^3}{(1-V_R)^2} \frac{r_f^2}{4k_z'} \qquad k_z' = k_o(Le/L)^2 \qquad (4)$$

where r_f is the radius of the fibers, V_R is the volume fraction of the liquid, k_o is a constant which is usually called the 'shape factor,' $(Le/L)^2$ is the 'tortuosity,' which was defined by Carman [11] as the square of the ratio of the effective average path length in the porous medium (Le) to the shortest distance measured along the direction of macroscopic flow (L), and k_z' is called the Kozeny constant. Using k_z' = 11 for flow perpendicular to the plane of the fibers, the data points fit well to the Kozeny-Carman equation for a liquid fraction range of 0.25 to 0.5. For liquid fractions smaller than 0.25, the value of k_z' decreases substantially with liquid fraction. This may be due to changes in the fiber direction during consolidation. When the consolidation pressure is applied, the fibers will move together until enough are in contact to bear some of the applied loads. Further increase of pressure will only distort the fiber tows and cause non-uniform packing. Non-uniform packing offers a less tortuous path for liquid flow. Thus, the value of the Kozeny constant k_z' could decrease with decreasing liquid fraction. Figure 4 also shows that within the error bars of three specimens per data point, there is no effect of either bed thickness or permeant surface tension.

B) Flow Parallel to the Fibers

For flow along the fiber direction in unidirectionally aligned fiber beds, Figure 5 shows that the Kozeny-Carman equation predicts the permeability well with a Kozeny constant of k_x' = 0.68 for water and k_x' = 0.35 for silicone oil as permeants for the liquid volume fraction range of 0.25 to 0.5. Although there is again no effect of sample thickness, there is an effect of surface tension. These values of k_x' are comparable to those obtained by Williams et al. [12] for silane-treated glass fiber beds with a constant liquid volume fraction, V_R = 0.486. Williams argued that the wetting forces can be comparable to the pressure forces as the liquid advances through a dry bed. In our case, however, only the first (highest V_R) data point was obtained on a dry bed, the remaining data arising as the wet bed consolidated. Furthermore, there seems to be no obvious reason why the permeant surface tension should affect the permeability for flow parallel to the fibers but not affect it for perpendicular flow. This anomaly remains to be resolved.

Thus, for unidirectionally aligned graphite fiber-reinforced resin prepregs, the ratio of the transverse to the axial fiber bed permeabilities is

$$\frac{K_z}{K_x} \cong \frac{1}{19} \tag{5}$$

Gutowski et al. reported a similar ratio of 0.7/17.9 for graphite fiber beds [13].

Off-Axis Fiber Orientations

The effect of fiber bed anisotropy in the flow direction on the permeability was determined for flow both transverse and parallel to the ply planes. Water was used as the permeant in all cases.

A) Transverse Permeability

Figure 6 shows that the transverse permeability is reasonably represented by the Kozeny-Carman equation, is a maximum for unidirectionally aligned fiber beds, and consistently decreases as the alternating plies are laid down at increasing angles to one another. The 0-90° layup provides the most tortuous path and therefore the lowest permeability.

Following the suggestions of Scheidegger [8] and Marcus [14] the anisotropic permeability K_z can be represented by

$$\frac{1}{K_z} = \frac{\cos^2\alpha}{K_{zuni}} + \frac{\sin^2\alpha}{K_{z90°}} \tag{6}$$

where K_{zuni} is the transverse permeability for a unidirectional fiber bed, $K_{z90°}$ is the transverse permeability for a 0-90° bed of fibers, and α is the angle between the fibers in successive plies. Using the Kozeny-Carman equation for permeability, equation (6) can be rewritten in terms of the Kozeny constants k'_{zuni} and $k'_{z90°}$,

$$k'_z = k'_{zuni} \cos^2\alpha + k'_{z90°} \sin^2\alpha \tag{7}$$

where k'_z is the measured Kozeny constant for the anisotropic fiber bed.

Using the experimental values for k'_{zuni} and $k'_{z90°}$, the dependence of k'_z on α can be predicted on a polar plot (Figure 7). The predicted values of k'_z for $\alpha = 30°$ and $45°$ agree reasonably well with the experimental values.

B) Axial Permeability

For flow parallel to the fiber planes, Figure 8 indicates that the permeability is again well described by the Kozeny-Carman equation. When flow is directly down the fiber axes (0° aligned fibers), the permeability is greatest as expected. As half of the fibers turn more and more "broadside" to the flow, the tortuosity increases and the permeability decreases.

The axial Kozeny constant, k_x', can also be described by Equation (5). Indeed, it was for this situation that the equation was initially derived [8]. Again, using the 0° (aligned) and 0-90° data points as fitting constants, the equation accurately predicts the 0-30° and 0-45° experimental results (Figure 9).

It is interesting, in comparing the transverse and axial permeabilities for the off-axis samples, to note that the ratio of the aligned (0°) to 0-90° permeabilities, $K_{uni}/K_{90°}$, is 4 in the case of axial flow and only 1.8 for transverse flow. This seems plausible since the placement of half of the fibers "broadside" to the flow in the case of axial flow will certainly create a more tortuous flow path than rotating half of the fibers but keeping their axes perpendicular to the flow, as would occur in the transverse case.

CONCLUSION

The major conclusion from this study is that the permeability of fiber beds in high performance composite laminates is highly anisotropic. This anisotropy (as much as a factor of 19 for a unidirectional laminate) must be taken into account in any accurate resin flow analysis and master processing model. Using an isotropic permeability will lead to totally erroneous and potentially costly processing mistakes.

ACKNOWLEDGMENT

We gratefully acknowledge the financial support of this work by the Materials Laboratory of the Air Force Wright Aeronautical Laboratories under contract No. F33615-83-C-5088, "Computer-Aided Curing of Composites," to the McDonnell Aircraft Company.

REFERENCES

1. A. C. Loos and G. S. Springer, J. Comp. Mat., 17, 135 (1983).
2. C. J. Bartlett, J. Elast. & Plast., 10, 369 (1978).
3. J. C. Halpin, J. L. Kardos and M. P. Duduković, Pure and Appl. Chem., 55, 893 (1983).
4. J. T. Lindt, SAMPE Quarterly, 4, 14 (1982).
5. T. G. Gutowski, J. Comp. Mat., 21, 172 (1987).
6. R. Dave, J. L. Kardos, and M. P. Duduković, Pol. Comps., 8, 29 (1987).
7. R. Dave, J. L. Kardos, and M. P. Duduković, Pol. Comps., 8 123 (1987).
8. A. E. Scheidegger, The Physics of Flow Through Porous Media, McMillan Co., New York (1974).
9. R. Lam and J. L. Kardos, Proc. Pol. Mat. Sci. Eng. (ACS), 57, 532 (1987).
10. A. M. Samarasinghe, Y. H. Huang, and V. P. Drenrich, J. Geotechnical Engineering Div., ASCE, 108, No. GT6, June 1982.
11. P. C. Carman, Trans. Inst. Chem. Eng., 15, 150 (1937).
12. T. G. Williams, C. E. Morris, and B. C. Ennis, Pol. Eng. Sci., 14, 413 (1974).

13. T. G. Gutowski, Z. Cai, S. Bauer, D. Boucher, J. Kingery, and
     S. Wineman, J. Comp. Mat., 21, 650 (1987).
14. H. Marcus, J. Geophys. Res., 67, 519 (1962).

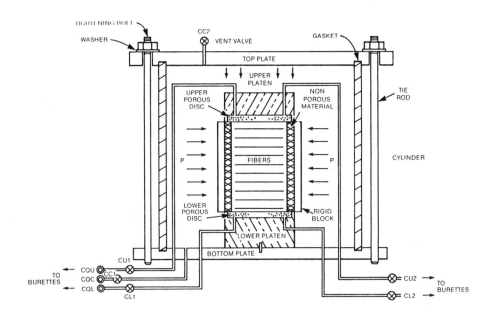

Fig. 1. Schematic of permeameter chamber (vertical section) for
         measurement of permeability perpendicular to the plane
         of the fibers during consolidation.

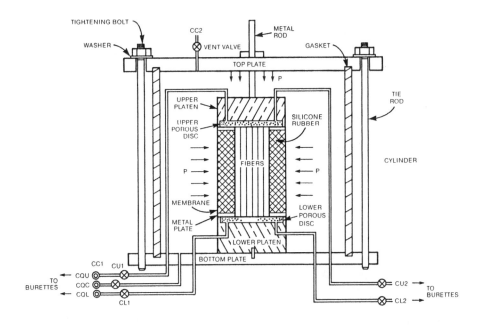

Fig. 2. Schematic of permeameter chamber (vertical section) for
         measurement of permeability parallel to the plane of the
         fibers during consolidation.

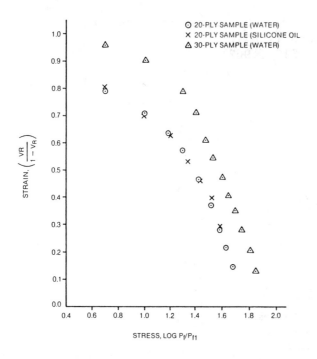

Fig. 3. Consolidation relationship for uniaxially aligned graphite fibers.

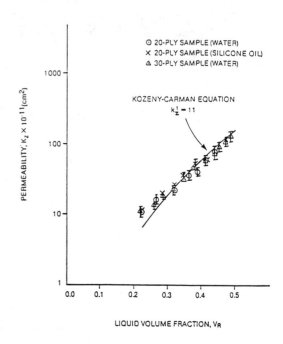

Fig. 4. Transverse permeabilities during consolidation for a uniaxially aligned graphite fiber bed.

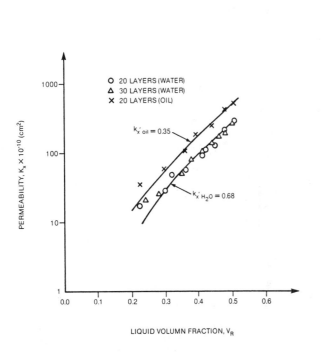

Fig. 5. Axial permeabilities during consolidation for a unidirectionally aligned graphite fiber bed.

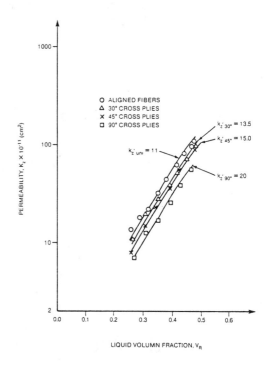

Fig. 6. Transverse permeabilities during consolidation as a function of ply orientation for a water permeant.

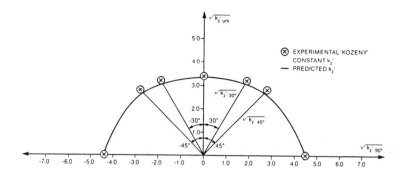

Fig. 7. Kozeny constant for transverse
flow as a function of off-axis
orientation angle, α, for a
water permeant.

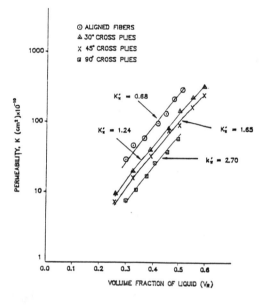

Fig. 8. Axial permeabilities during
consolidation as a function
of ply orientation for a
water permeant.

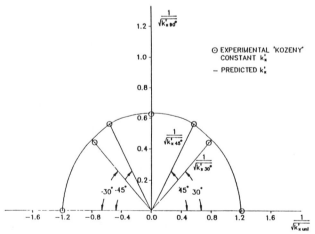

Fig. 9. Kozeny constant for axial flow
as a function of off-axis orientation
angle, α, for a water permeant.

11

Experimental Evaluation of the Prepregging Process

J. J. BREUKERS*, E. M. WOO AND J. C. SEFERIS
Polymeric Composites Laboratory
Department of Chemical Engineering
University of Washington
Seattle, WA 98195

ABSTRACT

The prepregging processing conditions for amine-cured epoxies, dicyanate and bismaleimide matrix resins, both unmodified and modified with elastomers and thermoplastics, were investigated in this study. Our laboratory-scale continuous hot-melt prepregger was utilized. The apparatus was modified in order to increase both quality of uniformity and processing flexibility. Finally the experimental conditions required for prepregging the different systems were compared on a dimensional analysis approach.

INTRODUCTION

The prepregging operation is the step in which the matrix resin and the reinforcing fibres are combined into a composite precursor. It is well established that the processing conditions influence both the structure and the final properties of the polymeric composite material [1]. Accordingly a study of the prepregging process as it influences both properties and structure of the resulting material may be a key to understanding composite manufacturing and usage [2].

There are several commercial methods of impregnating reinforcing fibres with a resinous matrix. Solution dip and solution spray techniques are performed with the resin matrix dissolved in a solvent. The low viscosity of the resin solution allows good impregnation of the reinforcing fibre bundles, but during subsequent processing the solvent should be removed to reduce void formation. In the hot-melt prepregging process, in principle the solvent removal step can be eliminated resulting in reduced void formation in the lamination step.

The melt impregnation technique was investigated experimentally with a laboratory-size prepregger, previously described and developed by Lee, Seferis and Bonner [3]. A schematic representation of the apparatus is shown in Figure 1. The fibres enter the process from the fibre creel and are collimated by a comb. In the original set-up, the top release paper is coated with a resin melt using a doctor blade to control the resin film thickness. Underneath the fibre bed another strip of release paper is added. In the heated impregnation zone resin and fibres are sandwiched between the release papers. Two compression rollers are used as the driving force for resin penetration into the fibre bed. The product then passes a chill plate to quench any possible reaction advancement. Finally the top release paper is removed and a product roller takes up the continuous prepreg.

---

* present address: DSM Research, P.O. Box 18, 6160 MD Geleen, The Netherlands

As in all high-performance production processes, product quality and especially uniformity are crucial. In terms of the prepregging process this means that the impregnation zone temperature should be constant throughout the whole batch. Temperature deviations of a few degrees might greatly affect the impregnation viscosity. Furthermore a homogeneous pressure distribution across the width of the prepreg is a necessity in order to avoid fibre misalignment.In addition the residence time in the impregnation zone must be adjusted so that minimum resin advancement is observed. The resin film thickness made on the knife-coater should be uniform both in time and across the width. A strict procedure for reproducible resin handling before prepregging should be developed in order to provide process consistency. Although this may not be required with resin systems currently in production, it may be expected to become of increase importance with new generation resin systems.

A central question in the evaluation of the prepregging process is the definition of what constitutes a good prepreg. Relatively simple physical and chemical tests on prepregs like resin content, resin flow, areal fibre weight etc., and poorly defined characteristics like tack and drape are specified by end-users that are difficult to quantify. Judgements are normally based on precursor properties (neat resin and fibres) and laminate properties in order to interpolate the intermediate (prepreg) properties.

From earlier experience the processing window for the prepregging process appeared to be relatively narrow. Temperature deviations of a few degrees were enough to deteriorate the prepreg quality at a certain combination of pressure and line speed. The experimental approach in this study was to set the impregnation temperature at a certain value using the same heater attenuation throughout the experiments to assure a minimal temperature fluctuation (due to the electrical heating). The line speed was then adjusted while applying a constant impregnation pressure until a visually good prepreg was obtained. Resin content, areal fibre weight and fibre filament wetting were checked after processing into prepreg tape.

Three thermosetting matrix resin systems were examined: an amine-cured epoxy, a dicyanate and a bismaleimide. Unmodified and modified formulations with reactive liquid rubbers and thermoplastics, were evaluated. They were chosen to provide a wide range in processing conditions and to study the effects of the modifiers on the processing conditions of the three chemically different basic systems. The impregnation requirements for the broad spectrum of impregnation viscosities were provided by several modifications made on the prepregger from the basic configurations described earlier. The data were used as an experimental input in an empirical [2,4] and a constitutive [5] prepreg model in order to investigate the relative ease of processability of the unmodified and modified resin systems.

EXPERIMENTAL

MATERIALS

Three basic thermoset resin systems were examined : an amine-cured epoxy, a dicyanate and a bismaleimide system. Model systems were formulated approximating commercial compositions They are listed in Figure 2 and Figure 3. In this study Hercules sized AS-4 W 3K carbon fibres were used.

The basic amine-cured epoxy system [6] was modified using carboxyl-terminated butadiene- acrylonitrile (CTBN) elastomers from BF Goodrich and Victrex 300P polyethersulphone (PES) from ICI [7]. The polyether- sulphone was dried at 80 °C overnight. After dissolving PES into TGDDM at 170 °C the resin was cooled down to 130 °C. At this temperature the NOVALAC and the DDS were added. Mixing continued for several minutes until a clear melt was obtained. For the CTBN, added at 130 °C, a similar procedure was used. Degassing under vacuum at 100 °C was performed for about 15 minutes. The dicyanate system used in this study was based on BADCy [8], supplied by Interez Inc. This resin was modified using CTBN rubbers [9] and Udel P-1800 polysulphone (PS). (PES appeared to be incompatible with BADCy).

The dicyanate system used in this study was based on BADCy [8], supplied by Interez Inc. This resin was modified using CTBN rubbers [9] and Udel P-1800 polysulphone (PS). (PES appeared to be incompatible with BADCy). After dissolving the PS into the BADCy at 170 $^{\circ}$C by mechanical stirring the mixture was cooled down to 120 $^{\circ}$C. At this temperature the catalyst solution was added. For the CTBN modified resin a similar procedure was used provided that the rubber was added at 130 $^{\circ}$C. Degassing under vacuum at 100 $^{\circ}$C was performed for about 15 minutes.

The bismaleimide system, supplied by Rhone-Poulenc Inc., was modified using CTBN rubber [10] and Victrex 300P polyethersulphone. Since the system already contained a catalyst, the PES was first dissolved in dichloromethane at room temperature. Both the PES - solution and the CTBN rubber were added at 120 $^{\circ}$C. Degassing under vacuum at 80 $^{\circ}$C was extended to 20 minutes in order to get rid of the dichloromethane.

After degassing all the resins were split into two parts: the major part for immediate prepregging while the rest was stored in the refrigerator for later viscosity measurements at impregnation temperature on a Brookfield viscometer.

PROCESSING

To meet impregnation requirements for several systems, several modifications to the apparatus were made. They are schematically shown in Figure 4. The fibre tows were first guided around several rollers and finally pre-spread on a roller just before entering the impregnation zone. The pre-spreaded fibres were impregnated with resin coated on the bottom paper coating instead of on the top paper. An additional comb was introduced just before the pre-spreading roller to guarantee a homogeneous fibre tow distribution across the width and to be able to vary the areal fibre weight. An additional pressure unit was placed right after the impregnation zone entrance to contribute to the driving force for resin penetration into the fibre bed. Weights were placed on the upper nip roll to perform the finishing impregnation touch. As a result of these modifications improvements in uniformity as well as in processing flexibility were observed for all the resin systems evaluated in this study.

CHARACTERIZATION

The prepregs were examined for resin content using the solvent extraction technique. For the amine-cured epoxies and the bismaleimide resins acetone was used as a solvent; for the dicyanate resins dichloromethane was used. By measuring the fibre tow weights per unit of length, the areal fibre weight could be calculated using the prepreg width, with the following equation:

$$areal\ fibre\ weight\ =\ \frac{\sum_{i=1}^{i=n} (ITW)}{PW} \qquad (1)$$

ITW = individual fibre tow weight per unit of length
PW = prepreg width
n = number of tows used in making the prepreg

Consistency checks were also performed using the solvent extraction technique. Differences were found to be within the range of experimental error.

A technique was developed utilizing an optical microscope for determining the extent of fibre filament wetting. A sample was prepared by first cutting a prepreg into small pieces. One piece was placed on a glass micro slide on which it was split along the fibre filament direction. Boardy prepreg pieces could be separated easily into two parts, resulting in a bad sample. Wet and tacky prepregs however did not cleanly separate: several individual filaments still connected the two pieces, schematically illustrated in Figure 5.

Samples for boardy prepregs were obtained by heating a small piece on a hot-stage before splitting. The technique is also applicable for thermoplastic matrix prepregs. A good sample can then be obtained by heating a small piece on a hot-stage above the melting temperature before splitting. Clear differences between wetted filaments and prepregs containing dry areas were found at 100 x magnifications. An example is shown in Figure 6.

RESULTS AND DISCUSSION

The processing conditions for the different resin systems are summarized in Figure 7. The materials were processed at an impregnation temperature of 100 °C (and in two cases at 130 °C because of extremely high viscosity at 100 °C) to minimize the effect of temperature fluctuations due to the electrical heating. The impregnation viscosities ranged from 230 mPa.s to 21,000 mPa.s , while line speeds varied from 4 to 57 mm/s. The impregnation viscosity of all modified resins increased and the line speed decreased at a constant weight on the impregnation rollers. This made a comparison of the effect of the modifier on the processing conditions very difficult. Initially it appeared that after modification processing became more difficult as was seen by the increase in residence time.

A recently developed analytical methodology, [2 - 5], was used to evaluate the influence of the impregnation zone temperature on the prepreg quality resulting from changes in impregnation viscosity.

The methodology was utilized both for the modified resins as well as for the unmodified resins, assuming that ageing and cure advancement are negligible.

The impregnation process may be viewed as a flow of a viscous liquid (the resin) through an anisotropic porous medium (the fibres). Empirical process modelling based on this idea has been attempted by Lee, Bravenec and Seferis [4]. Optimal impregnation requires an optimal residence time in the impregnation zone in order to penetrate completely and wet out fully the fibre bed but not to be squeezed out of the fibre bed. Accordingly a dimensionless flow time (FT) relating the real residence time in the impregnation zone to the optimal residence time may be defined as:

$$FT = \frac{real\ residence\ time}{optimal\ residence\ time} = \frac{k\ m\ g}{H_o^2\ \mu\ V\ W_o} \tag{2}$$

$$
\begin{aligned}
k &: \text{fibre bed permeability} & (3.10^{14}\ m^2) \\
m &: \text{mass of the impregnation rollers} & (6.6\ kg) \\
g &: \text{gravity constant} & (9.81\ m/s^2) \\
H_o &: \text{prepreg thickness} & (90\ m\ m^{**}) \\
V &: \text{line speed} & (m/s) \\
\mu &: \text{impregnation viscosity} & (Pa.s)
\end{aligned}
$$

** *based on a fibre density of 1800 kg/m³ , a resin density of 1300 kg/m³ , an areal fibre weight of 0.090 kg/m² , a resin content of 37 weight percent and a negligible void content.*

Since the flow parameter is normalized using a mathematically derived optimal reference condition, absolute values can not be utilized, but trends in the flow parameter can be used to compare the relative ease of processability of unmodified and modified resin systems. A higher value of the flow parameter means that the residence time in the impregnation zone was longer, implying a relatively more difficult processability.

Generalizing this methodology Ahn and Seferis [5] used continuity and momentum equations to derive a dimensionless number to characterize the impregnation process. This model also implies that under certain circumstances a superposition of impregnation viscosity, applied pressure and line speed seems to be possible, but compared to the empirical model the latter provides a more powerful tool, able to predict the optimal processing conditions as a function of the prepreg width.

The definition of this dimensionless viscous flow (VF) is:

$$VF = \frac{\text{resistance to viscous flow}}{\text{applied pressure}} = \frac{\mu \, V \, \sqrt{k \, e}}{m \, g} * 10^{12} \qquad (3)$$

μ	:	*impregnation viscosity*	*(Pa.s)*
V	:	*line speed*	*(m/s)*
k	:	*fibre bed permeability*	*($3.10^{14} m^2$)*
e	:	*fibre bed porosity*	*(0.4)*
m	:	*mass of the impregnation rollers*	*(kg)*
g	:	*gravity constant*	*($9.81 m/s^2$)*

In both models the permeability is assumed to be constant for all the system evaluated in this study. Both are dimensionless numbers that give a measure of the ratio of viscosity, applied pressure and residence time. They were both calculated from the experimental data. The results are shown in Figure 8. Since the flow time and the viscous flow number are inversely related to each other only the latter will be discussed.

A resin system shows relatively easier processability if the residence time in the impregnation zone at equivalent impregnation viscosity and applied pressure is shorter than the reference. In terms of the viscous flow number this implies that a higher value is obtained for a relatively easier processable resin system compared to a reference due to the relatively higher line speed.

For the amine-cured epoxy modifications a sharp decrease in the viscous flow number can be seen while impregnation viscosities are in the same order of magnitude implying a relatively more difficult processing compared to the unmodified resin. The same effect can be noticed for BMI/CTBN compared to the unmodified BMI.

The viscous flow numbers for the three dicyanate based resin systems predict a relatively easier processability for the modified resins. But since the impregnation viscosities differ two orders of magnitude, care should be taken when comparing the dimensionless numbers and translating them into relative ease of processability. For that reason an additional experiment was performed, processing the BADCy/cat at 48 °C instead of 100 °C.

The impregnation viscosity at this temperature was obtained from the viscosity-temperature profile shown in Figure 9. An extended comparison is made in Figure 10.From this figure it can be seen that although the sequence in relative ease of processability did not change, a different viscous flow number was obtained for the BADCy/cat system. This may imply that superposition of impregnation viscosity, applied pressure and line speed is only applicable in a certain viscosity range where the distribution of the total applied pressure between resin and fibre bed is balanced. Comparing BMI/PES and unmodified BMI the same effects as for the dicyanate based resins were seen. In this case not only the difference in impregnation viscosity of one order of magnitude was observed, but an additional effect should be considered as well. The unmodified BMI system already contained a catalyst. To lower the blending temperature and time in order to reduce ageing, the polyethersulphone was first dissolved in dichloromethane. The solvent might influence the impregnation viscosity, the surface tension and the flow resistance resulting in different impregnation characteristics.

CONCLUSION

Three basic thermosetting matrix resin systems: amine-cured epoxies, dicyanates and bismaleimides, unmodified and modified with elastomers (CTBN) and thermoplastics (PS and PES) were evaluated and provided a wide range in processing conditions. The effects of the modifiers on the processing conditions were investigated in this study. To meet appropriate processing requirements for the broad spectrum of viscosities, the impregnation step on the laboratory-scale prepregger was modified. Furthermore a technique was developed using optical microscopy which is capable of distinguishing wetting characteristics of the prepregs. Finally the experimental data were used in an analytical methodology of the prepregging process in order to describe the relative ease of processability.

ACKNOWLEDGEMENTS

The authors wish to acknowledge the financial and material support for this work which was provided by DSM Research, Interez Inc. and Rhone-Poulenc Inc. through project support at the Polymeric Composites Laboratory at the University of Washington.

REFERENCES
[1] : "The role of the Polymeric Matrix in the Processing and Structural Properties of Composite Materials", Ed. J.C. Seferis and L. Nicolais, Plenum Press,New York, (1983).
[2] : W.J. Lee,"Advanced Composite Prepreg Processing Science", Ph.D. Thesis in Engineering, University of Washington, february 1988.
[3] : W.J. Lee and J.C. Seferis, "Prepreg Processing Science", SAMPE Quarterly, Volume 17, No. 2, January 1986, pp. 58-68.
[4] : W.J. Lee, L.D. Bravenec and J.C. Seferis, "Hot-melt Prepreg Processing of Advanced Composites: A Comparison of Methods", SPE ANTEC '88 Pr. Accepted (1988).
[5] : K. Ahn and J.C. Seferis, in preparation.
[6] : H.S. Chu and J.C. Seferis, "Dynamic Mechanical Experiments for Probing Process-Structure-Property Relations In Amine-Cured Epoxies", Polym.Comp., 5, (2), p.124 (1984).
[7] : C.L.Loechelt, E.M. Woo and J.C. Seferis, "Process and Composition Induced Morphology in Thermosetting Matrix Systems", SPE ANTEC '88 Proc., Accepted (1988).
[8] : D.A. Shimp, "Thermal Performance of Cyanate Functional Thermosetting Resins", SAMPE Quarterly, Volume 19, No. 1, October 1987, pp. 41-46.
[9] : E.M. Woo, B.K. Fukai and J.C. Seferis, "Dicyanate Blends as Matrices for High Performance Composites", Amer.Soc. for Comp., submitted March (1988).
[10] : J-F Viot and J.C. Seferis, "Process-resolved Morphology of Bismaleimide Matrix Composites", Journal of Applied Polymer Science, Volume 34, pp.1459-1475.

figure 1   Schematic representation of the laboratory-size prepregger by Lee and Seferis.

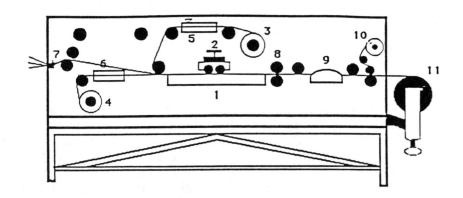

1. heated impregnation zone
2. pressure rollers
3. top paper let-off
4. bottom paper let-off
5. top coater
6. bottom coater

7. fibre comb
8. nip rolls
9. chill plate
10. top paper take-up
11. product wind-up

figure 2   Specification of the chemical components of the resin systems as used in this study.

amine cured epoxy

tetraglycidyl-4,4' -diamino-diphenylmethane	(TGDDM)
polyglycidyl ether of Bisphenol A Novalac	(NOVALAC)
4,4' - diamino-diphenyl sulphone	(DDS)

dicyanate

Bisphenol A dicyanate	(BADCy)
copperacetylacetonate catalyst in nonylphenol	(cat)

bismaleimide

diphenylmethane bismaleimide	
tolylene bismaleimide	
triallyl isocyanurate	(TAIC)
diphenyl silanediol	(DPS)
imidazole catalyst	

modifiers

carboxyl-terminated butadiene acrylonitrile	(CTBN)
polyethersulphone	(PES)
polysulphone	(PS)

figure 3   **Composition of the blends used in this study**
approximating commercial prepreg formulations

TGDDM/NOYALAC/DDS	(88.5 / 11.5 / 25)
TGDDM/NOYALAC/DDS/CTBN	(88.5 / 11.5 / 25 / 10)
TGDDM/NOYALAC/DDS/PES	(88.5 / 11.5 / 25 / 10)
BADCy/cat	(100 / 2.7)
BADCy/cat/CTBN	(100 / 2.7 / 10)
BADCy/cat/PS	(100 / 2.7 / 10)
BMI	(100)
BMI/CTBN	(100 / 10)
BMI/PES *	(100 / 10)

* dichloromethane used for blending

figure 4   **Schematic representation of the laboratory-size**
prepregger as used in this study. Modifications were
made to increase product uniformity and operating
flexibility.

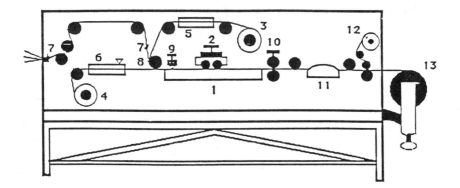

1. heated impregnation zone
2. pressure rollers
3. top paper let-off
4. bottom paper let-off
5. top coater
6. bottom coater
7. fibre comb

8. roller for fibre pre-spreading
9. consolidation of the pre-spreading
10. nip rolls
11. chill plate
12. top paper take-up
13. product wind-up

figure 5   **Difference between a bad and a good sample for the optical microscope to study fibre filament wetting.**

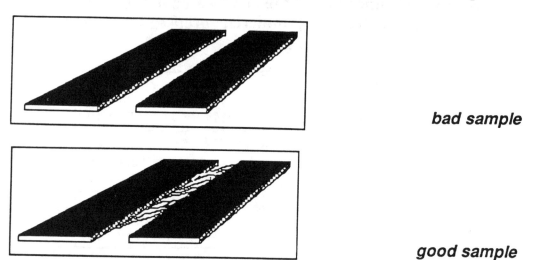

bad sample

good sample

figure 6   **Two examples of micrographs obtained for investigating fibre filament wetting. Number A shows wetting while number B shows dry filaments.**

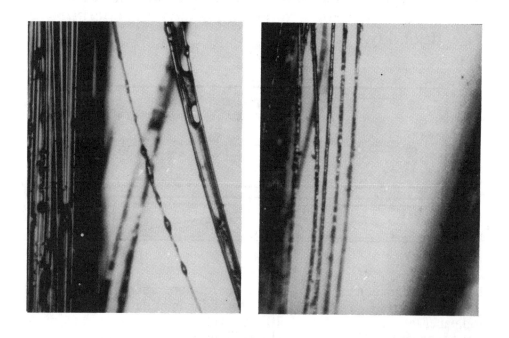

figure 7 **Prepreg processing conditions for the resin systems evaluated in this study.**

	IMPREGNATION TEMPERATURE (C)	IMPREGNATION VISCOSITY (mPa.s)	WEIGHT OF IMPREG-NATION ROLLERS (KG)	LINE SPEED (cm/s)	INITIAL RESIN FILM THICKNESS (μm)	AREAL FIBRE WEIGHT (gr/m^2)	PREPREG RESIN CONTENT (v %)	FILAMENT WETTING (qualitative)
TGDDM/NOYALAC/DDS	100	800	6.6	4.2	125	90	37	GOOD
TGDDM/NOYALAC/DDS/CTBN	100	1200	6.6	1.7	125	90	37	GOOD
TGDDM/NOYALAC/DDS/PES	130	1000	6.6	0.5	125	90	37	GOOD
BADCy/cat	100	230	6.6	4.2	125	90	37	GOOD
BADCy/cat/CTBN	100	17000	6.6	0.4	125	90	37	SUFFICIENT
BADCy/cat/PS	130	21000	6.6	0.4	125	90	37	SUFFICIENT
BMI	100	310	6.6	5.7	125	90	37	GOOD
BMI/CTBN	100	340	6.6	3.4	125	90	37	GOOD
BMI/PES	100	2800	6.6	1.0	125	90	37	GOOD

figure 8 **Viscous Flow Number and Flow Time Number calculated from the processing data.**

	IMPREGNATION VISCOSITY (mPa.s)	DIMENSIONLESS viscous flow number	dimensionless flow time number
TGDDM/NOYALAC/DDS	800	57	0.2
TGDDM/NOYALAC/DDS/CTBN	1,200	35	0.3
TGDDM/NOYALAC/DDS/PES	1,000	9	1.3
BADCy/cat	230	16	0.7
BADCy/cat/CTBN	17,000	115	0.1
BADCy/cat/PS	21,000	142	0.08
BMI	310	30	0.4
BMI/CTBN	340	20	0.6
BMI/PES *	2800	47	0.2

* dichloromethane used for blending

figure 9   **Viscosity-Temperature profile for the BADCy/cat resin system.**

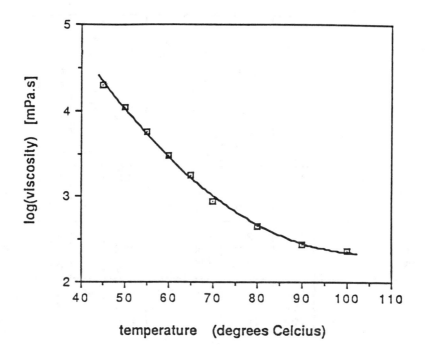

figure 10   **Processing conditions and Viscous Flow Number for the dicyanate based resin systems, processed at approximately the same impregnation viscosity.**

	IMPREGNATION TEMPERATURE (C)	IMPREGNATION VISCOSITY (mPa.s)	LINE SPEED (cm/s)	DIMENSIONLESS viscous flow number
BADCy/cat	48	13,000	0.4	88
BADCy/cat/CTBN	130	17,000	0.4	115
BADCy/cat/PS	130	21,000	0.4	142

Economic Issues in Composites Manufacturing

RICHARD M. McLANE

ABSTRACT

There are many issues, many technologies involved in the economics of manufacturing advanced composites. These issues start with the manufacturing of the raw materials, the types of materials, material forms and how they can be processed on automated equipment. In-house, the issues include the manufacturing technology (degree of automation) employed, CAD/CAM, quality consistency, cost of quality, non-touch costs, cost of capital, facilities cost and method of amortization. Finally, there is the economic issues of the design and what product is being manufactured.

The aim of this paper is not to cover all of these issues in depth, but rather to examine past composite programs, address the economic issues, and recommend a direction for the future.

1. INTRODUCTION

Composites has a negative image in many segments of the aircraft industry. The question is: Is this a technology of high performance, high cost, to be used in specialized applications where no other material will do the job as well? Or is this really one of the few new technologies which promises to deliver both improved performance and lower cost?

This author believes the latter statement to be true. It is a new technology, with cost growing pains. But the technology is sufficiently mature to where it can now be applied cost effectively, if we learn from our past experiences.

2. BACKGROUND

Advanced composites, referred to in this paper, are defined as aramid or carbon fiber reinforced thermoset polymer matrix composites. Boeing first started using advanced composites in production on the secondary structures of 757 and 767 airplanes. This production implementation followed a successful 737 Spoiler Flight Service Evaluation and a NASA-ACEE program in which five shipsets of 727 elevators and 737 horizontal stabilizers were certified and installed on production airplanes. See figure 1. The NASA program was considered a success in that the technical

R. M. McLane, MR&D Supervisor, Boeing Commercial Airplanes, P.O. Box 3707, M/S 79-85, Seattle, WA 98124-2207

objectives of weight reduction, structural performance and certification were met. Because it was a limited run of five shipsets, a realistic assessment of cost could not be made.

Boeing-Seattle Composites History

Year	Component	Quantity
1973	737 Spoilers	118 Parts
1979	727 Elevator	5½ Shipsets
1980	737 Horizontal Stabilizer	5½ Shipsets
1981	767 Secondary Structures	Production
1982	757 Secondary Structures	Production
1983	737 Secondary Structures	Production
1986	A-6 Wing	Production

Figure 1

Influenced by the NASA program, the decision to implement advanced composites on the 757 and 767 was primarily a technical one. Although cost projections were also considered, the prime motivation was weight savings on these new high technology aircraft designed on the heels of the oil shortage crisis and fears of ever escalating fuel prices. The designs and producibility verification for the 757 and 767 were done in-house, but much of the production fabrication was done by program participants and subcontractors.

The 737-300 was the next model to utilize advanced composites, again in secondary structures applications. This was the first attempt to design for automation, utilizing an Automated Tape Lay-up Machine (ATLM) for the skin panels of the rudder, elevators and spoilers. This was the first real test-case for the economics of advanced composites in that the composite parts were replacing metal or metal-fiberglass parts which had been used on the first 1200 737 airplanes.

The A-6 replacement wing program was the first production primary structure application of advanced composites at Boeing. In this instance, composites was virtually a program requirement to meet the objectives of increased strength and fatigue life with no weight increase. Again, the major skin and spar elements were designed for automation by ATLM's.

3.   THE PROBLEM

The 737 was the first opportunity to directly compare the cost of interchangeable structures made both by advanced composites and previous "conventional" structures such as bonded aluminum or aluminum/fiberglass construction. A direct cost comparison at a point in time showed an unfavorable picture for advanced composites, as shown in figure 2. However, this comparison is made between conventional structure at unit 1200 vs. advanced composites structures at unit 100.

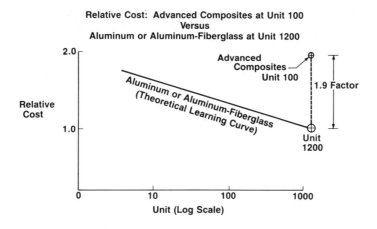

Figure 2 Flight Control Surfaces

On the A-6 program, design changes required due to work statement growth and a weight guarantee, severely complicated the design to where the original manufacturing automation plan could not be used (figure 3). This drove the program toward an entirely new approach to fabrication and, until some of these new approaches can be implemented, the program is facing costs higher than projected.

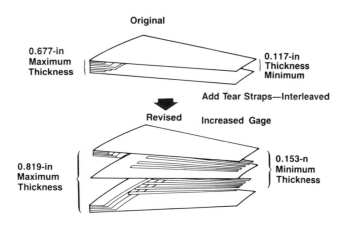

Figure 3 A-6 Outboard Upper Wing Skin

The response to the two situations described has had an unfavorable impact on the reputation of composites on programs where cost is of equal importance with performance. In fact, the 737 has changed some of the secondary structure back to the original materials to reduce costs (where tooling, designs, and certification already existed).

The A-6 re-wing program has had to replan their entire approach to fabrication. The A-6 now uses a machine assisted hand lay-up method, with 60-inch wide tape or fabric, automated ply nesting/cutting system, and automated ink jet ply locating/inspecting equipment. The actual placement of the plies is by hand. This method promises to reduce in half the lay-up flow time over the original approach of automated tape lamination of a very complex, multiple doubler design.

Because both programs failed to meet early cost projection, composites technology and the ability to forecast costs is experiencing a credibility gap in many circles.

4. ANALYSIS

In hindsight analysis of the cost problems, we can point to reasons why composites have not been cost effective. On the 737, the reasons varied, but the basic reason was that we were trying to compete in cost with a mature technology that was 1200 units down the learning curve.

The advanced composites designs were sometimes more complex than earlier designs. For example, the 737 outboard aileron design was changed from a full-depth honeycomb in metal/fiberglass to a panelized design with ribs to accommodate a subcontractor experience base, with the aim of lowering the risk in transferring the technology (figure 4). Not only were the costs higher in going to advanced composites, but the part actually gained two pounds in weight. Thus, the reputation of advanced composites further suffered in that not only are they more expensive, but they do not save any weight.

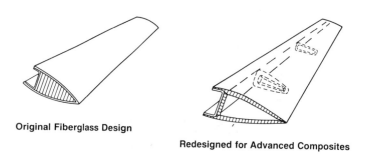

Original Fiberglass Design

Redesigned for Advanced Composites

Figure 4 Aileron

A condition that occurred on both the 737 and A-6 that affected costs is that the designs were restricted by having to duplicate existing structure in stiffness, strength, and hard points. The A-6 also had to resolve flutter and fatigue problems in the existing metal design.

5. LESSONS LEARNED

The experiences on the 737 and A-6 program have challenged the credibility of advanced composites. But they have also provided valuable lessons that should not be lost on future programs. A lot can be said about not applying composites in replacement applications. Also, a strong caution is warranted on small sized parts where automation is difficult.

However, the lessons can be boiled down to the following two basic needs:
   A.   Cost Effective Designs
   B.   Flexible Automation
Cost effective designs need not be 100 percent automatable. In fact,
that was one of our mistakes in the past. Early design concepts would
make an attempt at 100 percent automation, but inevitable tailoring or
"work statement growth" would creep in as test results came in, or
additional systems or structures interfaces were defined. As a result,
the costs would escalate from earlier projections.
   To illustrate what we mean by a cost effective design, figure 5 shows
an airfoil that might be envisioned by a manufacturing engineer. This
would be truly cost effective to build. However, the production product
is usually more like figure 6, a highly tailored, complex design that
would be difficult to automate and inherently expensive to fabricate.

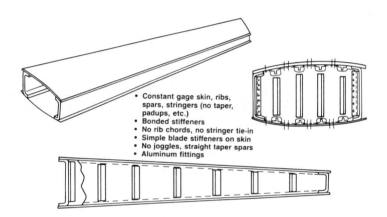

- Constant gage skin, ribs, spars, stringers (no taper, padups, etc.)
- Bonded stiffeners
- No rib chords, no stringer tie-in
- Simple blade stiffeners on skin
- No joggles, straight taper spars
- Aluminum fittings

**Figure 5 Composite Airfoil Designed by
Manufacturing**

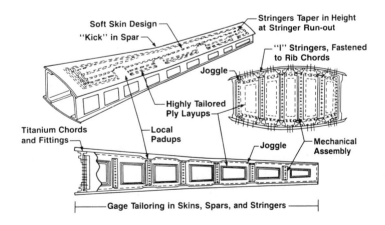

Soft Skin Design
"Kick" in Spar
Stringers Taper in Height at Stringer Run-out
"I" Stringers, Fastened to Rib Chords
Joggle
Highly Tailored Ply Layups
Titanium Chords and Fittings
Local Padups
Joggle
Mechanical Assembly
Gage Tailoring in Skins, Spars, and Stringers

**Figure 6 Composite Airfoil Designed by
Engineering**

A realistic cost effective design lies somewhere between the two illustrations. Compromises must be made between engineering and manufacturing. Composites technology is a leader in the area of producibility or design/build teams with manufacturing sign-off on the drawings. However, we still have a long way to go. The ideal of balance between manufacturing and engineering objectives has not yet been achieved.

The second basic need, flexible automation, is a corollary to cost effective design. We must realize that not every ply layup can be comparable to a sheet of plywood. Some tailoring of gages, padups, etc., is required to meet strength and weight objectives. These inevitable complexities must be planned for in developing an approach to automation. However, the design should recognize these non-automatable plies and group them in the stackup so alternate methods of fabrication can be used for these modules.

We have learned in the past that we cannot design toward the exclusive use of one piece of automated equipment. This might be acceptable for automating high volume parts, such as pultruded sections. However, it is not practical for large complex parts such as airfoil or fuselage skins. We have also learned that 100 percent automation is not necessarily required, nor desired. Flexible automation includes doing certain operations by hand, if it is more cost effective to do so.

The A-6 experience was an example of planning for 100 percent ATLM automation, only to experience a complete change in approach due to significant design changes. The depth of change experienced on the A-6 is hopefully not typical.

Figure 7 is a schematic of one such scheme for automating an airfoil skin panel, ribs and spars. It can handle formed parts such as ribs, spars and stringers via the drape form process. Most plies are laid by the ATLM, but the ability to accommodate small doublers or padups is provided by ply nesting/cutting, ink-jet ply location marking/inspection and semi-automated compaction features. Pre-plied doublers could be fabricated in a flexible laminating cell.

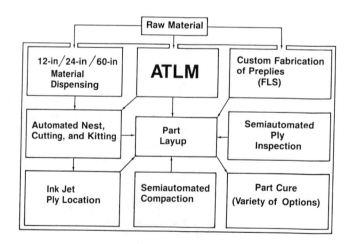

Figure 7 Flexible Composites Manufacturing Facility

Flexible automation is possible now, not only because we have gotten smarter, but the equipment capabilities for automating composites has grown significantly in recent years.

The effective use of this flexible automation approach is still dependent on cost effective designs. A highly tailored spar, with compound contoured flanges, pocketed web doublers, stiffener padups and joggles would still probably have to be hand laid up. The bottom line is that a complex design can ruin the best laid automation plans.

6. IMPLEMENTATION

To successfully implement the two basics of cost effective design and flexible automation is a challenge in itself. Without a concrete plan, these goals are nothing more than good intentions. The following four steps as essential ingredients in addressing the economics of advanced composites:

    A. <u>Get cost analysis tools in the hands of the designer.</u>

Currently, the designer has no real-time method of analyzing his design or various alternative designs for cost effectiveness. The current practice is for a designer to turn several alternatives for a rib design over to a Design Build Team (DBT) for analysis and get an answer back in three to six weeks. CAD tools need to be developed that allow the designer of an ATLM manufactured part to analyze the machine run time and material utilization for various design alternatives (see figure 8). These CAD tools have been developed by the machine tool builders and similar capabilities need to be implemented in the various CAD design systems. Design and staff engineers must be held accountable for structures cost, as well as weight and performance.

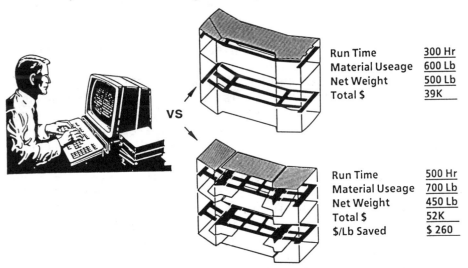

Run Time	300 Hr
Material Useage	600 Lb
Net Weight	500 Lb
Total $	39K

Run Time	500 Hr
Material Useage	700 Lb
Net Weight	450 Lb
Total $	52K
$/Lb Saved	$ 260

Figure 8 Cost Modeling

    B. <u>Challenge costly groundrules through testing.</u>

Many composites designs are overdesigned, overweight and too costly because of groundrules, assumptions, or basic design philosophy. Often, design criteria or concepts have evolved from earlier metal or fiberglass designs. Problems encountered on earlier developments become firmed up as groundrules or "sacred cows," which are adapted as absolute criteria. This can lead to designs costly to produce, overweight or not optimum from a composites view point.

One example of how different philosophies or criteria affect design can be seen in comparing the A-310 rudder with the 757/767 rudders (figure 9). Both are of similar construction and size; however, the A-310 is made with no internal ribs. Airbus claims to have substantiated their design with a vigorous testing program, including full-scale environmental testing. The question is, which design approach is better? The A-310 appears to be lower cost and probably lighter. But is it at risk from a reliability standpoint? Is the Boeing concept overdesigned?

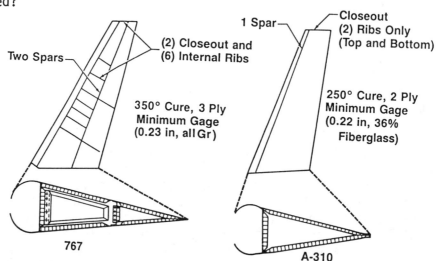

Figure 9 767 Versus A-310 Rudder Design Concept

Manufacturing also has groundrules that affect costs. Often, tools are overdesigned, or elaborate handling equipment has been built for parts that can be moved faster and easier by hand.

How do we challenge our sacred cows? Who is responsible? This leads to the next step.

C.   Assign focused responsibility for costs.

With cost analysis tools in the hands of the designer, and a willingness to challenge those ground rules which drive up costs, the next and most important step in implementing cost effective composites is to assign responsibility for costs. Historically, engineers have seen their primary job as producing a design which meets structural, environmental, weight, and damage tolerance requirements. If time permits and if manufacturing producibility personnel are available, then cost trades may be conducted to see if costs might be reduced within the above constraints. The engineer's primary job is perceived to design a durable, quality product. If it is also cost effective, that's a plus - as long as it does not interfere with "Job 1."

Quality properly should be the top priority, but too often, cost is so subordinate that it does not receive the necessary emphasis to make composites a cost success. Too often in the commercial world, composites has been a case of "the operation was a technical success, but the patient died an economic death."

Making composites an economic success is primarily a management responsibility. The organization must be structured so that an identifiable individual is responsible for cost. This is no easy task. Figure 10 illustrates the major cost centers ($) in the design-fab cycle. Each one of these value-added tasks contributes to the cost success (or failure) of a program.

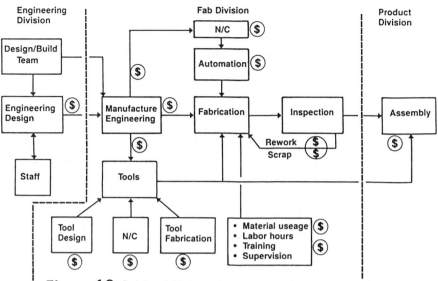

Figure 10 Critical Value-Added Tasks
Requiring Focused Management Responsibility

D.   Address all of the issues affecting costs.

The fourth step in implementing cost effective design and flexible automation is a solid development program that addresses all of the issues affecting cost. Figure 11 is a mind map of many of these economic issues. This paper has emphasized the manufacturing technology and cost effective design branches as being the key to controlling costs. However, the other branches cannot be ignored. Raw materials price is a significant portion of advanced composites structures (up to 25 percent of recurring cost). As can be seen from figure 11, there are a lot of issues that affect material price that need to be worked. Also, innovation through the use of novel material forms could provide a breakthrough in costs for fittings, frames, and similar complex structure.

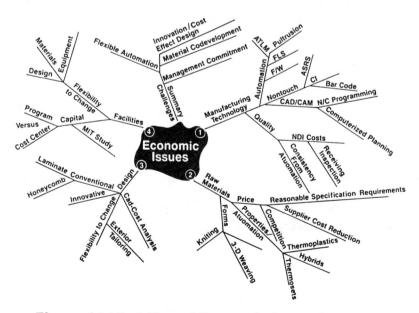

Figure 11 Mind Map of Economic Issues in
the Manufacturing of Advanced Composites

The issue of facilities cost is a real Catch 22 problem. Do you amortize the cost of automation against the parts being built, against a specific program or product, or add the cost to a burdened over- head rate and amortize it against all products made in a certain cost center?

A recent MIT study[1] involving a very simple economic model showed that facilities amortized against parts produced are not cost effective except for wet filament winding or wet pultrusion which involves low equipment costs. Amortizing costs against a certain model or product may push the breakeven point out to where a product manager may decide against the use of composites on his product. The last alternative, amortizing against an overhead burden rate, artificially drives up recurring costs to lead a manufacturer to decide to offload to a lower cost subcontractor who does not have automated capability. Thus, the design and development required to automate never takes place.

The solution is, again, flexible automation of cost effective designs. The designs should be automatable on generic, as opposed to point design, automated equipment applicable to a wide spectrum of parts. This will enable maximum utilization of equipment, and minimize types of automation and the overall facilities cost.

7.   SUMMARY

Thus far, in the history of advanced composites usage on major airframe structure, we do not have verifiable numbers that show that the cost of composite structure is competitive with aluminum. Experience on specific components that can be automated, such as rotor blades or filament wound tanks, shows that the potential for composites cost advantage is there, if the design and application is right. The question is, how do we ensure the right design and fabrication approach?

We have learned a lot from existing composites applications. It is difficult to compete cost-wise on applications where composites is replacing existing metal structure, both from a learning curve and design restriction standpoint. Although manufacturing automation capabilities have expanded significantly, the keys to successful cost effective composite structure are cost effective designs and flexible (adaptable to change) automation.

To facilitate implementation of cost effective designs and flexible automation, the four following steps are recommended:
    A.   Get cost analysis tools into the hands of the designer.
    B.   Challenge costly ground rules.
    C.   Assign focused responsibility for cost.
    D.   Address all of the technology issues.

The maturation of composites technology is now to the point where it can be applied cost effectively. As we apply the technology, it will further evolve in the materials, design and manufacturing areas, to where startling performance and cost breakthroughs will reward those who go forth in this exciting arena.

1   S. Krolewski and T. Gutowski, SAMPE Journal, May/June 1987.

SYMPOSIUM III

Biotechnology as Related to Composite Materials

Hierarchical Structure of the Intervertebral Disc

J. J. CASSIDY, A. HILTNER AND E. BAER

ABSTRACT

Optical microscope techniques are used to characterize the hierarchical structure of the collagenous components of the intervertebral disc. In the anterior annulus fibrosis, the thickness of lamellae increases abruptly 2 mm inward from the edge of the disc, dividing the annulus into peripheral and transitional regions. Lamellae in the lateral and posterior aspects of the disc have a broad distribution of lamellar thicknesses throughout the annulus. In alternating lamellae, fibers are inclined with respect to the vertical axis of the spine in a layup structure. From the edge of the disc inward to the nucleus, this interlamellar angle decreases from ± 62 to ± 45 degrees. Within lamellae, the collagen fibers exhibit a planar crimped morphology. The plane of the waveform is inclined with respect to the vertical axis by the interlamellar angle. From the edge of the disc inward, the crimp angle increases from 20 to 45 degrees and the crimp period decreases from 26 to 20 um. A hierarchical model of the intervertebral disc has been developed that incorporates these morphological gradients.

INTRODUCTION

The intervertebral disc is a complex hierarchical structure composed of the annulus fibrosis and nucleus pulposis which are anchored to the vertebral bodies by cartilagenous endplates. The annulus is made up of fibers of collagen type I arranged in concentric lamellae around the nucleus [1]. The fibers are organized in an alternating layup structure at an angle of about 65 degrees with respect to the spinal axis [2]. The ends of the fibers are bound to the cartilage endplates, thus anchoring the lamellae [1]. The nucleus pulposis is a amorphous gel of collagen type II, proteoglycans, and contains up to 88% water [3]. From the peripheral annulus to the nucleus there exists a smooth gradient of collagen types. At the edge of the annulus more than 95% of the collagen is type I. This percentage decreases along an approximately linear gradient to less than 5% in the nucleus. An opposing gradient exists for collagen type II, which comprises more than

Center for Applied Polymer Research and Department of Macromolecular Science, Case Western Reserve University, Cleveland, Ohio 44106

95% of the collagen of the nucleus and less than 5% of the annulus [4]. The morphology of these two collagen networks and their possible interaction is not understood.

The intervertebral disc in vivo serves to absorb shock and permit motion in the spinal column. It is loaded primarily in compression with elements of torsion, flexure, and extension. Mechanical testing in compression yields a non-linear stress-strain curve. Bulging of the disc during compressive loading indicates that axial loads are transmitted into radial and tangential tensile stresses in the lamellae of the annulus [5]. This supports the hypothesis that the disc behaves in compression as an incompressible fluid contained in a thick-walled cylinder [6]. More recently, investigators have explored the relationships among disc geometry, arrangement of collagen fibers in the annulus fibrosis, and compressive mechanical properties. A simple model based upon this work correlates mechanical properties with the arrangement of collagen fibers in the annulus by treating then as tendons and relating fiber extension to stress [7].

The object of this study is to characterize the hierarchical structure of the collagen in the intervertebral disc at the resolution of the optical microscope. These observations are incorporated into a model that correlates macroscopic mechanical behavior with the response of the disc structure to compressive deformation.

METHODS

Five fresh human lumbar spinal segments were removed at autopsy by transverse sectioning through the vertebral bodies. The subjects were ages 31, 36, 56, 59, and 80 years. The cause of death in all cases was unrelated to bone or connective tissue disease.

Specimens were double wrapped in plastic to maintain a moisture level of approximately 100 percent. If specimens were not used immediately, they were refrigerated at 5°C for not more than 24 hours. Individual segments were isolated by cutting through the vertebral bodies above and below the disc using a bandsaw. Posterior articulating facets and pedicles were removed so that the remaining test specimen consisted of an isolated intervertebral disc with approximately 1 cm of viable vertebral bone above and below. This specimen preparation involved exposing the disc to the air for a period of not more than 30 seconds which minimized water loss. The anterior-posterior length was approximately 30 mm and the lateral width was 40 mm in all cases and the height range 9.5 to 11.0 mm at the anterior periphery.

Discs for histomorphometry were immersed in formalin (40% formaldehyde solution) for 48 hours prior to sectioning. The disc was separated from the vertebral bodies by making a transverse cut through the cartilage endplate with a razor blade. Sections were taken along orthogonal planes related to the cylindrical geometry of the disc. These were in the transverse and radial (sagittal) anatomic directions and in the circumferential direction by cutting a chord across the periphery of the disc. In addition, sections were cut through the annulus fibrosis parallel to the long axis of the collagen fibers. Sections were mounted in paraffin, cut on a microtome, and mounted on slides. Three histological stains were employed: hemotoxylin and eosin for connective tissue, trichrome for collagen, and sirius red for collagen birefringence measurements. One disc from each of four subjects was used to measure lamellar thickness. A second disc was used to measure the interlamellar angle. A disc from each of two subjects was used to examine the morphology of the collagen fibers in the annulus fibrosis.

Transmission optical microscopy was used to examine the disc specimens over a range of magnifications from 10X to 1000X. A qualitative examination of the transverse, radial, and circumferential anatomy of the disc was made using brightfield and Nomarski differential interference contrast techniques. Quantitative measurements of the characteristic parameters of the collagen fibers were made using planar crossed polarized light as per the previously reported methodology [8].

RESULTS

A radial section through the disc, viewed between crossed polarizers in the optical microscope, reveals the relationship between the annulus fibrosis, nucleus pulposis, and cartilage endplates (Figure 1). While the nucleus and cartilages exhibit little optical activity, the dense fibrous lamellae are highly active. The lamellae of the annulus fibrosis are separate and distinct from one another with no interconnections visible at the optical microscope level.

In sections taken through the anterior annulus fibrosis, the thickness of the lamellae varies with radial location (Figure 2). At the periphery of the disc, lamellae average approximately 130 µm in thickness although at the very edge of the disc, many lamellae less than 100 µm in thickness were found. This region extends from the edge of the disc inwards for 2.2 mm and represents the outermost 18 lamellae. At this point, there is a sharp increase in lamellar thickness to approximately 260 µm. This thickness is maintained through the balance of the annulus inward to the nucleus pulposis from 2.2 to 7.5 mm in depth and comprises about 20 lamellae. This two part distribution in lamellar thickness divides the anterior annulus fibrosis into a peripheral region and a transitional region. In the vicinity of the nucleus, the boundaries between lamellae become poorly defined and the lamellae gradually are replaced by nuclear material. There is no clear demarcation between the annulus fibrosis and nucleus pulposis.

Lamellar thickness was also measured at the lateral and posterior aspects of the intervertebral disc. In sections through the lateral annulus, thickness values are widely scattered over a range from 80 to 400µm. In the posterior annulus the lamellar thickness is also scattered, however lamellae are generally thinner than in the lateral annulus with thicknesses ranging from 50 to 250µm. In both the lateral and posterior regions there is no change in the average lamellar thickness through the depth of the annulus fibrosis. A distinction between peripheral and transitional regions, found in the anterior annulus is not observed.

A circumferential section cut through the annulus fibrosis shows the organization of the fibrous layers (Figure 3). In successive lamellae, the orientation of the fibers alternates with respect to the spinal axis. This interlamellar angle is not constant through the thickness of the annulus, as reported previously [2]. The angle formed between the collagen fibers in the lamellae and the axis of the spinal column decreases from the edge of annulus inward.

Measurements of the fiber orientation were made on sections cut through the annulus at a point halfway between the anterior and lateral aspects on the disc (Figure 4) and halfway between the lateral and posterior.   In both positions, the angle of fiber orientation decreased from 62 degrees at a depth of 0.5mm to 47 degrees close to the nucleus. Since the lamellae in the anterior-lateral region are wider than in the posterior-lateral region, the 47 degree fiber orientation was observed at a depth of 7mm in the former and 5mm in the latter. Close to the nucleus pulposis it is difficult to differentiate between lamellae, and data points may be off by one lamella or approximately 300µm. Nevertheless, it appears that the angle of fiber orientation does not vary within a single lamellae as it circumscribes the disc, although the angle varies systematically from one lamella to the next. Through the depth of the annulus the decrease in fiber angle is gradual and no sharp transition is observed. The data are remarkably reproducible from specimen to specimen.

A section through the intervertebral disc cut parallel to the fibers in one set of alternating lamellae shows the concentric fibrous lamellae of the annulus fibrosis surrounding the nucleus. Viewed at higher magnification between crossed polarizers, characteristic extinction bands are observed along the collagen fibers. This pattern is similar to that observed in tendon and was analyzed using the same technique employed by previous investigators [8]. Periodic light and dark banding when viewed between crossed polarizers is indicative of periodic variation in orientation within a birefringent fiber. Dark bands are produced in those regions of the wavy fibers where the tangent to the wave is parallel to one of the polarization directions. When the fibers rotated about an axis normal to microscope stage, the banding pattern will shift. With the crossed polarizers oriented parallel and perpendicular to the overall fiber directions, the dark bands are evenly spaced along the fiber (Figure 5a). Upon rotation of the microscope stage, pairs of dark bands approach each other producing alternate narrow and wide light intervals. Upon further rotation to a particular angle 2ø, the dark bands bordering the narrow interval merge into a single dark band (Figure 5b). Further rotation causes the bands to become lighter until the fiber is uniformly bright where the polarizers are oriented at 45 degrees. Rotation of the microscope stage in the opposite direction yields the same effect in reverse (Figure 5c) and the dark bands are now located at positions where the collagen fibers were bright previously. This behavior denotes a planar crimped waveform configuration of the collagen fibers for which the crimp angle ø is readily obtained from the polarizer angle [8].

The crimp parameters at the anterior-lateral location were measured through the depth of two disc specimens. The crimp angle, ø, and crimp period, p, were measured and from these data, the crimp length, l, was calculated. The crimp angle increases from about 20 degrees at the periphery to 45 degrees in the lamellae closest to the nucleus while the period appears to decrease linearly through the depth from 26 to 20 µm (Figure 6). The calculated crimp length is constant through the depth of the annulus at about 13.5µm. This suggests that the length is the fundamental parameter of the crimp.

The suggested functional relationships are shown with solid lines in Figure 10: l is constant at 13.5µm, the period increases linearly through the depth of the annulus according to p = 0.70 x (depth) + 26.5, and the crimp angle increases as $\cos^{-1}$ (p/2l). This function expanded is:

$$\o = \cos^{-1} \{0.30 \text{ x (depth)} + 1\} \qquad (1)$$

Deviation of the data from this curve near the periphery of the disc indicates that p is not precisely linear with depth.

DISCUSSION

The intervertebral disc is a two part structure composed of the gelatinous nucleus pulposis surrounded by the concentric fibrous lamellae of the annulus fibrosis. A model for the fibrous organization of the annulus fibrosis is proposed based upon examination of the collagen fibers through the optical microscope. This model (Figure 7) incorporates the structural hierarchy of the collagen fibers on the macro- and microscopic scales. At the anterior aspect of the disc, the annulus fibrosis is divided into peripheral and transitional regions by the thickness of the lamellae. In the peripheral region, the lamellae are about 130µm or one-half the thickness of lamellae in the transitional region. The two regions are clearly delineated. At the lateral and posterior aspects of the disc, the lamellar thicknesses are distributed over a broad range through the depth of the annulus and the peripheral and transitional regions cannot be discerned. The average lamellar thickness is greater at the lateral aspect of the disc than at the posterior. Since the number of lamellae is the same at all locations, variations in lamellar thickness account for the annulus being thickest at the anterior region and thinnest at the posterior.

The lamellae that comprise the annulus fibrosis are continuous layers that form cylindrical walls around the nucleus pulposis. In each layer, the collagen fibers are arranged parallel to one another. In successive layers, the orientation of the fibers alternates with respect to the axis of the spinal column. For each lamellae, this interlamellar angle, θ, remains constant around the circumference of the disc. From the edge of the disc inward through the annulus, the interlamellar angle decreases in a linear manner.

Within each lamellae, collagen fibers are arranged in parallel arrays. The fibers display a planar crimped waveform with all fibers crimped in register. The plane of the waveform is perpendicular to the lamellar plane and inclined with respect to the spinal axis by the interlamellar angle. A gradient of crimp geometry exists in the annulus. The crimp angle, $\emptyset$, increases and the crimp period, p, decreases linearly from the periphery of the disc inward. The length of the crimp, l, remains constant.

The significance of this model and the complex gradient structure it represents can be appreciated when examining the mechanical response of the intervertebral disc to compression. Bulging of the disc during compression indicates that axial loads are translated to radial and tangential tensile stresses in the lamellae of the annulus fibrosis [5]. In this case, the expected response of the collagen fibers is reorientation in the radial direction and uncrimping of the waveform as is seen in the tendon and intestine. X-ray diffraction of the annulus under compression has demonstrated an increase in the interlamellar angle [9]. Preliminary experiments in this laboratory have shown a decrease in crimp angle and increase in crimp period as well as an increase in the interlamellar angle when the disc was fixed under a compressive load. This response of the hierarchical structure predicts a compressive stress-strain curve with a low modulus toe region during crimp straightening ans reorientation, a linear elastic region corresponding to extension of the fibers, a finally, a yield region as the structure undergoes irreversible damage. Compression of an isolated intervertebral disc specimen does show the three-part stress-strain curve characteristic of collagenous tissues in tension. As in the intestine, the lamellae will also be most resistant to tensile stress in the directions parallel to the collagen fibers. Uniaxial tensile tests of specimens cut through several layers of the peripheral annulus show maximum stiffness along the two fiber axes [10].

As a result of structural gradients through the annulus, the lamellae are not equally deformable. Peripheral lamellae, with a larger interlamellar angle and a smaller crimp angle, are less deformable than those closer to the nucleus. One study has attempted to measure the mechanical properties of the lamellae as a function of depth through the annulus. It was reported that specimens cut from peripheral lamellae were stiffer and had lower energy dissipation and residual deformation than those of more central lamellae [10].

In conclusion:

1. The collagen fibers that comprise the lamellae of the annulus have a planar crimped conformation similar to that seen previously in other collagenous tissues.

2. The structure of the intervertebral disc at the lamellar and fibrillar levels is characterized by gradients as lamellar thickness, interlamellar angle, and crimp angle vary systematically through the depth of the annulus fibrosis. A hierarchical model of the intervertebral disc that incorporates these observations has been devised.

3. The mechanical behavior predicted by this model is in qualitative agreement with experimental observations where the disc is tested in compression.

ACKNOWLEDGEMENT

The authors gratefully acknowledge the support of the U.S. Army Research Office (Grant No. DAAG29-84-K-0155).

REFERENCES

1. Inoue, H. and Takeda, T., "Three-dimensional observation of collagen framework of lumbar intervertebral discs", Acta Orthop. Scand., Vol. 46, 1975, pp. 949-956.

2. Hickey, D.S. and Hukins, D.W.L., "X-ray diffraction studies of the arrangement of collagenous fibres in human fetal intervertebral disc", J. Anat., Vol. 131, 1980, pp. 81-90.

3. Adams, P., Eyre, D.R., and Muir,; H., "Biochemical aspects of development and ageing of human lumbar intervertebral discs", Rheum. Rehab., Vol. 16, 1977, pp. 22-29.

4. Berthet-Columinas, C., Miller, A., Herbage, D., Ronziere, M., and Tocchetti, D., "Structural studies of collagen fibres from intervertebral disc", Biochim. Biophys. Acta, Vol. 706, 1982, pp. 50-64.

5. Markolf, K.L. and Morris, J.M., "The structural components of the intervertebral disc", J. Bone Joint Surg., Vol. 56-A, 1974, pp. 675-687.

6.   Nachemson, A., "The influence of spinal movements on the lumbar intradiscal pressure and on the tensile stresses in the annulus fibrosis", Acta Orthop. Scand., Vol. 33, 1963, pp. 183-207.

7.   Hickey, D.S. and Hukins, D.W.L., "Relation between the structure of the annulus fibrosis and the function and failure of the intervertebral disc", Spine, Vol. 5, 1980, pp. 106-116.

8. Diamant, J., Keller, A., Baer, E., Litt, M., and Arridge, R.G.C., "Collagen: ultrastructure and its relation to mechanical properties as a function of ageing", Proc. R. Soc. Lond., Vol. 180, 1972, pp. 293-315.

9.   Klein, J.A. and Hukins, D.W.L., "X-ray diffraction demonstrates reorientation of collagen fibres in the annulus fibrosis during compression of the intervertebral disc", Biochim. Biophys. Acta, Vol. 717, 1982, pp. 61-64.

10. Galante, J.O., "Tensile properties of the human annulus fibrosis", Acta Orthop. Scand., Suppliment 100, 1967, pp. 1-91.

Fig. 1. Radial section through the disc showing the annulus fibrosis, nucleus pulposis, and cartilage endplates. Note change in lamellar thickness between peripheral and transitional annulus fibrosis. (Sirius Red stain)

LAMELLAR THICKNESS VS. RADIUS

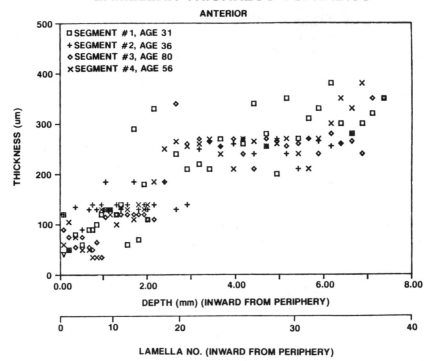

Fig. 2. The lamellar thickness gradient in the anterior annulus fibrosis: Plot of lamellar thickness vs. radius.

Fig. 3. Circumferential section through the annulus fibrosis showing orientation of collagen fibers in successive lamellae. (Sirius Red stain)

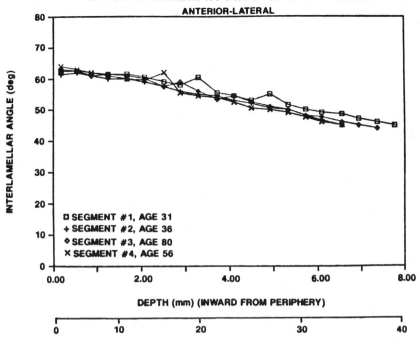

INTERLAMELLAR ANGLE VS. RADIUS

ANTERIOR-LATERAL

□ SEGMENT #1, AGE 31
+ SEGMENT #2, AGE 36
◆ SEGMENT #3, AGE 80
✕ SEGMENT #4, AGE 56

DEPTH (mm) (INWARD FROM PERIPHERY)

LAMELLA NO. (INWARD FROM PERIPHERY)

Fig. 4. The interlamellar angle gradient at the anterior-lateral aspect of the annulus fibrosis: Plot of interlamellar angle vs. radius.

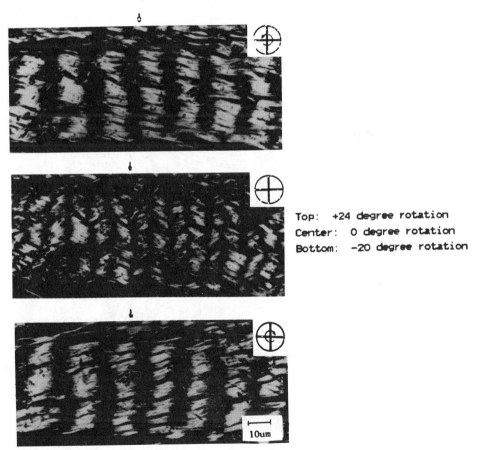

Top: +24 degree rotation
Center: 0 degree rotation
Bottom: -20 degree rotation

Fig. 5. Crimped collagen fibers within a single lamella stained with sirius red and viewed in crossed polarized light.

CRIMP ANGLE, PERIOD, LENGTH VS. RADIUS

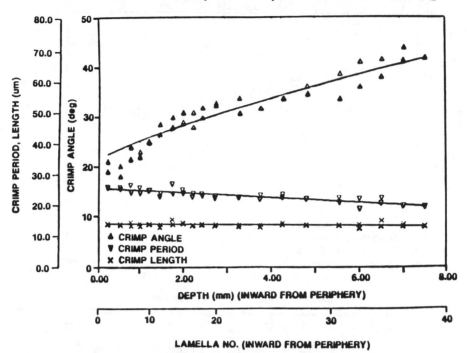

Fig. 6. The crimp angle and crimp period gradients in the annulus fibrosis: Scatterplot of crimp angle, period, and length vs. radius with suggested functional relationships highlighted with solid lines.

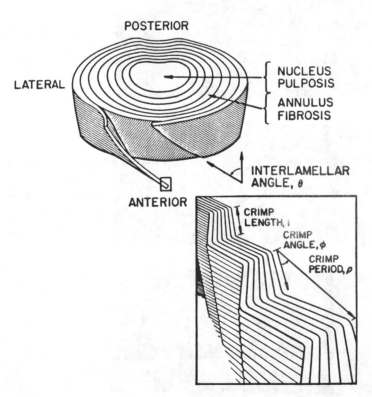

Fig. 7. Proposed hierarchical model of the intervertebral disc showing lamellar structure, fiber orientation, and crimp morphology in the annulus fibrosis.

Fracture Toughness Design in Horse Hoof Keratin

JOHN E. A. BERTRAM AND **JOHN M. GOSLINE**

ABSTRACT.

A fracture mechanics analysis was used to determine the relationship between fracture toughness, tissue microstructure and hydration level in horse hoof keratin. Fully hydrated material has greatest fracture resistance for cracks running parallel to the tubular component of the wall keratin (mean critical J-integral value 12 kJm^{-2}). This is two to three times greater than the value determined in the weakest direction, parallel to the distal contact surface of the hoof. Both stiffness and fracture toughness are profoundly affected by hydration level. Toughness is maximal (23 kJm^{-2}) at intermediate hydration (75% RH), which is very close to the hydration level found in the hoof in vivo.

INTRODUCTION.

The equine hoof-wall is an epidermal tissue which acts as an important component of the skeletal system. During locomotion the hoof-wall is subjected to repeated impact loads and abrasive interactions with the ground. In a galloping thoroughbred, for example, the vertical force acting on each foot is equal to or greater than the weight of the animal (about 9000 N). This load must be transmitted to the ground through the distal contact surface of the hoof, with a total area of about 20 cm^2. If this load is distributed uniformly over the contact area, the hoof-wall material experiences a stress of the order of 5 MPa. However, in the natural environment a horse would rarely encounter a substrate that allows uniform stress distribution, and it is likely that localized stresses reach an order of magnitude or more greater than this. To deal with these loads the hoof-wall material

1 J.E.A. Bertram, Dept. of Anatomy, Univ. of Chicago, Chicago, Il., 60637; J.M. Gosline, Dept. of Zoology, Univ. of British Columbia, Vancouver, B.C., Canada, V6T 2A9.

must be sufficiently rigid to prevent excess deformation, and
at the same time be tough enough to resist fracture.   In this
paper we consider how micro-structure and tissue hydration
effects provide the hoof-wall with mechanisms that lead to a
rigid and fracture-resistant hoof structure.   The details of
this study have already been published [1,2]

     Keratin is a fiber-reinforced composite material, formed
from long, slender α-helical protein fibers (ca. 10 nm in
diameter) embedded in a highly crosslinked, amorphous protein
matrix [3].   This composite is formed inside the cells that
synthesize these proteins, and the cells die in the final
stages of their differentiation when extensive inter- and
intra-molecular disulfide crosslinks are established between
the fibers and the matrix.   The mature keratin cells are
flattened, elliptical discs in which the α-helical fibers are
oriented parallel to the long axis of the cell.

     In the hoof, this molecular composite is arranged in a
complex pattern to form the micro-structure of the hoof-wall.
About half the keratin is arranged in cylindrical structures,
called tubules, that run the length of the hoof, parallel to
its external surface (Fig. 1). The axes of the cells and the
fibers they contain are arranged in a steep spiral around the
axis of the tubule and display a roughly alternating spiral

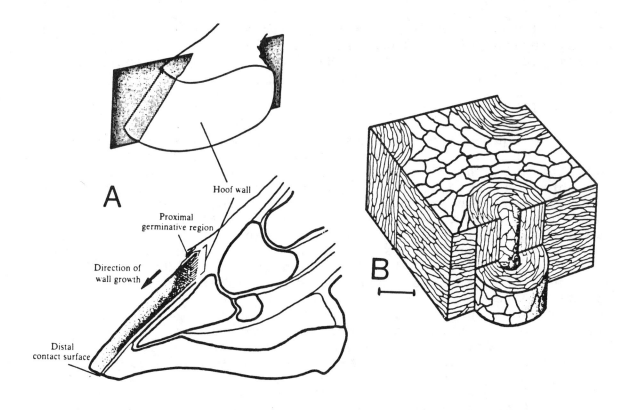

Fig. 1.   Diagram (A) of a whole hoof, and (B) a small
block of keratin, showing tubules, intertubular material
and cells. In (B), the top of the block is proximal;
the scale bar is 100 μm.

direction from layer to layer. The intertubular portion of
the wall keratin is composed of the same disc shaped cells,
but with cell axes and intracellular fiber orientations that
cross the tubule axes.

MATERIALS AND METHODS.

The toe region of the hoof-wall of freshly killed, adult
horses, obtained from an abattoir, was cut into rough blocks
and machined into samples (Fig. 2) for either tensile tests
(0.6 mm thick x 4.0 mm wide x 25 mm long) or compact-tension
(CT) tests (10 mm wide x 10 mm high x 3 mm thick). Samples
were machined on a Micromet saw, under a continuous flow of
water. Precut notches for CT specimens were cut with a
single-edged razor blade.

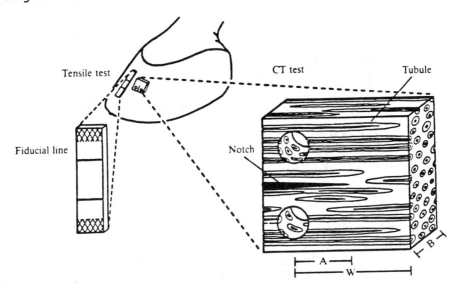

Fig. 2. Origin of tensile and compact tension samples.

The J-integral fracture toughness method assumes plane
strain conditions and requires fairly large specimens. The
specimen size in this study was the largest possible which
would ensure that only the central portion of the hoof-wall
was included in the specimen. Although smaller in overall
dimensions than recommended, the relative specimen dimensions
were those suggested for plane strain fracture analysis of
metals (ASTM standard E-399-80).
To evaluate the role of micro-structure in fracture, we
constructed four types of CT test samples, with the notch cut
at different orientations relative to the tubules and the
intertubular material (see Fig. 3). In orientation #1 the
notch was cut parallel to the axes of the tubules, and
therefore the notch would run in a proximal direction in the
hoof-wall. In orientation #3 the notch was cut normal to the
tubule axis. The intertubular orientation varies somewhat
between samples, and is expressed as a mean value (Table 2).

Orientations #2 and #4 were cut so that the notch ran at about
45° relative to the tubule axis, but in one case (#2) the
intertubular orientation ran roughly across the notch and in
the other (#4) the intertubular orientation ran nearly
parallel to the notch.

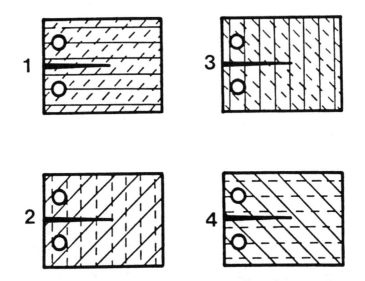

Fig. 3. Diagram of the orientation of the precut
notch relative to tubular orientation (solid lines) and
intertubular orientation (dashed lines)

       Tensile tests were conducted on an Instron Model 1122,
at an extension rate of 5 mm min^{-1}. Samples were clamped at
both ends with pneumatic grips, and in order to eliminate end-
effects, specimen strain was measured by following the
movement of fiducial marks using a video dimension analyzer
(Model 303, Instruments for Physiology and Medicine, La Jolla,
Calif.)
       Fracture toughness was evaluated using the J-integral
method [4], a technique that is useful for materials like
keratin that show a considerable degree of plastic
deformation. Crack growth measurements were obtained by the
compliance calibration method. This method involves measuring
the compliance of samples with known notch lengths, and then
inferring the effective notch length from changes in the
apparent compliance of CT specimens as the notch grows during
the fracture test. The details of the method and the
compliance calibration curves are given in [1,2].
       Fracture tests on fully hydrated material were carried
out with the specimen immersed in water at room temperature.
Tests at reduced hydration levels were carried out at 0%
relative humidity (RH), 53% RH and 75% RH in sealed chambers
which contained, respectively, phosphorus pentoxide drying
agent, saturated $Mg(NO_3)_2$ or saturated NaCl. Test samples
were allowed to equilibrate for 5 days in these chambers prior
to testing.

RESULTS.

   Tensile Tests.   As   shown   in   Table 1 and Fig 3, the
tensile properties of horse-hoof keratin are not significantly

RH (%)	E (GPa)	YIELD STRESS (MPa)	ORIENTATION RELATIVE TO TUBULES
100	0.41	9.2	PARALLEL
100	0.48	11.8	PERPENDICULAR
75	2.63	38.9	PARALLEL
53	3.63	--	PARALLEL
0	14.6	--	PARALLEL

TABLE 1.   Summary of tensile properties.   There is no
significant difference in stiffness or yield stress
between the parallel and perpendicular orientations at
100% RH   Other differences are significant.

altered by test orientation but are profoundly influenced by
hydration conditions.   Initial stiffness varies 40 fold
between fully saturated and fully dry material.   At higher
hydration levels hoof-wall keratin behaves much as other hard
keratins, displaying an initial linear elastic region after
which yield occurs.   Failure usually occurred outside the
fiducial marks, near to the grips, and for this reason

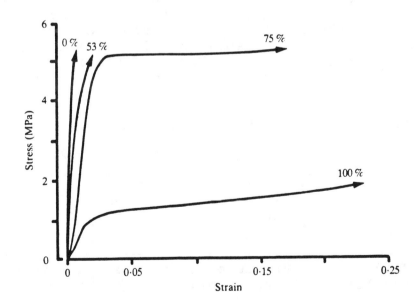

Fig. 4.   Typical tensile tests, stress parallel to
tubule axis, at four hydration levels.   Arrowhead
indicates sample failed at grips.

absolute strength and extensibility were not recorded. Stress
levels were, however, maintained in the post-yield region, and
some strain hardening behaviour was seen.

At lower hydration levels, little or no yield is seen,
and the material fractured in a "brittle" manner. This
fracture occurred, however, at stress levels several times
greater than the yield stress of the fully hydrated material.
The combination of high yield stress and high strain seen at
75% R.H. indicates a great deal of energy absorption at
intermediate hydration levels. This observation is confirmed
in fracture tests.

<u>Fracture Tests: Position and Orientation Effects.</u>
Initial fracture tests on orientation #1 samples at 100% RH
obtained from different vertical levels in the hoof wall
indicated that there were no significant differences in
fracture toughness, as expressed by the mean, critical J-
integral (J_{crit}), with location, with the exception that the
value for samples from the most distal position (i.e. about 1
cm from the contact surface of the hoof) was significantly
reduced to about half that at more proximal levels. This is
interesting because the distal location is the oldest material
present in the wall. It suggests that the hoof-wall keratin
accumulates fatigue damage, likely micro-cracks, with long-
term impact loading that weakens the distal portions of the
hoof. In all following tests only material from proximal
regions was used.

Table 2 summarizes the results of fracture tests carried
out at 100% RH on samples cut with notch orientations #1-#4,
as listed in methods above. These data demonstrate the
importance of the intertubular material in the fracture of
hoof-wall keratin. Orientation #1 shows the highest mean
fracture toughness (J_{crit}). In this case the notch runs
parallel to the tubules, in a proximal direction in the hoof,
and runs across the orientation of the intertubular material
at a mean angle of 59°. For orientation #3 the notch runs
perpendicular to the tubular direction. If hoof-wall keratin

ORIENTATION	MEAN ANGLE REL. TO NOTCH (°)		J_{crit}
	TUBULAR	INTERTUBULAR	(kJm^{-2})
#1	0	58.9	11.9
#2	39.3	75.5	8.0
#3	90	35.2	10.8
#4	43.8	14.1	4.6

Table 2. Fracture toughness in relation the to
morphological organization of the hoof-wall.

is a composite in which the tubules act as reinforcing fibers
and the intertubular portion acts merely as a matrix, then the
fracture toughness of orientation #1 should be lower than that

of orientation #3. This is obviously not the case (Table 2). Although not statistically different, the mean J_{crit} for #1 is somewhat larger than that for #3, suggesting that the intertubular material is important in resisting fracture.

Comparison of results for #2 and #4 illustrates this more clearly. Recall that in these samples the notches were cut at 45° to the tubules but either perpendicular to (#2) or parallel to (#4) the intertubular orientation. This comparison focuses on the role of the intertubular material. The J_{crit} for #2 is significantly greater than that of #4 by a factor of about two, clearly demonstrating the difficulty of propagating cracks across the intertubular material.

The dominance of intertubular orientation over crack propagation was further documented by measuring the direction of crack growth in CT specimens after fracture. Typically, cracks in samples with orientation #4 grew parallel to the pre-cut notch (i.e. parallel to the intertubular direction), whereas cracks in samples with orientation #2 diverged markedly from the notch direction and often grew at right angles to the notch. Linear regression analysis of all CT test specimens revealed a significant ($p < 0.05$) correlation between intertubular orientation and the direction of crack growth. No correlation was observed between crack growth and tubular orientation. Apparently, cracks can not be forced to cross the "grain" of the intertubular material. Rather they are diverted by and forced to propagate in between the layers of intertubular material, and in so doing propagate across the tubules.

Fracture Toughness: Hydration Effects. Table 3 summarizes the results of fracture tests carried out on orientation #1 samples equilibrated at various values of RH. As suggested in the tensile tests, intermediate hydration levels produced higher fracture toughness values than either

RELATIVE HUMIDITY (%)	MEAN J_{crit} (kJm^{-2})	WATER CONTENT (g H_2O/g protein)
100	11.9	0.40
75	22.8	0.18
53	5.6	0.12
0	8.7	0.05

Table 3. Compact tension fracture test results at different hydration levels. Water contents were determined by oven drying for 24 hr at 105°C. The residual water content at 0% RH probably indicates that some very tightly bound water was not removed by the phosphorus pentoxide drying agent or some other volatile material was removed with oven drying.

full hydration or very low levels of hydration. The value of 22.8 kJm^{-2} measured at 75% RH is significantly different from

the values at all other hydration levels.   The value   at 100%
RH was   significantly different from that at 53% RH, but could
not be distinguished from the value   at 0%   RH.   The means at
53% and 0% RH were not significantly different.

DISCUSSION.

   <u>Fracture   Toughness   of   Hoof-wall   Keratin.</u>   This study
provides the first documentation of the   fracture toughness of
hoof-wall keratin.   In   comparison to   other biomaterials for
which toughness data are available, hoof keratin appears to be
quite   exceptional   in   its   ability   to resist fracture.   For
example, the work of fracture for fresh bone is about   2 kJm^{-2}
[5]   and   wood   is   about   10 kJm^{-2} [6].   At optimum hydration
(i.e. 75% RH)   horse-hoof   keratin is   about 23   kJm^{-2}.   Thus,
hoof keratin   is   an   order of   magnitude tougher than bone and
about two times tougher than wood.   These   values   for hoof
keratin are similar to values reported recently for bovid horn
keratin [7], where it was   suggested   that   the   toughness may
approach 80 kJm^{-2} at the very tip of the horn.   Thus, keratins
are probably the toughest   materials in   the biological world,
and   their   toughness   compares   well   with   that   of man-made
composite materials.
   Both   the   stiffness   and   the   toughness   of hoof-wall
keratin are strongly affected by water content (Tables 1 & 3).
In addition, the water content of the keratin in the hoof of a
living horse   will vary   with location   and with environmental
humidity.   Thus,   it   is   possible,   indeed   likely,   that the
mechanical properties of the   keratin are modulated by shifts
in water content, so that this single biomaterial   can perform
a   variety   of   different   mechanical   functions   in different
regions of the hoof.   For example, The inner   surface of the
hoof-wall makes contact with living tissues and therefore will
be fully saturated.   The relatively   low stiffness,   low yield
stress   and   large   plastic   deformation   characteristic   of
saturated keratin may facilitate   the   transfer   of mechanical
stress from   the rigid bones of the internal skeleton, through
the soft basal membrane and living   tissues, to   the hoof-wall
[2].   The majority   of the hoof-wall thickness, however, must
deal with   the transmission of locomotor   forces in transient
impacts with the ground, and for this role adequate stiffness,
coupled with excellent   fracture   resistance,   seem essential.
The actual   water content   of the load-bearing, mid-portion of
the hoof-wall <u>in vivo</u> has been measured to be between 0.17 and
0.24 g H$_2$0 per g protein [8], and this range just brackets the
water content achieved at 75% RH   in the   present study (Table
3).   Thus, the bulk of the keratin in the hoof-wall functions
with a stiffness of   the order   of 3   GPa and   with a fracture
toughness that   is likely   to be   at or near an optimum value.
This leads us to conclude that the control of water content in
the hoof-wall is an important biological process that warrants
further attention.
   The data   presented here   provide sufficient information

to estimate the critical flaw size of hoof-wall keratin, Critical flaw size, a_{crit}, will be calculated as:

$$a_{crit} = E J_{crit} / \sigma_y^2$$

where E is modulus and σ_y is yield stress. Using values for 75% RH, the critical flaw size is about 4 cm. Since the whole hoof is only about 10 cm tall, flaws of this size are not expected and catastrophic fracture seems extremely unlikely. This is obviously a good thing for the horse.

The Fracture Design of the Hoof-wall. The hoof-wall is made from a material, keratin, that is a composite at the submicroscopic level. In addition, this molecular composite is arranged into larger scale structures that provide mechanisms which control the direction of crack propagation in the intact hoof. Because the hoof is part of the animals external epidermal layer, it must provide a barrier to separate the animal from its external environment as well as provide the supporting structure that transmits locomotor forces to the ground. Cracks that run parallel to the tubules, from the distal contact surface towards the proximal growth zone (see Fig. 1) are extremely dangerous, as they would expose the living tissues inside the hoof-wall to the outside world. It is not surprising, therefore, that the fracture toughness is greatest for cracks running in this direction (Table 2). In fact, we were unable to propagate cracks in this proximal direction in our tests because in all cases the cracks were diverted into the plane of the intertubular material. Thus, the value of 23 kJm^{-2} estimated for the toughness in this direction is really an underestimate of the true fracture toughness for cracks propagated parallel to the tubules.

The direction of least fracture toughness is in the plane of the intertubular material (orientation #4), and apparently cracks that start to grow in other directions are redirected into this plane. Thus, the intertubular material provides a "safety valve" that prevents dangerous, proximally-directed cracks by diverting them into a less dangerous, lateral direction. In addition, fracture in the plane of the intertubular material likely plays a role in determining the shape of the hoof. To a good first approximation the plane of the intertubular material is parallel to the ground contact surface, and if hoof wear occurs by fracture in this plane at the distal contact surface, then such programmed wear will maintain hoof shape [1]. Finally, controlled wear in this plane may actually reduce the chance of catastrophic fracture in the hoof-wall. The reduced fracture toughness of the most distal hoof-wall material indicates that the oldest tissue is subject to fatigue damage, most probably in the form of micro-cracks. The removal of older keratin, with its accumulated fatigue damage, and its replacement with newer material produced at the proximal growth zone may reduce the likelihood of major cracks occurring.

Since the intertubular material is largely responsible for preventing proximally directed cracks, we should ask, what is the role of the tubules? If all the keratin in the hoof-wall was oriented like the intertubular material, it is likely that the fracture toughness in the proximal direction would be even greater, as all the keratin in the hoof would then be oriented to prevent cracks from growing in this direction. It is equally likely, however, that the fracture toughness in the plane of the hoof contact surface would be very low and that the hoof would experience unacceptably rapid wear. Thus, we infer that the tubular micro-structure of the hoof-wall provides a higher-level composite structure that resists fracture in the lateral direction by obstructing the growth of cracks parallel to the intertubular direction. Thus, the tubules function in much the same way that knots increase the energy required to split a block of wood with the grain.

REFERENCES

1. Bertram, J.E.A. & Gosline, J.M. (1986) Fracture toughness design of horse-hoof keratin, J. Exp. Biol.,125:29-47.

2. Bertram, J.E.A. & Gosline, J.M. (1987) Functional design of horse hoof keratin: modulation of mechanical properties through hydration effects, J. Exp. Biol., 130: 121-136.

3. Fraser, R.D. & MacRae,T.P. (1980) Molecular structure and mechanical properties of keratins, in: Mechanical Properties of Biological Materials, J.F.V. Vincent & J.D. Currey, (eds.), Symp. Soc. Exp. Biol., XXXIV, 37-74.

4. Rice, J.R. (1968) A path independent integral and the approximate analysis of strain concentration by notches and cracks, J. Appl. Mech., 35: 379.

5. Wright, T.M. & Hayes, W.C. (1977) Fracture mechanics parameters for compact bone, J. Biomech., 10:410-430.

6. Jeronimidis, G. (1976) The fracture of wood in relation to its structure, in: Wood Structure in Biological and Technological Research, Baas, P, Bolton, A.J. & Catling, D.M. (eds.), Leiden: The University Press.

7. Kitchner, A. (1987) Fracture toughness of horns and a reinterpretation of the horning behaviour of bovids, J. Zool., Lond., 213:621-639.

8. Leach, D.H. (1980) The structure and function of the equine hoof-wall, PhD. Thesis, Dept. of Vet. Anatomy, University of Saskatoon, Saskatoon, Saskatchewan, Canada.

Development Program for the Microbial Conversion of Phenylacetylene to Meta-Hydroxy Phenylacetylene

PAUL E. SWANSON AND **RONALD J. HUSS**

ABSTRACT

Biological systems were evaluated for the ability to convert phenylacetylene to meta-hydroxy phenylacetylene. Two approaches were taken. One was a cell-free, enzymatic approach. Commercially available enzymes were evaluated for the ability to perform the bioconversion in the presence of oxygen and a variety of cofactors.

The second approach was a whole cell, microbial conversion. Ninety microorganisms were evaluated for the ability to specifically hydroxylate phenylacetylene.

Strategies for the continued development of this bioconversion system are discussed, along with critical aspects concerning the economic feasibility.

INTRODUCTION

The United States Air Force and the aeronautics industry are interested in the development of innovative and improved technology for the production of high performance, advanced aerospace composite resins. Characteristics of these resins include high temperature and moisture insensitivity and resistance to thermooxidative degradation.

One type of resin that may have potential for aerospace applications is the acetylene-terminated resin. A component of this type of resin is the end-capping group, meta-hydroxy phenylacetylene. Chemical synthesis of meta-hydroxy phenylacetylene is a multistep, expensive procedure. The current high cost for the chemical preparation of this compound has inhibited further development of acetylene-terminated resins.

Bio-Technical Resources was awarded a Phase I research contract by the Air Force through the Small Business Innovation Research Program to investigate biological approaches to the synthesis of meta-hydroxy phenylacetylene from phenylacetylene.

Paul E. Swanson, Research Fellow, and Ronald J. Huss, Director of Research, Bio-Technical Resources, Inc., 1035 South Seventh Street, Manitowoc, WI 54220

TECHNICAL FEASIBILITY

Enzymatic Conversion Program: Four enzymes, horseradish peroxidase, lactoperoxidase, chloroperoxidase, and tyrosinase, were evaluated under a variety of reaction conditions to determine the technical feasibility for the specific meta-hydroxylation of phenylacetylene. In addition to oxygen, these reactions require a hydrogen donating cofactor. A variety of cofactors were used. They included dihydroxy fumaric acid, nicotinamide adenine dinucleotide (reduced form), nicotinamide adenine dinucleotide phosphate (reduced form), flavin mononucleotide, ascorbic acid, erythorbic acid, cysteine, hydroquinone, resorcinol, and catechol. Reaction variables included time, temperature, pH, medium composition, concentrations of substrates, enzymes, and cofactors, and the addition of activating agents.

Two of the enyzmes, horseradish peroxidase and tyrosinase, converted phenylacetylene to meta-hydroxy phenylacetylene in the presence of oxygen and dihydroxy fumaric acid. However, reaction yields were very low. Other products, especially para-hydroxy phenylacetylene, were produced in higher concentrations than the meta-isomer. Varying reaction conditions afforded little improvement. Kinetic studies indicated that hydroxylation of phenylacetylene did not stop at monohydroxylation. None of the less expensive, more readily available cofactors could replace dihydroxy fumaric acid in the reaction. The inability of a commercial enzyme to specifically hydroxylate phenylacetylene using an inexpensive cofactor makes this system unattractive for further development.

Microbial Conversion Program: This program evaluated ninety microorganisms, including 49 fungi and 41 bacteria, for the ability to catalyze the bioconversion of phenylacetylene to meta-hydroxy phenylacetylene. Phenylacetylene is not a normal microbial substrate. Some microorganisms are unable to metabolize phenylacetylene, others are capable of partial degradation, still others are able to metabolize phenylacetylene completely and utilize it as a sole source of carbon and energy.

Microbial metabolism of aromatic compounds is typically oxidative degradation via hydroxylations of adjacent aromatic ring carbons followed by ring opening. Hydroxylated aromatic compounds such as meta-hydroxy phenylacetylene may be transient intermediates in this degradation pathway. These compounds are not expected to accumulate to high levels without genetic manipulation of the microorganism.

The technical feasibility for the microbial conversion of phenylacetylene to meta-hydroxy phenylacetylene was clearly demonstrated. Not only were several fungi and bacteria identified in the screening program that are capable of metabolizing phenylacetylene, but the desired product, meta-hydroxy phenylacetylene was also detected, suggesting that it may be an intermediate in the degradation pathway.

STRATEGIES FOR DEVELOPMENT OF THE PROCESS

The technical objectives for development of this bioconversion are to continue the microbial screening program to identify the best organism for the bioconversion and to begin development of the organism and the fermentation process. Microorganisms screened in the Phase I technical feasibility study were primarily from culture collections. Continued screening will focus on microorganisms isolated from environmental sources.

After the selection of the best organism, the pathway involved in the metabolism of phenylacetylene will be defined. An understanding of the pathway will aid in the development of selective strategies. The organism and process will be developed to achieve the following characteristics: 1) high molar yield of meta-hydroxy phenylacetylene (or a suitable precursor) from phenylacetylene; 2) high specificity for hydroxylation of phenylacetylene; 3) significant accumulation of product in fermentation broth; and 4) inability of the organism to completely metabolize phenylacetylene. These goals will be achieved through a combination of genetic manipulation of the selected organism, fermentation medium optimization, and optimization of process conditions, such as pH, temperature, and dissolved oxygen concentration.

ECONOMIC FEASIBILITY

The economics for the bioconversion of hydrocarbons may be analyzed using traditional fermentation economics. Preliminary estimates of the costs of bioconversion can be derived from these economic models and used to establish technical objectives. The analysis clearly demonstrated the economic feasibility for the microbial bioconversion. Fermentation costs in this system are dominated by the high cost of the substrate, phenylacetylene. To keep fermentation costs as low as possible, it is important to use phenylacetylene efficiently.

Composite Mechanics

Analysis of a Modified Free-Edge Delamination Specimen

HOWARD W. BROWN

An approximate energy release rate of a modified free edge delamination specimen is computed using a sublaminate modeling technique. A laminated plate theory is modified to model laminated composites as a system of sublaminates. The sublaminate displacement equations are written in cells that are used in assembling the displacement equations for the laminate. These expanded displacement equations are written in a form that considers the sublaminate surface tractions along with the displacement functions as dependent variables. Using sublaminate boundary conditions, the sublaminate field equations are assembled into a larger system of field equations representing the laminate. Boundary and continuity conditions that are consistent with those derived from variational methods are assembled using laminate layup, and sublaminate boundary conditions to solve for the displacements and intersublaminate stresses of specimens undergoing generalized plane strain. The displacement functions are then used in computing the energy release rate for the modified free-edge delamination specimen. The results of the method are compared with previous simpler methods and conclusions as to the accuracy of the simpler methods are made.

1. INTRODUCTION

Bonding laminae of unidirectional composite materials into an anisotropic laminate is presently one of the best approaches to designing high-strength, low-weight aircraft components. Delamination reduces the strength of the laminate. This delamination may occur and remain unseen and if it becomes extensive, the impaired strength may cause the laminate to fail catastrophically.

Delamination may be initiated by fatigue, an impact load, a fault in fabrication, or by the overloading of the laminate. Even if these delaminations are initially small and nonthreatening, the application of service loads may cause the delamination to grow and continue to undermine the integrity of the structure. In addition to these

Nonmetallic Materials Division, Air Force Materials Laboratory,
Wright-Patterson Air Force Base, Ohio 45433, U.S.A.

cracks it has been noted the interlaminar stresses near free edges may be high enough to initiate fracture even in the absence of any pre-existing cracks. Thus, it was necessary to develop theories that would approximate the interlaminar stresses that exist in laminates in order to understand the delamination of composite materials. It is apon these theories that the present research draws.

The method presented in the present paper is an extension of the higher order theory for laminated plates put forward by Whitney and Sun [1]. Whitney and Sun's theory is an extension of work originally done by Mindlin [2]. Mindlin developed a theory for dynamic response of homogeneous isotropic plates. The effect of transverse shear stress was included by adding terms linear in the thickness coordinate z to the assumed in-plane displacement field as follows:

$$u = u^0(x,y,t) + z\,\Psi_x(x,y,t) \ ,$$
$$v = v^0(x,y,t) + z\,\Psi_y(x,y,t) \ , \text{ and} \qquad (1)$$
$$w = w^0(x,y,t)$$

Mindlin's theory was applied to laminated plates by Yang, Norris, and Stavsky [3]. Whitney and Pagano [4] proposed correction factors that would match the approximate displacement solution to exact displacement solutions. Whitney and Sun extended the theory to include the first symmetric thickness shear and thickness stretch modes in the assumed displacement field.

The concept of applying laminate theory to individual plies was proposed by Pagano [5] while sublaminate models were employed by Valisetty and Rehfield [6] and by Armanios [7]. These theories utilize continuous displacement fields within the sublaminates and incorporate continuity conditions between sublaminates.

In the present paper the concept of sublaminates is extended to include elastic boundary conditions. These conditions are incorporated into the field equations such that the operator matrix for this set of equations is symmetric and that the sublaminate tractions as well as the displacement functions are dependent variables. By applying continuity and boundary conditions on the surfaces of the sublaminates the sublaminate field equations are easily assembled into a set of field equations representing the laminate . A deterministic set of boundary equations at the laminate edges and continuity equations at laminate interfaces is obtained by direct application of variational theory.

As an example, this technique is applied in computing the displacement field of a modified free-edge delamination specimen which is then used in computing the energy release rate of that specimen.

The idea of the edge delamination specimen under axial loading was proposed by Pagano and Pipes [8]. In their specimen no fiber breakage or transverse cracking was assumed and the delamination specimen was designed to avoid these failure modes. Failure was to be by interlaminar fracture only.

O'Brien [9] studied the onset and growth of delamination in graphite epoxy laminates. Using classical laminated plate theory, he derived the energy release rate by computing the strain energy as a function of the crack length and applying Griffith's equation for linear elastic fracture mechanics. Whitney and Knight [10] proposed a modified free-edge delamination specimen, the modification being the addition of starter cracks incorporated at the free-edges during the fabrication process. From photograghs of the fracture surfaces of the specimens the problems of fiber breakage, and crack jumping were greatly reduced.

2. HIGHER ORDER LAMINATE THEORY

The laminate theory used can be viewed as a higher order extension to the theory proposed by Yang, Norris and Stavsky [3] or as a simplication to Whitney and Sun's higher order theory for laminated composites, and is used to model a laminate undergoing generalized plane strain. For this class of problems the stresses are assumed to be independent of the axial coordinate, x, and time. To further simplify the theory with respect to Whitney and Sun's theory, the symmetric thickness shear i.e. the quadratic terms in z in u and v are neglected. Thus, the displacement field for this class of problems reduces to:

$$u = U(y) + z\,\Psi(y) + c_1\,x + c_2\,x\,y + c_3\,x\,z$$
$$v = V(y) + z\,\Omega(y) + c_4\,x - c_2\,x^2/2 + c_5\,x\,z \tag{3}$$
$$w = W(y) + z\,\Phi(y) + c_6\,x - c_3\,x^2/2 - c_5\,x\,y$$

where u, v, and w are displacement in the x, y, and z directions, and

U, V, W, Ψ, Ω, and Φ are functions of y.

The example problem will be a specimen undergoing uniform axial extension for which the displacement field is:

$$u = U(y) + x\,\epsilon + z\,\Psi(y)$$
$$v = V(y) + z\,\Omega(y) \tag{4}$$
$$w = W(y) + z\,\Phi(y)$$

The field equation for a laminate described by this displacement field is taken directly from Whitney and Sun [1] and written as follow:

$$
\begin{bmatrix}
L_{11} & L_{12} & 0 & L_{14} & L_{15} & L_{16} \\
L_{12} & L_{22} & 0 & L_{15} & L_{25} & L_{26} \\
0 & 0 & -L_{33} & -L_{34} & -L_{35} & -L_{36} \\
L_{14} & L_{15} & -L_{34} & L_{44} & L_{45} & L_{46} \\
L_{15} & L_{25} & -L_{35} & L_{45} & L_{55} & L_{56} \\
L_{16} & L_{26} & -L_{36} & L_{46} & L_{56} & -L_{66}
\end{bmatrix}
\begin{bmatrix}
U+x\epsilon \\ V \\ W \\ \Psi \\ \Omega \\ \Phi
\end{bmatrix}
=
\begin{bmatrix}
-q_x \\ -q_y \\ q \\ -m_x \\ -m_y \\ m
\end{bmatrix}
\tag{5}
$$

where $L_{11} = A_{66}\dfrac{\partial^2}{\partial y^2}$, $L_{12} = A_{26}\dfrac{\partial^2}{\partial y^2}$,

$L_{14} = B_{66}\dfrac{\partial^2}{\partial y^2}$, $L_{15} = B_{26}\dfrac{\partial^2}{\partial y^2}$,

$L_{16} = A_{13}\dfrac{\partial}{\partial x} + A_{36}\dfrac{\partial}{\partial y}$, $L_{22} = A_{22}\dfrac{\partial^2}{\partial y^2}$,

$L_{25} = B_{22}\dfrac{\partial^2}{\partial y^2}$, $L_{26} = A_{23}\dfrac{\partial}{\partial y}$,

$L_{33} = A_{44}\dfrac{\partial^2}{\partial y^2}$, $L_{34} = A_{45}\dfrac{\partial}{\partial y}$,

$L_{35} = A_{44}\dfrac{\partial}{\partial y}$, $L_{36} = B_{44}\dfrac{\partial^2}{\partial y^2}$,

$$L_{44} = D_{66} \frac{\partial^2}{\partial y^2} - A_{55} \quad , \quad L_{45} = D_{26} \frac{\partial^2}{\partial y^2} - A_{45} \quad ,$$

$$L_{46} = (B_{36} - B_{45}) \frac{\partial}{\partial y} \quad , \quad L_{55} = D_{22} \frac{\partial^2}{\partial y^2} - A_{44} \quad ,$$

$$L_{56} = (B_{23} - B_{44}) \frac{\partial}{\partial y} \quad , \quad L_{66} = D_{44} \frac{\partial^2}{\partial y^2} - A_{33} \quad ,$$

$$(A_{ij}, B_{ij}, D_{ij}) = \int_{-h/2}^{h/2} C_{ij} (1, z, z^2) \, dz \quad (i,j=1,2,3,\ldots,6), \tag{5a}$$

C_{ij} are components of the anisotropic stiffness matrix,

$$q_x = \tau_{xz}(h/2) - \tau_{xz}(-h/2) \quad ,$$
$$q_y = \tau_{yz}(h/2) - \tau_{yz}(-h/2) \quad ,$$
$$q = \sigma_{zz}(h/2) - \sigma_{zz}(-h/2) \quad , \tag{5b}$$
$$m_x = (\tau_{xz}(h/2) + \tau_{xz}(-h/2)) \, h/2 \quad ,$$
$$m_y = (\tau_{yz}(h/2) + \tau_{yz}(-h/2)) \, h/2 \quad , \text{ and}$$
$$m = (\sigma_{zz}(h/2) + \sigma_{zz}(-h/2)) \, h/2 \quad .$$

3.   SUBLAMINATE FIELD EQUATION

In this section a convenient form of representing the field equations associated with each sublaminate is proposed. In assembling a field equation representing a laminate the intersublaminate tractions must be included as field functions. Thus, it is convenient to rewrite Whitney and Sun's laminate field equation by incorporating the tractions σ_{zz}, τ_{xz}, and τ_{yz} of equation (5) as part of the dependent variables. It is also be convenient to incorporate the boundary and continuity conditions into the field equations in such a way that they can later be manipulated to represent the appropriate conditions. This can is done by introducing the following set of equations that assume direct relationships between the surface displacements and the surface tractions, akin to spring constants.

$$u(h/2) \; = U + x\epsilon + h \, \Psi/2 = K_1 \, \tau_{xz}(h/2)$$
$$v(h/2) \; = V + h \, \Omega/2 \qquad = K_2 \, \tau_{yz}(h/2)$$
$$w(h/2) = W + h \, \Phi/2 \qquad = K_3 \, \sigma_{zz}(h/2)$$
$$u(-h/2) = U + x\epsilon - h \, \Psi/2 = K_4 \, \tau_{xz}(-h/2) \tag{6}$$
$$v(-h/2) = V - h \, \Omega/2 \qquad = K_5 \, \tau_{yz}(-h/2)$$
$$w(-h/2) = W - h \, \Phi/2 \qquad = K_6 \, \sigma_{zz}(-h/2)$$

where K_i $(i=1,2,3,\ldots,6)$ are the elastic constants.

Thus for each sublaminate the field equation is:

$$\begin{vmatrix} [A_1] & [B] & \\ [B]^t & [L] & [C]^t \\ & [C] & [A_2] \end{vmatrix} \begin{vmatrix} \{\sigma_1\} \\ \{Y\} \\ \{\sigma_2\} \end{vmatrix} = \begin{vmatrix} \{0\} \\ \{f\} \\ \{0\} \end{vmatrix} \tag{7}$$

where $[L]$ is the symmetric form of the operator matrix in equation (5),

$$[A_1] = \begin{vmatrix} -K_1 & 0 & 0 \\ 0 & -K_2 & 0 \\ 0 & 0 & K_3 \end{vmatrix} \quad , \quad [A_2] = \begin{vmatrix} K_4 & 0 & 0 \\ 0 & K_5 & 0 \\ 0 & 0 & -K_6 \end{vmatrix} \quad ,$$

$$[B] = \begin{bmatrix} 1 & 0 & 0 & h/2 & 0 & 0 \\ 0 & 1 & 0 & 0 & h/2 & 0 \\ 0 & 0 & -1 & 0 & 0 & -h/2 \end{bmatrix} ,$$

$$[C] = \begin{bmatrix} -1 & 0 & 0 & h/2 & 0 & 0 \\ 0 & -1 & 0 & 0 & h/2 & 0 \\ 0 & 0 & 1 & 0 & 0 & -h/2 \end{bmatrix} ,$$

$$\{\sigma_1\} = \begin{Bmatrix} \tau_{xz}(h/2) \\ \tau_{yz}(h/2) \\ \sigma_{zz}(h/2) \end{Bmatrix} , \quad \{\sigma_2\} = \begin{Bmatrix} \tau_{xz}(-h/2) \\ \tau_{yz}(-h/2) \\ \sigma_{zz}(-h/2) \end{Bmatrix}$$

$$\{Y\} = \begin{Bmatrix} U \\ V \\ W \\ \Psi \\ \Omega \\ \Phi \end{Bmatrix} , \text{ and } \{f\} = \begin{Bmatrix} 0 \\ 0 \\ 0 \\ 0 \\ 0 \\ -A_{13}\epsilon \end{Bmatrix}$$

Next a method of assembling the laminate field equation by applying the boundary and continuity conditions that define the laminate to the sublaminate field equation must be defined.

4. ASSEMBLY OF LAMINATE FIELD EQUATION

The field equation for the laminate is set up first as a combination of the sublaminate field equations as follows:

$$\begin{bmatrix} [A_1]_1 & [B]_1 & & & & & \\ [B]_1^t & [L]_1 & [C]_1^t & & & & \\ & [C]_1 & [A_2]_1 & & & & \\ & & & \ddots & & & \\ & & & & [A_1]_n & [B]_n & \\ & & & & [B]_n^t & [L]_n^t & [C]_n \\ & & & & & [C]_n & [A_2]_n \end{bmatrix} \begin{bmatrix} \{\sigma_1\}_1 \\ \{Y\}_1 \\ \{\sigma_2\}_1 \\ \vdots \\ \{\sigma_1\}_n \\ \{Y\}_n \\ \{\sigma_2\}_n \end{bmatrix} = \begin{bmatrix} \{0\} \\ \{f\}_1 \\ \{0\} \\ \vdots \\ \{0\} \\ \{f\}_n \\ \{0\} \end{bmatrix} \qquad (8)$$

where the matrix operators of the sublaminate field equations are arranged along the diagonal of the matrix operator of the laminate field equation and there is no coupling of field functions of different sublaminates. Subscripts denote sublaminates and the matrices $[A_i]$ are yet to be defined.

Continuity of tractions across an interface between sublaminates couples the field dependent variables of the sublaminates as follows:

$$\begin{bmatrix} [A_1]_1 & [B]_1 & & & \\ [B]_1^t & [L]_1 & [C]_1^t & & \\ & [C]_1 & [A_2]_1+[A_1]_2 & [B]_2 & \\ & & [B]_2^t & [L]_2 & [C]_2^t \\ & & & [C]_2 & [A_2]_2 \end{bmatrix} \begin{bmatrix} \{\sigma\}_1^1 \\ \{Y\}_1 \\ \{\sigma\}_1^2 \\ \{Y\}_2 \\ \{\sigma\}_2 \end{bmatrix} = \begin{bmatrix} \{0\} \\ \{f\}_1 \\ \{0\} \\ \{f\}_2 \\ \{0\} \end{bmatrix} \qquad (9)$$

Continuity of displacements as well as tractions implies that $[A_2]_1$ and $[A_1]_2$ vanish thus equation (9) reduces to:

$$
\begin{bmatrix}
[A_1]_1 & [B]_1 & & & \\
[B]_1^t & [L]_1 & [C]_1^t & & \\
& [C]_1 & [0] & [B]_2 & \\
& & [B]_2^t & [L]_2 & [C]_2^t \\
& & & [C]_2 & [A_2]
\end{bmatrix}
\begin{bmatrix}
\{\sigma\}_1 \\
\{Y\}_1 \\
\{\sigma\}_2 \\
\{Y\}_2 \\
\{\sigma\}_3
\end{bmatrix}
=
\begin{bmatrix}
\{0\} \\
\{f\}_1 \\
\{0\} \\
\{f\}_2 \\
\{0\}
\end{bmatrix}
\tag{10}
$$

Fixed surfaces are defined by an elastic constant K_i equaling zero thus filling the $[A_i]$ matrices with zeroes. The tractions on surface where tractions are prescribed are known thus eliminating the need for the accompanying boundary equations and thus the K_i's of those equations need not be entered.

Thus K_i components in field equations for a laminate with no elastic boundary conditions either vanish, are zero, or are eliminated.

5. BOUNDARY AND CONTINUITY CONDITION EQUATIONS

The boundary condition equations for the laminate field equation depend on the vanishing variational equation of the potential energy of the laminate.

$$
\begin{aligned}
\delta P = \int_{x_0}^{x_1} &\left\{ \left(N_{xy}\delta U + N_y \delta V + Q_y \delta W + M_{xy} \delta \Psi + M_y \delta \Omega + R_y \delta \Phi \right) \Big|_{y_0}^{y_1} \right. \\
&+ \int_{y_0}^{y_1} \left\{ (-N_{xy'y} - q_x)\delta U + (-N_{y'y} - q_y)\delta V + (-Q_{y'y} - q)\delta W \right. \\
&\qquad + (-M_{xy'y} + Q_x - m_x)\delta \Psi + (-M_{y'y} + Q_y - m_y)\delta \Omega \\
&\qquad \left. + (-R_{y'y} + N_z - m)\delta \Phi \right\} dy \left. \right\} dx = 0
\end{aligned}
\tag{11}
$$

One member of each of the following pairs must be defined at a boundary: U, N_{xy}; V, N_y; W, Q_y; Ψ, M_{xy}; Ω, M_y; Φ, R_y (note these pairings) and all must be defined at a continuous interface. For the laminate field equations these pairs will change. For instance consider the case of a laminate with a fixed displacement w at z equal $-h/2$ (i.e. $W - h\Phi/2 = 0$). The first integral of equation (11) would reduce to:

$$
\int_{x_0}^{x_1} \left(N_{xy}\delta U + N_y \delta V + M_{xy}\delta \Psi + M_y \delta \Omega + (R_y + h Q_y/2)\delta \Phi \right) \Big|_{y_0}^{y_1} dx
\tag{12}
$$

Thus, there are now five pairs of boundary conditions instead of six.

6. MODIFIED FREE-EDGE DELAMINATION SPECIMEN

The modified free-edge delamination specimen shown in figure (1) is symmetrical about the midplane with starter cracks located at equal distances from the midplane on each side of specimen. The specimen is subjected to a uniform axial strain which it is assumed will cause the cracks to propagate similtaneously. This assumption was made in order to take advantage of the two planes for which either forces or displacements would be known due to symmetry conditions.

The first plane of known boundary conditions is the midplane of the laminate at which axial symmetry exists. Thus the surface boundary conditions are:

$$w = 0 \quad \text{and} \quad \tau_{xz} = \tau_{yz} = 0 \tag{13}$$

The second plane of known boundary conditions is the longitudinal yz plane cutting through the laminate perpendicular to the first plane. The boundary conditions are derived from the plane of symmetry conditions of the first plane, the 180° rotational plane of symmetry of the second plane, the assumed displacement field and the constitutive equations. Thus, these edge boundary conditions are:

$$U = V = \Psi = \Omega = 0 \quad \text{and} \quad Q_y = R_y = 0 . \tag{14}$$

The quadrant is itself divided into four elements which are dictated by the location of the crack tip as shown in figure (2). The plane of the crack defines the boundary between sublaminates and the plane perpendicular to the crack at the crack tip defines the boundary between sections of laminate. The boundary conditions for the free edge are:

$$N_{xy} = N_y = Q_y = M_{xy} = M_y = R_y = 0 \tag{15}$$

The boundary conditions for the top surfaces of elements one, two and three and for the bottom surface of element one are:

$$\sigma_{zz} = \tau_{xz} = \tau_{yz} = 0 \tag{16}$$

Displacements and interlaminar stresses are continuous at the interface of elements three and four.

At the interface between laminate sections displacement functions are continuous and forces are continuous.

These boundary and continuity conditions, along with coordinates, ply layup, axial strain, and material properties are input to a computer program that will set up the laminate field equations, solve for the generalized field solutions, set up the boundary and continuity equations and solve for the constants of the generalized displacement field.

Using the known sublaminate surface boundary conditions equation (8) can be reduced for each of the laminate sections. The laminate field equation for the laminate section containing the crack is:

$$\begin{bmatrix} [L]_1 & & \\ & [L]_2 & [C]_2^t \\ & [C]_2 & [0] \end{bmatrix} \begin{bmatrix} \{Y\}_1 \\ \{Y\}_2 \\ \{\sigma\}_3 \end{bmatrix} = \begin{bmatrix} \{f\}_1 \\ \{f\}_2 \\ \{0\} \end{bmatrix} \tag{17}$$

where $[C] = [\,0 \quad 0 \quad 1 \quad 0 \quad 0 \quad -h_2/2\,]$, and

$$\{\sigma_2\}_2 = \{\,\sigma_{zz}(-h_2/2)\,\} .$$

And, the laminate field equation of the uncracked laminate section is:

$$\begin{bmatrix} [L]_1 & [C]_1^t & & \\ [C]_1 & [0] & [B]_2 & \\ & [B]_2^t & [L]_2 & [C]_2^t \\ & & [C]_2 & [0] \end{bmatrix} \begin{bmatrix} \{Y\}_1 \\ \{\sigma\}_2 \\ \{Y\}_2 \\ \{\sigma\}_3 \end{bmatrix} = \begin{bmatrix} \{f\}_1 \\ \{0\} \\ \{f\}_2 \\ \{0\} \end{bmatrix} \tag{18}$$

where $[C]_1 = \begin{bmatrix} -1 & 0 & 0 & h_1/2 & 0 & 0 \\ 0 & -1 & 0 & 0 & h_1/2 & 0 \\ 0 & 0 & 1 & 0 & 0 & -h_1/2 \end{bmatrix}$,

$\qquad [B]_2 = \begin{bmatrix} 1 & 0 & 0 & h_2/2 & 0 & 0 \\ 0 & 1 & 0 & 0 & h_2/2 & 0 \\ 0 & 0 & -1 & 0 & 0 & -h_2/2 \end{bmatrix}$,

$\qquad [C]_2 = \begin{bmatrix} 0 & 0 & 1 & 0 & 0 & -h_2/2 \end{bmatrix}$

$\qquad \{\sigma_2\}_1 = \{ \tau_{xz}(-h_1/2),\ \tau_{yz}(-h_1/2),\ \sigma_{zz}(-h_1/2) \}^t$

$\qquad \{\sigma_2\}_2 = \{ \sigma_{zz}(-h_2/2) \}$.

The boundary and continuity equations are derived using variational theory as outlined in section 5, and are as follows:

at free edge $\qquad\qquad\qquad$ at interface $\qquad\qquad$ and at half laminate width

$N^1_{xy^1} = 0$, $\qquad\qquad\qquad U^1_1 = U^2_1$, $\qquad\qquad\qquad \Psi^2_1 = 0$,

$N^1_{y^1} = 0$, $\qquad\qquad\qquad V^1_1 = V^2_1$, $\qquad\qquad\qquad \Omega^2_1 = 0$,

$Q^1_{y^1} = 0$, $\qquad\qquad\qquad W^1_1 = W^2_1$, $\qquad\qquad R^2_{y^1} + h_1 Q^2_{y^1} = 0$,

$M^1_{xy^1} = 0$, $\qquad\qquad\qquad \Psi^1_1 = \Psi^2_1$, $\qquad\qquad\qquad U^2_2 = 0$,

$M^1_{y^1} = 0$, $\qquad\qquad\qquad \Omega^1_1 = \Omega^2_1$, $\qquad\qquad\qquad V^2_2 = 0$,

$R^1_{y^1} = 0$, $\qquad\qquad\qquad \Phi^1_1 = \Phi^2_1$, $\qquad\qquad\qquad \Psi^2_2 = 0$,

$N^1_{xy^2} = 0$, $\qquad M^1_{xy^1} + h_1 N^1_{xy^1}/2 = M^2_{xy^1} + h_1 N^2_{xy^1}/2$, $\qquad \Omega^2_2 = 0$, and

$N^1_{y^2} = 0$, $\qquad M^1_{y^1} + h_1 N^1_{y^1}/2 = M^2_{y^1} + h_1 N^2_{y^1}/2$, $\quad R^2_{y^2} + h_1 Q^2_{y^1} + h_2 Q^2_{y^2}/2 = 0$;

$M^1_{xy^2} = 0$, $\qquad R^1_{y^1} + h_1 Q^1_{y^1}/2 = R^2_{y^1} + h_1 Q^2_{y^1}/2$,

$M^1_{y^2} = 0$, and $\qquad\qquad\qquad U^1_2 = U^2_2$,

$R^1_{y^2} + h_2 Q_{y^2}/2 = 0$ $\qquad\qquad\qquad V^1_2 = V^2_2$,

$\qquad\qquad\qquad\qquad\qquad\qquad \Psi^1_2 = \Psi^2_2$,

$\qquad\qquad\qquad\qquad\qquad\qquad \Omega^1_2 = \Omega^2_2$,

$\qquad\qquad\qquad\qquad\qquad\qquad \Phi^1_2 = \Phi^2_2$,

$\qquad\qquad N^1_{xy^2} + N^1_{xy^1} = N^2_{xy^2} + N^2_{xy^1}$,

$\qquad\qquad N^1_{y^2} + N^1_{y^1} = N^2_{y^2} + N^2_{y^1}$,

$\qquad M^1_{xy^2} + h_1 N^1_{xy^1}/2 = M^2_{xy^2} + h_1 N^2_{xy^1}/2$,

$\qquad M^1_{y^2} + h_1 N^1_{y^1}/2 = M^2_{y^2} + h_1 N^2_{y^1}/2$, and

$R^1_{y^2} + h_1 Q^1_{y^1} + h_2 Q^1_{y^2}/2 = R^2_{y^2} + h_1 Q^2_{y^1} + h_2 Q^2_{y^2}/2$; $\qquad\qquad$ (19)

where superscripts indicate laminate section and subscripts indicate sublaminate.

7. ENERGY RELEASE RATE

The energy release rate for a modified free-edge delamination having fixed grips is

$$G = \frac{-\epsilon}{2} \frac{d}{da} \left\{ \int_0^a (N^1_{x^1} + N^1_{x^2})\, dy + \int_a^b (N^2_{x^1} + N^2_{x^2})\, dy \right\} \tag{20}$$

where b is half the specimen width.

Substituting the constitutive relation for N_x into equation (20), and integrating over y yields:

$$
\begin{aligned}
G = \frac{-\epsilon}{2} \frac{d}{da} \Big\{ &A_{12_1}\{ V^1_1(a) -V^1_1(0) +V^2_1(b) -V^2_1(a) \} + A_{12_2}\{ V^1_2(a) -V^1_2(0) +V^2_2(b) -V^2_2(a) \} \\
&+A_{13_1}\{ \underline{\Phi}^1_1(a) -\underline{\Phi}^1_1(0) +\underline{\Phi}^2_1(b) -\underline{\Phi}^2_1(a) \} + A_{13_2}\{ \underline{\Phi}^1_2(a) -\underline{\Phi}^1_2(0) +\underline{\Phi}^2_2(b) -\underline{\Phi}^2_2(a) \} \\
&+A_{16_1}\{ U^1_1(a) -U^1_1(0) +U^2_1(b) -U^2_1(a) \} + A_{16_2}\{ U^1_2(a) -U^1_2(0) +U^2_2(b) -U^2_2(a) \} \\
&+B_{12_1}\{ \Omega^1_1(a) -\Omega^1_1(0) +\Omega^2_1(b) -\Omega^2_1(a) \} + B_{12_2}\{ \Omega^1_2(a) -\Omega^1_2(0) +\Omega^2_2(b) -\Omega^2_2(0) \} \\
&+B_{16_1}\{ \Psi^1_1(a) -\Psi^1_1(0) +\Psi^2_1(b) -\Psi^2_1(a) \} + B_{16_2}\{ \Psi^1_2(a) -\Psi^1_2(0) +\Psi^2_2(b) -\Psi^2_2(a) \} \Big\}
\end{aligned}
$$

where $\quad \underline{\Phi}(y_1) - \underline{\Phi}(y_0) = \int_{y_0}^{y_1} \Phi\, dy$ (21)

8. RESULTS

Comparisons are made between the present method for computing the energy release rate and methods developed by O'Brien [9], Armanios and Rehfield [11 and 12], and a method using classical laminated plate theory. The laminate layups and material properties used in these comparisons are drawn from Armanios and Rehfield [12]. The material properties are:

E_{11}=134 GPa $E_{22}=E_{33}$=10.2 GPa $\nu_{12}=\nu_{13}$=0.3 ν_{23}=0.55

G_{12}=5.52GPa $G_{31}=G_{23}$=3.1 GPa

ply thickness (pt)=0.14×10^{-3} m   half specimen width (l) = 140 × pt

For purposes of comparison only the constant terms will be used. The sets of layups analyzed were selected from [12] The layups are defined by $[-(45+\theta)/-\theta/(45-\theta)/(90-\theta)]_s$, where θ indicates the rotation about a hole. Table 1 contains results of delamination occuring at the $-(45+\theta)/-\theta$ interface which yields symmetric sublaminates after delamination and table 2 contains results of delamination occuring at the $(45-\theta)/(90-\theta)$ interface which yields a nonsymmetric sublaminate after delamination.

TABLE 1 $10^6\ G_T$ (j/m)

layup	classical theory	A & R	YNS	present method	O'Brien
[50/-85/-40/5]	0.377	0.377	0.868	1.10	1.10
[40/85/-50/-5]	1.10	1.10	3.00	3.27	3.27
[30/75/-60/-15]	1.42	1.42	4.91	5.09	5.09
[20/65/-70/-25]	0.074	0.074	5.44	5.51	5.51
[10/55/-80/-40]	0.074	0.074	0.483	3.41	3.41

TABLE 2 $10^6\ G_T$ (j/m)

layup	classical theory	A & R	YNS	present method	O'Brien
[-45/0/45/90]	1.80	1.80	1.95	1.80	1.80
[-55/-10/35/80]	1.86	1.86	1.98	1.80	1.41
[-65/-20/25/70]	1.70	1.70	1.74	1.51	0.523
[-75/-30/15/60]	1.16	1.16	1.24	1.02	0.0497
[-85/-40/5/50]	1.07	1.07	1.52	1.46	1.10
[85/-50/-5/40]	1.60	1.60	3.14	3.30	3.27
[75/-60/-15/30]	1.83	1.83	4.91	5.55	5.09

9. CONCLUSIONS

A method has been derived that uses a higher order theory than those previously used to compute the energy release rate of a modified free-edge delamination specimen. The results were compared with those derived using lower order theories with the following conclusions. Classical theory and Armanios and Rehfield's method yield identical results which depending on the laminate may yield good or poor results. The energy release rate computed using Yang Norris and Stavsky's field assumptions yields greatly improved results but still vary greatly for some laminate layups when compared to the present method. Finally, the solution method derived by O'Brien appears to yield good results for symmetric laminates that yield symmetric sublaminates after delamination but for other cases the results may not be as reliable.

BIBLIOGRAPHY

1. Whitney, J. M., and Sun, C. T., "A Higher Order Theory For Extensional Motion of Laminated Composites," *Journal of Sound and Vibration*, Vol. 30, 1973, pp. 85-97.
2. Mindlin, R. D., "Influence of Rotatory Inertia and Shear on Flexural Motions of Isotropic, Elastic Plates," *Journal of Applied Mechanics*, American Society of Mechanical Engineers, Vol. 18, pp 31-38.
3. Yang, P. C., Norris, C. H., and Stavsky, Y., "Elastic Wave Propagation in Heterogenous Plates," *International Journal of Solids and Structures*, Vol. 2, pp. 665-684.
4. Whitney, J. M., and Pagano, N. J., "Shear Deformation in Heterogeneous Anisotropic Plates," *Journal of Applied Mechanics*, American Society of Mechanical Engineers, Vol. 37, pp. 1031-1036
5. Pagano, N. J., "Stress Fields in Composite Laminates," *International Journal of Solids and Structures*, Vol. 14, 1978, pp. 385-400.
6. Valisetty, R. R., and Rehfield, L. W., "A New Approach to Interlaminar Stress Analysis," ASTM Symposium on Delamination and Debonding of Materials, Pittsburgh, PA, 9-10, November 1983.
7. Armanios, Erian A., *New Methods of Sublaminate Analysis For Composite Structures and Applications to Fracture Processes*, Ph. D. dissertation, Georgia Institute of Technology, December 1984.
8. Pagano, N. J., and Pipes, R. B., "Some Observations on the Interlaminar Strength of Composite Laminates," International Journal of Mechanics and Science, Vol. 15, 1973, p. 679.
9. O'Brien, T. K., "Characterization of Delamination Onset and Growth in a Composite Laminate," *Damage in Composite Materials*, ASTM STP 775,
10. Whitney, J. M., and Knight, M., "A Modified Free-Edge Delamination Specimen," *Delamination and Debonding of Materials*, ASTM STP 876, W. S. Johnson, Ed., American Society for Testing and Materials, 1985, pp. 298-314
11. Armanios, E. A., and Rehfield, L. W., "Interlaminar Analysis of Laminated Composites Using A Sublaminate Approach," AIAA, 1986, 86-0969.
12. Armanios, E. A., and Rehfield, L. W., "Sublaminate Analysis of Interlaminar Fracture in Composites," NASA-CR-177228, N86-28130.

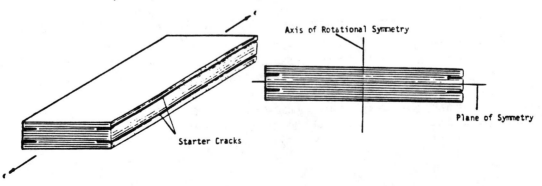

Figure (1)   Modified Free-Edge Delamination   Figure (2)   Division of Cross-Section

The Natural Frequencies of Symmetric Angle-Ply Laminates Derived from Eigensensitivity Analysis

ROBERT REISS, S. RAMACHANDRAN AND BO QIAN

ABSTRACT

In this paper, a new closed-form approximate solution for the natural frequencies of symmetric rectangular angle-ply laminates simply supported on all four edges is derived. The solution, obtained from eigensensitivity analysis, is expressed as a truncated Fourier series in the ply angle. Results show that the prediction for the fundamental frequency is quite accurate for engineering applications, often within 1-2% of the true frequency.

INTRODUCTION

The eigenvalue equation to determine the flexural natural frequencies for symmetric laminates seldom admits exact closed-form solutions. An exception is the well-known solution for specially orthotropic symmetric rectangular laminates simply supported on all sides [1]. However, when the coupling stiffness do not vanish, as is the case for symmetric angle-ply laminates, closed-form solutions are not available. Thus, apart from an approximate closed-form solution by Bert [2], researchers invariably resort to numerical procedures to obtain frequency response.

The determination of the natural frequencies of symmetric, simply-supported angle-ply laminates is the focus of this paper. The frequencies are presented in a truncated Fourier series in the ply angle θ. The Fourier coefficients are calculated from eigensensitivity analysis.

PROBLEM STATEMENT

Consider a symmetric laminate undergoing free flexural vibrations. The equation describing this motion can be expressed

R. Reiss, Professor of Mechanical Engineering, Howard University, Washington, D.C. 20059
S. Ramachandran, Senior Engineer Associate, Morton Thiokol, Inc., Brigham City, Utah 84302
B. Qian, Doctoral Student, Howard University, Washington, D.C. 20059

$$D_{11} \, w,_{xxxx} + D_{22} \, w,_{yyyy} + 2(D_{12} + 2D_{66}) \, w,_{xxyy}$$

$$+ D_{16} \, w,_{xxxy} + D_{26} \, w,_{xyyy} = \rho \, \omega^2 w \qquad (1)$$

where w is the mode shape, ω is the corresponding frequency, ρ is the mass density, D_{ij} are the flexural stiffnesses [1], x and y are the coordinates in the plane of the laminate, and indices following a comma denote differentiation with respect to the indicated argument.

For a balanced four-ply symmetric laminate $[\theta/-\theta]_s$, it is easy to show that the flexural stiffnesses appearing in Eq. (1) are given by

$$D_{11} = \frac{16}{3} \, h^3(U_1 + U_2 \cos 2\theta + U_3 \cos 4\theta)$$

$$D_{12} = \frac{16}{3} \, h^3(U_4 - U_3 \cos 4\theta)$$

$$D_{22} = \frac{16}{3} \, h^3(U_1 - U_2 \cos 2\theta + U_3 \cos 4\theta)$$

$$D_{16} = \frac{12}{3} \, h^3(\tfrac{1}{2} U_2 \sin 2\theta + U_3 \sin 4\theta) \qquad (2)$$

$$D_{26} = \frac{12}{3} \, h^3(\tfrac{1}{2} U_2 \sin 2\theta - U_3 \sin 4\theta)$$

$$D_{66} = \frac{16}{3} \, h^3(U_5 - U_3 \cos 4\theta)$$

where h is the common thickness of each ply and U_i (i=1,...,5) are invariant material properties of the identical constituent laminae [1].

Finally, for simply supported laminates, the appropriate boundary conditions are

$$x = 0, a \qquad\qquad w = D_{11} \, w,_{xx} + D_{12} \, w,_{yy} = 0,$$

$$y = 0, b \qquad\qquad w = D_{12} \, w,_{xx} + D_{22} \, w,_{yy} = 0. \qquad (3)$$

For $\theta=0$ or $\pi/2$, $D_{16} = D_{26}=0$, and the normalized eigensolutions to Eq. (1) with boundary values (3) is [1]

$$\omega^2_{mn} = \frac{\pi^4}{\rho a^4} \, [m^4 D_{11} + 2(D_{12}+2D_{66})m^2 n^2 R^2 + n^4 R^4 D_{22}]$$

$$\qquad (4)$$

$$w^{mn} = \frac{2}{\sqrt{\rho ab}} \, \sin \frac{m\pi x}{a} \sin \frac{n\pi y}{b}$$

where R is the plate aspect ratio

$$R = a/b. \qquad (5)$$

Clearly, the fundamental frequency occurs for m=n=1.

REVIEW OF EIGENVALUE DIFFERENTIATION

Before proceeding with the development of the solution to the stated problem, it is useful to review some general mathematical formulae that are essential to this approach.

Consider the class of eigenvalue problems which can be cast into the form

$$T*E(S)T \ w^{mn} = \lambda_{mn} M \ w^{mn} \qquad (6)$$

to which appropriate mixed boundary conditions must be appended. In Eq. (6), T and $T*$ are L_2 adjoint differential operators, S is a collection of parameters, and E and M are positive stiffness and mass operators, respectively. The subscripts m and n are included in order to facilitate a solution to Eq. (1).

The eigenvalue λ_{mn} is a functional of S. And if S changes by a small amount δS, λ_{mn} changes by a small amount $\delta\lambda_{mn}$ [3],

$$\delta\lambda_{mn} = (Tw^{mn}, \ \delta E \ T \ w^{mn}) \qquad (7)$$

where the normalized eigenfunctions satisfy

$$(w^{mn}, \ M \ w^{mn}) = 1 \qquad (8)$$

In Eqs. (7–8), $(\cdot,\cdot)$ denotes the usual L_2 inner product over the area spanned by the plate. Furthermore, the second variation of the eigenvalue λ_{mn} has also been derived [3]. Thus

$$\delta^2\lambda_{mn} = (T \ w^{mn}, \ \delta^2 E \ T \ w^{mn})$$

$$- 2 \sum_{i \neq m} \sum_{j \neq n} \frac{(T \ w^{ij}, \delta E \ Tw^{mn})^2}{\lambda_{ij} - \lambda_{mn}} \qquad (9)$$

Equations (7) and (9) are valid provided the eigenvalue λ_{mn} is not repeated.

DERIVATION OF THE NATURAL FREQUENCIES

It was previously stated that the object of this paper is to find an approximate expression for λ_{mn}. Since λ_{mn}, when treated as a function of θ, has a periodicity of π, it is convenient to expand the eigenvalues in the Fourier series

$$\lambda_{mn} = \tfrac{1}{2} a_o^{mn} + \sum_{p=1}^{\infty} a_p^{mn} \cos 2p\theta$$

$$+ \sum_{p=1}^{\infty} b_p^{mn} \sin 2p\theta. \qquad (10)$$

For orthotropic designs, $\theta = 0$ or $\pi/2$, and Eqs. (2) and (4) determine λ_{mn}. Thus $\lambda_{mn}(0)$ and $\lambda_{mn}(\pi/2)$ are known, i.e.

$$\lambda_{mn}|0,\pi/2 = \frac{16}{3} \frac{h^3\pi^4}{\rho a^4} \{m^4(U_1 \overset{+}{_-} U_2 + U_3)$$

$$+ 2m^2n^2R^2 (U_1 - 3U_3)$$

$$+ n^4R^4 (U_1 \overset{-}{_+} U_2 + U_3)\} \qquad (11)$$

Equation (11), however, is sufficient to determine only two of the Fourier coefficients appearing in Eq. (10). Knowledge of the derivatives of the eigenvalue will determine additional Fourier coefficients.

Toward this end, Eq. (1) is cast in the form of the abstract eigenvalue equation (6) by identifying $T, T*, E, M$, and λ_{mn}. It can easily be shown that

$$
T = \begin{bmatrix} \partial^2/\partial x^2 \\ \partial^2/\partial y^2 \\ 2\partial^2/\partial x \partial y \end{bmatrix}
\qquad
E = \begin{bmatrix} D_{11} & D_{12} & D_{16} \\ D_{12} & D_{22} & D_{26} \\ D_{16} & D_{26} & D_{66} \end{bmatrix}
\tag{12}
$$

and

$$
T* = T^T, \qquad M = \rho, \qquad \lambda_{mn} = \omega^2_{mn}, \qquad S = \theta.
\tag{13}
$$

After substitution of the specific operators (12) into Eq. (7), the following expression for the derivative of the eigenvalues is obtained:

$$
\frac{d\lambda_{mn}}{d\theta} = D'_{11} \ (w,^{mn}_{xx}, \ w,^{mn}_{xx}) + 2D'_{12} \ (w,^{mn}_{xx}, \ w,^{mn}_{yy})
$$

$$
+ 4 \, D'_{16} \ (w,^{mn}_{xx}, \ w,^{mn}_{xy}) + D'_{22} \ (w,^{mn}_{yy}, \ w,^{mn}_{yy})
$$

$$
+ 4 \, D'_{26} \ (w,^{mn}_{yy}, \ w,^{mn}_{xy}) + 4D'_{66}(w,^{mn}_{xy}, \ w,^{mn}_{xy})
\tag{14}
$$

where ( )' denotes differentiation with respect to θ.

The right-hand side of Eq. (14) can be evaluated, exactly, at $\theta = 0$ and $\pi/2$. For either orthotropic case, D'_{11}, D'_{12}, D'_{22} and D'_{66} vanish identically. The inner products associated with the terms D'_{16} and D'_{26} also vanish. It may therefore be concluded that

$$
\frac{d\lambda_{mn}}{d\theta} \bigg|_{\theta=0,\pi/2} = 0
\tag{15}
$$

Reiss and Ramachandran reached the same conclusion using a somewhat different argument [4]. In fact, by using arguments based upon symmetries and skew-symmetries, they concluded that all odd derivatives of λ_{mn}, when evaluated at $\theta=0$ or $\pi/2$, vanish identically. Consequently, b^{mn}_p vanish for all p, m and n. It remains to calculate the second derivative of the eigenvalues for $\theta=0$ and $\pi/2$. While these computations are algebraically messy, they are nevertheless straightforward. The results are

$$
\frac{d^2\lambda_{ij}}{d\theta^2} \bigg|_{\theta=0} = \frac{64\pi^4 h^3}{3\rho a^4} \{ U_2(j^4 R^4 - i^4) + 4U_3(6i^2 j^2 R^2 - j^4 R^4 - i^4) \}
$$

$$
- \frac{6144h^3}{\rho a^2 b^2} \sum_{m \neq i} \sum_{n \neq j} \frac{m^2 n^2 i^2 j^2}{(m^2-i^2)^2(n^2-j^2)^2} \{(U_2 + 4U_3)(m^2+i^2) +
$$

$$+ (U_2-4U_3)\ (n^2+j^2)R^2\}^2/\{(U_1+U_2+U_3)(m^4-i^4)$$

$$+ 2(U_1-3U_3)(m^2n^2-i^2j^2)R^2 + (U_1-U_2+U_3)(n^4-j^4)R^4\} \quad (16)$$

and

$$\left.\frac{d^2\lambda_{ij}}{d\theta^2}\right|_{\theta=\pi/2} = \frac{64\pi^4h}{3\rho a^4}\ \{\ U_2(i^4-j^4R^4) + 4U_3(6i^2j^2R^2-j^4R^4-i^4)\}$$

$$- \frac{6144h^3}{\rho a^2b^2}\ \underset{m\neq i}{\Sigma}\ \underset{n\neq j}{\Sigma}\ \frac{m^2n^2i^2j^2}{(m^2-i^2)^2(n^2-j^2)^2}\ \{(U_2-4U_3)(m^2+i^2)$$

$$+ (U_2+4U_3)\ (n^2+j^2)R^2\}^2/(U_1-U_2 + U_3)\ (m^4-i^4)$$

$$+ 2(U_1-3U_3)(m^2n^2-i^2j^2)R^2+(U_1+U_2+U_3)(n^4-j^4)R^4\} \quad (17)$$

The boundary values (11),(16) and (17) are sufficient to determine the Fourier coefficients $a_0^{mn}, a_2^{mn}, a_4^{mn}$ and a_6^{mn}, i.e.

$$\tfrac{1}{2}\ a_0^{mn} = \tfrac{1}{2}[\lambda_{mn}(0) + \lambda_{mn}(\tfrac{\pi}{2})] + [\lambda_{mn}''(0) + \lambda_{mn}''(\pi/2)]/32$$

$$a_2^{mn} = \tfrac{9}{16}\ [\lambda_{mn}(0) - \lambda_{mn}(\pi/2)] + [\lambda_{mn}''(0)- \lambda_{mn}''(\pi/2)]/64$$

$$\qquad\qquad\qquad\qquad\qquad (18)$$

$$a_4^{mn} = \qquad\qquad\qquad - [\lambda_{mn}''(0) + \lambda_{mn}''(\pi/2)]/32$$

$$a_6^{mn} = \tfrac{-1}{16}\ [\lambda_{mn}(0) - \lambda_{mn}(\pi/2)] - [\lambda_{mn}''(0) - \lambda_{mn}''(\pi/2)]/64$$

NUMERICAL RESULTS

While higher order derivatives of λ_{mn} with respect to θ may be calculated using the methodology developed in Ref. [3], the resulting algebraic complexity negates the simplicity of Eqs. (18). Further, numerical illustrations demonstrate that a Fourier series truncated at a_6^{mn} is sufficiently accurate for most engineering purposes.

In the following tables, numerical calculations for the fundamental frequency obtained by setting m=n=1 in Eq. (10) are validated. This validation is facilitated by defining the non-dimensional frequency

$$k_1 = \frac{\omega_{11} \sqrt{\rho} \; b^2}{\sqrt{U_1} \; h^3} \tag{19}$$

In Table 1, comparisons are made for a typical boron–epoxy laminate with the following material properties:

$$E_2/E_1 \quad = \quad 0.100$$

$$G_{12}/E_1 \quad = \quad 0.025$$

$$\nu_{12} \quad = \quad 0.300$$

The columns labelled $\bar{k}_1$ are the values of the frequency calculated from a Rayleigh–Ritz procedure, while the columns labelled k_1 contain the fundamental frequency calculated from Eq. (10). The largest error occurs for the square plate when $\theta = \pi/4$ and is 3.6%. For R=2, calculated values for $\bar{k}_1$ and k_1 differ by less than 1%, and for R=5, the accuracy is within 0.5%.

TABLE 1 RESULTS FOR BORON EPOXY

θ°	R=1		R=2		R=5	
	$\bar{k}_1$	k_1	$\bar{k}_1$	k_1	$\bar{k}_1$	k_1
0	38.89	38.89	15.59	15.59	11.39	11.39
15	41.01	40.81	18.12	18.05	12.15	12.14
30	45.51	44.39	23.14	22.92	15.18	15.22
45	47.79	46.07	27.78	27.57	20.83	20.93
60	45.51	44.39	31.56	31.46	27.54	27.58
75	41.01	40.81	34.36	34.34	32.80	32.80
90	38.89	38.89	35.45	35.45	34.78	34.78

In Table 2, results for the fundamental frequencies are evaluated for a T300/5208 graphite epoxy laminate. The material constants, in this case, are

$$E_2/E_1 \quad = \quad 0.0569$$

$$G_{12}/E_1 \quad = \quad 0.0396$$

$$\nu_{12} \quad = \quad 0.28$$

The slightly greater E_1 to E_2 stiffness ratio is reflected by slightly larger discrepancies between the calculated fundamental frequency and the one obtained from a Rayleigh–Ritz approximation. Again, the greatest difference, 4.1%, is observed in Table 2 for R=1, $\theta = 45°$. For R=2 and 5, the maximum differences are 1.4% and 1%, respectively.

TABLE 2 RESULTS FOR GRAPHITE EPOXY

θ°	R=1		R=2		R=5	
	$\bar{k}_1$	k_1	$\bar{k}_1$	k_1	$\bar{k}_1$	k_1
0	39.26	39.26	14.36	14.36	9.04	9.04
15	40.94	40.73	16.94	16.85	10.34	10.35
30	44.72	43.50	22.23	21.93	14.40	14.53
45	46.73	44.82	27.38	27.08	20.88	21.02
60	44.72	43.50	31.70	31.56	27.95	27.99
75	40.94	40.73	34.85	34.82	33.32	33.32
90	39.26	39.26	36.05	36.05	35.30	35.30

In Table 3, a similar comparison is made for a high modulus graphite-epoxy laminate. Typical material values were selected to be

$$E_2/E_1 \quad = \quad 0.0250$$

$$G_{12}/E_1 \quad = \quad 0.0125$$

$$\nu_{12} \quad = \quad 0.2500$$

The greatest error, again at $\theta = \pi/4$ and R=1, is 6.4%. Representative values for k_1 and $\bar{k}_1$ are tabulated in Table 3. The discrepancies for R=2 and 5 are less than 3% and 2%, respectively.

TABLE 3 RESULTS FOR HIGH MODULUS GRAPHITE EPOXY

θ°	R=1		R=2		R=5	
	$\bar{k}_1$	k_1	$\bar{k}_1$	k_1	$\bar{k}_1$	k_1
0	37.95	37.95	11.69	11.69	6.21	6.21
15	39.81	39.49	14.95	14.73	7.86	7.89
30	44.34	42.39	21.24	20.61	12.71	12.98
45	46.78	43.77	27.02	26.45	20.24	20.46
60	44.34	42.39	31.80	31.56	28.24	28.30
75	39.81	39.49	35.34	35.30	34.23	34.23
90	37.95	37.95	36.71	36.71	36.44	36.44

In order to use Eq. (10) to calculate higher order natural frequencies, special care must be taken in order to insure that the comparisons made are for like eigenvalues. While the fundamental frequency is always λ_{11}, the next lowest frequency may be λ_{12} or λ_{21} depending upon R and θ. Equation (10) does not provide directly any specified natural frequency, other than the fundamental frequency. Several computations for different values of m and n must be made.

Table 4 shows higher order frequency results for the square boron epoxy laminate considered earlier. The eigenvalues are symmetric about $\theta=45°$. The columns labelled $\bar{k}_2, \bar{k}_3$ and $\bar{k}_4$ contain, respectively, the second, third and fourth natural frequency, calculated from a Rayleigh-Ritz scheme. Similarly, the column k_2, k_3 and k_4 are the corresponding non-dimensional eigenvalues calculated from Eq. (10); in parentheses next to each k_i are the values of m and n that were used to compute the specific eigenvalue. While these calculated frequencies are slightly less accurate than the fundamental frequencies shown in Table 1, they still provide a reasonable engineering estimate.

TABLE 4 HIGHER ORDER NATURAL FREQUENCIES FOR R=1

$\theta°$	$\bar{k}_2$	$k_2(m,n)$	$\bar{k}_3$	$k_3(m,n)$	$\bar{k}_4$	$k_4(m,n)$
0	62.36	62.36(1,2)	112.49	112.49(1,3)	141.81	141.81(2,1)
20	76.26	76.06(1,2)	131.20	132.83(1,3)	135.14	134.13(2,1)
30	87.62	88.72(1,2)	129.21	125.39(2,1)	148.64	157.62(1,3)
40	95.81	101.92(1,2)	124.42	114.40(2,1)	162.01	182.09(2,2)

In Table 5, the same boron-epoxy plate is considered, but with an aspect ratio R=5. Just as Eq. (10) provides increasingly better estimates of the fundamental frequency as R increases, it also provides increasingly better estimates of the higher natural frequencies as R increases. Indeed, the accuracy of k_2, k_3 and k_4 is within 1.1%, 1.2% and 0.6%, respectively. Moreover, the four lowest curves, as functions of θ, generated from Eq. (10) do not intersect and, therefore, each of the first four natural frequencies is associated with a specific λ_{ij}, i.e.

$$k_1 = \lambda_{11} b^2 \sqrt{\rho/h^3 U_1}$$

$$k_2 = \lambda_{21} b^2 \sqrt{\rho/h^3 U_1}$$

$$k_3 = \lambda_{31} b^2 \sqrt{\rho/h^3 U_1}$$

$$k_4 = \lambda_{41} b^2 \sqrt{\rho/h^3 U_1}$$

ACKNOWLEDGMENT

This study was supported by the NASA Langley Research Center through Grant No. NAG-1-383.

TABLE 5 HIGHER ORDER NATURAL FREQUENCIES FOR R=5

$\theta°$	$(m,n)=(2,1)$		$(m,n)=(3,1)$		$(m,n)=(4,1)$	
	$\bar{k}_2$	k_2	$\bar{k}_3$	k_3	$\bar{k}_4$	k_4
0	13.47	13.47	18.57	18.57	27.10	27.10
20	16.95	16.91	23.28	23.24	31.65	31.46
40	23.00	23.26	29.16	29.50	36.74	36.66
50	26.34	26.55	31.29	31.57	37.63	37.53
70	32.53	32.52	34.55	34.51	37.47	37.31
90	35.14	35.14	35.86	35.86	37.05	37.05

REFERENCES

1. Jones, R.M., Mechanics of Composite Materials, Scripta Book Company, Washington, D.C., 1975.

2. Bert, C.W., "Optimal Design of a Composite-Material Plate to Maximize its Fundamental Frequency," Journal of Sound and Vibration, Vol. 50, 1977, pp. 229-237.

3. Reiss, R., "Design Derivatives of Eigenvalue and Eigenfunctions for Self-Adjoint Distributed Parameter Systems," AIAA Journal, Vol. 24, No. 7, 1986, pp. 1169-1172.

4. Reiss, R. and Ramachandran, S., "Maximum Frequency Design of Symmetric Angle-Ply Laminates," Composite Structures 4, Volume 1. Analysis and Design Studies, edited by I.H. Marshall, Elsevier Applied Science Publishers LTD, 1987, pp. 1.476-1.487.

An Improved Procedure for Free Edge Stress Analysis in Composite Laminates

SOM R. SONI AND DAVID K. CHU

Abstract

Free edge stress analysis studies have been conducted by a number of investigators and are reviewed in reference [1]. Recently Pagano [2] developed a unified model to predict accurately response of multi-layered composites using layer equilibrium concept. Attempts for calculating interlaminar stress components for thick composites, with the help of this model, were hampered by the overflow/underflow computational difficulties associated with the mathematical method pertaining to the solution of the free edge boundary value problem. The advent of the Global-Local Variational Model [3] helped solving free edge effect problems for a number of thick laminates, but still the applicability of the model was limited for the same reasons as in [2]. The present paper describes an alternative approach to solve the boundary value problem avoiding the above mentioned difficulties. This approach consists of re-arranging the intermediate equations in an eigenvalue and eigenvector matrix form for the determination of roots of the polynomial encountered in the solution procedure. This approach has been demonstrated to be advantageous as compared to the plynomial expansion form solution used in [2,3]. The present paper demonstrates the applicability of the alternative approach and the significant effect of the stacking sequence and ply orientations on the response of the laminate.

Introduction

A number of studies have been conducted [1] investigating the stress fields in composite laminates and free edge delamination mode of failure. The present study demonstrates three main aspects of the present procedure of stress analysis of free edge problem. The first aspect deals with the markedly improved problem solution capability developed for the investigation of the free edge problem, the second aspect deals with the demonstration of alternative approach of stacking laminates to avoid interlaminar failure by reducing magnitude and changing sign of stimulated stress components (Figures 2-7, Tables 2-7), and the third pertains to the fact that the computation time for the present procedure is considerably lower than that for the other method (Figure 10).

The results given here are based upon the extended Reissner's Variational Principle [2]. The solution technique given in [2] has been modified to facilitate the computation of stress fields in composite laminates with large number of layers. The technique used earlier was based upon the expansion of a determinant containing unknown values of a parameter into a polynomial for determining the roots, intermediate quantities, of the polynomial. In this procedure we encounter digital values of very high and/or low manitude. This creates computational overflow

AdTech Systems Research Inc, 1342 N. Fairfield Road, Dayton, Ohio 45432

overflow/underflow problems and provide unacceptable rounding of numbers in approximation of the results. In the present approach the procedure is developed such that the governing equations are aligned to provide a standard eigenvalue-eigenvector formulation. In the polynomial solution approach, the application of the theory was limited to a laminate with maximum of eight subdivisions that amounts to a laminate with maximum of sixteen layers (because, in symmetric laminates only half of the laminate thickness is considered for analysis of free edge problem under inplane applied strain). In the eigenvalue-eigenvector approach the laminate with half thickness containing 40 or more layers/subdivisions can be modeled (depending upon the computer memory). The authors' experience is with an IBM RT PC computer with 40 subdivisions of the half laminate thickness.

Problem Description

Figure 1 shows the laminate geometry, coordinate axes, and the direction of the applied strain. The magnitude of stress components stimulated by the application of the given strain or stress is dependent upon the stacking sequence and ply orientation. The model developed in reference [2] has been used to conduct the stress analysis of the laminate for constant inplane applied strain ε in x direction (ε_x). The numerical procedure of solution as described in reference [2] has been modified such that the values of intermediate parameters, such as eigenvalues and eigenvectors, are computed by different approach. In the new approach, the equilibrium equations and interfacial continuity conditions are re-arranged to written in the standard eigenvector-eigenvalue form with required number of eigenvalues. This set of equations is used to obtain the final solution of the problem.

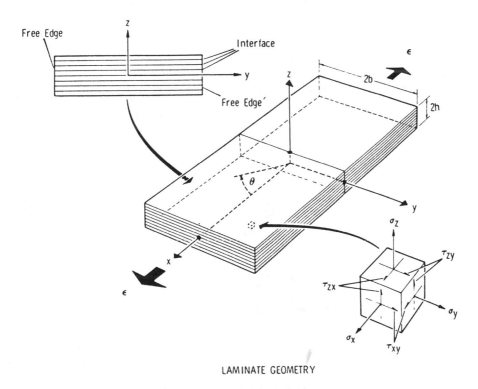

LAMINATE GEOMETRY

Figure 1

Results and Discussion

The computer program is written for IBM RT PC computer. A large number of test cases were run for ascertaining the accuracy of the results obtained by this program. Comparisons were made with the results obtained by existing method of polynomial expansion form. An excellent agreement between the results by the two procedures was observed. All the laminates considered for this study are given in table 1.

In this study the laminate configuration $(\pm 30)_s$ was considered for demonstrating the effect of variation of the number of subdivisions in the half thickness in the analysis. The laminate is made of T300/C-1034 graphite epoxy material. Four different cases of subdivision N=4,8,20,40 are considered. All the three interlaminar stress components in the region of free edge at the mid-plane and 30/-30 interface are given in figures 8 and 9. A significant difference in the results for various values of N has been observed. For examle, figure 8 shows the stress component σ_z at the mid-plane of the laminate for all values of N. It can be seen that the stress

Table 1
Table of Laminates Considered for Analysis

Table	Figure	Laminate	Objective
2	2	$(0/\pm 45/90)_s$ * $(0/90/\pm 45)_s$	High Modulus Graphite Epoxy Reverse sign of σ_z at midsurface
3	3	$(0/45/90/-45_2/90/45/0)_s$ * $(0/90/\pm 45)_{2s}$	T300/1034-C graphite epoxy Reduced Interlaminar stresses
4	4	* $(0/90/\pm 45)_s$ $(0/45/90/-45)_s$	Reduced Interlaminar stress components
5	5	* $(\pm 45/0/90)_s$ $(45/0/-45/90)_s$	Reduced interlaminar stress comp.
6	6	* $(\pm 45/0/90)_s$ $(45/0/-45/90_2/45/0/-45)_s$	Reduced interlaminar stress comp.
7	7	* $(0/90/\pm 45)_s$ * $(\pm 45/0/90)_s$	Reduced interlaminar stress comp.
	8-10	$(\pm 30)_s$	N=4,8,20,40; applicbility of the approach and solution efficiency.

* Laminates marked with star are stacked according to a conventional laminates used in research and development activities. The other laminates in the table are alternative better choices without compromising the inplane strength capabilities.

distribution for N=40 is considerably different from that for other values of N. The present study is going to affect a number of investigations conducted earlier with less accurate models. Further, that will help us to obtain better understanding of failure mechanisms in composite lamintes. The failure criterion developed in references [4,5] may also have to be re-investigated.

Example Demonstrating Change of Direction of σ_z at Midsurface:

This example is given to show that the sign of the stress component σ_z at the free edge stimulated by the application of an inplane tensile strain $\varepsilon_x=1.\times10^{-6}$ can be reversed just by re-arranging the stacking sequence of the ply layup. Figures 2a and 2b show the stacking sequences and interface numbers for $(0/\pm45/90)_s$ and $(0/90/\pm45)_s$-laminates. As shown in table 2, $(0/\pm45/90)_s$ laminate exhibits tensile σ_z at the mid-plane and $(0/90/\pm45)_s$-laminate exhibits compressive σ_z at the midsurface. In classical lamination theory the influence of interlaminar stress components is neglected.

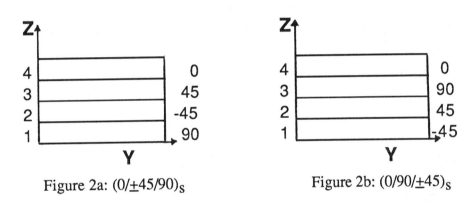

Figure 2a: $(0/\pm45/90)_s$ Figure 2b: $(0/90/\pm45)_s$

Figure 2: Laminate stacking sequence and interface number for two laminates.

The model for these results contains eight subdivisions of half the laminate thickness with each layer subdivided into two sublayers. The maximum interlaminar stress components are given at four interfaces shown in figure 2. According to the mathematical nature of the problem, the stress component σ_z and τ_{xz} are definite at the mid surface whereas singular at other interfaces named 2 to 4. However, the current model gives nonsingular results at all interfaces and the singular nature is observed from the trend of the variation of the stress components near the free edge. These results are given for same order of model discretization for both cases of the laminate. τ_{yz} values are given at a location away from the free edge where maxima occurs. The maximum value is determined on the basis of a comparison of stress component calculated at different equidistant locations. The actual maximum value and location may be different than the value given here. Material properties used for the computation of results for this case are given below:

E_{11}=35.7 (Msi), E_{22}=1.03 (Msi), E_{33}=1.03 (Msi)
G_{12}=0.7 (Msi), G_{13}=0.7 (Msi), G_{23}=0.55 (Msi)
υ_{12}=0.33, υ_{13}=0.33 υ_{23}=0.45

Table 2

Out-of-plane stress components in Psi at four interfaces, for two laminates, for $\varepsilon_x=1.\times10^{-6}$

Interface #	σ_z	τ_{xz}	τ_{yz}	Interface #	σ_z	τ_{xz}	τ_{yz}
		$(0/\pm45/90)_s$				$(0/90/\pm45)_s$	
1	2.86	0.	0.	1	-1.86	0.	0.
2	2.56	-.45	1.7	2	-2.41	3.57	-.74
3	-.87	2.9	0.7	3	0.25	-0.8	-1.57
4	-.4	-.82	-.27	4	0.83	-.12	0.35

Examples for reducing interlaminar stresses by re-arranging ply sequence:

A number of laminates given in table 1 and shown in figures 3 to 7 are analyzed to demonstrate the effect of stacking sequence on the interlaminar stress components stimulated by the applied in-plane strain of $\varepsilon_x=1.\times10^{-6}$. The results are calculated at different interfaces and are given in Psi. The material properties used for these calculations and other forthcoming calculations are those for T300/1034-C, as given below:

$$E_{11}=20.0 \text{ (Msi)}, \qquad E_{22}=1.4 \text{ (Msi)}, \qquad E_{33}=1.4 \text{ (Msi)},$$
$$G_{12}=0.8 \text{ (Msi)}, \qquad G_{13}=0.8 \text{ (Msi)}, \qquad G_{23}=0.6 \text{ (Msi)},$$
$$\upsilon_{12}=0.3, \qquad \upsilon_{13}=0.3, \qquad \upsilon_{23}=0.6$$

Laminates $(0/45/90/-45_2/90/45/0)_s$ and $(0/90/+45/-45)_{2s}$:

In this study two cases of laminate $(0/90/\pm45)_{2s}$ with different stacking sequences are considered. In the first case the ply orientation is in an order such that the adjacent ply angle is minimum. These two cases with their interface numbers are given in figure 3 and the corresponding maximum interlaminar stress components are given in table 3.

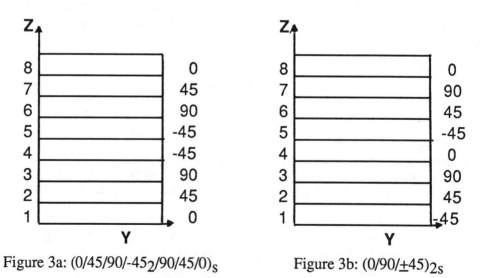

Figure 3a: $(0/45/90/-45_2/90/45/0)_s$ Figure 3b: $(0/90/\pm45)_{2s}$

Figure 3: Laminate stacking sequence and interface number for two laminates.

Table 3

Out-of-plane stress components in Psi at eight interfaces for two laminates, for $\varepsilon_x = 1 \times 10^{-6}$.

Interface #	σ_z	τ_{xz}	τ_{yz}	Interface #	σ_z	τ_{xz}	τ_{yz}
	$(0/45/90/-45_2/90/45/0)_s$					$(0/90/\pm45)_{2s}$	
1	-0.05	0	0	1	1.56	0	0
2	-0.23	0.61	0.14	2	-1.14	2.41	0.44
3	0.83	-1.37	-.99	3	0.71	-.58	-1.16
4	0.77	-1.42	-.98	4	0.94	-.31	0.54
5	-.34	0.0	0.01	5	-.68	-0.73	0.65
6	0.78	1.42	-.99	6	-1.22	2.4	-.21
7	0.87	1.36	0.98	7	0.36	-.5	1.04
8	-.15	-.61	-.14	8	0.44	-.11	0.30

A careful inspection of these results shows that the stacking sequence of figure 3a provides lower stress components σ_z, τ_{xz} and τ_{yz} as compared to those in figure 3b.

Laminates $(0/90/+45/-45)_s$ and $(0/45/90/-45)_s$:

The results given in table 4 are for laminates $(0/90/\pm45)_s$ and $(0/45/90/-45)_s$ as shown in figures 4a and 4b. In the model the half thickness is divided into eight subdivisions, each layer has been subdivided into two sublayers. Here again as in the case of laminate of figure 2a and 2b the maximum interlaminar stress components are higher for laminate $(0/90/\pm45)_s$ as compared to that for laminate $(0/45/90/-45)_s$. Thus the mode of failure in the former case can be different than that of the latter.

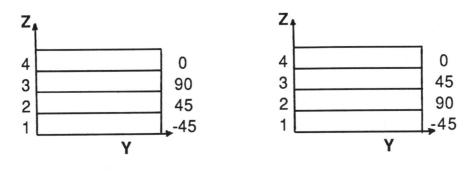

Figure 4a: $(0/90/\pm45)_s$ Figure 4b: $(0/45/90/-45)_s$

Figure 4: Laminate stacking sequence and interface number for two laminates.

Table 4

Out-of-plane stress components in Psi at four interfaces, for two laminates for $\varepsilon_x = 1. \times 10^{-6}$.

Interface #	σ_z	τ_{xz}	τ_{yz}	Interface #	σ_z	τ_{xz}	τ_{yz}
		$(0/90/\pm45)_s$				$(0/45/90/-45)_s$	
1	-1.27	.0	.0	1	-.51	.0	.0
2	-1.58	3.1	-.48	2	.7	1.5	-1.03
3	0.33	-0.73	-1.06	3	1.48	1.5	1.06
4	0.97	-.1	0.36	4	-.34	-.84	-.16

Laminates $(45/-45/0/90)_s$ and $(45/0/-45/90)_s$:

In figure 5 two stacking sequences $(\pm45/0/90)_s$ and $(0/90/\pm45)_s$ of a laminate are considered. The half laminate thickness has been subdivided into eight mathematical sublayers, i.e. each layer has been subdivided into two sublayers. The inspection of the stress components given in table 5 shows that the laminate τ_{yz} values are given at a location away from the free edge where maxima occurs. The maximum value is determined on the basis of a comparison of stress component calculated at different points. The actual maximum value and location may be different than the given value.

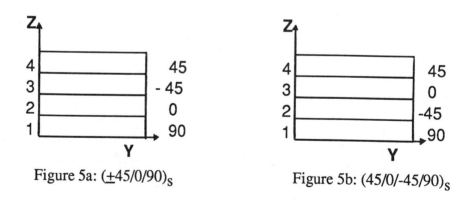

Figure 5a: $(\pm45/0/90)_s$ Figure 5b: $(45/0/-45/90)_s$

Figure 5: Laminate stacking sequence and interface number for two laminates.

Table 5

Out-of-plane stress components in Psi at four interfaces for two laminates for $\varepsilon_x = 1. \times 10^{-6}$.

Interface #	σ_z	τ_{xz}	τ_{yz}	Interface #	σ_z	τ_{xz}	τ_{yz}
		$(\pm45/0/90)_s$				$(45/0/-45/90)_s$	
1	2.53	0.	0.	1	2.48	0.	0.
2	3.01	-.07	0.76	2	2.32	-.46	1.11
3	0.3	-.59	0.8	3	0.17	1.38	0.41
4	-.29	2.23	0.38	4	0.04	1.44	0.4

Laminates $(45/-45/0/90)_{2s}$ and $(45/0/-45/90_2/45/0/-45)_s$:

Table 6 shows the interlaminar stress components for two cases of ply stacking sequences $(\pm45/0/90)_{2s}$ and $(45/0/-45/90_2/45/0/-45)_s$ for same ply orientations. It can be seen that stepwise gradual increase of ply angle in adjacent layers, figure 6b, stimulate lower interlaminar stress components as compared to conventional method, figure 6a. In these models each layer has been subdivided into two sublayers. The total number of subdivisions was sixteen. The results show that laminate at the left hand side, figure 6a, may fail due to interlaminar shear at 45/-45 interface whereas that situation is avoided in the case of a laminate given in the right hand side, figure 6b. There are two advantages in the laminate in figure 6b, the first one is: the magnitude of σ_z is less than that in the laminate in figure 6a and the second one is: the magnitude of the τ_{xz} is considerably lower than the other one. This value in $(\pm45/0/90)_{2s}$ laminate is higher than that in $(45/0/-45/90_2/45/0/-45)_s$ laminate by about about 50%.

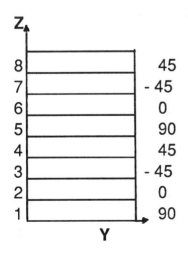

Figure 6a: $(\pm45/0/90)_{2s}$

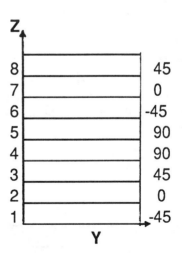

Figure 6b: $(45/0/-45/90_2/45/0/-45)_s$

Figure 6: Laminate stacking sequence and interface number for two laminates with same ply orientations and different stacking sequence.

Table 6

Out-of-plane stress components in Psi at eight interfaces for two laminates for $\varepsilon_x = 1. \times 10^{-6}$.

Interface #	σ_z	τ_{xz}	τ_{yz}	Interface #	σ_z	τ_{xz}	τ_{yz}
		$(\pm 45/0/90)_{2s}$				$(45/0/-45/90_2/45/0/-45)_s$	
1	2.26	0	0	1	-.57	0	0
2	2.69	-.06	0.68	2	-.49	-1.67	-.43
3	0.01	-.62	0.67	3	0.002	-1.5	-.45
4	-.58	2.41	-.16	4	1.35	0.5	-1.08
5	1.37	-.63	-1.21	5	2.55	-.003	-.06
6	2.85	-.28	0.58	6	2.31	-.46	1.03
7	0.21	-.76	0.66	7	0.16	1.38	0.38
8	-.31	2.17	0.27	8	0.04	1.43	0.38

Laminates $(0/90/45/-45)_s$ and $(45/-45/0/90)_s$

In this example two cases of stacking sequence considered are $(0/90/\pm 45)_s$ and $(\pm 45/0/90)_s$ with the same ply orientations. The half thickness has been subdivided into eight subdivisions, each layer subdivided into two sublayers. The inspection of these results shows that out-of-plane normal stress component σ_z stimulated due to inplane applied load is compressive in laminate $(0/90/\pm 45)_s$ whereas that for laminate $(\pm 45/0/90)_s$ is tensile. Also the magnitude is also considerably reduced. Similarly τ_{xz} is higher in former case than that in the latter case. The key factor to demonstrate here is that by re-arranging the stacking sequence one can change the magnitude and sign of out of plane stress components to one's advantage. τ_{yz} values are given at a location away from the free edge where maxima occurs. The maximum value is determined on the basis of a comparison of stress component calculated at different points. The actual maximum value and location may be different than the given value.

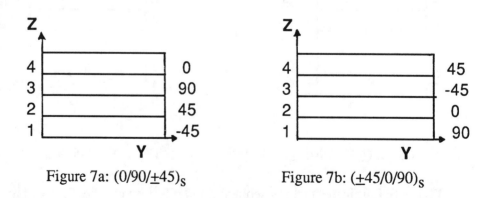

Figure 7a: $(0/90/\pm 45)_s$ Figure 7b: $(\pm 45/0/90)_s$

Figure 7: Laminate stacking sequence and interface number for two laminates.

Table 7

Out-of-plane stress components in Psi at four interfaces for two laminates for $\varepsilon_x=1.\times10^{-6}$.

Interface #	σ_z	τ_{xz}	τ_{yz}	Interface #	σ_z	τ_{xz}	τ_{yz}
		$(0/90/\pm45)_s$				$(\pm45/0/90)_s$	
1	-1.27	.0	.0	1	2.53	0.	0.
2	-1.58	3.1	-.48	2	3.01	-.07	0.76
3	0.33	-0.73	-1.06	3	0.3	-.59	0.8
4	0.97	-.1	0.36	4	0.3	-.59	0.8

Conclusions

This work shows an advanced capability in studying the free edge effects in thick composite laminates. The method of solution used has dramatically increased the problem solving capability of the theory of reference [2]. This has opened up new avenues to apply this theory for solving other boundary value problems, such as: laminate with transverse crack and interfacial crack. The results shown here demonstrate the benefit of spiral stacking of the plies as compared to conventional way of stacking the plies in composite laminates. Also it has been demonstrated how to suppress a specific mode of failure by re-arranging the stacking sequence.

Acknowledgement:

A part of this work was sponsored by the US Air Force Materials Laboratory, through a subcontract from Martin Marietta Energy Systems Inc. The authors acknowledge the help from Dr. S. W. Tsai and Dr. N. J. Pagano for the successful execution of this program.

References:

1. S. R. Soni and N. J. Pagano,'Models for Studying Free Edge Effects', AdTech Systems Research Report Number ASR-1-TR-1987, Dayton, OH.

2. N. J. Pagano,'Stress Fields in Composite Laminates', Int. Jl. Solids & Structures, Vol 14, 1978, pp.385-400.

3. N. J. Pagano and S.R. Soni,'Global-Local Laminate Variational Model', Int. Jl. Solids & Structures, Vol. 19,No. 3, pp. 207-228, 1983.

4. R. Y. Kim and S. R. Soni,'Experimental and Analytical Studies on the Onset of Delamination in Laminated Composites', Jl. Composite Materials, Vol. 18, January 1984.

5. S. R. Soni and R. Y. Kim,'Delamination of Composite Laminates Stimulated by Interlaminar Shear', Composite Materials: Testing and Design (Seventh Conference), ASTM STP 893, J.M. Whitney, Ed., American Society for Testing and Materials, Philadelphia, 1986, pp.286-307.

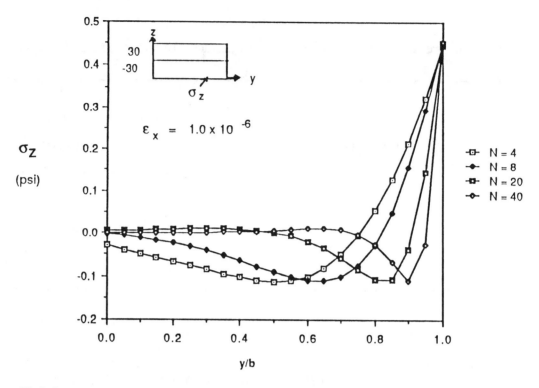

Fig 8: Stress component σ_z for different subdivisions of laminate thickness, at the mid plane

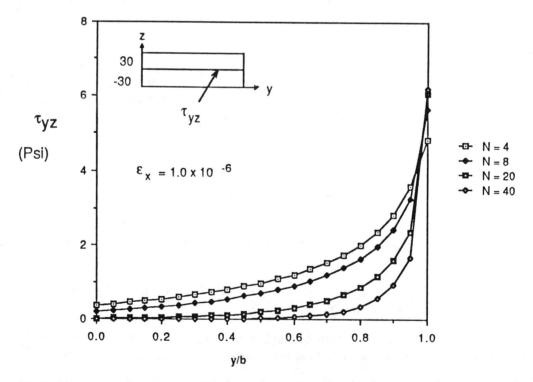

Fig 9a: Stress components τ_{yz} at 30/-30 interface for different subdivisions of laminate thickness

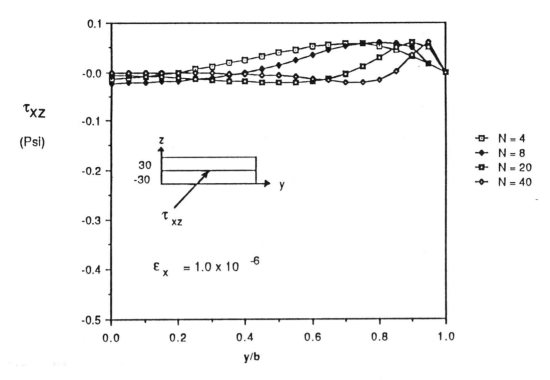

Fig 9b: Stress components τ_{xz} at 30/-30 interface for different subdivisions of laminate thickness

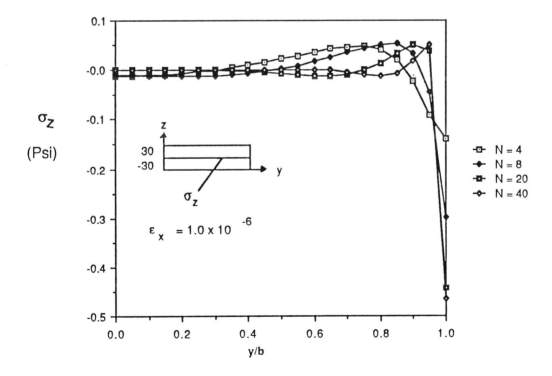

Fig 9c: Stress components σ_z at 30/-30 interface for different subdivdisions of laminate thickness

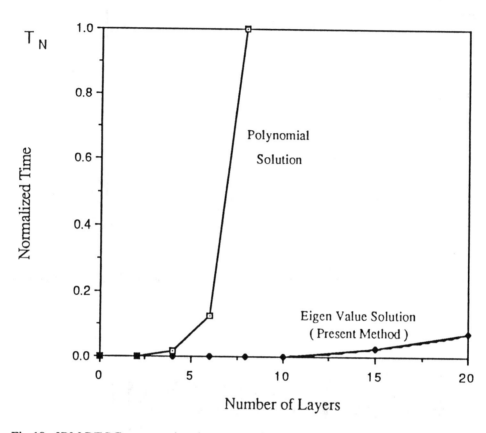

Fig 10: IBM RT PC computation time comparision for solving problems by two approaches.

T_N = 1 correspods to 15 hours execution time by polymomial method of solution

Effects of Microstructure on Failure in Peek and its Short-Fiber Composites

J.-N. CHU, G.-M. WU AND J. M. SCHULTZ

ABSTRACT

The effects of spherulite size, orientation and molecular weight on the mechanical failure of PEEK and its carbon and glass short-fiber composites were investigated. A self-seeding method was used to produce PEEK samples of different average spherulite size, and a permanganic etching method was used to reveal the spherulites. It was found that spherulite size increased with melting temperature and decreased with the molecular weight of the material. The failure behavior of these samples was studied using a compact tension (CT) test. It was found that spherulite size plays a very important role in the fracture mechanism of PEEK. Preferred chain orientation in the neat resin was imposed using a rolltrusion technique. Short-fiber orientation was varied using an inverted extrusion method. The fracture toughness K_Q of oriented neat PEEK was found to be much higher for a crack propagating perpendicular (T-type) to the machine direction (MD) than for a crack propagating parallel to it (L-type). Fractography shows that little plastic deformation is associated with T-type fracture, whereas there is profuse plastic deformation for L-type fracture. These results indicate a major role of orientation-enhanced modulus in fracture improvement. For the composites, the crack propagates along fiber ends, in a fiber-avoidance mode, for T-type failure; for L-type fracture, the fracture path is preferentially along the fiber/matrix interface.

INTRODUCTION

PEEK, or poly(ether-ether ketone), is a newly developed engineering thermoplastic with potentially vast application in advanced composites, because of its exceptional performance. Under many circumstances the properties of a matrix material can profoundly influence the bulk properties of composite materials.[1] In turn, the failure behavior of

Center for Composite Materials, University of Delaware
Newark, DE 19716

a semicrystalline, thermoplastic matrix material depends on microstructural details such as spherulite size, orientation, degree of crystallinity, and degree of transcrystallinity. This work will deal principally with the effect of spherulites and orientation on fracture.

It has been well established that melt history and molecular weight are two major factors affecting the spherulite size of a semicrystalline thermoplastic material.[2] Thus, if the crystallization temperature is fixed and the crystallization time is long enough, the spherulite size will vary with the melt temperature and molecular weight. Using materials whose microstructure has been controlled in this way, the effect of spherulite size on failure can be defined.

Polymeric solids with a high degree of molecular alignment possess markedly anisotropic mechanical properties. Both the modulus and stiffness are greater in the direction of drawing than in the transverse direction. Likewise, the mechanical properties of composites also show a strong dependence on the orientation of short fibers. Some degree of orientation always accrues to the finished article during the commercial process of formation, e.g., by injection molding. This orientation will also affect the practical applications of such materials. Thus, the strength and ductility of polymeric composites can be greatly modified by preferred orientation of the characteristic structural units (molecular chains or reinforced short fibers) within the workpiece.

SPHERULITE SIZE CHARACTERIZATION[3]

Both low molecular weight 150P and high molecular weight 450G PEEK materials were used in this study. Specimens were prepared by compression molding. Their spherulite sizes were controlled by using different melting temperatures, i.e. 380°C, 400°C and 420°C, respectively. After the specimens were molded, they were quenched to an isothermal heat treatment temperature of 310°C for 4 hours in order to develop a fully spherulitic microstructure. After that, the spherulite size was determined by microscopic observation of permanganic-etched surfaces.[4] The effect of melting temperature on spherulite size is shown in Fig. 1. In this figure, each data point is based on the average spherulite size of the 5 largest spherulites. This figure shows that the spherulite size varies substantially with melting temperature in 150P PEEK. This is because for 150P PEEK, the nucleation density (number of nuclei or spherulites per unit volume) decreases with increasing melting temperature. Thus, after crystallization, larger spherulites can be found in the samples with higher melt temperatures.

The spherulite size of 450G PEEK did not change dramatically with melt temperature.(As shown in Fig. 1) This is because the nucleation density of this material is very high. The growth of the spherulites in the high nucleation density 450G PEEK was stopped by other spherulites before the development of the long lamellae bundles characteristic of 150P PEEK. The diameter obtained from these samples was essentially the size of a spherulitic nucleus.

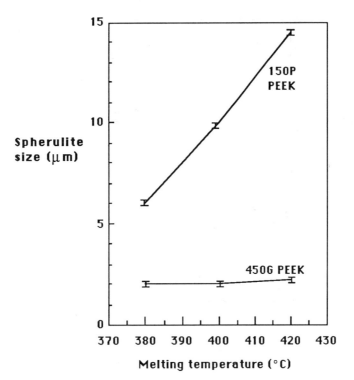

Fig. 1 Effect of melting temperature on spherulite size.

FRACTURE MECHANISM STUDY [3]

Compact tension testing was conducted to measure the fracture toughness of 150P and 450G PEEK samples. The effect of melting temperature on fracture toughness is shown in Fig. 2. Each data point represents the measurement of at least three different specimens. The error bar is from 0.5% to 7.8%. This Figure shows that the fracture toughness decreases with melting temperature. Since spherulite size increases with increasing melt temperature, the fracture toughness decreases with spherulite size.

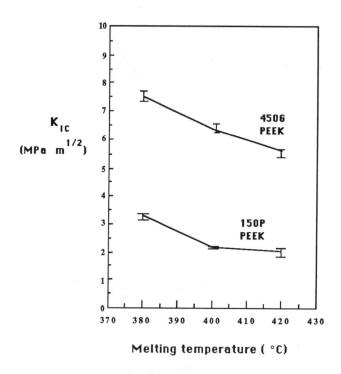

Fig. 2   Effect of melting temperature on fracture toughness
         (sample thickness, 0.32 cm, or 1/8 in)

An SEM analysis of the initial crack fracture surface was performed on
the fractured 150P PEEK specimens.   Fracture surface micrographs of the
region of slow crack growth for 150P 380°C specimen (as shown in Fig.
3) showed that nuclei were pulled out from the surface by the applied
force.   This   implies   that   the   crack   propagated   through   the
spherulite.[5]   As for the higher melting temperature 150P specimens,
because the spherulites are larger it is assumed they exhibit a smaller
density gradient from high density nuclei to low density boundary.
Thus a crack propagates more easily through these specimens, the
fracture toughness of them become lower.   Fig. 4 and 5 are SEM
micrographs of the slow crack region of melting temperature 400°C and
420°C 150P PEEK, respectively.   The micrographs show circular-drawn
structures.   The dimensions of these structures are very similar to
those of spherulites in the same material.   The dimension of the
fracture structure increases with increasing processing temperature.
Furthermore, the amount of subsurface structure appears to decrease at
higher temperatures, due to the larger subsurface distances needed to
be traversed in order for the crack to join the spherulite centers via
interspherulite fracture.

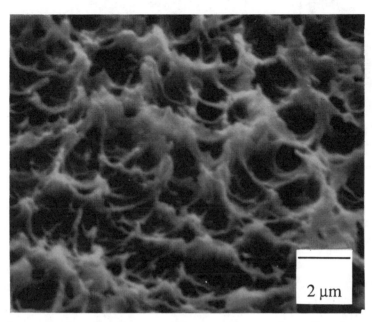

Fig. 3 SEM micrograph of slow crack growth region fracture surface in
105P PEEK (melt processed at 380°C). Micrograph shows nucleus
pull-out.

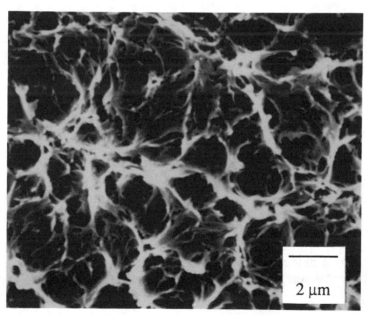

Fig. 4 SEM micrograph of slow crack growth region fracture surface in
150P PEEK (melt processed at 400°C). Picture shows relatively
less amount of nuclei.

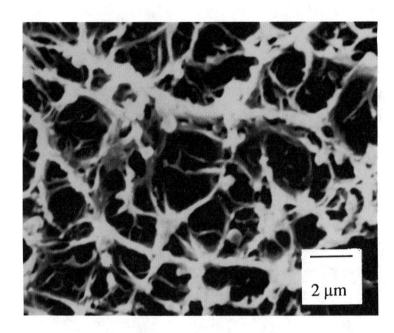

Fig. 5 SEM micrograph of slow crack growth region fracture surface in
       150P PEEK (melt processed at 420°C). Picture shows relatively
       fewer nuclei.

EFFECT OF MOLECULAR CHAIN ORIENTATION

Unoriented samples were compression-molded from ICI Victrex (R) 450G
PEEK pellets at 400°C under a moderate pressure (3.5 MPa). The cooling
rate, measured in the press platten, was about 20°C/min throughout the
crystallization region.   Oriented   samples   were   prepared   using   a
rolltrusion technique at 300°C[5,6]   The degree of crystallinity was
deduced from density measurement.   The degree of orientation was
determined qualitatively by wide-angle x-ray scattering diffraction
patterns and quantitatively by further measurement using an image
analysis system.

The densities, crystallinities, Hermans orientation indices, and
fracture toughness K_Q of the specimens are listed in Table 1.   The
crystallinities of the oriented specimens are higher than those of the
unoriented specimens.   This increase of the crystallinity likely
results from the annealing of the specimens during the rolltrusion
process. Figure 6 shows WAXS photographs of (a) the unoriented and (b)
the oriented PEEK. A strong preference for the crystallographic c-axis
of the PEEK crystals along the machine direction (MD) is seen. The
Hermans orientation index f_H is calculated from the optical density
along the (110) arcs measured by an image analyzer (Cambridge
Quantiment 970). The unoriented sample gives f_H=0, while the oriented
samples give f_H~0.4.

TABLE 1

Sample	Density	Crystallinity	Hermans Orientation	K_Q (T)	K_Q (L)
	g/cm^3	0-100(%)	Index	MPa$\sqrt{m}$	
Raw Material	1.295	27	–	–	–
DR=1.0	1.305	36	0	9	–
DR=3.0	1.307	38	0.42	13	7
DR=3.5	1.309	39	0.42	13	6
DR=4.0	1.310	40	0.43	13	6

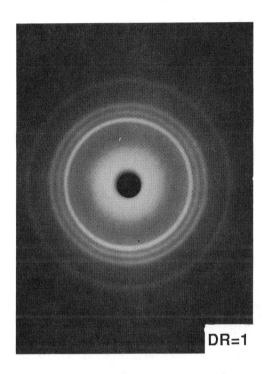

DR=1   DR=3.5

Fig. 6 WAXS photographs of unoriented (DR=1) and oriented (DR=3.5) PEEK
samples. For DR=1, the four major Debye rings ((110), (111),
(200) and (211) appear to be of uniform intensity originating
from randomly oriented crystallities. For DR=3.5 the (110)
reflection shows an obvious enhanced intensity on the equator,
while (111) shows a less obvious enhancement of intensity on the
±30° diagonals.

The effect of orientation on K_Q is investigated in two directions.
T-type refers to the case in which the initial notch and crack growth
is normal to the preferred molecular chain direction, while L-type
refers to the parallel case. The results indicate a remarkable
increase of 45% in K_Q for the oriented specimens (T-type). However,
the K_Q value drops about 30% for the L-type oriented specimens. The
high anisotropy of the oriented PEEK is clearly reflected by a factor
of two in stress intensity factor between T- and L- type specimens.
The modulus and strain energy relase rate (neither shown) also show the
same trend. A previous study of the effects of thickness, initial

crack length ratio, and strain rate on K_Q suggests a good applicability of linear elastic fracture mechanics (LEFM) concepts to these materials.[7]

EFFECT OF SHORT FIBER ORIENTATION

Again, unoriented short fiber reinforced composites (PEEK/carbon and PEEK/glass) were compression-molded, from pellets of PEEK 450CA30 and 450GL30. The preferred orientation of short fibers in oriented composites was introduced, using an inverted extrusion method. Figure 7 shows the microstructure at the surface of the molding, indicating a random orientation of spherulitic nuclei. The fiber orientations were characterized from polished surfaces, using an image analyzer.

TABLE 2

Sample	Hermans Orientation Index 0-10	K_Q/K_O
PEEK/Carbon	0	1.00
PEEK/Carbon (T)	0.46	1.38
PEEK/Carbon (T)	0.24	1.16
PEEK/Carbon (L)	0.46	0.88
PEEK/Carbon (L)	0.24	0.97
PEEK/glass	0	0.83

* K_O is the stress intensity factor of randomly oriented PEEK/Carbon.

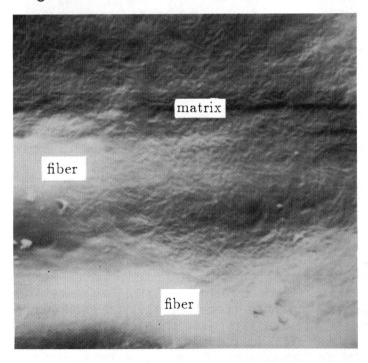

Fig. 7

The microstructure at the surface of oriented composites. A random orientation of spherulitic nuclei is seen within the PEEK matrix. The fibers shown are covered by a thin layer of matrix.

The mechanical characterization results are summarized in Table 2. An increase of 40% in K_Q when a crack propagates from parallel (L-type) to normal (T-type) to the preferred orientation of fibers. Figure 8 describes fracture surfaces in (a) the T-type and (b) the L-type PEEK/carbon specimens. The crack seems to propagate preferentially along fiber ends (in a fiber-avoidance mode) in T-type failure, and parallel to the fiber/matrix interface in L-type fracture. The large amount of PEEK matrix residue left on carbon fiber surfaces implies a good bonding between carbon fibers and PEEK matrix. But this kind of good fiber-matrix bonding cannot be detected in PEEK/glass composites. The poorer bonding apparently leads to the lower stress intensity factor K_Q observed in this system.[8]

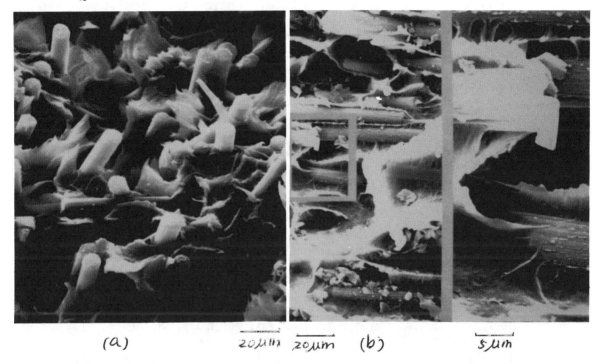

(a)   20μm   20μm   (b)   5μm

Fig. 8   Fracture surfaces of (a) the T-type and (b) the L-type
         PEEK/Carbon composites. The crack propagates along fiber ends
         in the T-type fracture and along fiber/matrix interface in the
         L-type failure. Most of the fibers are surrounded with highly
         deformed matrix.

CONCLUSIONS

The following conclusions are obtained in this study:
(1) The spherulite size of PEEK is greatly influenced by the polymer grade. 450G PEEK, which has a relatively high molecular weight, exhibits only nuclear spherulites, whereas the lower molecular weight 150P PEEK exhibits fully developed spherulites. It is presumed that the difference in behavior of these two grades is due to additives, more than to the molecular weight difference.
(2) For 150P PEEK, spherulite size of isothermally crystalized material increases dramatically with melting temperature.

(2)   In unfilled PEEK, the fracture toughness data obtained by compact tension testing indicates that the fracture toughness decreases with increasing melt temperature.

(3)   The fracture surface morphology studies of 150P PEEK indicate that intraspherulite fracture occurs in this material. A common morphological feature in these samples is nucleus pull-out.

(4)   Oriented PEEK tends to exhibit high specific strength (strength/density) and stiffness (toughness/density). The alignment of the covalently bonded polymer molecules in the machine direction leads to a considerable strength when deformed in that direction. This indicates the ease of intermolecular cleavage (where the resistance is a Van der Waals force) as opposed to molecular fracture (where the resistance is a covalent bond).

(5)   Preferred orientation of short fibers in PEEK composites can be made using a simulating technique of injection-molding. Fractography indicates a major role of orientation-enhanced toughness in the fracture improvement.

ACKNOWLEDGMENT

This work was supported by the U.S. Army Research Office under Grant No. DAAG29-85-K-0042

REFERENCES

1.   Seferis, J. C., 1986, Polymer Composites, 7 (3), 158.

2.   Wunderlich, B., 1976, "Macromolecular Physics", Academic Press, New York.

3.   Chu, Jia-Ni, 1988, M.S. Thesis, University of Delaware.

4.   Olley, R. H., Bassett, D. C., and Blundell, D. J., Polymers, 127, 344.

5.   Way, J. L. Atkinson, J. R., and Nutting, J. 1974, J. Mater. Sci., 9, 293.

6.   Shanker Narayanan, M. J., and Magill, J. H., 1986, J. Mater. Sci. Lett., 5, 267.

7.   Wu, G.-M. and Schultz, J. M., submitted to Polym. Eng. Sci.

8.   Heike Motz, 1987, Ph.D. Dissertation, University of Delaware.

Analysis of the End Notch Flexure Specimen Using a Higher Order Beam Theory Based on Reissner's Principle

JAMES M. WHITNEY

ABSTRACT

A higher order beam theory based on second order displacements in the thickness coordinate is derived in conjunction with Reissner's variational principle. Homogeneous, orthotropic materials are considered. Both the inplane normal stress and the interlaminar shear stress distributions exactly satisfy the equilibrium equations of classical theory of elasticity. The resulting field equations are applied to an analysis of the end notch flexure specimen for measuring interlaminar Mode II critical strain energy release rate. Numerical results from the present theory are compared to those obtained from finite elements. Excellent agreement is obtained.

INTRODUCTION

The end notch flexure (ENF) test is performed in conjunction with a 3-point bend specimen containing a mid-plane starter crack of desired length, a, at the end (see Fig.1). A Mode II critical strain energy release rate, G_{IIC}, can be determined by measuring the load and center deflection at the instant the starter crack propagates. Classical beam theory can be utilized in deriving a relationship between center deflection and load with the result

$$\delta = \frac{(2L^3 + 3a^3)P}{8E_x bh^3}$$

(1)

where δ is the beam deflection under the load nose and E_x is the effective bending modulus in the x- direction of a beam of thickness 2h. Combining eq. (1) with the definition of strain energy release rate for fixed load, we obtain the result

$$G_{IIC} = \frac{P_C}{2b} \frac{d\delta_C}{da} = \frac{9a^2 P_C^2}{16E_x b^2 h^3}$$

(2)

where the subscript C denotes critical values associated with the onset of crack propagation.

Materials Research Engineer, Air Force Materials Laboratory, Wright-Patterson Air Force Base, Ohio 45433. Also Adjunct Professor of Materials Engineering, Universtiy of Dayton, Dayton, Ohio.

Although the ENF test has been utilized with both unidirectional and multidirectional laminates, difficulties have been encountered in the latter case [1] due to the crack not being self-similar (crack not remaining in the mid-plane). This often results in high values of G_{IIC} as compared to unidirectional composites. Thus, the ENF test is currently being utilized in conjunction with unidirectional specimens only.

Finite element analysis of the ENF specimen have been performed by a number of investigators [2-5]. In each of these studies significant departure from beam theory, eq. (2), was observed for a certain range of values of a/L. In the work performed by Mall and Kochhar [3] and by Salpekar, Raju, and O'Brien [5], the largest departure from eq. (2) was observed for small crack lengths with beam theory providing accurate results for a/L $\geq$ 0.5. The same trend was observed by Barrett and Foschi [2] with significant departure from eq. (2) observed over the range 0.1 $\leq$ a/L $\geq$ 0.8. However, their finite element results were applied to wood beams with specimen geometries which differ considerably from those commonly utilized with graphite/epoxy unidirectional composites. The analysis performed by Gillespie, Carlsson, and Pipes [4] shows significant departure from eq.(2) with increasing values of a/L, which is opposite to the trends shown by other investigators.

A modified version of the Whitney-Sun plate theory [6] has been applied to the analysis of a homogeneous ENF specimen [7]. The deformation state is assumed to be cylindrical bending with the following displacement field

$$u = u^0(x) + z\psi(x) + \frac{z^2}{2}\phi(x)$$
$$w = w(x)$$

(3)

where u is the inplane displacement (x direction) and w is the transverse deflection (z direction). This displacement field is then utilized in conjunction with the principle of stationary potential energy to obtain the governing equations. In this approach it is common to employ a k factor, or factors, in conjunction with the interlaminar shear constitutive relations. For the modified whitney-Sun theory these constitutive relations are of the form

$$Q_x = k_1 G_{13} h\left(\psi + \frac{dw}{dx}\right)$$
$$R_x = k_2 G_{13}\frac{h^3}{12}\phi$$

(4)

where k_1 and k_2 are the shear correction factors, G_{13} is the shear modulus relative to the x-z plane, h is the plate thickness, and

$$Q_x, R_x = \int_{-h/2}^{h/2} \tau_{xz}(1, z)\, dz$$

In these integrals τ_{xz} denotes the shear stress in the x-z plane.

A unique determination of the shear correction factors cannot be obtained. However, both the compliance and the energy release rate are a function of k_1 and k_2 when this approach is utilized in analyzing the ENF specimen. In the present paper a higher order beam theory based on Reissner's variational principle [8] is developed. With this approach the constitutive relations are determined directly without the need for shear correction factors. The resulting field equations are applied to the analysis of the ENF specimen. Numerical values for Mode II strain energy release rate are then compared to existing finite element results.

A HIGHER ORDER BEAM THEORY BASED ON REISSNER'S PRINCIPLE

Consider a homogeneous, orthotropic beam of width b and thickness h with the coordinate system in the middle surface. The x coordinate denotes the axial direction of the beam and the z coordinate denotes the thickness direction. Coordinates at the ends of the beam are given by x_0 and x_1, respectively. A surface traction $q = \tau_{xz}(x, -h/2)$ is assumed to be applied to the bottom surface of the beam. The displacements are assumed to be of the form given by eq. (3). The inplane normal stress, σ_x, is assumed to have the following distribution through the beam thickness:

$$\sigma_x = \frac{3}{4bh}\left[3 - 20\left(\frac{z}{h}\right)^2\right]N + 12\frac{z}{bh^3}M - \frac{30}{bh^3}\left[1 - 12\left(\frac{z}{h}\right)^2\right]S \tag{5}$$

where

$$N, M, S = \int_{-h/2}^{h/2} \sigma_x\left[1, z, \frac{1}{2}\left(\frac{z}{h}\right)^2\right]dz$$

From classical theory of elasticity

$$\frac{\partial\sigma_x}{\partial x} + \frac{\partial\tau_{xz}}{\partial z} = 0 \tag{6}$$

Substituting eq. (5) into eq. (6) and integrating with respect to z, we obtain the interlaminar shear stress distribution through the plate thickness

$$\tau_{xz} = \frac{3}{2bh}\left[1 - 4\left(\frac{z}{h}\right)^2\right]Q + \frac{30z}{bh^2}\left[1 - 4\left(\frac{z}{h}\right)^2\right]R - \left[1 - 6\left(\frac{z}{h}\right)\right.$$
$$\left. - 12\left(\frac{z}{h}\right)^2 + 40\left(\frac{z}{h}\right)^3\right]\frac{q}{4} \tag{7}$$

where

$$Q = bQ_x, \quad R = bR_x$$

In addition it is assumed that

$$\sigma_y = \tau_{xz} = \tau_{yz} = 0 \tag{8}$$

The effects of transverse normal stress, σ_z, are neglected in the formulation of the field equations. Thus, the expression for σ_z which satisfies equilibrium is not displayed.

The principle of stationary potential energy provides a well recognized approach to the development of field equations in terms of displacements. Reissner [8], however, developed a variational theorem of elasticity that simultaneously provides constitutive relations, equilibrium equations, and the boundary conditions. For our current applications this variational principle can be written in the form

$$\int_{x_0}^{x_1} \int_{-h/2}^{h/2} [(\frac{\partial u}{\partial x} - \frac{\sigma_x}{E_1}) \delta\sigma_x + (\frac{\partial w}{\partial x} + \frac{\partial u}{\partial z}) \delta\tau_{xz} - (\frac{\partial \sigma_x}{\partial x} + \frac{\partial \tau_{xz}}{\partial z}) \delta u$$

$$- \frac{\partial \tau_{xz}}{\partial x} \delta w] \, dz \, dx - \int_{x_0}^{x_1} [\tau_{xz}(x, -h/2) - q] \delta u \, dx + \int_{-h/2}^{h/2} [\sigma_x(\tilde{x}, z)$$

$$- \sigma] \delta u \, dz + \int_{-h/2}^{h/2} [\tau_{xz}(\tilde{x}, z) - \tau] \delta u \, dz + \int_{-h/2}^{h/2} \sigma_x(\hat{x}, z) \delta u \, dz$$

$$+ \int_{-h/2}^{h/2} \tau_{xz}(\hat{x}, z) \delta w \, dz = 0 \tag{9}$$

where $\tilde{x}$ denotes a beam end(s) on which one or both of the stresses σ_x or τ_{xz} are prescribed and $\hat{x}$ denotes a beam end(s) on which one or both of the displacements u and w are prescribed. Prescribed values of σ_x and τ_{xz} at beam ends are denoted by σ and τ, respectively. Vanishing of terms inside the double integral which are multiplied by stress variations yield the constitutive relations, while vanishing of terms multiplied by displacement variations yield the equilibrium equations.

Substituting eqs. (3), (5), and (7) into eq. (9), and integrating the results with respect to z, we obtain the required constitutive relations, equilibrium equations, and required boundary conditions. The inplane constitutive relations are given by

$$N = E_1 \, bh \, (\frac{du^0}{dx} + \frac{h^2}{24} \frac{d\phi}{dx})$$

$$M = E_1 \frac{bh^3}{12} \frac{d\psi}{dx}$$

$$S = E_1 \frac{bh^3}{24} (\frac{du^0}{dx} + \frac{3h^2}{40} \frac{d\phi}{dx}) \tag{10}$$

while the interlaminar shear constitutive relations are of the form

$$Q = \frac{bh}{6} [5G_{13} (\psi + \frac{dw}{dx}) + \frac{q}{2}]$$

$$R = \frac{7bh^2}{120} (G_{13} h \, \phi - \frac{3q}{2}) \tag{11}$$

For the case of vanishing surface traction, $q = 0$, eqs. (4) and (11) can be made identical by letting $k_1 = 5/6$ and $k_2 = 7/10$. The equilibrium equations are as follows

$$\frac{dN}{dx} - bq = 0$$

$$\frac{dM}{dx} - Q + \frac{bh}{2} q = 0$$

$$\frac{dQ}{dx} = 0$$

$$\frac{dS}{dx} - R - \frac{bh^2}{8} q = 0 \tag{12}$$

The vanishing of the integrals taken at the ends of the beam require that one term of each of the products $(Nu^0, M\psi, S\phi, Qw)$ be prescribed at $x = x_0, x_1$.

APPLICATION TO THE ENF SPECIMEN

For purposes of the present analysis it is only necessary to consider the upper half of the ENF specimen as shown in Fig. 2. Over the crack interval, $- a \leq x \geq 0$, the surface traction vanishes, $q = 0$, while in the cracked region, $0 \leq x \geq (2l - a)$, $q \neq 0$. In addition pure bending deformation is assumed in the uncracked region, i.e. $u (- h/2) = 0$. Thus

$$u^0 = \frac{h}{2} \psi - \frac{h^2}{8} \phi, \quad 0 \leq x \geq (2L - a) \tag{13}$$

and the surface traction q now becomes a dependent variable. In the interval $- a \leq x \geq 0$, $q=0$, and the four kinematic functions in eqs. (3) are the dependent variables.

The final analysis consists of a particular solution and a complementary solution. The particular solution satisfies the conditions

$-a \leq x \geq 0$:

$$N = 0$$

$$Q_p = -\frac{P}{4}, \quad q_p = 0$$

$$M_p = -\frac{P}{4} (x + a) \tag{14}$$

$0 \leq x \geq (L - a)$:

$$Q_p = -\frac{P}{4}, \quad q_p = -\frac{3P}{8bh} \tag{15}$$

$(L - a) \leq x \geq (2L - a)$:

$$Q_p = \frac{P}{4}, \quad q_p = \frac{3P}{8bh} \tag{16}$$

where the subscript p denotes the particular solution. The first three equilibrium relationships in eqs. (12) are exactly satisfied by eqs. (14) - (16). If we assume ϕ_p = constant the constitutive relations, eqs. (10) and (11) can be integrated to obtain u^0_p, ψ_p, and w_p. The third equilibrium relationship in eqs. (12) can then be solved directly for ϕ_p.

We now require the complementary solution to satisfy the conditions

$- a \leq x \geq 0$:

$$N_c = M_c = 0 \tag{17}$$

$-a \leq x \geq (2L - a)$:

$$Q_c = 0 \tag{18}$$

where the subscript c denotes the complementary solution. Combining eqs. (10) - (13) with the conditions (17) and (18), we arrive at the following field equations for the homogeneous solution

$-a \leq x \geq 0$:

$$\frac{\partial u^0_c}{\partial x} + \frac{h^2}{24} \frac{\partial \phi_c}{\partial x} = 0$$

$$\frac{du^0_c}{dx} + \frac{h^2}{24} \frac{d\phi_c}{dx} = 0$$

$$\frac{d\psi_c}{dx} = 0$$

$$E_1 \frac{d^2 u^0_c}{dx^2} + E_1 \frac{3h^2}{40} \frac{d^2 \phi_c}{dx^2} - \frac{7}{5} G_{13} \phi_c = 0 \tag{19}$$

$0 \leq x \geq (2L - a)$:

$$E_1 h^2 \frac{d^2 \psi_c}{dx^2} - \frac{E_1 h^3}{6} \frac{d^2 \phi_c}{dx^2} - 2q_c = 0$$

$$E_1 h^2 \frac{d^2 \psi_c}{dx^2} + 6q_c = 0$$

$$10 \, G_{13} (\psi + \frac{dw_c}{dx}) + q_c = 0$$

$$E_1 h^2 \frac{d^2 \psi_c}{dx^2} - E_1 \frac{h^3}{10} \frac{d^2 \phi_c}{dx^2} - \frac{14}{5} G_{13} h \phi_c - \frac{9}{5} q_c = 0 \tag{20}$$

The total solution is subjected to the boundary conditions

at x = 0, (2L - a):

$$N = M = S = w = 0 \tag{21}$$

In addition, continuity of u^0, ψ, ϕ, w, N, and M are required at $x = 0$. Continuity of Q is automatically satisfied by conditions imposed on the particular and complementary solution.

The resulting solutions for the deflection under the load nose, denoted by δ, and the Mode II strain energy release rate are of the form

$$\delta = \frac{P}{8E_1 bh} \{ 2\bar{L}^3 + 3\bar{a}^3 + \frac{(704\bar{L} + 51\bar{a})}{340} (\frac{E_1}{G_{13}}) + \frac{9\bar{a}}{\lambda^2} [1$$

$$+ \frac{1}{\sinh \lambda(2\bar{L}-\bar{a})} (\lambda\bar{a} \cosh \lambda(2\bar{L}-\bar{a}) - 2 \sinh \lambda\bar{L})]$$

$$+ (\frac{E_1}{G_{13}}) \frac{\sinh \lambda\bar{L} (\lambda\bar{a} + 2 \sinh \lambda(\bar{L} - \bar{a}))}{352 \lambda \sinh \lambda(2\bar{L} - \bar{a})} \} \tag{22}$$

$$G_{II} = \frac{9P^2 \bar{a}^2}{16E_1 b^2 h} \{ 1 + \frac{1}{\lambda^2 \bar{a}^2} [1 + \frac{\lambda^2}{60} (\frac{E_1}{G_{13}})] + \frac{1}{\sinh^2 \lambda(2\bar{L}-\bar{a})} [1$$

$$+ \frac{2}{\lambda\bar{a}} (\sinh \lambda(2\bar{L}-\bar{a}) - \sinh \lambda\bar{L}) \cosh \lambda(2\bar{L}-\bar{a})$$

$$- \frac{2}{\lambda^2 \bar{a}^2} \sinh \lambda(2\bar{L}-\bar{a}) \sinh \lambda\bar{L}] + (\frac{E_1}{G_{13}}) \frac{\lambda^2 \sinh \lambda\bar{L}}{396 \bar{a}^2} [\lambda\bar{a} \cosh \lambda(2\bar{L}-\bar{a})$$

$$+ \sinh \lambda(2\bar{L}-\bar{a}) - 2 \sinh \lambda\bar{L}] \} \tag{23}$$

where

$$\lambda = 4 \sqrt{\frac{14}{5} (\frac{G_{13}}{E_1})}, \quad \bar{L} = L/h, \quad \bar{a} = a/h$$

It should be noted that the current solution is perfectly compatible with deformations that occur in the bottom half of the beam. In particular, a solution can be obtained to the lower half of the beam which will produce the same distribution of q(x) in the uncracked region as the current solution. Since u = 0 along the centerline of the uncracked region and w is independent of z, complete compatibility of the upper and lower halves of the beam will be assured.

NUMERICAL RESULTS

As a numerical example we consider a unidirectional composite with the following properties

$$E_1/E_3 = 11.9, \quad E_1/G_{13} = 25.7, \quad \nu_{13} = 0.3, \quad L/h = 22.4$$

where E_3 and ν_{13} denote Young's modulus in the z-direction and Poisson's ratio as measured from contraction in the z-direction during a uniaxial tensile test in the x-direction, respectively. These properties and dimensions are typical of state-of-the-art graphite/epoxy unidirctional beams currently utilized in ENF specimens.

Ratios of G_{II} / G_{II}^{BT}, where the superscript BT denotes the beam theory solution as given by eq. (2), are shown in Table 1 as a function of a/L. Results from eq. (23) are compared to finite element results obtained by Gillespie, Carlsson, and Pipes [4] and by Salpekar, Raju, and O'Brien [5]. Resuls in the last column are obtained from the relationship

$$G_{II} = \frac{9P^2\bar{a}^2}{16E_1 b^2 h} [1 + \frac{0.2}{\bar{a}^2} (\frac{E_1}{G_{13}})] \tag{24}$$

This expression was developed by Carlsson, Gillespie, and Pipes [9] and is based on shear deformation theory in the uncracked region ($0 \leq x \geq 2L-a$) and on an approximate elasticity solution for an end loaded cantilever beam [10] in the cracked region ($-a \leq x \geq 0$). Continuity of displacement at the crack tip, $x = 0$, is not attained, however, with this formulation.

DISCUSSION AND CONCLUSIONS

The numerical results in Table 1 show a good correlation between eq. (23) and the finite element results reported in Ref. [5]. The finite element results obtained by Gillespie, Carlsson, and Pipes [4] display a large discrepancy compared to the other results. Furthermore, as previously pointed out, their results show an increasing departure from beam theory with increasing values of a/L, while the other results show the opposite trend. Relatively close agreement is also obtained between the current approach, eq. (23), and the beam analysis developed by Carlsson, Gillespie, and Pipes [9].

The use of Reissner's principle in developing the higher order beam theory presented in the current paper allows the direct development of constitutive relations without the need for shear correction factors. This approach allows one to obtain a more accurate solution for the determination of Mode II strain energy release rate in conjunction with the ENF specimen than classical beam theory. Since the results are in closed form, they can be used more readily than results obtained from the finite element approach.

REFERENCES

1. A. J. Russell and K. N. Street, "Factors Affecting the Interlaminar Fracture Energy of Graphite/Epoxy Laminates," *Proceedings of the Fourth International Conference on Composite Materials*, Elsevier, North Holland (1984), pp. 279-286.

2. S. Mall and N. K. Kochhar, "Finite-Element Analysis of End-Notch Flexure Specimens," *Journal of Composite Technology and Research*, Vol. 8 (1986), pp. 54-57.

3. J. D. Barrett and R. O. Foschi, "Mode II Stress Intensity Factors for Cracked Wood Beams," *Engineering Fracture Mechanics*, Vol. 9 (1977), pp. 371-378.

4. J. W. Gillespie, Jr., L. A. Carlsson, and R. B. Pipes, "Finite Element Analysis of the End Notched Flexure Specimen for Measuring Mode II Fracture Toughness," *Composites Science and Technology*, Vol. 27 (1986), pp. 177-197.

5. S. A. Salpekar, I. S. Raju, and T. K. O'Brien, "Strain-Energy Release Rate Analysis of the End Notch Flexure Specimen Using the Finite Element Method," *Journal of Composites Technology and Research*, to be published.

6. J. M. Whitney and C. T. Sun, "A Higher Order Theory for Extensional Motion of Laminated Composites," *Journal of Sound and Vibration*, Vol. 30 (1973), pp. 85-91.

7. J. M. Whitney, "Stress Analysis of the End Notch Flexure Specimen for Mode II Delamination of Composite Materials," *AIAA/ASME/ASCE/AHS 28th Structures, Structural Dynamics, and Materials Conference*, Part I (1987), pp. 177-197.

8. E. Reissner, "On a Variational Theorem in Elasticity," *Journal of Mathematics and Physics*, Vol. 29 (1950), pp. 90-97.

9. L. A. Carlsson, J. W. Gillespie, Jr., and R. B. Pipes, "On the Analysis and Design of End Notch Flexure (ENF) for Mode II Testing," *Journal of Composite Materials*, Vol. 20 (1986), pp. 594-604.

10. Timoshenko and J. N. Goodier, *Theory of Elasticity*, Second Edition, McGraw-Hill (1951), pp. 35-39.

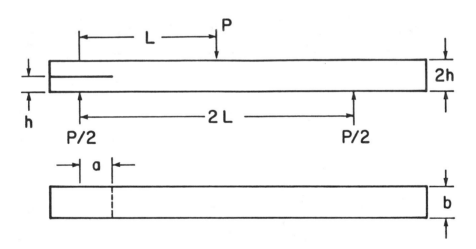

Fig. 1   End notch Flexure Specimen

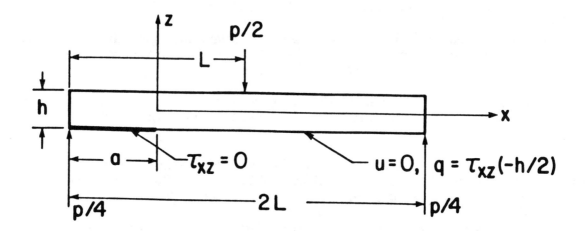

Fig. 2   Stress analysis model

Table 1   Mode II strain energy release rate

	G_{II}/G_{II}^{BT}			
a/L	EQ. 23	Ref. 4	Ref 5	EQ. 24
0.2	1.388	---	1.305	1.256
0.3	1.247	1.252	---	1.114
0.4	1.181	---	1.142	1.064
0.5	1.143	1.371	---	1.041
0.6	1.118	---	1.090	1.028
0.7	1.101	1.472	---	1.021
0.8	1.088	---	1.064	1.016
0.9	1.079	---	1.050	1.013

SYMPOSIUM IV—SESSION B

Composite Mechanics

Torsion of Laminates Containing Orthotropic Layers

R. D. KURTZ AND J. M. WHITNEY

Abstract

The exact elasticity solution for simple torsion of cross–ply laminate is presented. The particular geometry investigated is that of a rectangular plate. Several examples are shown in detail and the solution is compared with the existing theory for the torsion of homogeneous orthotropic bars. It is demonstrated that the homogeneous solution is sufficiently rigorous for most practical applications.

Introduction

In many applications, composite laminates have become so thick that the loading can no longer be considered as in–plane. These thick structures have intensified the need for practical methods for determining out–of–plane deformation, stress, and failure. This was the motivation for the recent paper by R.D. Kurtz and C.T. Sun [1]. The paper presented a simplified and accurate experimental testing procedure to determine the transverse modulus 'G_{23}' and the interlaminar shear strength of composite materials. This approach relies on the fundamental assumption that under simple torsion the deformation of a cross–ply laminate can be predicted by the existing solution for the torsion of a homogeneous orthotropic bar. The present paper will focus on this assumption. The exact elasticity solution for the simple torsion of symmetric cross–ply laminates will be developed and compared to the homogeneous case.

Materials Research Engineers, Air Force Wright Aeronautical Laboratories, Wright–Patterson AFB, OH 45433.

The approach is not limited to symmetric cross–ply laminates and can be extended to unsymmetric laminates as well. The only criteria is that the plies which make up the laminate must be orthotropic relative to the x–y axis system parallel to the sides of the rectangular plate under consideration.

Theory

For the special case of a symmetric laminate with only 0° and 90° plies laying in the x–y plane as shown in Figure 1, an exact elasticity solution can be developed to describe the deformation produced by torsion. The geometry considered is of rectangular planform and cross–section of width 'a' and thickness 'b'. The origin of the coordinate system is located at the middle of the left–hand free edge of the laminate. Under an applied torque, the cross–sections of the laminate twist with respect to another along with the simultaneous warping of the cross–sectional planes. In this state of deformation there are only two non–zero stress components. These are the shear stresses acting in the plane perpendicular to the center of twist. The state of stress can be written as :

$$\sigma_x = \sigma_y = \sigma_z = \tau_{xy} = 0 \; ; \; \tau_{xz} = \tau_{xz}(y, z) \quad \tau_{yz} = \tau_{yz}(y, z) \tag{1}$$

The equilibrium equations for the kth ply then reduces to :

$$\frac{\partial\left(\tau_{yz}^k\right)}{\partial y} + \frac{\partial\left(\tau_{xz}^k\right)}{\partial z} = 0 \tag{2}$$

Using u, v, and w to denote the deformations in the x, y, and z directions, the boundary conditions become :

$$\tau_{xy}^k = 0 \text{ at } y = 0, a \tag{3}$$

$$\tau_{xz}^k = 0 \text{ at } z = \pm b/2 \tag{4}$$

u, v, and w continuous between plies $\tag{5}$

τ_{xz}^k continuous between plies $\tag{6}$

Assuming a displacement field of :

$$u = \beta [z (y - 2) + f^k(y,z)]$$
$$v = - x z \beta$$
$$w = x y \beta \tag{7}$$

where,

β = the angle of twist

We note that continuity between plies is assured for v and w as given by Equation 7. Using Hooke's Law for an orthotropic material in conjunction with Equation 7, we obtain the shear stresses:

$$\tau^k_{xy} = G^k_{xy} \left[\frac{\partial u}{\partial y} + \frac{\partial v}{\partial x} \right]$$

$$\tau^k_{xz} = G^k_{xz} \left[\frac{\partial u}{\partial z} + \frac{\partial w}{\partial x} \right] \tag{8}$$

where $G_{xy}{}^k$ and $G_{xz}{}^k$ are the shear moduli of the kth ply relative to the x–y and x–z planes, respectively. Combining Equations 7 and 8 with the equilibrium condition, Equation 2, we obtain the result :

$$G_{xy} \frac{\partial^2 f^k(y,z)}{\partial y^2} + G_{xz} \frac{\partial^2 f^k(y,z)}{\partial z^2} = 0 \tag{9}$$

A solution to Equation 9 which satisfies the boundary conditions, Equation 3 and 4, is of the form :

$$f^k(y,z) = \sum_{m=1,2,3...}^{\infty} \beta \left[A^k_m \cosh(\mu^k \lambda_m z) + B^k_m \sinh(\mu^k \lambda_m z) \right] \cos(\lambda_m y) \tag{10}$$

with,

$$\mu^k = (G_{xy}{}^k / G_{xz}{}^k)^{1/2}$$

$$\lambda_m = m \pi / a$$

$$A_m^k, B_m^k = \text{constants}$$

Substituting Equation 9 into Equations 6 and 7, we obtain the following results :

$$u^k = z \beta (y - 2) + \sum_{m=1,3,5\ldots}^{\infty} \beta \left[A_m^k \cosh(\mu^k \lambda_m z) + B_m^k \sinh(\mu^k \lambda_m z) \right] \cos(\lambda_m y)$$

(11a)

$$\tau_{xy}^k = \sum_{m=1,3,5\ldots}^{\infty} - \lambda_m \beta \left[A_m^k \cosh(\mu^k \lambda_m z) + B_m^k \sinh(\mu^k \lambda_m z) \right] \sin(\lambda_m y)$$

(11b)

$$\tau_{xz}^k = 2 \beta (y - 1) + \sum_{m=1,3,5\ldots}^{\infty} \mu^k \beta \left[A_m^k \sinh(\mu^k \lambda_m z) + B_m^k \cosh(\mu^k \lambda_m z) \right] \cos(\lambda_m y)$$

(11c)

Equation 11b exactly satisfies the boundary condition given by Equation 3. Expanding the first term of Equation 11c into a fourier sine series, we obtain the result :

$$\tau_{xz}^k = \sum_{m=1,3,5\ldots}^{\infty} \frac{-8 \lambda_m^2}{a} + \mu^k \beta \left[A_m^k \sinh(\mu^k \lambda_m z) + B_m^k \cosh(\mu^k \lambda_m z) \right] \cos(\lambda_m y)$$

(11d)

Now consider a symmetric laminate constructed of ' n ' plies, the top and the bottom ply must satisfy the conditions given by Equations 4, 5, and 6 while the interior plies need only to satisfy the conditions given by Equations 5 and 6. By substituting Equation 11 into the conditions 4, 5, and 6, a system of ' 2n ' equations is obtained for each value of 'm'. The first two equations obtained from Equation 4 are :

$$\sum_{m=1,3,5\ldots}^{\infty} \left\{ \frac{-8 \lambda_m^2}{a} + A_m^k \mu^k \sinh(\alpha_m^k z^k) + B_m^k \mu^k \cosh(\alpha_m^k z^k) \right\} = 0$$

$$k = 1, n+1 \quad (12a)$$

where,

$$\alpha_m^k = \mu^k \lambda_m$$

$$z^k = b/2 - t (k - 1)$$

$$t = \text{ply thickness}$$

Substituting Equation 11a into Equation 5, we obtain the following 'n–1' equations :

$$\sum_{m=1,3,5...}^{\infty} \left\{ A_m^k \cosh(\alpha_m^k z^k) + B_m^k \sinh(\alpha_m^k z^k) \right\}$$

$$- \left\{ A_m^{k+1} \cosh(\alpha_m^{k+1} z^k) + B_m^{k+1} \sinh(\alpha_m^{k+1} z^k) \right\} = 0$$

$$k = 2, 3 ... n \qquad (12b)$$

Substituting Equation 11d into Equation 6, the final ' n – 1 ' equations are obtained :

$$\sum_{m=1,3,5...}^{\infty} \left\{ G_{xz}^k \left[\frac{-8 \lambda_m^2}{a} + A_m^k \mu^k \sinh(\alpha_m^k z^k) + B_m^k \mu^k \cosh(\alpha_m^k z^k) \right] \right\}$$

$$- \left\{ G_{xz}^{k+1} \left[\frac{-8 \lambda_m^2}{a} + A_m^{k+1} \mu^{k+1} \sinh(\alpha_m^{k+1} z^k) + B_m^{k+1} \mu^{k+1} \cosh(\alpha_m^{k+1} z^k) \right] \right\} = 0$$

$$k = 2, 3 ... n \qquad (12c)$$

For symmetric laminates, an examination of Equation 11d reveals that τ_{xz} must be symmetric about the mid–plane, z=0. Thus, the constant A must be zero in the mid–ply. In addition, the system of equations can now be reduced from '2n' to ' n–1' by letting ' k ' range from ' 1 ' to ' n/2 – 1 ' in Equations 12a – 12c and setting :

$$A_m^{n/2} = 0, \quad A_m^{n/2+k} = - A_m^{n/2-k} \qquad (13)$$

One final equation is needed to determine the angle of twist, ' β '. This is obtained by recognizing that the total torque applied to the composite plate must be equal to the total moment produced by the shear stresses which is written as :

$$\int \int (y \, \tau_{xz}^k - z \, \tau_{xy}^k) \, dy \, dz = T \qquad (14)$$

where,

T = the applied torque

Substituting Equations 11 into 14 and integrating over the thickness, we find :

$$T = 4\beta\sum_{k=1}^{n}\sum_{m=1,3,5\ldots}^{\infty}\frac{zG_{xy}}{\mu^{k}\lambda_{m}}\left[A_{m}^{k}\sinh(\mu^{k}\lambda_{m}z) + B_{m}^{k}\cosh(\mu^{k}\lambda_{m}z)\right] -$$

$$\frac{G_{xy}^{k}}{(\mu^{k}\lambda_{m})^{2}}\left[A_{m}^{k}\cosh(\mu^{k}\lambda_{m}z) + B_{m}^{k}\sinh(\mu^{k}\lambda_{m}z)\right] -$$

$$\lambda_{m}^{2}G_{xz}^{k}\left[A_{m}^{k}\cosh(\mu^{k}\lambda_{m}z) + B_{m}^{k}\sinh(\mu^{k}\lambda_{m}z) - \frac{8z}{a\lambda_{m}^{2}}\right]\Bigg|_{z^{k}}^{z^{k+\cdot}}$$

$$(14)$$

Equations 12 and 14 represent a complete system of equations which can be solved with the aid of a computer. Because the equations are in the form of a summation, the system is solved with a term by term procedure. The solution converges rapidly and only a small number of terms are needed to obtain an accurate solution. A convergence study has shown that only 25 terms are needed to achieve three place accuracy.

Numerical Results

The solution for a homogeneous orthotropic plate subjected to torsional loading has been solved previously by Lekhnitskii in [2]. The solution is good for any material provided the state of stress is equivalent to conditions given by Equation 1 and the principle material axis are aligned with the coordinate system chosen (i.e. there can be no shear extension coupling). For the sake of brevity, the homogeneous solution will not be repeated here. However, it can be shown that the solution presented in this paper will reduce to the solution obtained by Lekhnitskii for the special case of a homogeneous, orthotropic plate subjected to torsional loading.

The particular interest in this problem is the investigation of $[0°/90°]_s$ laminates. Therefore, we will restrict our discussion to only these types of lay–ups. In order to apply homogeneous solution, the cross–ply laminate must be replaced with an equivalent homogeneous elastic solid. The shear moduli of the equivalent solid are obtained from the individual ply properties. Methods for deriving the effective moduli for composite laminates have been developed by Pagano [3] and Sun [4]. For the simple case of a cross–ply laminate, the equivalent moduli obtained from both these methods are given by :

$$G_{xy} = G_{12} \tag{15a}$$

$$\frac{1}{G_{xz}} = \frac{1}{2} \left[\frac{1}{G_{12}} + \frac{1}{G_{13}} \right] \tag{15b}$$

where the subscripts 1, 2, and 3 refer to the principle material properties. Equation 15a can be derived from the fact that the in–plane shear modulus in a cross–ply laminate does not vary with the thickness and is equal to the principal modulus 'G_{12}'. Equation 15b is more difficult to derive and is based on the application of a constant shearing stress to the laminate.

Using the definitions in 15a – b and the solution for a homogeneous plate, the exact solution for the cross–ply laminate presented in this paper can be compared to that of an equivalent homogeneous solid. The larger the number of repeated [0/90] sublaminates which make up the cross–ply laminate, the closer the laminate approximates a homogeneous material. As a result, we chose to analyze the extreme case of a four ply laminate. Both $[0/90]_s$ and $[90/0]_s$ laminates are considered. The individual ply thickness is normalized with respect to the width of the laminate. The normalized shear moduli are chosen as :

$$G_{12} = 1.0 \ \ (\text{force / length}^2)$$
$$G_{13} = 0.5 \ \ (\text{force / length}^2) \tag{16}$$

which represents one of the more severe cases found in current materials. Figure 2 is a comparison between the angles of twist of the three laminates. As the width to ply thickness ratio ('a/t') becomes large, the angle of twist in both the $[0/90]_s$ laminate and the $[90/0]_s$ laminate approach that of the homogeneous solid. This is because as 'a/t' becomes large the in–plane modulus which is equivalent in each case begins to dominate the response. Figures 3 and 4 show plots of the shear stress distribution along the free edges of the laminated bar for a width to ply thickness ratio of four. Because of symmetry, only half the free edge is considered. The effect of the variable ply properties are clearly shown in Figure 3. In this plot a dramatic change occurs in the slope of the $[0/90]_s$ and $[90/0]_s$ curves at the interface between the $0°$ and $90°$ plies ($a/t = .25$). In both Figures 3 and 4, the shear stresses of the equivalent homogeneous solid give a good approximation of the shear stresses for the $[0/90]_s$ and $[90/0]$ laminates (a maximum of 15% difference). Analysis shows that the addition of more plies significantly improves this comparison. In most applications of composite material, the number of plies and width to thickness ratio of the laminates are sufficiently large so that the the use of the homogeneous theory causes no significant error.

Conclusions

In this paper the exact elasticity solution for the simple torsion of a cross–ply laminate has been developed. A comparison between existing theory and the exact solution has demonstrated that the deformation of the cross–ply laminate can be predicted by the solution for a homogeneous orthotropic bar. In most composite laminates where the ratio between the width to ply thickness is large, the homogeneous solution has proven to be very accurate. This is of some practical importance because the exact solution for the cross–ply laminate is complex and inconvenient to use.

References

1. Kurtz, R.D. and Sun, C.T. " Torsion of Thick Laminates", ASTM – 9th Symposium on Composite Matls., April 1988.

2. Lekhnitskii, S.G., *Theory of Elasticity of Anisotropic Body*, Mir Publishers, USSR, 1981.

3. Pagano, N.J., "Exact Moduli of Anisotropic Laminates," in *Composite Materials*, Vol. 2, (G.P. Sendeckyj, Editor), Academic Press, New York, 1974, pp. 23–45.

4. Sun, C.T., and Li, Sijian, "Three–Dimensional Effective Elastic Constants for Thick Laminates," to appear in *Journal of composite Materials*.

Figure 1 – Coordinate System

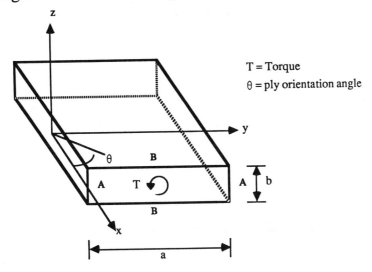

Figure 2 – Angle of Twist Normalized by the
Homogenous Angle of Twist

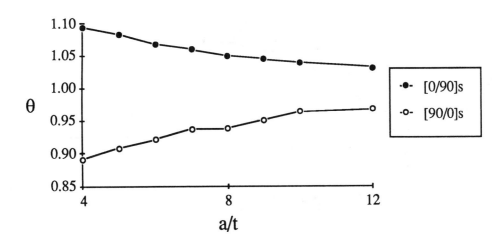

Figure 3 – Transverse Shear Stress
at the free edge y = 0 and a/t = 2

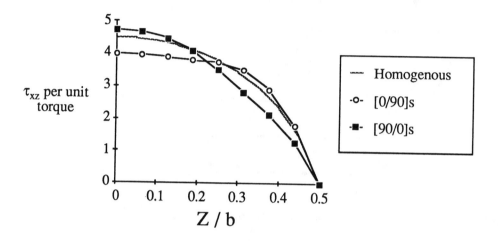

τ_{xz} per unit torque

Z / b

Homogenous
-o- [0/90]s
-■- [90/0]s

Figure 4 – In–plane Shear Stress
at the free edge z = b/2 and a/t = 2

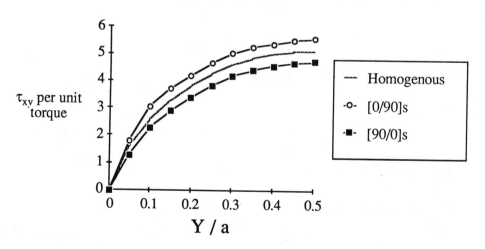

τ_{xy} per unit torque

Y / a

Homogenous
-o- [0/90]s
-■- [90/0]s

Correction Factors for Finite-Width Anisotropic Plates Containing Elliptical Openings

SENG C. TAN

ABSTRACT

The finite-width correction factors for anisotropic and orthotropic plates containing an elliptical opening are derived and given in a tractable and closed-form solution. Comparison with experimental data and finite element solutions reveals that the present theory is very accurate for a broad range of opening-to-width ratio, 2a/w, and opening aspect ratio, a/b. The isotropic finite-width correction factors can only be used in a narrow region without causing significant error.

1. INTRODUCTION

By definition, a FWC (finite-width correction) factor is a scale factor which is applied to multiply the notched infinite-plate solution to obtain the notched finite-plate result. The FWC factor can be discussed in the following aspects: (1) the stress or strain concentration such as the maximum tangential stress at the opening edge; (2) the stress intensity factor of a plate with a crack and; (3) the interpolation of finite plate testing data (such as strength) to infinite plate result; this concept has been utilized for the strength analysis of isotropic and anisotropic plates. In this aspect, most of the strength analyses are based on infinite plate solution because a compact form stress distribution is available. Therefore, there is a need for the data reduction between finite and infinite-width plates. The first two aspects have been discussed a great deal for isotropic material [1-9] and orthotropic material [10-12]. The first and the third aspects are the inspirations for the present paper. The third aspect has seldom been discussed in the literature.

The FWC factors for isotropic plate with a hole or a crack can be determined accurately using a curve fitting technique [5]. Whereas for finite-width anisotropic or orthotropic plates, the stress analyses have been proceeded mainly by using finite element methods. Due to the lack of a closed-form solution for an anisotropic finite-width plate containing a cutout, the isotropic FWC factors have been applied for the data reduction of notched orthotropic finite-width laminates to notched infinite plates. In some cases, the isotropic FWC factors can be applied for anisotropic plates, for instance, the stress intensity of an infinite length composite laminate with a crack [10]. This is because the stress distribution around the crack is independent of the material properties. However the application of an anisotropic FWC factor to transform the maximum stress concentration and the notched strength data between a finite and an infinite-width plate with an opening

Seng C. Tan, Research Scientist, AdTech Systems Research Inc., Air force Wright Aeronautical Laboratories, AFWAL/MLBM, Wright-Patterson AFB OH 45433-6533.

other than a crack has seldom been discussed in the literature.

In the present study, a compact form solution is derived for the FWC factors of anisotropic and orthotropic plates containing a central elliptical opening. Published numerical and experimental results for respective orthotropic and isotropic plates are applied to examine the present theory.

2.   FUNDAMENTAL THEORY

An anisotropic FWC factor can be derived by proper utilization of the stress distribution around the opening of an infinite anisotropic plate. Two derivations are presented in the following: one is by using the exact two-dimensional anisotropic normal stress distribution; another one is by considering an approximate orthotropic stress distribution for an infinite plate.

2.1   Anisotropic FWC Factor

The stress distribution of an infinite anisotropic laminate containing a central elliptical opening, Figure 1, can be derived using a complex variable method [13-15]. The normal stress, σ_y^∞, along the x-axis of the laminate, under the applied stress, $\overline{\sigma}_y$, is

$$\sigma_y^\infty(x,0) = \overline{\sigma}_y + \overline{\sigma}_y \, \mathrm{Re} \left\{ \frac{1}{\mu_1 - \mu_2} \left[\frac{-\mu_2(1 - i\mu_1\lambda)}{\sqrt{\gamma^2 - 1 - \mu_1^2\lambda^2} \, (\gamma + \sqrt{\gamma^2 - 1 - \mu_1^2\lambda^2})} \right. \right.$$
$$\left. \left. + \frac{-\mu_1(1 - i\mu_2\lambda)}{\sqrt{\gamma^2 - 1 - \mu_2^2\lambda^2} \, (\gamma + \sqrt{\gamma^2 - 1 - \mu_2^2\lambda^2})} \right] \right\} \tag{1}$$

where $\gamma = x/a$, $\lambda = b/a$ and μ_1 and μ_2 are the solutions of the characteristic equation:

$$a_{22}\mu^4 - 2a_{26}\mu^3 + (2a_{12} + a_{66})\mu^2 - 2a_{16}\mu + a_{11} = 0 \tag{2}$$

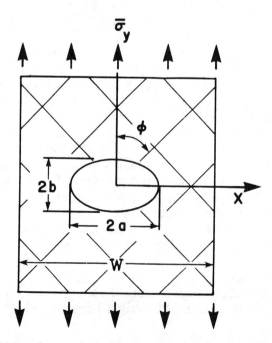

Figure 1. A finite-width anisotropic laminate containing a central elliptical opening.

where a_{ij}, $i, j = 1, 2, 6$ are the compliances of the laminate with 1 and 2 parallel and transverse to the loading direction respectively. Only the principal roots of Equation (2) should be chosen, i.e., two of the four roots that have positive imaginary part.

According to the definition of the FWC factor stated in Section 1; and an assumption that the normal stress profile for a finite plate is identical to that for an infinite plate except for a FWC factor, the following relation is obtained:

$$\frac{K_T}{K_T^\infty} \, \sigma_y^\infty(x,0) = \sigma_y(x,0) \tag{3}$$

where K_T/K_T^∞ is the FWC factor and K_T and K_T^∞ denote the stress concentration at the opening edge on the axis normal to the applied load for a finite plate and an infinite plate, respectively. The parameter σ_y is the y-component of normal stress for a finite-width plate. Justification of the above assumption is made in Section 4. The comparison of the present solution with the experimental result [2] in Figure 2; the finite element result in Figure 4 and those in reference 16 show that this assumption is acceptable for $a/b \geq 1$ with $2a/W < 0.5$ and for a smaller $2a/W$ ratio with $a/b < 1$. Outside these regions, the stress distribution tends to deviate from the assumption. Therefore, an improved theory is also developed in Section 3. The solution of the improved theory agrees very well with the experimental and numerical results.

From the consideration of equilibrium condition (force resultant) in the loading direction, the following equation is obtained from the integration of Equation (3):

$$\frac{a\,K_T}{K_T^\infty} \int_1^{W/2a} \sigma_y^\infty(x,0)d\gamma = a \int_1^{W/2a} \sigma_y(x,0)d\gamma = \overline{\sigma}_y \bullet W/2 \tag{4}$$

Substitution Equation (1) into Equation (4) yields

$$\frac{K_T^\infty}{K_T} = 1 - \frac{2a}{W} + \mathrm{Re}\left\{ \frac{1}{\mu_1 - \mu_2} \left[\frac{\mu_2}{1 + i\mu_1\lambda} \left(1 - \frac{2a}{W} - i\mu_1\lambda(\frac{2a}{W}) - \sqrt{1 - (1 + \mu_1^2\lambda^2)(2a/W)^2} \right) \right.\right.$$

$$\left.\left. - \frac{\mu_1}{1 + i\mu_2\lambda} \left(1 - \frac{2a}{W} - i\mu_2\lambda(\frac{2a}{W}) - \sqrt{1 - (1 + \mu_2^2\lambda^2)(2a/W)^2} \right) \right]\right\} \tag{5}$$

Equation (5) is the reciprocal of the anisotropic FWC factor. The imaginary part of the solution for the term with square root should keep the same sign as that for the parameter under the square root. This also holds for Equation (10).

For orthotropic laminate, the characteristic equation, Equation (2), reduces

$$a_{22}\mu^4 + (2a_{12} + a_{66})\mu^2 + a_{11} = 0 \tag{6}$$

which has the solution:

$$\mu^2 = -\frac{2a_{12} + a_{66}}{2a_{22}} \pm \sqrt{(\frac{2a_{12} + a_{66}}{2a_{22}})^2 - \frac{a_{11}}{a_{22}}} \tag{7}$$

Solving Equation (7) yields

$$\mu_{1,2,3,4} = \sqrt{r} \, [\cos(\theta/2 + k\pi) + i\sin(\theta/2 + k\pi)], \quad k = 0,1 \tag{8}$$

where

$$r = \sqrt{a_{11}/a_{22}} \tag{9a}$$

$$\theta = \tan^{-1} \{\pm \sqrt{\frac{a_{11}}{a_{22}} - (\frac{2a_{12} + a_{66}}{2a_{22}})^2} \Big/ \frac{-(2a_{12} + a_{66})}{(2a_{22})}\} \tag{9b}$$

Again, only the two principal roots of Equation (8) should be used for computing the FWC factors. Note that if θ, Equation (9b), lies in the second or third quadrant, it has to be subtracted from π before substituting into Equation (8).

2.2   Approximate Orthotropic FWC Factor

In an earlier work [15], an approximate stress distribution for an infinite orthotropic laminate containing an elliptical opening has been derived. The solution, which was found very accurate within the range $0 \le b/a \le 1$, is

$$\frac{\sigma_y^\infty(x,0)}{\sigma_y} = \frac{\lambda^2}{(1-\lambda)^2} + \frac{(1-2\lambda)\gamma}{(1-\lambda)^2\sqrt{\gamma^2 - 1 + \lambda^2}} + \frac{\lambda^2\gamma}{(1-\lambda)(\gamma^2 - 1 + \lambda^2)^{3/2}}$$

$$- \frac{\lambda^7}{2}(K_T^\infty - \frac{2}{\lambda}) \, [\frac{5\gamma}{(\gamma^2 - 1 + \lambda^2)^{7/2}} - \frac{7\lambda^2\gamma}{(\gamma^2 - 1 + \lambda^2)^{9/2}}] \tag{10}$$

where the parameters have all been defined and

$$K_T^\infty = 1 + \frac{1}{\lambda}\sqrt{\frac{2}{A_{66}}\left(\sqrt{A_{11}A_{22}} - A_{12} + \frac{A_{11}A_{22} - A_{12}^2}{2A_{66}}\right)} \tag{11}$$

where A_{ij}, $i, j = 1, 2, 6$ denote the effective laminate in-plane stiffnesses with 1 and 2 parallel and transverse to the loading direction respectively.

Substituting Equation (14) into (4) yields

$$\frac{K_T^\infty}{K_T} = \frac{\lambda^2}{(1-\lambda)^2} + \frac{(1-2\lambda)}{(1-\lambda)^2}\sqrt{1 + (\lambda^2 - 1)(2a/W)^2} - \frac{\lambda^2}{(1-\lambda)}\frac{(2a/W)^2}{\sqrt{(1 + (\lambda^2 - 1)(2a/W)^2}}$$

$$+ \frac{\lambda^7}{2} (\frac{2a}{W})^6 (K_T^\infty - 1 - \frac{2}{\lambda})\{[1 + (\lambda^2 - 1)(\frac{2a}{W})^2]^{-5/2} - (\frac{2a}{W})^2[1 + (\lambda^2 - 1)(\frac{2a}{W})^2]^{-7/2}\} \qquad (12)$$

When $\lambda = b/a = 1$ (a circular hole), the first part of Equation (12) is undetermined. After applying L'Hospital rule twice w.r.t. λ and substituting $\lambda = 1$ we obtain

$$K_T^\infty/K_T = [2 - (2a/W)^2 - (2a/W)^4]/2 + (2a/W)^6(K_T^\infty - 3)[1 - (2a/W)^2] / 2 \qquad (13)$$

When $\lambda = b/a = 0$ (a crack), Equation (12) reduces to

$$K_T^\infty/K_T = \sqrt{1 - (2a/W)^2} \qquad (14)$$

which is the same as the Dixon formula [7] and is independent of the material properties.

2.3   Isotropic FWC Factor

The FWC factors for an isotropic plate with an elliptical opening can be obtained by substituting $K_T^\infty = 1 + 2/\lambda$, isotropic stress concentration factor, into Equation (12), i.e.

$$\frac{K_T^\infty}{K_T} = \frac{\lambda^2}{(1-\lambda)^2} + \frac{(1 - 2\lambda)}{(1-\lambda)^2}\sqrt{1 + (\lambda^2 - 1)(\frac{2a}{W})^2} - \frac{\lambda^2}{(1-\lambda)}(\frac{2a}{W})^2[1 + (\lambda^2 - 1)(\frac{2a}{W})^2]^{-1/2} \qquad (15)$$

Equation (15) can also be obtained by substituting $\mu_1 = \mu_2 = i$ into Equatio (5). When $\lambda = b/a = 1$ (a circular hole), the solution can easily be seen from Equation (13) as

$$K_T^\infty/K_T = [2 - (2a/W)^2 - (2a/W)^4] / 2 \qquad (16)$$

When $\lambda = b/a = 0$ (a crack), the FWC factor is given in Equation (14).

3.   IMPROVED THEORY (FOR a/b < 4)

A parametric study in Section 4, Figure 2, shows that the maximum finite-width stress concentration predicted by Equations (15) and (16) are more accurate for $a/b \geq 4$ than for $a/b < 4$. The predictions using the fundamental theory for $a/b < 4$ are too low compared to the experimental data. Therefore, a magnification factor M is developed to improve the previous basic theory. Knowing the fact that the accuracy of Heywood formula [8] is very good for isotropic plate with a circular opening, we derive an anisotropic solution that will recover to the Heywood formula under the isotropic and same opening condition. This can be accomplished by multiplying the 2a/W ratio of Equation (16) by a magnification factor M; equating it to the Heywood formula, Equation (22), and solving for M. Finally, multiplying the opening-to-width ratio, 2a/W, of Equation (5) by the magnification factor M, we obtain the improved anisotropic FWC factor as:

$$\frac{K_T^\infty}{K_T} = 1 - \frac{2a}{W}M + Re\{\frac{1}{\mu_1 - \mu_2}[\frac{-\mu_2}{1 + i\mu_1\lambda}(\frac{2a}{W}M - 1 + i\mu_1\lambda(\frac{2a}{W}M) + \sqrt{1 - (1+\mu_1^2\lambda^2)(\frac{2a}{W}M)^2})$$

$$+ \frac{\mu_1}{1 + i\mu_2\lambda}(\frac{2a}{W}M - 1 + i\mu_2\lambda(\frac{2a}{W}M) + \sqrt{1 - (1 + \mu_2^2\lambda^2)(\frac{2a}{W}M)^2}\ )]\} \tag{17}$$

where

$$M^2 = \frac{\sqrt{1 - 8[\frac{3(1 - 2a/W)}{2 + (1 - 2a/W)^3} - 1]} - 1}{2(2a/W)^2} \tag{18}$$

The approximate orthotropic FWC factor, Equation (12), becomes

$$\frac{K_T^\infty}{K_T} = \frac{\lambda^2}{(1 - \lambda)^2} + \frac{(1 - 2\lambda)}{(1 - \lambda)^2}\sqrt{1 + (\lambda^2 - 1)(\frac{2a}{W}M)^2 - \frac{\lambda^2}{(1 - \lambda)}(\frac{2a}{W}M)^2[1 + (\lambda^2 - 1)(\frac{2a}{W}M)^2]^{-1/2}}$$

$$+ \frac{\lambda^7}{2}(\frac{2a}{W}M)^6(K_T^\infty - 1 - \frac{2}{\lambda})\{[1 + (\lambda^2 - 1)(\frac{2a}{W}M)^2]^{-5/2} - (\frac{2a}{W}M)^2[1 + (\lambda^2 - 1)(\frac{2a}{W}M)^2]^{-7/2}\} \tag{19}$$

When $\lambda = b/a = 1$, Equation (19) reduces to

$$\frac{K_T^\infty}{K_T} = \frac{3(1 - 2a/W)}{2 + (1 - 2a/W)^3} + \frac{1}{2}(\frac{2a}{W}M)^6(K_T^\infty - 3)[1 - (\frac{2a}{W}M)^2] \tag{20}$$

For isotropic plate the FWC factor, Equation (15), becomes

$$\frac{K_T^\infty}{K_T} = \frac{\lambda^2}{(1 - \lambda)^2} + \frac{(1 - 2\lambda)}{(1 - \lambda)^2}\sqrt{1 + (\lambda^2 - 1)(\frac{2a}{W}M)^2 - \frac{\lambda^2}{(1 - \lambda)}(\frac{2a}{W}M)^2[1 + (\lambda^2 - 1)\frac{2a}{W}M)^2]^{-1/2}} \tag{21}$$

When $\lambda = 1$, Equation (21) is simplified to

$$\frac{K_T^\infty}{K_T} = \frac{3(1 - 2a/W)}{2 + (1 - 2a/W)^3} \tag{22}$$

Equation (22) recovers the familiar form of Heywood formula [8], which has been widely used for isotropic plate.

4.   EXPERIMENTAL AND NUMERICAL COMPARISONS

Using the experimental data presented in Reference [2], the prediction of the maximum tangential stress concentration of a finite-width isotropic plate containing an elliptical opening under uniaxial loading is examined in Figures 2a-b. Comparison shows that the present basic approach, Equations (15) and (16), is highly accurate for a/b > 1 with 2a/W < 0.5 and the improved theory Equations (21) and (22), has excellent agreement with the data for a/b < 4. Neuber solution [9] is too high for a/b > 1 while Isida solution [6] is quite accurate for the cases considered. When a/b = 1, the solutions by Heywood, Wahl-Beeuwkes and Isida are practically the same.

The predictions for the FWC factors of orthotropic laminates containing a circular hole are compared to the finite element solution [16] in Figures 3a-b. The material properties, Example 1, are listed in Table 1. It is shown that both the improved anisotropic theory, Equation (17), and the improved orthotropic solution, Equation (20), agree very well with the finite element solution. Even when the 2a/w ratio is over 90 percent , the predictions are still in excellent agreement with the finite element solution, Table 2.

The present approach assumes that the normal stress, σ_y (x,0) (Figure 1), across the net section of a finite-width plate is proportional to that of an infinite-width plate by a factor. The solution of this assumption can be examined using a typical finite element solution [17]. In Figure 4, comparisons are made for: (1) the result of FWC factor obtained from Equation (5) multiplied by the infinite plate solution; (2) the FWC factor obtained from Equation (13) multiplied by the approximate orthotropic solution, Equation (10), and; (3) the isotropic FWC factor multiplied by the infinite orthotropic plate solution. The material properties for the graphite/epoxy (Gr/Ep) T300/5208 [0/90]$_{4s}$ laminate being used is listed in Table 1, example 2. The result appears that the normal stress distribution for the finite element solution correlates better with the solutions obtained by using the orthotropic FWC factor, Eqs (5) and (13), than by using the isotropic FWC factor. In this comparison, Heywood formula, Equation (22), was used for the isotropic FWC factor.

5. ANISOTROPIC PLATE WITH AN ELLIPTICAL OPENING

Using the material properties shown in Table 1, some graphite-epoxy T300/5208 laminates, [±75]$_s$, [0$_6$/90]$_s$, [0/45]$_s$ and an isotropic plate, containing an elliptical opening with aspect ratio, a/b, equal to 1/2 and 2 were studied in Figures 5a and 5b respectively.

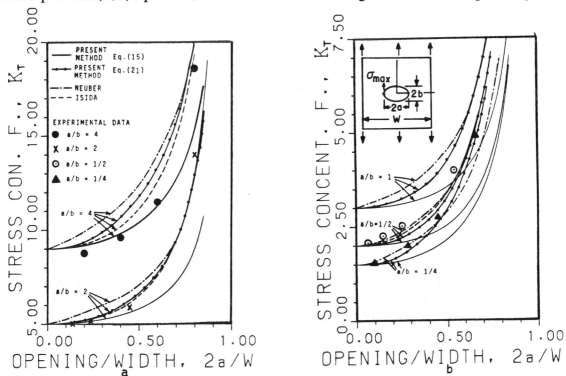

Figure 2.   The maximum tangential stress concentration of an isotropic finite-width plate containing an elliptical opening.

The result shows that the FWC factors are in descending order for the $[\pm 75]_S$, isotropic, $[0/45]_S$ and $[0_6/90]_S$ laminates. This order suggests that the FWC factor is more significant for matrix dominated laminates than for fiber dominated laminates. A comparison of the curves, $a/b = 1/2, 1, 2$ and 5, show that the difference in FWC factors of these laminates is in descending order with increasing ratio of a/b. When $a/b \geq 5$, the effect of anisotropy upon the FWC factors is practically negligible. If a/b is less than unity, anisotropy can be very significant for the FWC factors even when the opening-to-width ratio is small. An example for $a/b = 1/2$ and $2a/w = 1/3$ for these laminates is shown in Table 3.

6.   DISCUSSIONS AND CONCLUSIONS

The finite-width correction factors of anisotropic, orthotropic and isotropic plates containing an elliptical opening are presented in a closed-form solution. The basic theory can be improved significantly by multiplying the opening-to-width ratio, $2a/W$, by a magnification factor M. Comparisons with experimental data and some finite element solutions show that the basic solution has good accuracy for $a/b \geq 4$ while the improved theory has excellent accuracy for the domain $a/b < 4$.

Illustration with Figures and Tables reveals that within the region of $a/b > 1$ with $2a/W < 1/3$, FWC factors of most orthotropic laminates can be estimated by using the isotropic FWC factors with less than 6 percent error. Beyond this range, however, significant error can be caused by estimating the anisotropic and orthotropic FWC factors using the isotropic FWC factors, Figures 3 and 5.

The result of the present study also concludes that:

1. Matrix dominated laminates, e.g., $[\pm 75]_S$, have higher FWC factors than those of the isotropic plate (for a same opening length) while the fiber dominated laminates, e.g., $[0_6/90]_S$, are just the opposite, Figures 3 and 5;

2. For each ratio of $2a/W$, the FWC factors are increasingly sensitive and the values are higher for the decreasing ratio of a/b;

3. The FWC factors of matrix dominated laminates are extremely sensitive to the opening aspect ratio, a/b, whereas the fiber dominated laminates are somewhat less sensitive than that of the matrix dominated laminates;

4. For $a/b \geq 5$ the influence of the anisotropy upon the FWC factors vanishes (for infinite length).

It is important to point out that the FWC factors can be used to transform the laminate notched strength (finite-width) to infinite plate notched strength only if the stress distribution at the neighboring points around the maximum stress concentration agrees with the actual solution. This is because, in the failure analysis, the maximum stress concentration cannot explain the notched strength due to the width effect. The component of normal stress at a very small distance (usually within 2.54 mm) away from the maximum stress point, however, has been applied successfully to explain the width and the hole size effects [14, 15, 17, 18] for the notched strength of composite laminates. For a finite-width plate, if the stress profile at a small distance around the maximum tangential stress point (usually at the opening edge) agrees with the stress profile of an infinite plate multiplied by the FWC factor, then the notched strength can be correlated to that of an infinite plate. In Figure 4, the normal stress distribution of a finite-width plate with a central hole predicted using the present theory agrees very well with a typical finite element solution. This result supports the application of the FWC factors to interpolate a finite-width-specimen data to an infinite plate result, especially the solutions for the maximum tangential stress and the notched strength. The correction factors for the width effect and the hole size effect can all be computed utilizing the present analysis if their opening-to-width ratio, $2a/W$, is known.

It is interesting to note that the magnification factor M is only a function of 2a/W and is independent of a/b ratio. Therefore, this factor, M, could be applied as well to some other opening shapes such as rectangular, oval, etc. Part of this paper will appear in [19], which also includes the second stress concentration, K_{Tn}, and second stress concentration factor, K_{Tn}/K_T^∞.

REFERENCES

1. Wahl, A. M. and Beeuwkes, R., "Stress Concentration Produced by Holes and Notches," Trans. American Society for Testing and Materials, Vol. 56, 1934, pp.617-625.

2. Durelli, A. J., Parks, V. J. and Feng, H. C., "Stresses Around an Elliptical Hole in a Finite Plate Subjected to Axial Loading," J. Applied Mechanics, Trans. American Society for Testing and Materials, Vol. 88, 1966, pp. 192-195.

3. Jones, N. and Hozos, D., "A Study of the Stresses Around Elliptical Holes in Flat Plates," J. Eng. for Industry, Trans. American Society for Testing and Materials, Vol. 93, 1971, pp. 688-695.

4. Paris, P. C. and Sih, G. C., "Fracture Toughness Testing and Its Applications," American Society for Testing and Materials, STP 381, 1965, pp. 84-113.

5. Brown, JW. F., Jr., and Srawley, J. E., "Plane Strain Crack Toughness Testing of High Strength Metallic Materials," American Society for Testing and Materials, STP 410, 1969.

6. Peterson, R. E., Stress Concentration Factors, John Wiley & Sons, New York, 1974.

7. Dixon, J. R., "Stress Distribution Around a Central Crack in a Plate Loaded in Tension; Effect of Finite Width of Plate," J. Royal Aeronautical Society, Vol. 64, March 1960, pp. 141-145.

8. Heywood, R. B., Designing by Photoelasticity, Chapman and Hall, London, 1952, p.163.

9. Neuber, H., Kerbspannungslehre, Julius Springer, Berlin, Germany, 1937. Translation, Theory of Notched Stresses, Office of Technical Services, Dept. of Commerce, Washington, D. C, 1961.

10. Mar, J., "Fracture and Fatigue of Bi-Materials," Mechanics of Composites Review, Air Force Materials Laboratory and Air Force Office of Scientific Research Technical Report, Oct. 26-28, 1976, pp. 117-122.

11. Konish, H. J., Jr., "Mode I Stress Intensity Factors for Symmetrically-Cracked Orthotropic Strips," Fracture Mechanics of Composites, American Society for Testing Materials, STP 593, 1975, pp. 99-116.

12. Atluri, S. N., Kobayashi, A. S. and Nakagaki, M., "A Finite-Element Program for Fracture Mechanics Analysis of Composite Material," Fracture Mechanics of Composites, American Society for Testing and Materials, STP 593, 1975, pp. 86-98.

13. Lekhnitskii, S. G., Anisotropic Plates, Translated from the second Russian edition by Tsai, S. W. and Cheron, Gordon and Breach, 1968.

14. Tan, S. C., "Notched Strength Prediction and Design of Laminated Composites Under In-Plane Loadings," J. of Composite Materials, Vol. 21, August 1987, pp. 750-780.

15. Tan, S. C., "Laminated Composites Containing an Elliptical Opening. I. Approximate Stress Analyses and Fracture Models," J. of Composite Materials, Vol. 21, Oct. 1987, pp. 925-948.

16. Hong, C. S. and Crews, J. H., Jr., "Stress-Concentration Factors for Finite Orthotropic Laminates with a Circular Hole and Uniaxial Loading," NASA Technical Paper 1469, 1979.

17. Nuismer, R. J. and Whitney, J. M., "Uniaxial Failure of Composite Laminates Containing Stress Concentrations," Fracture Mechanics of Composites, American Society for Testing and Materials, , STP 593, 1975, pp. 117-142.

18. Tan, S. C., "Effective Stress Fracture Models For Unnotched and Notched Multidirectional Laminates," to appear in J. Composite Materials, April 1988.

19. Tan, S. C., "Finite-Width Correction Factors For Anisotropic Plate Containing A Central Opening," to appear in J. of Composite Materials, September 1988.

Table 1. Elastic properties of typical graphite/epoxy laminae.

Parameter	Example 1	Example 2 (T300/5208)
Longitudinal Young's Modulus, E_{11}	146.900 GPa	147.548 GPa
Transverse Young's Modulus, E_{22}	10.890 GPa	11.032 GPa
In-Plane Shear Modulus, G_{12}	6.412 GPa	5.309 GPa
In-Plane Poisson's Ratio, ν_{12}	0.380	0.29

Table 2. Comparison of the improved theory, Equations (17) and (20), and the finite element solution [16] for orthotropic laminates, Example 1 in Table 1, with a circular hole and $2a/W = 0.91$.

Laminate Layup	$[0/\pm45/90]_S$	$[0]$	$[90]$	$[0/90]_S$
K_T, Equation (20)	7.4101	3.9393	8.5554	5.0844
K_T, Equation (17)	7.4101	4.6263	8.8426	5.5033
K_T, Finite Element	7.5556	4.5084	8.9068	5.4393

Table 3. The FWC factors for Gr/Ep T300/5208 laminates with an elliptical opening, $a/b = 1/2$, and opening-to-width ratio, $2a/W = 1/3$, using the improved solutions.

Lay-Up	isotropic	$[\pm75]_S$	$[0_6/90]_S$	$[0/45]_S$
Equation (19)	1.2121	1.3225	0.9237	-
Equation (17)	1.2121	1.3268	1.1599	1.1606

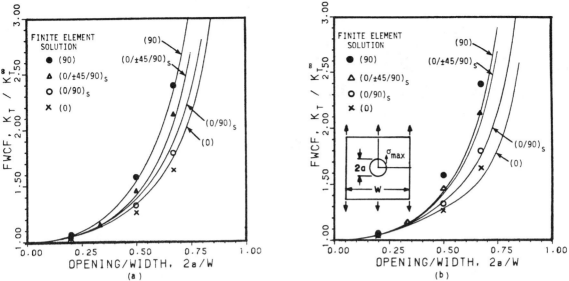

Figure 3. Comparison for the predictions of FWC factors of maximum tangential stress of orthotropic laminates with a circular hole using the improved solutions: (a) Equation (17); (b) Equation (20) and the finite element solution [16].

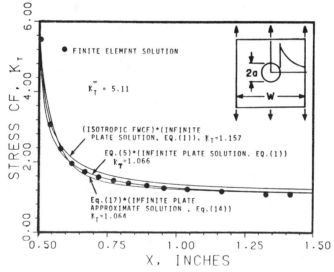

Figure 4. The normal stress distribution of a Gr/Ep T300/5208 $[0/90]_{4s}$ laminate containing a 1 in. diameter hole, 2a/W = 1/3.

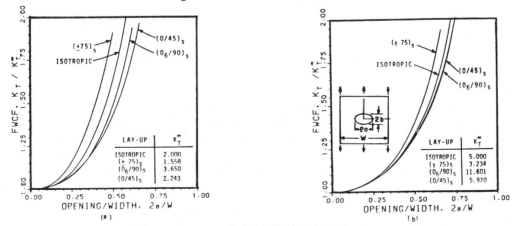

Figure 5. The FWC factors for some Gr/Ep T300/5208 laminates containing an elliptical opening with (a) a/b = 1/2 and; (b) a/b = 2 using the improved solution (17).

Yield Criteria for Fiber Composites Based on Energy Principles

P. S. THEOCARIS

ABSTRACT

Yield criteria developed for isotropic materials depending on pressure and dilatancy were extended to transversely isotropic (transtropic) materials by convenient transformations which maintain the direction of the axis of symmetry of the yield surface and its general form derived from considerations of energy principles. It has been shown that an elliptic paraboloid quadric surface with its axis of symmetry parallel to the hydrostatic axis in the principal stress space describes satisfactorily the yield behavior of a large spectrum of fiber composites and transtropic polymers. Moreover it yields an insight in the strength behavior of these materials. Two typical examples one concerning the Ca-Epoxy composite and the other an oriented polycarbonate exemplify the potentialities of the criterion.

INTRODUCTION

With the advent of new materials and especially fiber composites it became imperative to remodel the existing yield criteria which were based on crude simplifications like the continuing general isotropy of the material during loading and the absence of any strength differential effect (Bauschinger effect). Thus many new and more sensitive criteria better adapted to the real mechanical behavior of composites during loading were introduced.

These criteria are based on Hill's well known theory for anisotropic metals [1] which is based on Mises' first attempt to formulate failure in anisotropic solids [2]. While Hill's criterion does not take into account the strength differential effect, apparent in all solids, Hoffman's criterion [3] presents a further improvement by incorporating this effect in Hill's criterion by adding linear terms in the quadratic expression of Hill's criterion.

The tensor polynomial criterion, introduced by Tsai and Wu [4], constitutes a flexible and mathematically elegant version of a criterion, formulated by means of the Cartesian components of the stress tensor. However, it is represented by hypersurfaces in the six-dimensional stress space, impossible to be readily visualized geometrically in a physical stress space. Only plane sections of this hypersurface were possible to be

P.S. Theocaris, Professor Department of Engineering Science, Athens National Technical University, P.O.Box 77230 Athens (175-10) GREECE

studied, representing quadric surfaces in the $(\sigma_x, \sigma_y, \sigma_{xy})$ parametric space. However, even these subspaces do not yield a direct interrelation with the directions of the externally imposed loading and the material strength directions, a drawback causing the necessity of meticulous and delicate experiments for its definition.

Theocaris [5] has presented recently a paraboloid of revolution failure criterion for isotropic bodies, which, later on, was extended to an elliptic paraboloid, convenient for anisotropic materials [6,7]. The physical basis and properties of this elliptic paraboloid criterion is presented in this paper when adapted to transtropic materials, as they are the fiber-reinforced composites.

PROPERTIES OF THE ELLIPTIC PARABOLOID YIELD SURFACE (EPFS)

The shape properties and position in the principal stress space of the proposed yield surface for transtropic materials will be summarized here.

Consider a transversely isotropic (transtropic) body with σ_{Ti} and σ_{Ci} its tension and compression yield or failure strengths, where the index i runs for i=1,2,3 and the 3-axis represents the strong axis of the material, whereas $\sigma_{T1}=\sigma_{T2}$ and $\sigma_{C1}=\sigma_{C2}$ are the transversely isotropic strengths. In other words we assume that the 3-axis is the fiber axis of the composite. We assume further that the principal stress-axes coincide with the material principal strength axes.

The failure surface intersects the σ_i-axes at points A_i $(\delta_{1i}\sigma_{T1}, \delta_{2i}\sigma_{T2}, \delta_{3i}\sigma_{T3})$ and $B_i(-\delta_{1i}\sigma_{C1}, -\delta_{2i}\sigma_{C2}, -\delta_{3i}\sigma_{C3})$, where δ's are the kronecker deltas and its axis is parallel to the hydrostatic axis $\sigma_1=\sigma_2=\sigma_3$.

If the 3-axis is the strong axis of the transtropic material, the σ_{T3}- and σ_{C3}-stresses are the longitudinal failure strengths in tension and compression parallel to the direction of the fibers. Then, it is valid that $\sigma_{T1}=\sigma_{T2}$ and $\sigma_{C1}=\sigma_{C2}$ and the expression for the yield surface takes the form [6,8]:

$$\frac{\sigma_1^2+\sigma_2^2}{\sigma_{T1}\sigma_{C1}} + \frac{\sigma_3^2}{\sigma_{T3}\sigma_{C3}} - \left(\frac{2}{\sigma_{T1}\sigma_{C1}} - \frac{1}{\sigma_{T3}\sigma_{C3}}\right)\sigma_1\sigma_2 - \frac{\sigma_2\sigma_3+\sigma_3\sigma_1}{\sigma_{T3}\sigma_{C3}} +$$

$$+\left(\frac{1}{\sigma_{T1}} - \frac{1}{\sigma_{C1}}\right)(\sigma_1+\sigma_2)+\left(\frac{1}{\sigma_{T3}} - \frac{1}{\sigma_{C3}}\right)\sigma_3 = 1 \qquad (1)$$

It has been shown [8] that the quadric surface (1) is an elliptic paraboloid surface, whose axis of symmetry is parallel to the hydrostatic axis, $\sigma_2=\sigma_2=\sigma_3$, and it is symmetric to the diagonal plane, $\sigma_3\delta_{12}$, containing the σ_3-axis and passing through the diagonal δ_{12} lying in the σ_1, σ_2-plane.

Figure 1 presents the formation of the elliptic paraboloid yield surface, (EPFS), whose axis of symmetry KO'z' is parallel to the hydrostatic axis, Oz, in the principal stress space $O\sigma_1\sigma_2\sigma_3$. The Cartesian system Oxyz is formed by the Ox-axis which is the bisector of the $O\sigma_1\sigma_2$ right angle, the Oz-axis coinciding with the hydrostatic axis and the Oy-axis forming a tri-orthogonal system of reference. The Cartesian O'x'y'z-system is another tri-orthogonal system, whose origin O' is the intersection of $Oy\equiv Oy'$-axis and the axis O'z' of symmetry of the (EPFS). The O'x'-axis is parallel to the Ox-axis. While the Oxyz-system corresponds to an equivalent but isotropic material, the O'x'y'z'-system corresponds to the actual (EPFS). It is obvious that the Oxy- and O'x'y'-axes belong to the deviatoric plane

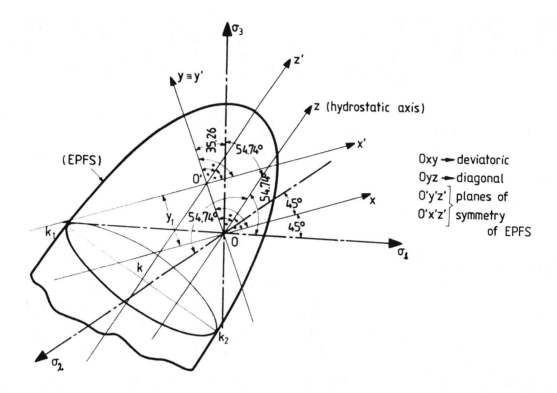

Fig.1 Three-dimensional form of the (EPFS) in the principal stress space
     $\sigma_1,\sigma_2,\sigma_3$ and the Cartesian $0'x'y'z'$ coordinate system where $0'z'$ is
     the axis of symmetry of (EPFS).

passing through the origins 0 and $0'$ of the coordinate systems.
     The distance y_1 between the hydrostatic axis and the axis of symmetry
of the (EPFS) is given by [6,8]:

$$y_1 = \frac{\sqrt{6}\,\sigma_{T3}\sigma_{C3}}{9}\left[\frac{\sigma_{C1}-\sigma_{T1}}{\sigma_{T1}\sigma_{C1}} - \frac{\sigma_{C3}-\sigma_{T3}}{\sigma_{T3}\sigma_{C3}}\right] \qquad (2)$$

We further assume that $\sigma_{C1}=\sigma_{C2}>\sigma_{T1}=\sigma_{T2}$, condition which is always satisfied
in transtropic materials. However, it is worthwhile indicating a peculiar
behavior for fiber reinforced materials and other transtropic ones. While
it is customary to have for all isotropic materials $\sigma_C>\sigma_T$, for the
transtropic materials the reverse situation happens, that is we have always
$\sigma_{T3}\geqq\sigma_{C3}$. This phenomenon may be explained by internal instabilities of the
transtropic material due to weaknesses along the interfaces of composites,
or along the drawing direction of oriented materials, which make their
yielding in tension to be higher than the respective value for compression.
     Besides the principal diagonal $\sigma_3\delta_{12}$-plane, which is a plane of
symmetry for the (EPFS), there is also the $0'x'z'$-plane, which is also
symmetric to the (EPFS). This plane has as equation the relation:

$$y = y_1 \qquad (3)$$

     Thus, the (EPFS) possesses a double symmetry along the $0'y'z'$- and
$0'x'y'$-planes. This is valid only for transtropic materials.
     The intersection of (EPFS) by the principal diagonal $\sigma_3\delta_{12}$-plane
creates a parabola [8], whose equation is given by:

$$y' = \pm \left(\frac{2}{3} \sigma_{T3}\sigma_{C3}\right)^{\frac{1}{2}} (\alpha+\beta z')^{\frac{1}{2}} \tag{4}$$

with

$$\alpha = 1 + \frac{\sigma_{T3}\sigma_{C3}}{9}\left\{\frac{\sigma_{C1}-\sigma_{T1}}{\sigma_{T1}\sigma_{C1}} - \frac{\sigma_{C3}-\sigma_{T3}}{\sigma_{T3}\sigma_{C3}}\right\} \tag{5}$$

$$\beta = -\frac{2}{\sqrt{3}}\left\{\frac{\sigma_{C1}-\sigma_{T1}}{\sigma_{T1}\sigma_{C1}} + \frac{\sigma_{C3}-\sigma_{T3}}{2\sigma_{T3}\sigma_{C3}}\right\} \tag{6}$$

While the parameter α is always positive, the parameter β may be positive or negative depending on the particular values of the material strength parameters. Since the vertex of the parabola (9) has a z'-coordinate given by:

$$z'_v = -\frac{\alpha}{\beta} \tag{7}$$

the vertex of the (EPFS) lies on the positive z'-semi-axis of the O'x'y'z'-frame, if β is negative and vice versa. Then, for β negative we have compression strong paraboloids, whereas for β positive tension strong ones. We designate these two main categories of (EPFS) as belonging to C-strong and T-strong materials respectively. For $\beta=0$ the paraboloid degenerates into an elliptic cylinder, as this described by Hill's criterion [9].

Another important condition which always is satisfied by fiber composites and transtropic materials is the following

$$4\sigma_{T3}\sigma_{C3} > \sigma_{T1}\sigma_{C1} \tag{8}$$

which implies that a>0 and the quantity $(4\sigma_{T3}\sigma_{C3}-\sigma_{T1}\sigma_{C1})$ is always positive. Condition (8) deserves the name of a *stability condition* [8]. Failing to satisfy condition (8) a composite has a condition of strengths σ_{C3} and σ_{T1} satisfying the inequality:

$$\sigma_{C3} \ll \sigma_{T1}/3 \tag{9}$$

condition, which indicates a very poor compressive strength of the composite along its strong axis and a consequent rejection of the material as unacceptable, because of internal weaknesses during the preparation of the composite, or because of excesses in the formation of the anisotropy during rolling or other cold or hot treatment.

The importance in defining from the beginning the type of transtropic material before any further study of its particular properties was clearly demonstrated in ref.[8]. Thus, it was found that although the majority of transtropic materials are of the C-strong type, there are also exceptions where the material presents a T-strong behavior. Typical T-strong examples of materials are the oriented polypropylene and all the types of paper sheets.

INTERSECTIONS OF (EPFS) BY CHARACTERISTIC PLANES

One of the most important intersections of the (EPFS) is its intersection with deviatoric plane $(\sigma_1+\sigma_2+\sigma_3)=0$ passing through the origins O and O' and lying on the plane O'x'y', coincident with the Oxy-plane. Its equation is given by:

$$ax'^2 + by'^2 = \alpha \tag{10}$$

where α is given by relation (5) and a,b are constants expressed by:

$$a = \frac{4\sigma_{T3}\sigma_{C3}-\sigma_{T1}\sigma_{C1}}{2\sigma_{T1}\sigma_{C1}\sigma_{T3}\sigma_{C3}} \quad \text{and} \quad b = \frac{3}{2\sigma_{T3}\sigma_{C3}} \tag{11}$$

The intersection of the deviatoric plane and the (EPFS) is proved to be again an ellipse whose major semi-axis a_2 is given by:

$$a_2 = \left\{ \frac{2\sigma_{T3}\sigma_{C3}}{9}\left[1 + \frac{\sigma_{T3}\sigma_{C3}}{9}\left\{ \left(\frac{1}{\sigma_{T1}} - \frac{1}{\sigma_{C1}}\right)-\left(\frac{1}{\sigma_{T3}} - \frac{1}{\sigma_{C3}}\right)\right\}^2 \right] \right\}^{\frac{1}{2}} \tag{12}$$

The intersection of the (EPFS) and the O'x'z'-plane yields another parabola similar to the parabola expressed by relation (10) with a similarity factor λ_0 given by:

$$\lambda_0 = \left(\frac{3\sigma_{T1}\sigma_{C1}}{4\sigma_{T3}\sigma_{C3}-\sigma_{T1}\sigma_{C1}} \right)^{\frac{1}{2}} \tag{13}$$

so that the minor semi-axis a_1 of the elliptic cross-section on the deviatoric plane is equal to $a_1=\lambda_0 a_2$.

Considering now the intersection of the (EPFS) and the principal $(\sigma_3\delta_{12})$-diagonal plane we state that this intersection is a parabola expressed by:

$$y' = \pm\left(\frac{2\sigma_{T3}\sigma_{C3}}{3}\right)^{\frac{1}{2}}\left\{ 1 + \frac{\sigma_{T3}\sigma_{C3}}{9} I^2 - \frac{2}{\sqrt{3}}\left[\left(\frac{1}{\sigma_{T1}} - \frac{1}{\sigma_{C1}}\right) + \frac{1}{2}\left(\frac{1}{\sigma_{T3}} - \frac{1}{\sigma_{C3}}\right]z'\right\}^{\frac{1}{2}} \tag{14}$$

where I is given by:

$$I = \left[\left(\frac{1}{\sigma_{T1}} - \frac{1}{\sigma_{C1}}\right)-\left(\frac{1}{\sigma_{T3}} - \frac{1}{\sigma_{C3}}\right)\right] \tag{15}$$

The distances d and d' between the deviatoric plane and either the point of piercing of the (EPFS) by the hydrostatic axis, or the apex of the paraboloid are given by [6]:

$$d = \frac{\sqrt{3}}{\left[2\left(\frac{1}{\sigma_{T1}} - \frac{1}{\sigma_{C1}}\right)+\left(\frac{1}{\sigma_{T3}} - \frac{1}{\sigma_{C3}}\right)\right]} \tag{16}$$

and:

$$d' = \frac{\sqrt{3}\left[1 + \frac{\sigma_{T3}\sigma_{C3}}{9}\{I\}^2\right]}{\left[2\left(\frac{1}{\sigma_{T1}} - \frac{1}{\sigma_{C1}}\right) + \left(\frac{1}{\sigma_{T3}} - \frac{1}{\sigma_{C3}}\right)\right]} \tag{17}$$

Distances d and d' define immediately the relative position of the whole paraboloid surface referred to the origin 0 of coordinates. These distances together with the y_1-distance characterize the degree of anisotropy of the material.

Finally it is important to define the intersection of the (EPFS) with the principal plane $(\sigma_1\sigma_3)$ (or $\sigma_2\sigma_3$) since the position and the form of this intersection is the basic checking for the validity of the criterion because all experimental tests are executed on this principal plane.

It can be readily shown [8] that this intersection is an ellipse whose equation is given by:

$$\frac{\sigma_1^2}{\sigma_{T1}\sigma_{C1}} + \frac{\sigma_3^2}{\sigma_{T3}\sigma_{C3}} - \frac{\sigma_1\sigma_3}{\sigma_{T3}\sigma_{C3}} + \left(\frac{1}{\sigma_{T1}} - \frac{1}{\sigma_{C1}}\right)\sigma_1 + \left(\frac{1}{\sigma_{T3}} - \frac{1}{\sigma_{C3}}\right)\sigma_3 = 1 \tag{18}$$

The center of this ellipse is given by the coordinates [8]:

$$\sigma_{1M} = \frac{(\sigma_{T3} - \sigma_{C3})\sigma_{T1}\sigma_{C1} + 2\sigma_{T3}\sigma_{C3}(\sigma_{T1} - \sigma_{C1})}{4\sigma_{T3}\sigma_{C3} - \sigma_{T1}\sigma_{C1}} \tag{19}$$

and

$$\sigma_{3M} = \frac{\sigma_{T3}\sigma_{C3}[(\sigma_{T1} - \sigma_{C1}) + 2(\sigma_{T3} - \sigma_{C3})]}{4\sigma_{T3}\sigma_{C3} - \sigma_{T1}\sigma_{C1}} \tag{20}$$

and therefore the angle θ_M subtended by the polar radius of the center of the ellipse and the σ_1-axis is given by:

$$\theta_M = \frac{1}{2}\tan^{-1}\left(\frac{\sigma_{T1}\sigma_{C1}}{\sigma_{T1}\sigma_{C1} - \sigma_{T3}\sigma_{C3}}\right) \tag{21}$$

The major and minor semi-axes of this ellipse are expressed by:

$$a_{3M} = \left\{\frac{2\sigma_{T3}\sigma_{C3}[K+L+M]}{(N-P)R}\right\} \tag{22}$$

$$a_{1M} = \left\{\frac{2\sigma_{T3}\sigma_{C3}[K+L+M]}{(N+P)R}\right\} \tag{23}$$

where the quantities K to R are given by:

$$K = \sigma_{T1}^2(\sigma_{T3}\sigma_{C3}+\sigma_{C1}\sigma_{T3}-\sigma_{C1}\sigma_{C3}-\sigma_{C1}^2)$$

$$L = \sigma_{C1}^2(\sigma_{T3}\sigma_{C3}+\sigma_{T1}\sigma_{C3}-\sigma_{T1}\sigma_{T3})$$

$$M = \sigma_{T1}\sigma_{C1}(\sigma_{T3}^2+\sigma_{C3}^2) \tag{24}$$

$$N = (\sigma_{T3}\sigma_{C3}+\sigma_{T1}\sigma_{C1})$$

$$P = [(\sigma_{T1}\sigma_{C1}-\sigma_{T3}\sigma_{C3})^2+\sigma_{T1}^2\sigma_{C1}^2]^{\frac{1}{2}}$$

and

$$R = (4\sigma_{T3}\sigma_{C3}-\sigma_{T1}\sigma_{C1})$$

As an example of the validity of the (EPFS) criterion it was applied to two typical transtropic materials that is a strong fiber composite, the graphite-epoxy and, for comparison, to a weak anisotropic polymer, the oriented polycarbonate. The yield strengths of these materials are given by [10,11]:

Graphite-Epoxy: σ_{T3} = 1,043MPa, σ_{C3} = 43.10MPa

σ_{T1} = 698MPa, σ_{C1} = 142.24MPa

Oriented PC: σ_{T3} = 65.20MPa, σ_{C3} = 42.70MPa
σ_{T1} = 35.20MPa, σ_{C1} = 45.20MPa

Figures 2 to 5 present the form of the intersections of their respective (EPFS)'s with the principal $\sigma_3\delta_{12}$-diagonal and the $(\sigma_1\sigma_3)$-principal plane.

The parabolas and ellipses shown in these figures corroborate the experimental evidence with tests executed in the $(\sigma_1\sigma_3)$-plane better than any other criterion [8].

Considering also the superiority of this criterion vis-a-vis all other empirical criteria concerning the well defined shapes of the yield loci in various sections, obeying physical laws of conservation of energy, and the better approximation with existing experimental results, it may be concluded that (EPFS) constitutes a reliable phenomenological yield criterion.

REFERENCES

1 R. Hill,"A Theory of the Yielding and Plastic Flow of Anisotropic Metals", *Proc. Roy. Soc. Lond.*, Ser.A, Vol.**193**, 1948, pp.281-297.

2 R. von Mises, "Mechanik der plastischen Formänderung von Kristallen", *Zeit. ang. Math. and Mech.*, Vol.**8**, 1928, pp.161-185.

3 O. Hoffman, The Brittle strength of Orthotropic Materials", *Jnl. Comp. Mat.*, Vol.**1**, 1967, pp.200-206.

4 S.W. Tsai and E.M. Wu, "A General Theory of Strength for Anisotropic Materials", *Jnl. Composite Mat.*, Vol.**5**, 1971, pp.58-80.

5 P.S. Theocaris, "Generalized Failure Criteria in the Principal Stress Space", *Theoret. and Appl. Mech.*, Bulgarian Academy of Sciences, Vol.**19**(2), 1987, pp.74-104.

6 P.S. Theocaris, "Failure Characterization of Anisotropic Materials by Means of the Elliptic Paraboloid Failure Criterion", *Uspechi Mechanikii* (Advances of Mechanics), Vol.**10**(3), 1987, pp.83-102.

7 P.S. Theocaris and Th. Philippidis, "The Paraboloidal Failure Surface of Initially Anisotropic Elastic Solids", *Jnl. of Reinf. Plastics and Comp.*,

Vol.**6**(4), 1987 pp.378-395.

8 P.S. Theocaris, "Failure Criteria for Transtropic, Pressure Dependent Materials: The Fiber Composites", *Rheologica Acta*, Vol.27, 1988, pp.

9 P.S. Theocaris, The Elliptic Paraboloid Failure Surface for 2D-Transtropic Plates (Fiber Laminates), *Engng. Fract. Mech.*, Vol. , 1988, pp.

10 R. Narayanaswami and H.M. Adelman, "Evaluation of the Tensor Polynomial and Hoffman Strength Theories for Composite Materials", *Jnl. Comp. Mat.*, Vol.**11**, 1977, pp.366-377.

11 R.M. Caddell and A.R. Woodliff, "Macroscopic Yielding of Oriented Polymers", *Jnl. Mat. Sci.*, Vol.**12**, 1977, pp.2028-2036.

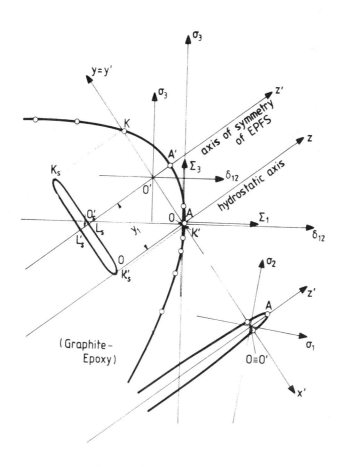

Fig.2 Intersection of (EPFS) for graphite epoxy with the principal diagonal plane $(\sigma_3 \delta_{12})$.

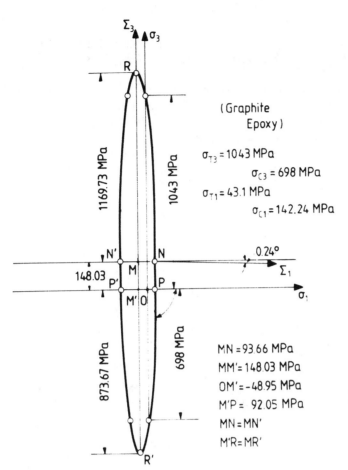

(Graphite Epoxy)

$\sigma_{T3} = 1043$ MPa

$\sigma_{C3} = 698$ MPa

$\sigma_{T1} = 43.1$ MPa

$\sigma_{C1} = 142.24$ MPa

1169.73 MPa

1043 MPa

873.67 MPa

698 MPa

148.03

0.24°

MN = 93.66 MPa

MM' = 148.03 MPa

OM' = -48.95 MPa

M'P = 92.05 MPa

MN = MN'

M'R = MR'

Fig.3 Intersection of (EPFS) for graphite epoxy with the principal $(\sigma_1 \sigma_3)$-plane.

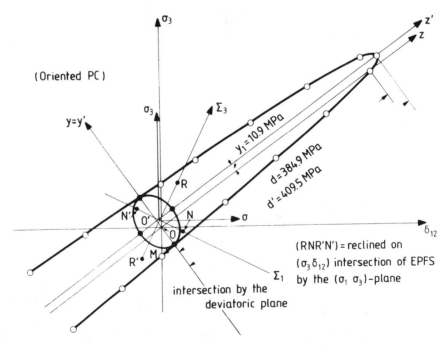

(Oriented PC)

$y_1 = 10.9$ MPa

d = 384.9 MPa

d' = 409.5 MPa

(RNR'N') = reclined on $(\sigma_3 \delta_{12})$ intersection of EPFS by the $(\sigma_1 \sigma_3)$-plane

intersection by the deviatoric plane

Fig.4 Intersection of (EPFS) for oriented polycarbonate with the principal diagonal plane $(\sigma_3 \delta_{12})$.

144

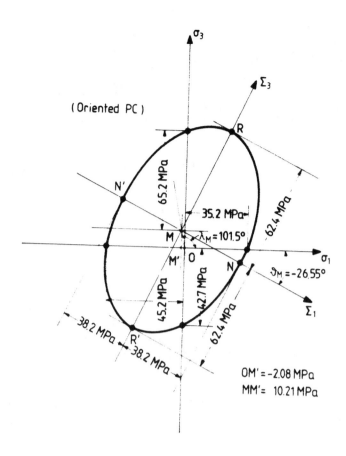

Fig.5 Intersection of (EPFS) for oriented polycarbonate with the principal $(\sigma_1\sigma_3)$-plane.

Transverse Stress Calculations for Laminated Composite Shell Structures Using Plate/Shell Finite Element Formulations

JOHN J. ENGBLOM, JOSEPH. P. FUEHNE AND **JAMIL M. HAMDALLAH**

ABSTRACT

Shear deformable plate and shell finite element formulations are utilized to predict transverse stresses in composite shell structures. In-plane stresses are computed by solving the constitutive equations at the layer level. These stress variations then serve as input to the equilibrium equations which are integrated to obtain the transverse stresses. The approach improves upon the 'classical' laminate theory and thus provides an efficient numerical procedure for thin to moderately thick shell structures.

INTRODUCTION

Laminated fiber-reinforced composite materials add a new dimension to the design process in the sense that the engineer can 'design' the fundamental directional material properties needed in given design applications. The macroscopic material behavior of the laminate is, therefore, a function of how the plies, i.e., the individual laminae, are specified and bonded together. While composites have the advantage of a higher strength-to-weight ratio than many of the monolithic materials, their failure modes are more complex and, presently, not as well understood. A failure mode of great interest involves the delamination of adjacent plies within a laminate. This particular type of failure is related to the fact that composite plate and shell structures have lower strength in the thickness direction. Thus it is essential to determine the magnitude of the stresses developed in the thickness direction due to the prescribed design loads. These particular stress quantities are generally denoted as either the 'transverse' or as the 'through-the-thickness' stresses.

J.J. Engblom, Associate Professor, Mechanical Engineering Dept., Texas A&M University, College Station, Texas 77843
J.P. Fuehne, Ph.D. Student, Mechanical Engineering Dept., Texas A&M University, College Station, Texas 77843
J.M. Hamdallah, Engineer, General Dynamics Corporation, Fort Worth, Texas 76101

There are many applications of composites for thin to moderately thick shell structures, e.g., pressure vessels, aircraft skins, jet engine components such as fan and compressor blades, etc. Even for relatively thin structures, it has been shown that the inclusion of transverse shear deformation is essential in being able to quantify the transverse stress variation. For this reason, the classical lamination theory can not always provide the stress information needed in designing such structures.

The plate and shell elements utilized in this work include shear effects to some degree by relaxing the Kirchhoff-Love hypothesis. This is accomplished by allowing the neutral surface displacements to be independent of the rotations. Stresses in the plane of a layer are calculated using the constitutive equations at the layer level. The term 'layer' herein refers to a set of adjacent plies of equal fiber orientation. The transverse stresses, however, are obtained by integrating the equilibrium equations in the thickness direction. This approach yields a transverse stress variation of higher order than would have otherwise been obtained via solution of the constitutive equations.

Examples presented herein include helically wound pressure vessels having different winding geometries and end constraints. The element formulations are shown to have the ability to predict the significant interlaminar stresses in the boundary layer region near the cylinder constraints. It is emphasized that the approach taken is applicable to much more general structures, e.g., doubly curved airfoil shapes.

DISPLACEMENT FIELD

Both the plate and shell element formulations are based on eight-noded quadrilateral geometries. There are four corner and four midside nodes located at the midsurface of the element as depicted in Fig. 1. Note that the shell element degenerates into a plate element in the special case when all eight nodes lie in the x-y plane of the element. Displacements are defined for any point in the element as a function of midsurface displacements and midsurface rotations as given below

$$u(x,y,z) = u_0(x,y) + z[n_z \theta_x(x,y) - n_y \theta_z(x,y)]$$

$$v(x,y,z) = v_0(x,y) + z[n_z \theta_y(x,y) + n_x \theta_z(x,y)]$$ (1)

$$w(x,y,z) = w_0(x,y) - z[n_x \theta_x(x,y) + n_y \theta_y(x,y)]$$

where u_0, v_0, and w_0 are the midsurface displacements and θ_x, θ_y, and θ_z are the midsurface rotations, all of which are prescribed in the element cartesian coordinate system. Also shown in Fig. 1 are the curvilinear coordinates ξ, η, and ζ, where ξ and η lie in the element midsurface and ζ defines the transverse (thickness) direction. Curvature of the element is accomodated by definition of surface normals η_x, η_y, and η_z which represent components of the normal to the midsurface in element coordinates. When η_z equals 1 and both η_x and η_y are identically zero, the element geometry then simply reduces to that of a plate as mentioned above. A bi-quadratic interpolation (shape) function is utilized to specify the location of nodal points and to specify displacement variation. Since both geometry and displacements are defined via the same functions, the formulation can be considered isoparametric. The shape functions are written in terms of the curvilinear coordinates and the displacement variation in u_o can be represented as

$$u = u[1\ \xi\ \eta\ \xi^2\ \xi\eta\ \eta^2\ \xi^2\eta\ \xi\eta^2 \quad \zeta(1\ \xi\ \eta\ \xi^2\ \xi\eta\ \eta^2\ \xi^2\eta\ \xi\eta^2)] \tag{2}$$

also, it is noted that the variation in v_o, w_o, θ_x, θ_y and θ_z are each represented in the same manner. These shape functions can be transformed in the traditional way to relate the displacements at any point in the element to nodal point motion. Denoting terms in the shape function as N_i, Equations (1) can be given in terms of these functions and nodal point displacements/rotations as

$$\begin{pmatrix} u \\ v \\ w \end{pmatrix} = \sum_{i=1}^{8} N_i \left(\begin{pmatrix} u_i \\ v_i \\ w_i \end{pmatrix}_0 + \frac{\zeta t_i}{2} \begin{bmatrix} n_{z_i} & 0 & -n_{y_i} \\ 0 & n_{z_i} & n_{x_i} \\ -n_{x_i} & -n_{y_i} & 0 \end{bmatrix} \begin{pmatrix} \theta_x \\ \theta_y \\ \theta_z \end{pmatrix}_i \right) \tag{3}$$

where subscript 'i' denotes the ith node and t_i represents the thickness of the element at the ith node. The shape function utilized in this formulation assures that both constant strain states and rigid body modes can be represented. Furthermore, the chosen displacement variation provides the basis for the interlaminar stress calculations. The matrix form of the elemental displacements can be further simplified as

$$\begin{pmatrix} u \\ v \\ w \end{pmatrix} = \sum_{i=1}^{8} [N_i'] \begin{pmatrix} u_{0_i} \\ v_{0_i} \\ w_{0_i} \\ \theta_{x_i} \\ \theta_{y_i} \\ \theta_{z_i} \end{pmatrix} \tag{4}$$

where the matrix $[N_i']$ combines the N_i and ζ terms given in Equation (3) above.

STRAIN FIELD

Having defined both geometry and displacement variation for the element, appropriate derivatives of Equation (3) provide the relationship between strains in the element to the nodal point displacements/rotations. In short notation, this strain-to-nodal-deformation relationship is given as

$$\{\varepsilon\} = [B]\{\delta\} \tag{5}$$

where $\{\varepsilon\}$ represents the element strains and $\{\delta\}$ the element displacements (including translations and rotations of the midsurface at element node points), each in elemental coordinates. [B] is the matrix of partial derivatives taken with respect to the element coordinates. This, of course, requires use of the Jacobian to transform the derivatives from the ξ, η, ζ coordinates to element coordinates for both plate and shell formulations. Since volume integrals must be obtained which involve numerical integration in the plate/shell 'thickness' direction, a relationship between strain and displacement in reference coordinates 'in the plane of' and 'normal' to the shell is required. For the plate element, [B] satisfies this requirement; however, in the case of the shell element, [B] must be transformed in a two step process. Firstly, the derivatives in [B] of elemental displacements u,v,w with respect elemental coordinates x,y,z are transformed to derivatives of u',v',w' with respect to x',y' and z'. Here, the prime superscripts indicate that the displacements and coordinates are defined in the desired coordinates, i.e., tangent and normal to the shell. These coordinates are herein denoted as the 'local' coordinate system. This transformation is represented as

$$
\begin{bmatrix}
\frac{\partial u'}{\partial x'} & \frac{\partial v'}{\partial x'} & \frac{\partial w'}{\partial x'} \\[2mm]
\frac{\partial u'}{\partial y'} & \frac{\partial v'}{\partial y'} & \frac{\partial w'}{\partial y'} \\[2mm]
\frac{\partial u'}{\partial z'} & \frac{\partial v'}{\partial z'} & \frac{\partial w'}{\partial z'}
\end{bmatrix}
= [L]^T
\begin{bmatrix}
\frac{\partial u}{\partial x} & \frac{\partial v}{\partial x} & \frac{\partial w}{\partial x} \\[2mm]
\frac{\partial u}{\partial y} & \frac{\partial v}{\partial y} & \frac{\partial w}{\partial y} \\[2mm]
\frac{\partial u}{\partial z} & \frac{\partial v}{\partial z} & \frac{\partial w}{\partial z}
\end{bmatrix}
[L]
\qquad (6)
$$

where [L] contains the direction cosines defined for any particular integration point on the shell. This transformation provides a strain-displacement relationship of the form

$$
\{\varepsilon'\} = [B']\{\delta\}
\qquad (7)
$$

where $\{\varepsilon'\}$ contains the strains and [B'] the appropriate derivatives in the local coordinate system. Secondly, the element displacements are transformed into local coordinates as represented below

$$
\{\delta\} = [TR]\{\delta R\}
\qquad (8)
$$

where [TR] is a matrix of direction cosines and $\{\delta R\}$ contains the nodal point displacements in the local coordinate system. The desired strain-displacement form is now given as

$$
\{\varepsilon'\} = [B'][TR]\{\delta R\} = [BR]\{\delta R\}
\qquad (9)
$$

where [BR] is the modified form of [B] needed to calculate the stiffness matrix for the shell formulaton. Note again that the plate formulation does not require the double transformation described above. Both plate and shell element stiffness matrices are obtained in the traditional manner by integrating a triple matrix product of the form

$$[B]_t[D][B] \tag{10}$$

for the plate, and

$$[BR]_t[D][BR] \tag{11}$$

for the shell. Here [D] represents the material matrix which
is defined on a layer-by-layer basis as integration proceeds
in the 'thickness' direction for the element. Gauss quadrature
is utilized to perform the integration within each layer. See
[1,2] for a more in-depth discussion of both plate and shell
formulations.

STRESS VARIATION

Stresses within each layer are computed on the basis of
the constitutive equations as given below for the k_{th} layer

$$\{\sigma'\}^k = [D]^k\{\varepsilon'\}^k \tag{12}$$

These stresses are defined in the primed 'local' coordinate
system. As noted, these are simply the element coordinates
when the shell reduces to a plate geometry. The transverse
stresses determined on the basis of (12) would yield only a
uniform variation. This stress distribution is significantly
improved in both the plate and shell formulations by using the
equilibrium equations.

To appreciate this approach, consider first the
variation in stresses in the x',y' directions, including
shear, as calculated using the constitutive equations in (12).
Variation in the x' direction has the functional form

$$\sigma'_x = \sigma'_x[x',y',x'y',x'^2,y'^2, z'(x',y',x'y',x'^2,y'^2)] \tag{13}$$

while the normal stresses in y' and shear in x'y' vary in the
same manner. These stresses within each layer can be
calculated at each Gauss point and approximated by use of a
'smoothing function'. In the absence of body forces then, the
equilibrium equations define the through-the-thickness stress
variation as

$$\frac{\partial \tau'_{xz}}{\partial z'} = -\left(\frac{\partial \sigma'_x}{\partial x'} + \frac{\partial \tau'_{xy}}{\partial y'}\right)$$

$$\frac{\partial \tau'_{yz}}{\partial z'} = -\left(\frac{\partial \tau'_{xy}}{\partial x'} + \frac{\partial \sigma'_y}{\partial y'}\right) \tag{14}$$

$$\frac{\partial \sigma'_z}{\partial z'} = -\left(\frac{\partial \tau'_{xz}}{\partial x'} + \frac{\partial \tau'_{yz}}{\partial y'}\right)$$

where the derivatives on the right hand side of (14) are

obtained directly on the basis of the previously discussed smoothing functions. The transverse stresses are thus determined by numerically integrating the expressions in (14) in the z' direction. The through-the-thickness variations have the form

$$\tau'_{xz} = \tau'_{xz}[x'z', y'z', z'^2(1, x', y')]$$

$$\tau'_{yz} = \tau'_{yz}[x'z', y'z', z'^2(1, x', y')] \qquad (15)$$

$$\sigma'_z = \sigma'_z[z'^2, z'^3]$$

which gives a parabolic transverse stress distribution. The numerical procedure for computing these transverse stresses from equilibrium has been presented in [3,4].

RESULTS

The present plate and shell element formulations are utilized here to calculate the transverse shear stress variation that occurs in the 'boundary layer' region of a laminated cylindrical shell. Some of the results are compared to those given in [5], which presented an analytcal solution specifically for determining the interlaminar stresses in laminated cylindrical shells. Plate/shell solutions are compared to those given in [5] for a three-layer, cross-ply wound cylinder. Fiber orientation is defined as [$0°, 90°, 0°$] and results are obtained for two different sets of boundary conditions including both simply-supported and clamped ends. The cylinder geometry and loading is the same for all results presented herein. The cylinder has a length L = 50 inches, radius R = 25 inches and wall thickness t = 0.25 inches. Also, the cylinder is subjected to a uniform internal pressure of 100 psi. Two material systems are considered as defined below

Boron-Epoxy:

$$E_{11} = 32.5 \text{ MPSI}$$
$$E_{22} = 1.84 \text{ MPSI}$$
$$G_{12} = 0.642 \text{ MPSI}$$
$$G_{23} = 0.361 \text{ MPSI}$$
$$\nu_{12} = 0.256$$

Glass-Epoxy:

$$E_{11} = 6.00 \text{ MPSI}$$
$$E_{22} = 1.50 \text{ MPSI}$$
$$G_{12} = 0.80 \text{ MPSI}$$
$$G_{23} = 0.60 \text{ MPSI}$$
$$\nu_{12} = 0.25 \text{ MPSI}$$

To be completely consistent with the properties given in [5], it is also assumed that $G_{13} = G_{12}$.

Since the laminated geometry under consideration is

symmetric, only one quadrant of the cylinder with suitably specified boundary conditions need be modelled. Two levels of mesh refinement have been used to assure that convergent solutions have been obtained. These include both 10X10 and 16X16 rectangular meshes. These meshes along with the geometry are shown in Figure 2. Note that all of the solutions presented herein are for the more refined mesh. Calculated values for the longitudinal transverse shear stress, τ_{yz}, occuring at the interface of the 0° and 90° layers are given in Figures 3-6. The first two of these Figures relate to the Boron-Epoxy material system, while the latter two Figures provide results for the Glass-Epoxy system. Regardless of which boundary condition is considered, the results indicate essentially zero transverse stresses away from the end constraint. Whereas in the vicinity of the cylinder end, a 'boundary layer' develops in which the transverse stresses become quite significant. As demonstrated in Figures 3-6, the calculated transverse stresses are in excellent agreement with the analytical solutions given by Waltz [5]. Note that the shell element model gives better agreement with the analytical solution than does the plate element model, as would be expected. While the 10X10 mesh results are not presented herein for these cases, the solutions compare well with the 16X16 results and do demonstrate convergence.

Finally, calculated transverse stresses are presented for a cylindrical shell with a quasi-isotropic layup having ply sequence $[0^{\circ},+45^{\circ},-45^{\circ},90^{\circ}]_s$. The material system is assumed to be the Glass-Epoxy previously defined, while the geometry and loading are unchanged. Figures 7 and 8 present the transverse stress 'τ_{yz}' variation calculated using the plate and shell element formulations, again for both clamped and simply-supported boundary conditions. The 'boundary layer' stresses are similar to those shown in the previous cases. Based on the previous examples, it is believed that the shell element provides the more accurate results. All stress results represent average values at selected node points on the model. It is interesting to note that the variation in these stresses, i.e., at a particular node shared by more than one element, is much less pronounced in the shell than in the plate formulation. It seems this is simply a further indication that the shell formulation is more suitable in modelling structures having curvature.

CONCLUSIONS

Good results have been obtained for the transverse stresses in laminated cylinders using both plate and shell finite element formulations. The approach is based on an assumed displacement variation but the equilibrium equations are solved to obtain these transverse (through-the-thickness) stresses. While the formulaton is much more efficient than the use of fully 3-dimensional elements, it is limited to thin to moderately thick structures. The shell elements yield improved results over those obtained via the plate formulation, and it should be emphasized that the shell elements can be used to model general doubly curved geometries whereas the plate element does not have such a general capability.

REFERENCES

1. Hamdallah, J.M., and Englom, J.J.,"Finite Element Plate Formulation Including Transverse Shear Effects for Composite Shell Structures," Journal of Reinforced Plastics and Composites, Accepted for publication.

2. Fuehne, J.P., and Engblom, J.J.,"A Shear Deformable, Doubly Curved Finite Element for the Analysis of Composite Structures," Composite Structures, Accepted for publication.

3. Engblom, J.J., and Ochoa, O.O.,"Finite Element Formulation Including Interlaminar Stress Calculations,"Computers and Structures, Vol. 23, No. 2, 1986, pp. 241-249.

4. Englom, J.J., and Ochoa, O.O.,"Through-the-Thickness Stress Predictions for Laminated Plates of Advanced Composite Materials,"International Journal for Numerical Methods in Engineering, Vol. 21, 1985, pp. 1759-1776.

5. Waltz, T.L., "Interlaminar Stresses in Laminated Cylindrical Shells of Composite Materials," Master's Thesis, Aerospace and Mechanical Engineering Dept., University of Delaware, 1975.

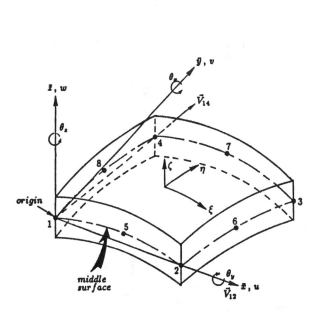

Fig. 1 A typical doubly curved element showing nodes 1–8, the elemental coordinate system, and the six degrees of freedom per element, $u, v, w, \theta_x, \theta_y, \theta_z$.

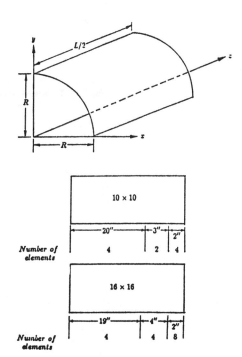

Fig. 2 The reference coordinate system and the two meshes including the dimensions of the elements.

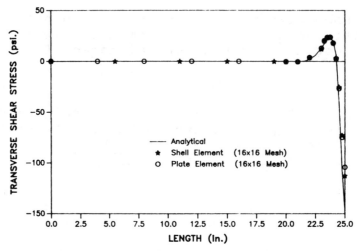

Fig. 3 Transverse shear stress at the interface of the 0° and 90° layers for a boron-epoxy cylinder with simply-supported edges.

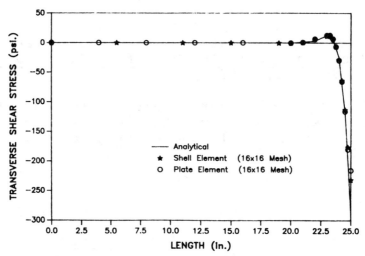

Fig. 4 Transverse shear stress at the interface of the 0° and 90° layers for a boron-epoxy cylinder with clamped edges.

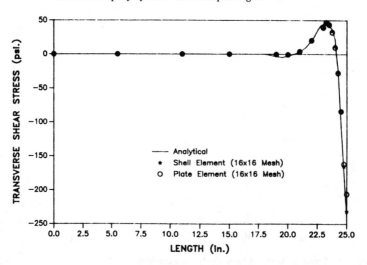

Fig. 5 Transverse shear stress at the interface of the 0° and 90° layers for a glass-epoxy cylinder with simply-supported edges.

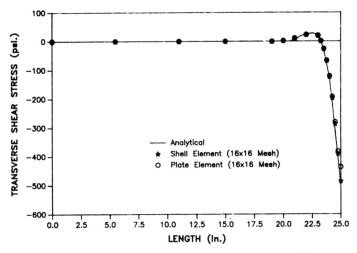

Fig. 6 Transverse shear stress at the interface of the 0° and 90° layers for a glass-epoxy cylinder with clamped edges.

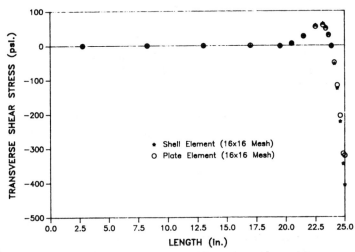

Fig. 7 Transverse shear stress at the interface of the −45° and 90° layers for a quasi-isotropic, glass-epoxy cylinder with simply-supported edges.

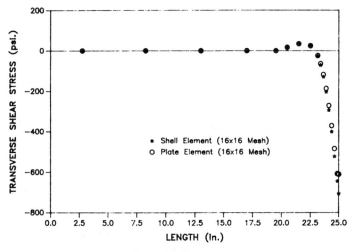

Fig. 8 Transverse shear stress at the interface of the −45° and 90° layers for a quasi-isotropic, glass-epoxy cylinder with clamped edges.

A Procedure for Accurate Determination of Elastic Properties for Unidirectional Composite Materials

J. M. STARBUCK[1] AND Z. GÜRDAL[1]

Abstract

A combined experimental analytical procedure for accurate determination of the elastic properties for unidirectional composite materials is proposed. The procedure is based on conducting on-axis and off-axis compression and/or tension tests and using a finite element method to analyze the specimen configuration. A use of an optimization algorithm is proposed to obtain the correct material properties by minimizing an objective function expressed in terms of the experimentally measured apparent values and the values calculated using the finite element method. The values of the material properties obtained for various off-axis coupon configurations for aramid/epoxy and graphite/epoxy material systems are presented. These results are compared with the axial modulus values measured from 0° and 90° test coupons.

Introduction

There are a number of different test configurations that can be used for mechanical property characterization of unidirectional composite materials and choosing the right one can be a difficult task because of the lack of established methodology for comparison. Also some test configurations may yield more accurate results for certain material properties compared to others and the degree of accuracy may be dependent on the values of the properties themselves.

In general, three different test configurations are required to determine the elastic material properties in the principal material directions. Two relatively simple tension or compression test coupons, one with longitudinal fibers and one with transverse fibers, are needed to obtain E_1 and E_2. The determination of the shear modulus, G_{12}, either requires an off-axis coupon [1] or a special test fixture with appropriate specimen geometry to obtain a pure shear stress state [2]. It has been demonstrated in Ref. 1 that off-axis tension tests can be used to obtain accurate values for G_{12} if the aspect ratio of the coupon is greater than 10. This was necessary to assure the uniformity of the stress state at the gage section of the specimen. For specimen aspect ratios smaller than 10, appropriate corrections are required to account for the shear-coupling and end-constraint effects.

Conducting a series of uniaxial tension tests with very large specimen aspect ratios would require large amounts of material. In order to reduce the material requirement one can either calculate the minimum value of the aspect ratio that would produce a uniform

---

[1] Graduate Student and Assistant Professor, respectively, Department of Engineering Science and Mechanics, Virginia Polytechnic Institute and State University, Blacksburg, VA 24061

stress state or include appropriate correction factors to calculate the actual material property, both of which require the correct values of all the elastic material properties. For a material characterization test these values may not be readily available. In order to solve this dilemma, a combined analytical experimental procedure is proposed.

The basic premise behind the procedure is to iterate the unknown material properties until the calculated properties from the analysis of a given specimen configuration agree with the measured properties obtained during testing of the same configuration. A formal minimization algorithm is used to reduce the error between the calculated and measured quantities. Various apparent elastic properties of two material systems, graphite/epoxy and aramid/epoxy, are measured by using short aspect ratio compression specimens. Analysis of the specimen configuration is performed by using a two-dimensional finite element program.

Experimental Work

The experimental work presented in this paper was conducted using a recently designed compression test fixture [3] shown in Fig. 1. The fixture uses an end-loaded coupon supported by four circular side pins along the unloaded edges of the specimen. This configuration makes it possible to use specimens with a longer gage section than what is commonly used in compression testing. The longer gage section allows room for the mounting of a rosette strain gage in the test section that may be used to obtain shearing strains required for the calculation of shear modulus, G_{12}. This is important since now a simple compression test may be used for determining the shear modulus as long as the apparent value, resulting from the nonuniformity of the stresses in the gage section, measured during the test is modified to take into account the end-effects. Additional features of the new compression fixture are ease of specimen preparation and alignment, reduction of stress concentrations, and simple boundary conditions that make analysis of the test specimen easier. The boundary conditions are such that the specimen can be modelled as a plate simply-supported along all four sides. The specimen was supported along the loaded edges by using quarter-circle cross-section grips, and the circular side pins are used to prevent out-of-plane buckling. Also, the four corners of the specimen are restrained against transverse displacements along the loaded edges to prevent brooming of the ends and slipping of the off-axis specimens in a transverse direction. A more detailed description of this fixture is given in Ref. 3.

The specimens tested were 1.0 in. wide and 1.5 in. long, and were instrumented with a three-arm rosette for measuring longitudinal, transverse, and shearing strains. A uniaxial strain gage was mounted on the back surface to monitor and correct the data for out-of-plane bending, if there was any. Nominal thicknesses of the graphite/epoxy and aramid/epoxy material were 0.10 and 0.09 in., respectively. Several orientations of each material have been tested.

The apparent values of the elastic material properties ($\bar{E}_x$, $\bar{G}_{12}$, $\bar{\nu}_{xy}$) were obtained from the stress-strain response curves generated from the experimental measurements based on their simple definitions.

$$\bar{E}_x = \frac{\sigma_x}{\varepsilon_x}; \quad \bar{G}_{12} = \frac{\tau_{12}}{\gamma_{12}}; \quad \bar{\nu}_{xy} = \frac{\varepsilon_y}{\varepsilon_x} \tag{1}$$

where

$$\sigma_x = \frac{P}{A}; \quad \tau_{12} = -\cos\theta \sin\theta \sigma_x; \quad \gamma_{12} = 2\cos\theta \sin\theta \, (\varepsilon_y - \varepsilon_x) + (\cos^2\theta - \sin^2\theta) \, \gamma_{xy} \tag{2}$$

The strains ε_x, ε_y, and γ_{xy} were measured with the rectangular rosette strain gage, the applied load P was obtained from the load cell, and A is the cross-sectional area of the specimen. A typical stress-strain curve used for determining the axial Young's modulus of aramid/epoxy is shown in Fig. 2. The results are given in Table 1 for the two material systems, aramid/epoxy and graphite/epoxy, for different fiber orientations tested.

If the stress/strain state at the gage section of an off-axis coupon was uniform and uniaxial, one could determine the properties in the principal material directions by using the classical transformation equations. However, due to the small specimen size (aspect ratio of 1.5) the conditions imposed at the boundaries will propagate into the gage section and produce a nonuniform stress state. The nonuniformity of the stresses is more pronounced for an off-axis specimen because of the coupling between the extensional and shearing deformations. The degree of nonuniformity in the stress state is also a function of the material properties because of the effects of material anisotropy on boundary layer stress distribution.

Finite Element Analysis

A remedy for the inability to achieve a uniform stress state is to perform an analysis of the test coupon configuration in order to evaluate the effects of the nonuniform stress state on the experimentally determined material properties. A two-dimensional anisotropic finite element analysis of the specimens was performed to obtain the effects of the material anisotropy and boundary conditions on the stress distribution within the gage section. The finite element mesh used consisted of 6-noded isoparametric triangular elements. The applied loading was introduced by specifying a uniform compressive displacement at one end of the discretized coupon that resulted in 1% overall strain in the specimen. The stresses and strains were determined at the nodal point corresponding to the location of the strain gage, and the apparent finite element elastic properties (E_x^{fe}, G_{12}^{fe}, v_{xy}^{fe}) were calculated in the same manner (Eqs. 1 and 2) used in the post-processing of the experimental data. The applied axial stress, σ_x, in the equations was calculated by summing the nodal forces across the end of the coupon and dividing it by the cross-sectional area. This was equivalent to measuring the applied load from a load cell during the test.

For a known set of principal material properties, the finite element method and the transformation equations that neglect the boundary effects are used to calculate the local properties at the center of an off-axis specimen in the global xy-coordinate system. The ratio of the finite element results to the transformation results for the axial Young's modulus, E_x^{fe}/E_x^{tr}, of aramid/epoxy and graphite/epoxy are shown in Fig. 3 as a function of fiber orientation, θ. As stated previously, if a uniform stress state existed with no end-constraint effects, then this ratio would be equal to unity. However, it is observed that although for large fiber orientations the ratio is close to unity, for small fiber orientations the modulus calculated from the finite element analysis can become significantly larger than the value calculated from the transformation relations. Furthermore, the difference is a strong function of the material anisotropy as depicted by the difference of the ratio between the graphite/epoxy and aramid/epoxy specimens.

The ratios of the Poisson's ratio, v_{xy}^{fe}/v_{xy}^{tr}, and shear modulus, G_{12}^{fe}/G_{12}, calculated by the finite element procedure and transformation relations are shown in Figs. 4 and 5, respectively, for the two different material systems. These results indicate that the nonuniform stress state affects the finite element calculation of Poisson's ratio for all off-axis fiber orientations. Even for the on-axis 0° specimen there is a slight difference between the finite element and transformation relation calculations. The results for the shear modulus also show differences for all fiber orientations, but the difference is more pronounced for angles less than 45°.

Based on the results presented in Figs. 3 through 5, it seems it is feasible to find the modification factor to correct the apparent properties obtained from various off-axis orientations in order to account for the effects of shear-extension coupling and load-introduction. This is, of course, assuming that we have a good representation of the stress distribution in the specimen gage section with our finite element analysis. For example, it is clear from Fig. 5 that if one performs a test on a 35° off-axis aramid/epoxy specimen, the value obtained for the shear modulus will correspond to the actual shear modulus of the material. However, the flaw behind this reasoning is that, one has to have the correct material properties to feed into the finite element program to be able to generate curves similar to the ones shown in Figs. 3 through 5. The shape of these curves is a strong function of all the material properties. In the case of a shear modulus curve, for example, even if we have the correct moduli in the principal material directions, the angular orientation that corresponds to the G_{12}^{fe}/G_{12} ratio of unity changes as a function of the unknown value of G_{12}. Therefore, the dilemma faced here is that, correct values of properties are needed in order to calculate the accurate values of the properties...!

Experimental - Analytical Procedure

The iterative approach proposed is to vary the unknown material property, or properties, along the principal material directions (E_1, E_2, G_{12}, v_{12}) until the values of the apparent properties ($E_x^{fe}, G_{12}^{fe}, v_{xy}^{fe}$) of an off-axis specimen calculated from the finite element program agree(s) with the experimentally obtained elastic properties ($\bar{E}_x, \bar{G}_{12}, \bar{v}_{xy}$) for the same off-axis configuration. In order to make such an iterative procedure tractable, initially only one of the properties, G_{12}, is treated as an unknown material property during analyses. The other fixed material properties (E_1, E_2, v_{12}) which are used for the finite element analyses are taken to be the values obtained from the on-axis test configurations. In general, the axial modulus values measured from specimens having 0° and 90° fiber orientations provide reliable results for the principal moduli E_1 and E_2, respectively. The reason for using the value of the shear modulus, G_{12}, as the unknown quantity is the erratic nature of the experimentally obtained values for different off-axis angles. For six different off-axis angles, 10°, 15°, 30°, 45°, 60°, and 75°, six different values of the shear modulus ranging from a low of 0.196 Msi for the 45° orientation to a high value of 1.57 Msi for the 10° orientation was obtained.

The effect of the variation in the shear modulus, G_{12}, on the calculated value of the axial modulus is shown in Fig. 6 as a function of the off-axis angle for various values of the shear modulus. The filled triangular symbols in the figure represent the experimentally obtained values of the axial modulus. Clearly, the values of the axial modulus, E_x, can vary significantly from the experimental values depending on the value of the shear modulus used in the analysis.

The iteration procedure proposed in the first paragraph of this section was, therefore, used to minimize the difference between the calculated value of the axial modulus and the experimentally obtained modulus for each of the off-axis orientations. The resulting values for the moduli were 0.321, 0.389, 0.479, 0.234, 0.226, and 0.482 Msi for the 10°, 15°, 30°, 45°, 60°, and 75° off-axis specimens, respectively, with an average value of 0.355 Msi. There is still some variation in the value of the shear modulus from one off-axis orientation to another, and the average seems to be higher than the values reported elsewhere for this material. Nevertheless, compared to the uncorrected apparent values, there is a substantial improvement in the scatter for the shear modulus. It is suspected that the boundary conditions used in the finite element modeling of the specimen may not correspond to the actual conditions during testing because of internal slipping of the loaded specimen edges, especially for small off-axis angles. A more in-depth study is needed to determine a reliable off-axis angle for determination of the shear modulus based on small coupon compression testing.

Optimization Procedure

Potentially, the procedure described above can be used to find more than one material property based on a single test. For example, even if the modulus along the fiber direction is unknown along with the shear modulus of the material, based on the apparent experimental values measured from an off-axis test, the two unknown quantities can be iterated until the finite element analysis produce the same results as the test. There are two difficulties, however, associated with this problem. First of all, the iterative procedure outlined above is extremely cumbersome and only one of the input properties could efficiently be treated as a variable. Second, deciding on which of the measured quantities ($\bar{E}_x$, $\bar{G}_{12}$, $\bar{v}_{xy}$, or combinations of these) to be used for comparison with the analytical ones can be a difficult task. As a possible solution to these difficulties, it is proposed that the unknown material properties be varied automatically and systematically by using a mathematical optimization algorithm which will minimize an objective function defined in terms of the experimentally measured and the analytically predicted apparent properties.

Initially, an objective function defined in the form of a root-mean square error expression

$$\text{Minimize} \quad \sqrt{(\bar{E}_x - E_x^{fe})^2 + (\bar{v}_{xy} - v_{xy}^{fe})^2 + (\bar{G}_{12} - G_{12}^{fe})^2} \tag{3}$$

is considered. Preliminary work indicated that there is a need for normalizing the design variables and different terms in the objective function to be minimized because of large differences in the magnitudes of the potential design variables. It is observed that the rate of convergence (and sometimes the resulting values) to be very sensitive to the type of normalization used. In order to implement a more physically meaningful weighting between different design variables, another candidate for the objective function in the form of strain energy density expressed in terms of the material properties and local strains

$$\left(\bar{W} - W^{fe}\right) \tag{4}$$

where

$$W = \frac{1}{2}\left\{\bar{Q}_{11}\varepsilon_x^2 + \bar{Q}_{22}\varepsilon_y^2 + \bar{Q}_{66}\gamma_{xy}^2 + 2\bar{Q}_{12}\varepsilon_x\varepsilon_y + 2\bar{Q}_{16}\varepsilon_x\gamma_{xy} + 2\bar{Q}_{26}\varepsilon_y\gamma_{xy}\right\} \tag{5}$$

is under consideration. The $\bar{Q}$ terms in the above expression are the transformed reduced stiffnesses.

References

1. Pindera, M. J., and Herakovich, C. T., "Shear Characterization of Unidirectional Composites with the Off-Axis Tension Test", Experimental Mechanics, Vol. 26, No. 1, March, 1986, pp. 103-112.

2. Iosipescu, N., "New Accurate Procedure for Single Shear Testing of Metals", Journal of Materials, Vol. 2, No. 3, Sept., 1967, pp. 537-666.

3. Gürdal, Zafer, and Starbuck, J. M., "Compressive Characterization of Unidirectional Composite Materials", Proc., International Conference on Analytical and Testing Methodologies for Design with Advanced Materials, (G. C. Sih, ed.), Montreal, Aug.26-28, 1987.

Table 1: Average Apparent Material Properties

Fiber Angle θ	Aramid/Epoxy			Graphite/Epoxy		
	E_x (msi)	ν_{xy}	G_{12} (msi)	E_x (msi)	ν_{xy}	G_{12} (msi)
0	8.29	0.418	---	20.13	0.438	---
15	5.09	0.550	0.761	12.52	0.335	2.97
45	0.630	0.652	0.196	3.14	0.539	1.02
90	0.673	0.036	---	1.65	0.022	---

Figure 1: End-loaded-coupon compression test fixture, Ref. 3.

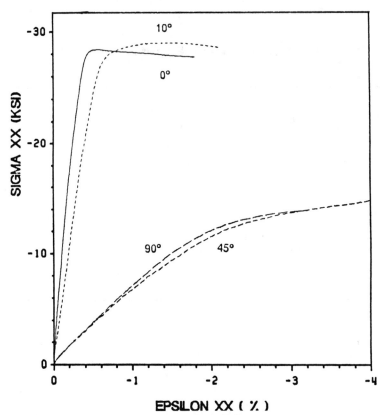

Figure 2: Typical stress-strain curves for aramid/epoxy specimens.

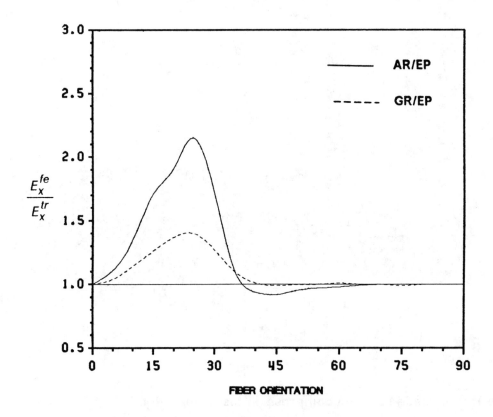

Figure 3: Ratio of the calculated finite element axial modulus to the modulus calculated from transformation equations.

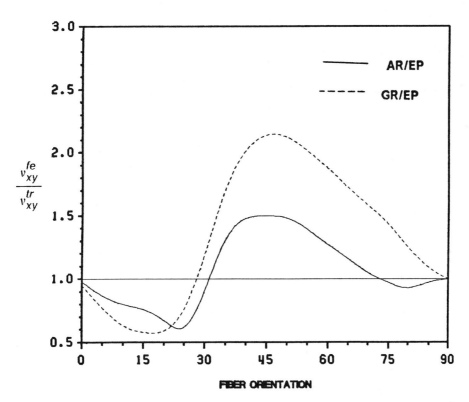

Figure 4: Ratio of the Poisson's ratio calculated from the finite element results to the Poisson's ratio calculated from transformation relations.

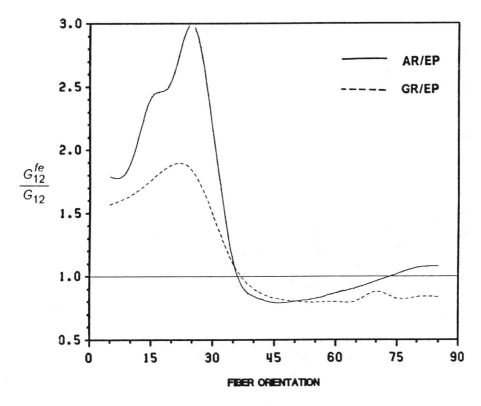

Figure 5: Ratio of the shear modulus calculated from the finite element results to the actual shear modulus.

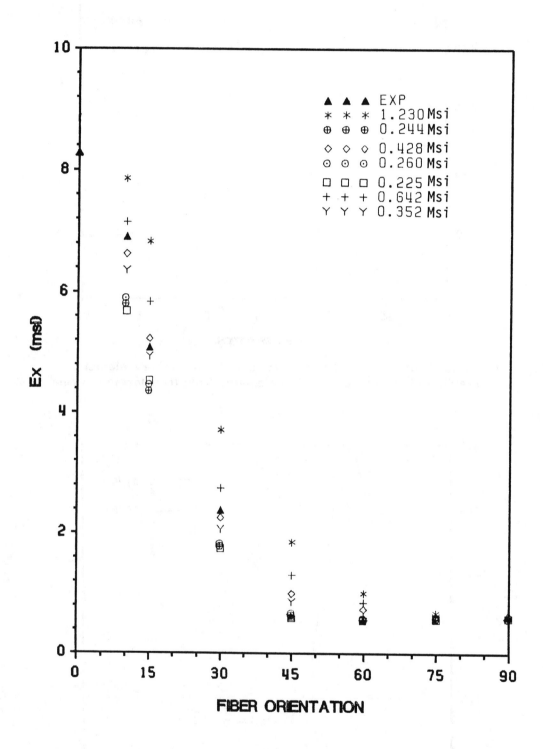

Figure 6: Axial modulus as function of fiber orientation for different G_{12} values used in the finite element calculations.

SYMPOSIUM V

High Performance Composites

Bismaleimids: A Historical Perspective

MARC E. CARREGA

ABSTRACT

   Bis-maleimid, bis(4-maleimidophenyl)methane, has been
polymerized on itself in the year 1964, the product was very
brittle. By copolymerizing it with a diamine a useful thermoset was obtained which is the base of Kerimid ® 601.
The development of powder compounds with fillers was slow and
the main application has been resins for multilayer electronic
circuit boards. Since the early 80's new Kerimid ® are developed for electronic and prepregging use. These products are
hot melt processable like epoxies but with much higher thermal
properties, the newest ones being tougher than K 601.

   The interest of RHONE-POULENC (RP) for thermostable
polymers started during the mid 60's when its textile division
Rhodiaceta was looking for heat resistant fibers. On november
13, 1964 The Battelle Institute and RP filed for a Swiss
patent covering the reaction of a bismaleimid (BMI) on itself
or with an other mono or bi-functionnal maleimid.

   On September 1966, RP was working with the BMI bis-
(4-maleimidophenyl)methane (coded name M 3) :

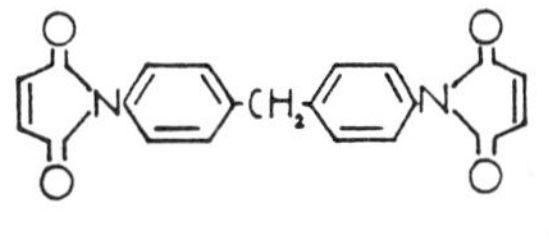

   Parts could be moulded from M 3 powder polymerized on
itself but developed cracks after a 150 hours exposure at
a temperature of 250° C.

   Working on the Michael reaction of the maleic double-
bond it was discovered by Bargain and Combet that the copolymer of M 3 with the bis(4-aminophenyl)methane (MDA)

Director of Polymer Research – RHONE-POULENC CHIMIE – Speciality Chemicals Division
Les Miroirs – 92097 PARIS LA DEFENSE – France

at the molar ratio M 3/MDA = 1,3 gave moulded parts unmodified after a 475 hours exposure at 250°C. The product coded M 33 was patented on july 13, 1967 [2] and was the basis of the future Kerimid 601 which was developed by optimizing the M 3/MDA ratio. The toughness of moulded parts was much higher than these made of poly-M 3. The decision to go ahead with the development of this type of BMI was taken during the year 1968.

RP approched General Electric in the US and they decided to jointly evaluate potential applications of these products. The market of M 33 type moulding powders and their compounds with fillers were extensively investigated. Compounds reinforced with glass fibers, alumina, etc ... and lubricating additives (graphite, MoS_2) were tested as bearings, structural parts, etc ... After two years GE aknowledged the heat and chemical resistance of M 33 but the toughness was not high enough to make for instance blades of turbomachines. The cost of the compounds was to high to let hope a very large market in the very near future. GE decided not to go further. 54 patents have been filed, of which 29 during the collaboration with RP.

By this time an improved process to synthesize M 3 had been worked out and also a continuous process for the preparation of M 33 prepolymers. The goal was to lower the cost of the Kinel ® compounds and to enlarge the market. Many applications were studied. Some have been successful like diamond emery-wheels or bases for automotive cigar lighters were ceramics were cracking. These uses needed all the thermostability of BMI. Other were technically sucessful but the part could not accept the price : RP worked with the Michelin Cy to make car wheels for the Citroen-Maseratti car. M 33 compounds were also tried with success to make disc breaks pads; however the advantage of higher abrasion resistance and a better "fading" behaviour was not enough to offset the disadvantage in price of Kinel compared to phenolics.

A small racing car used an engine made by Polymotor and fitted with pistons made of Kinel. To-day this company is developping "plastic" engines with pistons made of Torlon ® (a more expensive polymer).

Since Kinel could not been developed for large markets it was decided to turn to those markets were the cost of the polymer is a small fraction of the cost of the finished part. This is the case for multilayered electronic circuit boards where the BMI represents 2 % of the value of a layer circuit.

It was found that when glass cloth was impregnated with a solution of Kerimid ® 601 (a brand of M 33) and copper claded the adhesion measured by the peel-strength was very good after immersion in the soldering bath and that the thermal expansion coefficient in the tickness direction fitted very well the one of the connecting metal. This is a very important property for very complex and costly electronic systems which can be subjected to important tempe- rature variations. Kerimid 601 has stayed ursurpassed for more than a decade for the making of multi-layered circuit boards used for military and aerospace applications.

Kerimid was not resistant enough to glue the tiles of the future space shuttle but it was tried in the early seven- ties. Kerimid prepregs and Kinel are used in jet engines in contact with oil or at a high temperature.

Kerimid 601, a solid powder at room temperature, is normally used as solutions in such solvents as NMP which is largely accepted in electronic industry. Since Kerimid 601 is delivered as a prepolymer, a small percentage of MDA can be extracted by solvents.

The possible carcinogenicity of this diamine was a concern in the case of extensive use by manufacturers in civil electronic applications, although in the cured board no diamine is left unreacted that can be leached out.

In order to solve the problem of the few per cent of extractable diamine of Kerimid 601 and to improve its tough- ness the research on BMI was reactivated in the early eighties. This is demonstrated in Fig.1 where the number of patents filed annualy by R.P. on the subject of BMI's is shown.

Some epoxy/BMI products had been studied in the seventies which solved a large part ot the diamine problem but at the expense of the maximum service temperature [3]. New avenues have been open by the use of other non-carcinogenic products in place of MDA and the introduction of phenyl-silanediol type oligomers [4,5]. Other studies have shown that Kerimid with a very low glass transition before cure can be formulated which offer the possibility of preparing resins for casting without solvents and also to prepare prepregs with carbon cloth and unidirectionnal tape. These prepregs can be handled like standard epoxy systems.

Two parallel developments are underway :
- resins for electronic circuit board with no toxic components,
- resins for prepregging with no toxic components and an improved toughness for the same service temperature.

Since it is well known that the higher the thermosta- bility the lower the toughness in an homologous series.

A number of experimental Kerimid ® FE 70 xxx have been prepared during the last 3 years. $G_{1}c$ values for the neat resins and the corresponding T 300 carbon fiber compo- sites are depicted in table 1. with the Tg measured by DMA at 3,5 °C/min.

Even when G 1 c are lower than this of K 601, the FE 70 xxx products offer the advantage of "hot melt" processing and less internal stress. For some of them a higher resilience is obtained.

CONCLUSION

Although not so high as polyimides, BMI have a higher thermal resistance than epoxies with the same type of processing. The new Kerimid have an even higher resistance than K 601 without some of the drawbacks and keep the advantage of those properties which dominate over the epoxies. This allows to be optimistic about the future of these products on a larger basis of application in electronic industry as well as for some aerospace structural parts. The improvement of the resin properties will give way to new molding compounds.

The knowledge that RP has developed with the Kermel ® fibers and Rhodeftal ® which is related to the poly-imid amid chemistry offers other possibilities.

REFERENCES

1.   Swiss patent n°14838/64 growing patent n° 430220

2.   French patent n°1 555 564

3.   Woo E.M., Chen L.B., Seferis J.C.
     J. Mat'ls Sci. vol. 22, 1987, pp. 3665

4.   Lopez P.R. "Characterization and Processing of Polyimide Matrix Composites" M.S. Thesis in Engineering.
     Dept. of Chemical Engineering, Uty of Washington, Seattle, 1986.

5.   Viot J.F., Seferis J.C.
     J. of Appl. Poly. Sci. vol. 84, 1987, p.1459

TABLE 1 – Thermal and mechanical properties of Kerimid and composites made of them

KERIMID	Neat resin		Composite (CF- T 300)		
	Tg(DMA) °C	G$_{1c}$ J/m2	Tg(DMA) °C	ILSS(unidir.) MPa	G$_{1c}$(cloth) J/m2
601	290	70	290	110	–
FE 70 003	400	75	360	61	200
FE 70 006	380	65	370	–	–
FE 70 011	360	75	360	105	200
FE 70 015	330	60	320	90	200
FE 70 020	240	80	240	116	300
FE 70 023	310	65	320	108	220
FE 70 024	240	–	240	115	300
FE 70 026	310	65	300	108	220

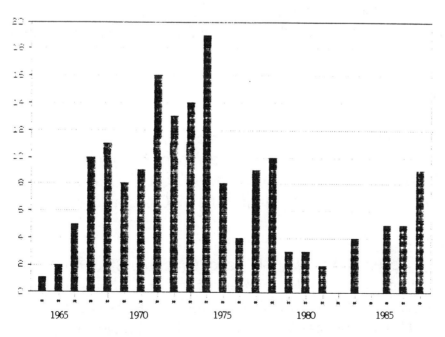

Fig. 1 – Number of patents filed annually by RP on the subject of BMI.

The Processing of Ceramic-Ceramic Composites

D. DORAISWAMY, I.-L. TSAO, S. C. DANFORTH, A. N. BERIS AND A. B. METZNER

ABSTRACT

Highly concentrated ceramic suspensions are used to manufacture whisker reinforced ceramic-ceramic composites by the injection molding technique. A detailed understanding of the rheology of these systems is crucial to the success of this operation. A simple constitutive equation (based on recoverable strain) is presented which describes experimental data for concentrated suspensions of silicon particles in polyethylene. These materials display both fluid-like and elastic responses but do not possess viscoelasticity of the classical kind. In general, these materials exhibit yield stresses and a strong dependence of the dynamic properties on the strain to which the material is subjected. Both of these features are described by the model. The results indicates that it is possible to correlate the dynamic response with the shearing behavior in steady state flows which are usually encountered in injection molding.

INTRODUCTION

Whisker reinforced ceramic composites have a number of appealing characteristics such as light weight, high strength, high fracture toughness, and high temperature resistance. Typical applications which exploit these properties are in the manufacture of IC engine components

D. Doraiswamy (Research Fellow), I-L. Tsao (Graduate Student), S.C. Danforth (Associate Professor), Department of Ceramic Engineering, Rutgers University, Piscataway, NJ 08854-0909; A.N. Beris (Assistant Professor), A.B. Metzner (Professor), Department of Chemical Engineering, University of Delaware, Newark, DE 19716.

and turbine blades. The need for a processing technology for manufacturing superconducting ceramic cables could overshadow these areas in
relative importance in the not too distant future. An important route for
the manufacture of ceramic-ceramic composites is by the injection
molding of concentrated ceramic suspensions, as this technique has the
potential for rapid manufacture of parts of complex shape.

The ceramic particles and fibers are mixed with a binder (usually a low
molecular weight polymer) and the formulation is then used to mold the
rquired part. The binder is used merely as a vehicle in order to fabricate
the part and lend strength and processability to the system. The binder
must then be burned off in a controlled manner before the residual
ceramic can be sintered [1].

In this paper a model is presented which describes the rheology of
concentrated ceramic suspensions and can be used as a framework to
study the injection molding of these materials. The goal is to relate
pressure levels to injection rates and also to determine the whisker
orientation, residual stresses, and unwanted deformations in the
molded part. The results of this work should also be applicable to other
related systems such as foams, printing inks and mine tailings.

CONSTITUTIVE MODEL

A constitutive equation which relates stresses to deformation rates is
presented in order to describe the important rheological characteristics of
these materials: the occurence of yield stresses, shear thinning behavior
and a non-linear strain dependence of the dynamic properties. An analysis
is then developed which enables a determination of the model parameters
from conventional linear viscoelastic properties in dynamic measurements.

Highly concentrated ceramic suspensions might be expected to exhibit a
Hookean elastic response below a critical strain and a viscous fluid-like
response above the critical strain:

$$
\underline{\underline{\tau}} = \begin{cases} (\frac{\tau_y}{\gamma} + K\dot{\gamma}^{n-1})\underline{\underline{\dot{\gamma}}} & |\hat{\gamma}| = \gamma_c \\ G\underline{\underline{\hat{\gamma}}} & |\hat{\gamma}| < \gamma_c \end{cases} \tag{1}
$$

The critical strain γ_c is related to the yield stress τ_y at which the
material first starts to exhibit viscous flow:

$$\tau_y = G\gamma_c \tag{2}$$

in which G is the elastic modulus.   The yield stress is a parameter of interest because it is likely to be related to the green strength of the ceramic.   The effective strain $\hat{\gamma}$ incorporates a recoverable strain in the model:

$$\frac{\partial \hat{\gamma}}{\partial t} = \begin{cases} 0 & |\hat{\gamma}| = \gamma_c \\ \dot{\gamma} & |\hat{\gamma}| < \gamma_c \end{cases} \tag{3}$$

Equation (3) indicates that below the yield stress (or critical strain) the material behaves like a purely elastic solid and the entire strain is recoverable; above the yield stress, however, the maximum recoverable strain is limited to the critical strain. Physically, this is equivalent to stating that the elastic deformation is recoverable while the viscous deformation is not, *i.e.* , the material is only capable of remembering the strain it undergoes as an elastic solid. Further details of the model are provided elsewhere [2,3]. Eqs. (1) - (3) are simple, but significant, generalizations of the classical Herschel-Bulkley fluid [4] and are the proposed equations for concentrated ceramic suspensions.   A more restrictive form of these equations has also recently been proposed by Yoshimura and Prud'homme [5] in order to analyze the rheological behavior of emulsions. The parameters in the model are the critical strain γ_c, the elastic modulus G, and the power law constants K and n.

ANALYSIS FOR OSCILLATORY FLOW

In typical dynamic measurements a sinusoidal strain of the form

$$\gamma = \gamma_0 \sin(\omega t) \tag{4}$$

is imposed on the material and the resultant stress, which is also periodic, is studied:

$$\tau = \gamma_0 \sum_{k=1}^{\infty} [G_k' \sin(k\omega t) + G_k'' \cos(k\omega t)] \tag{5}$$

For linear viscoelastic materials the first harmonics are sufficient to describe the stress response while for non-linear materials the higher harmonics could also become significant. For linear materials the in-phase component of the resultant stress is characterized by the storage modulus G'_1 (abbreviated to G') which is a measure of the elastic response of the material while the out-of-phase component is characterized by the loss modulus G''_1 (abbreviated to G").

For the kinematics outlined in Eq. (4), it is possible to obtain analytic expressions for G' and G" in terms of the dimensionless strain S, the dimensionless stress D, and the phase shift θ [2,3]:

$$G' = \frac{2G}{\pi} \left[\frac{2-S}{S} \sin\theta + \frac{D}{S} \frac{(\sin\theta)^{n+1}}{(n+1)} + \frac{\theta}{2} + \frac{\sin 2\theta}{4} \right] \tag{6}$$

$$G'' = \frac{2G}{\pi} \left[\frac{2-S}{S} \cos\theta + \frac{D}{S} F(\theta,n) + \frac{3}{4} + \cos 2\theta \right] \tag{7}$$

In the above equations, θ is the phase shift introduced as a result of the elastic nature of the fluid given by

$$\theta = \cos^{-1}\left(1 - \frac{2}{S}\right) \quad 0 \le \theta \le \pi \text{ for } S \ge 1 \tag{8}$$

The function $F(\theta,n)$ is an infinite converging series in terms of sine and cosine functions of θ and gamma functions of n. Additional details are presented elsewhere [2,3]. From Eqs. (6) and (7) it can be seen that the first harmonics G' and G" alone can be used to determine the material parameters of non-linear ceramic suspensions which can then be used to determine the higher harmonic coefficients. The material parameters, once determined, can subsequently be used to evaluate the steady state behavior of these materials. This feature is of considerable practical significance since, in general, dynamic data are far more easily obtained in the laboratory than steady shear data in the range of shear rates commonly encountered in injection molding.

EXPERIMENTAL

The experimental system investigated in this work was a 70 vol % suspension of silicon particles (supplied by Globe Metal Co.) in polyethylene (suppied by Allied Chemicals). The polyethylene exhibited

Newtonian behavior with a viscosity of 6 Pa.s at a temperature of 140°C. In the future this work will be extended to concentrated suspensions containing SiC carbide whiskers in addition to Si particles.

Experimental dynamic as well as steady state data were obtained on a Rheometrics RMS 800 mechanical spectrometer in a cone and plate geometry. The plate diameter was 25 μm and the cone angle was 5.7°. The average particle size of the silicon powder was 8 μm. Additional experimental details are provided in reference [6].

RESULTS AND DISCUSSION

The dynamic viscosities η' and η'', and the complex viscosity η^* are often used in place of the moduli G' and G'' and are defined as follows:

$$\eta' = \frac{G''}{\omega} \tag{9}$$

$$\eta'' = \frac{G'}{\omega} \tag{10}$$

$$\eta^* = \sqrt{\eta'^2 + \eta''^2} \tag{11}$$

Typical complex viscosity data are plotted in Fig. (1) in terms of an effective shear rate defined as the product of frequency, ω and the maximum imposed strain, γ_0. It can be clearly seen that the curves for all the different strains (ranging from 1% to 20 %) are superimposed when the abscissa is defined in this manner. An initial slope of -1 is clearly visible (corresponding to an elastic solid-like response) followed by a gradual transition to a viscous shear thinning response.

The complex viscosity predicted by Eqs. (6) and (7) are plotted in Fig. (2) in terms of the same abscissa as in Fig. (1) for the following parameter values: G = 50,000 Pa, K = 400 Pa.s^n, n = 0.5 and γ_c = 0.005. The parameter values were determined by fitting the individual dynamic moduli (G' and G'') using the 1 % strain curve and the other curves were generated using the same values [2,3]. It can be clearly seen that the model predicts the proper superposition of the curves for the various strains in terms of the effective shear rate. The model is thus able to predict the proper non-linear dependence of the material properties on the strain to which the material is subjected.

It is worth mentioning here that the magnitudes of both moduli are finite over a range of frequencies which is not the case for the elastic Bingham model or the Herschel-Bulkley fluid.   The modified Herschel-Bulkley model proposed in this work, however, predicts finite values for both moduli with a maximum error within 25%.   Finally, it should also be noted that these preliminary data have not been corrected for slip effects which could introduce additional errors.   Initial investigations, however, indicate that slip effects might not be very severe (causing an error in the apparent viscosity of less than 15 % [6]).

Typical experimental steady state cone and plate data are also plotted in Fig. (1) and the corresponding theoretical predictions in Fig. (2). The parameter values used were the same as those used for the prediction of the dynamic behavior. It can be seen that the agreement between theory and experiment is very good.   It thus appears that that the model enables a prediction of the steady state behavior from dynamic measurements. It is planned to further check the validity of the model by obtaining high shear rate capillary data in the range of shear rates commonly encountered in injection molding.

Some of the important features of the model are summarized below:

1) The model is able to describe the non-linear strain dependence of the dynamic properties of these materials.   The observed elastic, viscous and yielding phenomena are described by the constitutive equation.

2) The model incorporates the concept of a recoverable strain which is limited to a maximum value equal to the critical strain at which the material first starts to exhibit viscous flow. Although the value of the critical strain is typically very small, it can severely influence the flow patterns (and, consequently, the whisker orientation) in molds of complex shape (like turbocharger molds) where abrupt changes in flow direction are possible.

3) The current constitutive equation assumes that the transition from an elastic to a viscous response occurs at a point value of the strain rather tnan over a range of strains. It is intended in an extension of this work to incorporate the more realistic assumption of a smooth transition in behavior from an elastic to a viscous response.

4) The model parameters are determined solely using conventional dynamic measurements (or linear viscoelastic properties). In the future the analysis outlined in this work will be presented in a more useful graphical form in order to facilitate estimation of the parameter values in the model.

5) The model enables a determination of the yield stress at which the transition from an elastic solid-like response to a viscous response occurs.   The yield stress is a parameter of interest in processing applications because it determines the size of the largest part which will not slump under its own weight. It is therefore likely to be related to the green strength of the unfired ceramic.

6)   In some molding applications a random orientation of whiskers may be desirable while, in other applications, oriented whiskers may be preferred. The magnitude of the yield stress (or the critical strain) is related to the size of the plug-flow region in the mold. Since the whiskers would tend to maintain a random orientation in the plug-flow or unyielded region, a good mold design could exploit this feature in order to obtain some control over the whisker orientation.

7) The model enables a correlation between the dynamic response and the steady state shearing behavior. This feature is of considerable practical significance since dynamic data are relatively easy to obtain in the laboratory as compared to steady shearing data in the range of shear rates commonly encountered in injection molding.

CONCLUSIONS

A non-linear model has been presented for describing the rheological behavior of concentrated ceramic suspensions which combines elastic, viscous and yielding phenomena.   A major feature of the formulation is the incorporation of a limiting recoverable strain. Although the magnitude of the critical strain is quite small it can severely influence the flow patterns in molds of complex geometry. The model parameters are determined from conventional linear viscoelastic properties in dynamic measurements. A useful practical feature of the model is that it permits a prediction of the steady state behavior from dynamic measurements. Experimental data for a 70 vol % suspension of silicon particles in polyethylene indicate good agreement between data and experiment for both dynamic as well as steady state shearing behavior.

NOMENCLATURE

Roman Letters

D = viscous stress/elastic stress = $K\gamma_0{}^n\omega^n/(G\gamma_C)$

$F(\theta,n)$ = function defined in Eq. (7)

G = elastic modulus, Pa

G'_k, G''_k = k^{th} harmonic coefficients

G'_1 = storage modulus (abbreviated to G'), Pa

G''_2 = loss modulus (abbreviated to G''), Pa

K = consistency in power law model, Pa.s^n

n = power law index

S = dimensionless strain = γ_0/γ_C

Greek letters

$\underline{\dot{\gamma}}$ = deformation rate tensor, 1/s

$\dot{\gamma}$ = second invariant of deformation rate tensor, 1/s

$\hat{\underline{\gamma}}$ = effective strain defined in Eq. (2)

γ_C = critical strain

γ_0 = maximum imposed strain defined in Eq. (4)

η',η'' = dynamic viscosities defined in Eqs. (9) and (10), Pa.s

η^* = complex viscosity defined in Eq. (11), Pa.sτ = stress tensor, Pa

$\underline{\tau}$ = stress tensor, Pa

τ = second invariant of stress tensor, Pa

τ_y = yield stress defined in Eq. (2), Pa

ω = frequency, 1/s

REFERENCES

[1] Tsao, I-L., Metzner, A.B. and Danforth, S.C., "Rheological Study of Highly Filled Ceramic-Ceramic Composites," Proceedings of the American Society for Composites, Second Technical Conference, September 23-25, Technomic, Lancaster, 1987, p. 316.

[2] Doraiswamy, D., Tsao, I-L., Beris, A.N., Danforth, S.C. and Metzner, A.B., "The Rheology of Ceramic Suspensions," paper to be presented at the Xth World Congress on Rheology, Sydney, August, 1988.

[3] Doraiswamy, D., Tsao, I-L., Beris, A.N., Danforth, S.C. and Metzner, A.B., "A Rheological Model for Concentrated Ceramic Suspensions," in preparation.

[4] Herschel, W.H. and Bulkley, R., 39, 291 (1926) as cited by Bird, R.B., Dai, G.C. and Yaruso, B.J., <u>Rev. Chem. Engg.</u>, Vol. 1, 1983, pp. 1.

[5] Yoshimura, A.S. and Prud'homme, R.K., <u>Rheologica Acta</u>, Vol. 26, 1987, pp. 428.

[6] Tsao, I-L., Doraiswamy, D. and Danforth, S.C., "Rheological Behavior of Injection Moldable Ceramic-Ceramic Composites," paper to be presented at the 3rd International Symposium on Ceramic Materials and Components for Engines, Las Vegas, November, 1988.

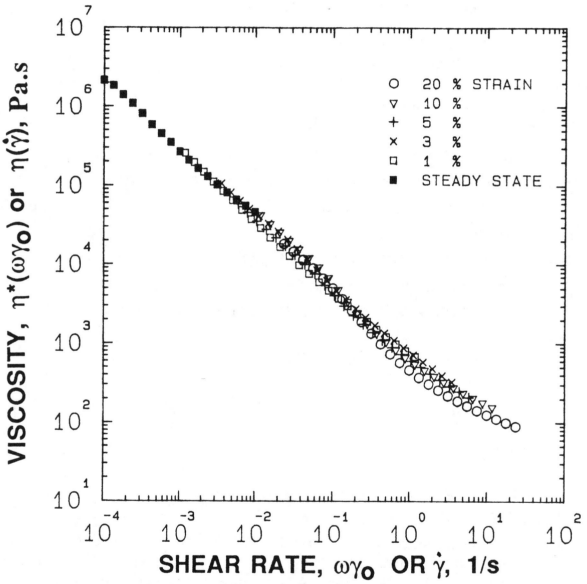

Fig.1. Experimental complex viscosity data and steady state shear viscosity data for 70 vol % Si in polyethylene as a function of the "effective shear rate," $\gamma_o \omega$, and the steady state shear rate, $\dot{\gamma}$, respectively.

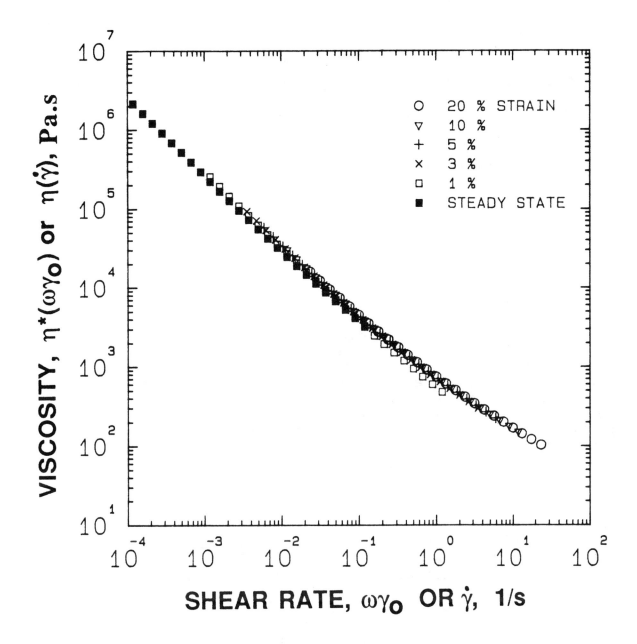

Fig. 2. Theoretical predictions for complex and steady state shear viscosities as a function of the "effective shear rate," $\gamma_0 \omega$, and the steady state shear rate, $\dot{\gamma}$, respectively.

Squeeze Flow Approach to Rheology and Consolidation of Fiber Reinforced Thermoplastic Systems

K. M. NELSON, J.-A. E. MANSON AND J. C. SEFERIS[1]

ABSTRACT

A squeeze flow rheometer uses two mutually parallel plates moving closer together as a means of measuring the viscosity of the contained material. Such a system is not only ideal for rheological characterization of thermoplastic resins but also for studying the consolidation process of the corresponding fiber reinforced matrices. This study addresses the application of squeeze flow rheology in measuring the viscosity of amorphous polymers, and in particular, high temperature thermoplastic systems used as matrices in high performance composites. By the nature of thermomechanical analysis, a significant advantage of this rheological technique is its applicability to high temperature polymers. Both dynamic (increasing temperature) and isothermal data were represented by the WLF (Williams, Landel, and Ferry) and Andrade equations. In a classical application of time-temperature superposition, a master curve was generated from a series of isothermal shear-stress shear-rate data. Moreover, the activation energy of flow was calculated from the WLF and Andrade equations and was compared with values reported in the literature. Essentially, any instrument capable of applying constant force and controlling the temperature, while at the same time measuring displacement, is capable of squeeze flow measurement. However, geometrical and apparatus dependent factors must be properly accounted for in the calculations.

INTRODUCTION

The squeeze flow rheometer, as an accessory to a thermomechanical analyzer, has the distinct advantage of applicability over a broad temperature range, making it

K. M. Nelson, J.-A. E. Manson, J. C. Seferis, Polymeric Composites Laboratory, Department of Chemical Engineering, University of Washington, Seattle, Washington 98195.
[1] Author to whom correspondence should be addressed.

particularly attractive to the rheological study of high temperature thermoplastics. Squeeze flow rheology has been limited in traditional polymer applications due to the fact that only low rates of shear are obtainable. However, for composite consolidation, it is the low shear rate conditions that are applicable during processing. Dynamic (increasing temperature) or isothermal experiments are conducted to measure the temperature dependence of viscosity, in this case the viscosity of an amorphous, high temperature thermoplastic.

The WLF expression adequately describes the viscosity behavior of a polymer as a function of temperature at a given shear rate. In addition, for temperatures greater than 100°C above the glass transition, the viscosity can be represented by an Arrhenius approach to the activated process (Andrade equation) [1]. The rheology of polymers can also be characterized by shifting isothermal shear-rate shear-stress data to produce a master curve about a single temperature, encompassing a broader shear rate spectrum.

EXPERIMENTAL

The Du Pont 943 Thermomechanical Analyzer (TMA) with the parallel plate rheometer attachment was used to study the rheology of high temperature amorphous thermoplastics. The squeeze flow rheometer attachment consisted of two cylindrical, stainless steel disks housed in a frame which kept them mutually parallel. The sample (0.5 to 1 mm thick) was placed between the plates, each 5.08 mm thick and 9.53 mm in diameter, as shown in Fig. 1. This was then placed upon the thermomechanical analyzer (TMA) stage with the probe resting on top.

The polymer used in this study is part of a broad family of polyetherimide thermoplastics whose basic chemical structure is similar to that depicted schematically in Fig. 2. American Cyanamid's CYPAC resin (CYPAC X7005) is the amorphous polyetherimide [2] used in this research and is part of a family of polyetherimide resins manufactured by General Electric under the trade name of Ultem^tm. For comparison purposes, standard samples of well-characterized polymers were also investigated with the squeeze flow system. The viscosity of polycarbonate Makrolon (Bayer AG) and polystyrene Vestyron 114 (CW Huls AG) was measured.

The neat thermoplastic samples were molded into 1 mm thick billets using a matched die quench mold developed by Seferis et al [9]. Wafers were cut from the slowly-cooled billet using a 9.53 mm. diameter punch. Alternately, a 0.3 mm thick film was produced using a hydraulic press (MTP14 Tetrahedron Associates Inc., Materials Test Presses). Wafers punched from this stock were stacked upon each other to equal the approximate 1 mm thickness that was found to be ideal for this experiment. Dynamic experiments were performed at heating rates of 5 and 10°C/min. Typical data are presented in Fig. 3.

Of primary importance in a squeeze flow experiment is a

precise measurement of the separation of the two plates as a function of time and temperature. The TMA is "zeroed" with no sample between the plates. The recorded dimension is then an accurate measurement of the actual distance between the plates. To account for the thermal expansion of the stainless steel plates the dimensional change of the empty apparatus was measured with respect to temperature and subtracted from the corresponding experimental values.

RESULTS AND DISCUSSION

The procedure followed for calculating viscosity is outlined in work by Cessna and Jabloner [3]. For flow between two parallel plates where the surface area of contact is constant, and thus the applied normal pressure is constant, the viscosity is given by:

$$\mu = \frac{(4F/3\pi r^4)}{d(1/h^2)/d(1/t)} \tag{1}$$

$$\dot{\gamma} = (3rh/2)\frac{d(1/h^2)}{d(1/t)} \tag{2}$$

The shear stress is the product of the apparent viscosity and shear rate; or the apparent viscosity is the ratio of the shear stress to the shear rate:

$$\mu = \tau / \dot{\gamma} \tag{3}$$

$$\tau = \mu\dot{\gamma} = 2Fh/\pi r^3 \tag{4}$$

Where:
μ = Apparent Viscosity (Pa·sec.)
$\dot{\gamma}$ = Shear Rate (1/sec.)
τ = Shear Stress (Pa)
F = Force (N)
h = Plate Separation (m)
t = Time (sec.)
r = Plate Radius (.004765 m)

The simplicity of these working equations makes them ideal for evaluation using spread sheet type programs. Ease of graphical representation of viscosity models in terms of temperature and shear rate dependence and rapid manipulation of the data is readily accomplished. All the results presented in this text have been processed using spread sheets [4].
Viscosity profiles of polyetherimide, polystyrene, and polycarbonate are presented in Fig. 4, 5, and 6, and they compare favorably to values reported in the literature [5]. As represented by the squeeze flow method, the viscosity of the amorphous polymers are well-behaved. Work done by E. Macho et al uses the Vogel-Falcher expression [6] to express the temperature dependence of the viscosity of several

amorphous thermoplastics. The Vogel-Falcher expression for viscosity is essentially a hybrid of the two models employed in this study.

In general, polymers behave as Newtonian fluids provided the shear rate is small enough. For a Newtonian fluid above the glass transition temperature (or the melting point), the viscosity may be modelled by the Andrade equation, vis.:

$$\mu = K e^{Ea/RT} \tag{5}$$

Where:
K = A constant for a given shear stress
Ea = Activation energy for the flow process
T = Absolute temperature
R = Ideal gas constant

Values for the activation energy Ea depend on the polymer, but generally range from 5,000 to 50,000 cal/mole. The activation energy of the polyetherimide resin was calculated from the slope of a log viscosity, reciprocal temperature plot. This value was calculated as 32.0 Kcal/mole.

It has been well established for amorphous polymers that the temperature dependence of viscosity may be modelled by the WLF equation [1]. The WLF equation intrinsically predicts a temperature dependence of the apparent activation energy for the polymer, while the Andrade expression is limited by an assumption of a constant activation energy. The WLF equation [7,8] is written as:

$$Log_{10}\mu/\mu_g = \frac{-C_1(T-Tg)}{C_2 + (T-Tg)} \tag{6}$$

Where:
T, Tg = Temperature and glass transition temperature of the polymer respectively (Kelvin).
μ, μ_g = Viscosity and interpolated viscosity at glass transition respectively (poise).
C_1, C_2 = WLF Constants

Quite often, general values for the constants C_1 and C_2 are used and are sometimes assumed to be independent of the polymer used. The general values of $C_1 = 17.44$ and $C_2 = 51.6$ Kelvin are referred to as the universal constants. However, for better representation of the data, specific values of μ_g, C_1, and C_2 are calculated for the polyetherimide, using a least squares fit. These best fit values are $C_1 = 9.6$, $C_2 = 59.6$ Kelvin, and Log $\mu_g = 11.7$ poise. The WLF equation with these values closely predicts the viscosity of the polyetherimide; for example see Fig. 7.

The activation energy can be calculated from the WLF parameters C_1, C_2, μ_g. Simply stated, it is proportional to the change in the log of viscosity with respect to reciprocal temperature, producing the desired results:

$$Ea = \frac{R(2.303)C_1 C_2 T^2}{(C_2 + T - T_g)^2} \tag{7}$$

Where T is in Kelvin

As the temperature approaches Tg, the activation energy increases according to equation 7. At the glass transition, the activation energy is a maximum and is calculated as:

$$Ea = \frac{R(2.303)C_1 Tg2}{C_2} \tag{8}$$

Measurements from isothermal experiments are more representative of the actual viscosity at a given temperature partly because kinetic effects are eliminated. Constant temperature experiments allow a closer look at the effects of shear rate on viscosity. Fig. 8 is a plot of isothermal viscosity measurement showing each individual data point. Each experiment begins at high shear rates and proceeds to lower ones as the plates come together, allowing only a narrow range of shear rates. Using time-temperature superposition, multiple sets of isothermal shear-stress shear-rate data can be manipulated to represent a single set at a single temperature, encompassing a broader shear rate range.

Since most of the processing of polyetherimide is done at 320°C, this temperature was chosen as a reference temperature about which shifting was done. Fig. 9 shows the unshifted shear rate data. The shifted data, referred to as the master curve, forms one continuous line at the reference temperature. The horizontal amount an isothermal curve is shifted on a log scale is the shift factor:

$$a_t = \dot{\gamma}_{Tg}/\dot{\gamma}_T \tag{9}$$
$$a_t = \mu/\mu_g \tag{10}$$

Where   at = Shift factor

Four sets of data, taken at 285, 307, 320, and 330°C are shifted and displayed in Fig. 9. The shift factors can be expressed mathematically by applying the relations of equations 9 and 10 above.

$$Log_{10}\mu/\mu_r = Log_{10}a_t = \frac{-C_1(T - Tr)}{(C_2 + T - Tg)} \tag{11}$$

Thus the shifting is described by the WLF constants C_1, C_2, and μ_g. Most polymer systems behave as Newtonian fluids, provided that the shear rate is low enough. At low shear rates, typically less than 1.0 sec^{-1}., the viscosity approaches its asymptotic limit μ_o. This value is different for each temperature. Shifting of the data produces a master curve predicting Newtonian behavior, covering a small spectrum of shear rates (Fig. 10). Essentially no information is gained by shifting a Newtonian fluid, since the viscosity is unaffected by shear rate.

The WLF equation can be written with respect to any reference temperature by recalculating the parameters using the following relationship.

$$C_1{}^* = \frac{C_1 C_2}{C_2 + (Tr-Tg)} \tag{12}$$

$$C_2{}^* = C_2 + (Tr-Tg) \tag{13}$$

$$\mu_r = \mu_g - \frac{C_1(Tr-Tg)}{C_2 + (Tr-Tg)} \tag{14}$$

Finally, the WLF equation and the Andrade equation were plotted and compared to the experimentally determined isothermal shift factors (generated from isothermal shifts, see Fig. 11). Both the fitted WLF constants and the Andrade constants were calculated from dynamic data, generating best fit curves which were independent of the isothermal data. If these constants accurately model the dynamic data (solid lines in Fig. 11), they would be expected to model the isothermal data (squares in Fig. 11) equally well, and they do.

Independent measurement of the viscosity of the polyetherimide resin is nominally 3 x 10^5 poise at 383°C and 1 x 10^6 poise at 360°C, using a Rheometrics parallel plate viscometer at steady shear. This value compares to 7 x 10^4 and 1 x 10^5 poise respectively as measured using the squeeze flow method. Although further investigation is required to determine the cause of this discrepancy, it can possibly be attributed to scaling and geometric effects. These issues will be further investigated, as will applications to fiber reinforced systems.

CONCLUSIONS

The squeeze flow rheometer has been successfully utilized to model the viscosity of an amorphous polyetherimide polymer (CYPAC X7005). The WLF parameters for viscous flow were calculated and reported as: C_1 = 9.6, C_2 = 59.6 (Kelvin), and log μ_g = 11.7 (poise) and the activation energy for the flow process was calculated as 32.2 kcal/gmole (320°C), as measured using the Andrade equation. Most polymers will exhibit Newtonian properties at the shear rates of the squeeze flow method.

Experimental shifting of isothermal shear stress - shear rate data resulted in shift factors predicted by the WLF equation, using the parameters calculated from dynamic experiments. The squeeze flow technique is particularly applicable to amorphous thermoplastic polymers, since the viscosity changes predictably over a given temperature range. The squeeze flow viscometer quickly gives viscosity temperature data at high temperatures by doing a single experiment.

The same squeeze flow principle can be applied to the study of laminate consolidation during processing. Understanding the rheological behavior of the prepreg during the consolidation process is an important step in linking the processing to the properties of the laminate itself.

ACKNOWLEDGMENT

The authors would like to express their appreciation to Mr. Alan Fenstermaker and Dr. Steven Peake of American Cyanamid for their helpful assistance and collaboration. Financial assistance for this work was provided by American Cyanamid through corporate support to the Polymeric Composites Laboratory at the University of Washington. Instrument support and assistance was provided to the Polymeric Composites Laboratory by Du Pont Instruments and Tetrahedron Associates and is gratefully acknowledged.

REFERENCES

1.   L.E. Nielson; Polymer Rheology, Mortel Dekker Inc., New York, (1977)

2.   S. L. Peake, A. Maranci, 32nd Intrnl. SAMPE Symposium Preprints, April 6-7 (1987)

3.   Cessna, Jabloner, J. Elas. Plas., 6, p. 103 (1974)

4.   L.S. Reich, S.H. Patel, Am. Laboratory, p. 23, (1987)

5.   E. Macho, A. Alegria, J. Colmenero, Polym. Eng. and Sci. Vol. 27(11), (1987)

6.   A.K. Doolittle, J. Appli. Phy. 21., p. 1471 (1951)

7.   J.D. Ferry, Viscoelastic Properties of Polymers, 2nd Ed., John Wiley & Sons, New York, (1970)

8.   M. L. Williams, R. F. Landel, J. D. Ferry, J. Amer. Chem. Soc., 77, p. 3701, (1955).

9.   J. C. Seferis, Polym. Comp. (7)3, p. 158, (1986).

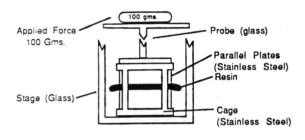

Fig. 1. Squeeze Flow
Experimental Apparatus.

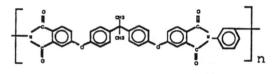

Fig. 2. Typical
Polyetherimide Structure.

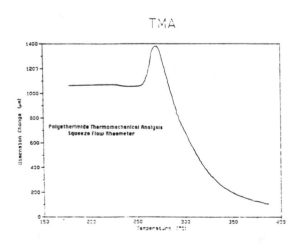

Fig. 3. Thermomechanical
Analysis Scan of
Polyetherimide with Squeeze
Flow Attachment.

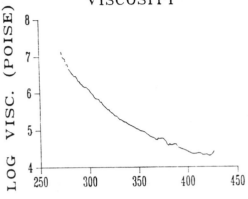

Fig. 4. Log Viscosity /
Temperature Plot of
Polyetherimide.

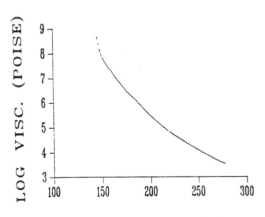

Fig. 5. Log Viscosity /
Temperature Plot of
Polystyrene.

POLYCARBONATE VISCOSITY

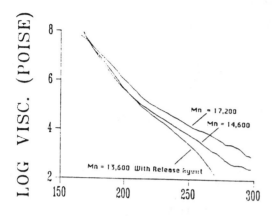

Fig. 6. Log Viscosity /
Temperature Plot of
Polycarbonate.

WLF EQUATION
POLYETHERIMIDE VISCOSITY

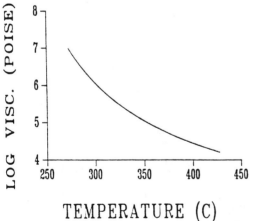

Fig. 7. WLF Equation
Representation of
Polyetherimide Viscosity
Using Best Fit Parameters.

POLYETHERIMIDE VISCOSITY
VISCOSITY — SHEAR RATE

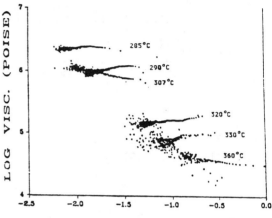

Fig. 8. Log Viscosity / Log
Shear Rate Plot of
Polyetherimide Showing
Newtonian Behavior of
Isothermal Data.

POLYETHERIMIDE VISCOSITY
SHEAR STRESS—SHEAR RATE

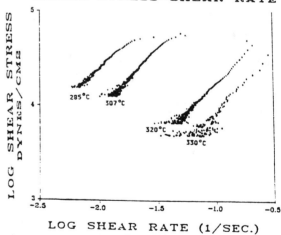

Fig. 9. Log Shear Stress /
Log Shear Rate Plot of
Polyetherimide at Four
Isothermal Temperatures.

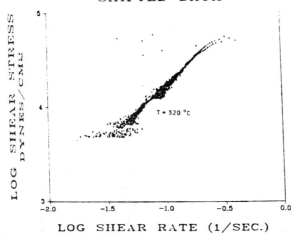

Fig. 10. Shifted Isothermal Shear Stress / Shear Rate Data for Polyetherimide. Shifting Temperature is 320 C.

Fit of Shift Factors

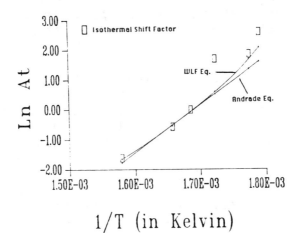

Fig. 11. WLF and Andrade fit of Isothermal Shift Factors (At).

Dicyanate Blends as Matrices for High Performance Composites

E. M. WOO, B. FUKAI AND J. C. SEFERIS[1]

ABSTRACT

Prepregs and composites were prepared from toughened thermoset resins as matrices derived from the cyclotrimerization of aryl dicyanates modified with specific amounts of thermoplastic polysulfone (PS) and/or reactive rubber (CTBN). The thermoplastic phase separated from the thermoset phase during cure as the resin molecular weight increased and more polar cyanate functional groups were transformed to less polar polycyanurate linkages. Rubber component helps to produce coalesced spherical particles. The morphology was found to be affected by compositions, as well as gel temperature. Fracture measurements with the composites were performed using double-edge-notch (DEN) geometry while neat resin toughness was determined with the compact tension (CT) geometry. Translation of matrix resin toughness to composite toughness was found to be strongly influenced by prepregging and composite processing parameters in general. The correlation between processing parameters and composite properties are discussed in this study.

INTRODUCTION

Thermosetting matrices for high performance composites have been shown in recent years to be comparable to thermoplastic ones if properly blended with appropriate rubber and thermoplastics. Recent developments[1-6] primarily focused on epoxy based systems, have concentrated in providing improvements in the fracture toughness of the composites without sacrificing resin dominated property performance. However, the use temperature of these systems is controlled by the maximum temperature of the epoxy employed in the formulation. Although bismeleimides and polyimides are most often utilized for thermosetting resin-based composites when higher temperature performance is

E. M. Woo, B. Fukai, J. C. Seferis, Polymeric Composites Laboratory, Department of Chemical Engineering, University of Washington, Seattle, WA 98195

[1] Author to whom correspondence should be addressed.

demanded, there has traditionally been a gap in use temperature between high temperature epoxies and bismaleimides. Such a gap may be filled with the use of dicyanate based polymers while maintaining the processing ease commonly found with epoxy and bismaleimide based systems. In previous studies we have been investigating processing-structure-property interrelations of epoxy and bismaleimides systems, both unmodified and modified for toughness improvements[7-8]. Furthermore, our blending studies of epoxies with BMI's have demonstrated the possibility of controlling use temperature, as well as utilizing toughness improvement schemes that have proved beneficial with epoxy based systems. Accordingly, we have begun evaluating the processing-structure-property interrelations of dicyanate resins in an effort to understand toughness improvements at different temperatures for matrix systems used in high performance composites. Specifically, this paper focuses on a series of dicyanate resins, including bisphenol-A dicyanate (BADCy), that were modified with the addition of polysulfone (PS) and/or carboxyl terminated butadiene rubber (CTBN). A particulate phase-separated morphology was observed when dicyanate resins were blended with both PS and CTBN additives. In addition, prepregging and lamination of both unmodified and modified formulations further established the importance of matrix morphology in translating toughness improvements from neat resins to composites.

EXPERIMENTAL

Materials

All dicyanate resins, both unmodified and modified with polysulfone, were supplied by Interez, Inc. However, all compositions containing CTBN's were formulated in this laboratory. A liquid reactive rubber, CTBN (B. F. Goodrich Hycar 1300 X8), was used as a morphology modifier.

Bisphenol-A dicyanate (NC-O-Ar-O-CN) is a crystalline material with a melting temperature of 79°C. Crystallinity varies for other dicyanate resins depending on the chemical structure. To reduce the crystallinity, it must be partially trimerized. It cures slowly at high temperatures above 250°C, but generally it is cured at lower temperatures in the presence of a copper acetylacetonate/nonyl phenol solution. Upon curing, the more polar Ar-(O-CN) cyanate structure is transformed through triazine ring formation to a less polar polycyanurate structure without giving out any volatile by-products. The trimerization reaction is given as follows[9-10]:

Also, a formulation containing all or part of a liquid dicyanate (LDICy) resin was supplied by Interez, Inc. The liquid dicyanate resin was used to modify the viscosity of resins.

Prepreg Processing

The dicyanate resins and polysulfone-modified formulations at various thermoplastic contents were processed into continuous carbon fiber prepregs using the scaled down continuous hot-melt prepregger of Lee, Seferis, and Bonner[11]. Sixteen spools of Hercules AS-4 carbon fibers were collimated and impregnated with the resins to form a prepreg approximately 3.5 cm wide. The prepregging parameters, such as temperature and pressure applied, were carefully controlled in accordance with our earlier prepreg modeling study[11]. Viscosity-temperature profiles for the resin formulations were determined as a requirement for setting appropriate impregnation parameters. The impregnation temperatures were between 80-140°C so that the viscosities for the different formulations were approximately 1,000 centi-poise during impregnation.

Cure Cycle

The unidirectional prepregs were cut, laminated, and cured in a Lipton high performance autoclave. The laminates were processed using a commercially available Zip-Vac vacuum bag. Experimental cure cycles were designed according to the available information of curing the neat dicyanate resins[8]. They were first heated to 135°C at a 3 - 5°C/min heating rate and held at that temperature for 40 min before a pressure of 85 psi was applied. Heating was then resumed at the same heating rate to 177°C, at which the sample was cured for 2 hours. The laminate was subsequently post-cured at 200 - 220°C in a hot-air circulation oven for 2 hours. The post-curing step was necessary to develop optimum mechanical properties in the dicyanate resins as has been suggested by Shimp[8].

Morphology

Morphological studies were performed on neat resin and composite samples using a Nikon optical polarized light microscope. Selected samples were also examined with the novel scanning acoustic microscopy technique with an ELSAM microscope from E. Leitz of West Germany (E. Leitz, Inc.).

Mechanical Measurements

Dynamic mechanical modulus and glass transition temperatures (Tg's) were determined by measuring the dynamic mechanical properties at 1 Hz using the DuPont DMA-983 apparatus. The temperatures of maximum change of loss flexural modulus were used as a measure of Tg's of the samples. Fracture toughness measurements were performed on both neat and composite specimens using a double edge notch geometry (DEN)[12] or compact tension geometry (CT). Measurements were performed with an Instron (Model 1035) hydraulic tester interfaced via a Harrier interface to an IBM personal computer for data calculation and analysis.

RESULTS AND DISCUSSION

Neat, unmodified dicyanate resins could be prepregged easily at temperatures 80 - 100°C. In this temperature range, the resins exhibited viscosities under or near 1000 centi-poise. Viscosities above or under this value gave difficulties for hot-melt prepregging. Low viscosities caused excessive resin flow resulting into dry fibers, while too high viscosities resulted in the incomplete wetting of fiber.

For the polysulfone-modified formulations, the resins became not only highly viscous but also rubbery, especially at high polysulfone contents. For example, at a 20% polysulfone content, the BADCy/polysulfone blend at 100°C had a viscosity of 30,000 cps. Its viscosity dropped to 1000 cps only at 160°C, which was close to the curing temperature. It became difficult to process these formulations since hot-melt prepregging at high temperatures where reactions take place is not recommended. Therefore, for the BADCy/PS.25 blend, a small quantity of methylene chloride was incorporated in the formulations to reduce the viscosity and solution elasticity. Most of the solvent in the formulations was driven off during impregnation. Subsequent drying of the resulting prepreg was done at 60°C whenever appropriate.

Table I summarizes the prepregging parameters that were used to produce good quality prepregs from the formulations studied and the resin contents of the resulting prepregs. All prepregs were quite uniform and the fiber was properly wetted through as observed by optical microscopy. Resin contents were measured on some of the prepreg formulations using a methylene chloride solvent extraction technique. The resin contents were found between 32 to 35% by weight (which is equivalent to 68 - 65% fiber content by weight). In terms of volume percentages, the resin contents are equivalent to 40.5 - 45%, which compared favorably with commercial epoxy or bismaleimide prepregs.

TABLE I
Prepreg Processing Parameters and Resin Contents

Formulations Designation	Manufacturer's Designation	PS Content (%)	Impreg. Temp. (C)	Resin Content (% by weight)
BADCy	RDX 80352	0	85-90	32.2
BADCy/PS.10	ESR 315-A	10	115	34.1
BADCy/PS.20	ESR 315	20	NT	NT
BADCy/PS.25	ESR 315-B	25	125	NT
BADCy/LDICy/ PS.20	ESR 321	20	140	32.5
LDICy/PS.20	ESR 322	20	130	NT

NT - not tested

The laminates after the cure cycles and post-cure treatment were examined under the optical light microscope. The samples were cut across the fiber direction and polished gradually to 0.3-micron grid. Figure 1 shows the micrographs of the cross sections of laminates of three different BADCy/polysulfone compositions: 100/0 (BADCy), 90/10 (BADCy/PS.10), and 75/25 (BADCy/PS.25). The fiber distribution in the high-viscosity matrix material, BADCy/ PS.25, is as good as that in the low-viscosity matrix material, BADCy. Good fiber distribution in cured laminates usually reflects good prepregs. The micrographs also confirm impregnation of the fiber was well done with the dicyanate formulations.

Figure 2 shows the dynamic mechanical properties of the cured neat BADCy samples. BADCy exhibits a single Tg of 282°C, while the 10% polysulfone-modified formulation has two Tg's, indicating a two-phase system. The Tg of the dicyanate phase is almost the same as that of the neat BADCy, indicating very little plasticization effect by the polysulfone component. The polysulfone phase exhibits two partially-overlapped, small peaks at 190 and 210°C. This is probably related to incomplete phase-separation of polysulfone from the curing network.

Figure 3 shows the dynamic mechanical properties of the BADCy/PS.25 laminate at the transverse direction. Since prepregging of this formulation was made possible by the aid of methylene chloride solvent, it is interesting to examine the effect of the solvent. The Tg's of the two phases move closer to each other, suggesting that the plasticization effect is greater as a result of solvent processing.

The sizes of the phase domains of the precipitated polysulfone in the cured dicyanate resins are generally small. Although distinct phases can be detected using the DMA method, the phase domains may be too small to be observed using an optical or electron microscope. It has been suggested by Shimp et al[13] that the agglomerates had cross-sectional diameters of about 0.05 microns. However, in order to have a synergistic effect in enhancing the fracture toughness, phase-separation has to produce agglomerates which generally are in the order of microns.

Figure 4 shows the optical micrographs of BADCy/PS.10 and the same resin modified with 15 PHR of CTBN. The cured BADCy/PS.10 exhibits barely-visible tiny agglomerates of irregular geometries. Addition of CTBN to the blend produced a morphology with coalesced distinct particles. Figure 5 shows the acoustic microscopy results for the same resin formulations. Since the acoustic microscopy reflects the mechanical properties directly, the darker regions are the lower modulus (rubber) and the lighter regions are the higher modulus (polysulfone or polycyanurate). The micrograph indicates that the rubbers are the dispersed phase in the continuous matrix but inside the rubber phase domains, there may be discrete, small polysulfone or polycyanurate dispersed particles.

Figure 6 shows the DMA results of the 15 PHR CTBN-modified BADCy/PS.10. At the 15 PHR composition, the

CTBN significantly plasticizes the polycyanurate phase, as indicated by the lower Tg and highly depressed relaxation peak of this phase. However, the CTBN contents can be further optimized to produce desirable agglomerate sizes without sacrificing high temperature properties. By comparison, the CTBN does not plasticize the polysulfone phase. The slightly, more-pronounced relaxation of the polysulfone phase in this sample indicates that precipitation of the polysulfone component becomes more complete as a result of the addition of CTBN to the formulation. The result that CTBN facilitates phase- separation agrees with the microscopic observation.

Finally, Figure 7 compares the fracture toughness in terms of K_{Ic}'s for the three systems: BADCy, BADCy/PS.10, and BADCy/LDICy/PS.20 in their resin and laminate forms. All laminates had the same resin content of 30% by weight. While the fracture toughness of the neat resins increases in the order of BADCy/LDICy/PS.20> BADCy/PS.10> BADCy, the K_{Ic} of the composites decreases in that order. Possibly, this behavior is the result of the composite samples not being cured in an optimal manner providing quite distinct morphologies from their corresponding neat resin values. Further study is under way to establish morphological features of matrices for toughness improvement in the composites.

CONCLUSION

This study has demonstrated hot-melt prepregging as a viable processing technique for dicyanate resins including high-viscosity polysulfone-modified tough formulations. Although phase-separation of polysulfone from the thermosetting resin during cure does not always result in well-defined agglomerates of optimal sizes, the morphology can be modified and controlled by adding a rubber component. However, neat resin fracture toughness improvements may not be directly translated to composite fracture toughness. This may be a result of further modification of the resin morphology resulting from the incorporation of carbon fiber. Thus, this study has demonstrated that for direct translation of toughness improvements from neat resin to composite, the morphological features of a matrix system must be identified and controlled through processing for optimum composite performance.

ACKNOWLEDGMENTS

The authors would like to acknowledge D. Shimp of Interez Inc. for introducing them to this class of resins and for his guidance and input in selecting an appropriate dicyanate system for study and evaluation; to M. Potter formally of Interez for catalyzing this study; and to Dr. R. Graver of Interez for support and expert coordination of activities. Financial assistance for this work was provided in part by National Science Foundation Presidential

Young Investigators award to Prof. J. C. Seferis and matching fund support to the Polymeric Composites Laboratory Consortium provided by the following industrial sponsors: Interez Inc., Boeing Commercial Airplane Co., Lockheed Space & Missiles Co. and Rhone Poulenc, Inc. The instrumental and material support provided to the Polymeric Composites Laboratory at the University of Washington by Zip Vac Co. and the Acoustic Microscope by E. Leitz Co., extensively utilized in this study, were greatly appreciated.

REFERENCES

1.   Boll, D. J., W. D. Bascom, J. C. Weidner and W. J. Murri, J. Mater. Sci., 2667, (1986).

2.   Woo, E. M., J. C. Seferis and L. D. Bravenec, Proc. of Am. Soc. Comp., 569 (1987).

3.   Bauer, R. S., Proc. of 18th Int. SAMPE Tech. Conf., 510, (1987).

4.   Diamant, J. and R. J. Moulton, Proc. of 29th Natl SAMPE Symp., 422, (1984).

5.   Bucknall, C. B. and I. K. Partridge, Polym. Eng. Sci., 26, 54, (1986).

6.   Jabloner, H., B. J. Swetlin, and S. G. Chu, U. S. Patent, 4,656,207, assigned to Hercules Inc., (1987).

7.   Viot, J. F. and J. C. Seferis, J. Appl. Polym. Sci., 34, 1459, (1987).

8.   Woo, E. M., L. B. Chen and J. C. Seferis, "Characterization of Epoxy-Bismaleimide Network Matrices," J. Mater. Sci., in press (1987).

9.   Shimp, D. A., SAMPE Qtrly, 19(1), 41, (1987).

10. Sultan, J. N., R. C. Laible, and F. J. McGarry, J. Appl. Polym. Sci., 6, 127, (1971).

11. Lee, W. J., J. C. Seferis, and D. C. Bonner, SAMPE Qtrly, 17(2), 58, (1986).

12. Coxon, B, T. Walker, L. Ilcewicz, and J. C. Seferis, Proc. of Soc. Exper. Mech. (SEM), 144, (1987).

13. Shimp, D. A., F. A. Hudock and W. S. Bobo, Proc. of 18th Intnatl SAMPE Tech. Conf., 851, (1987).

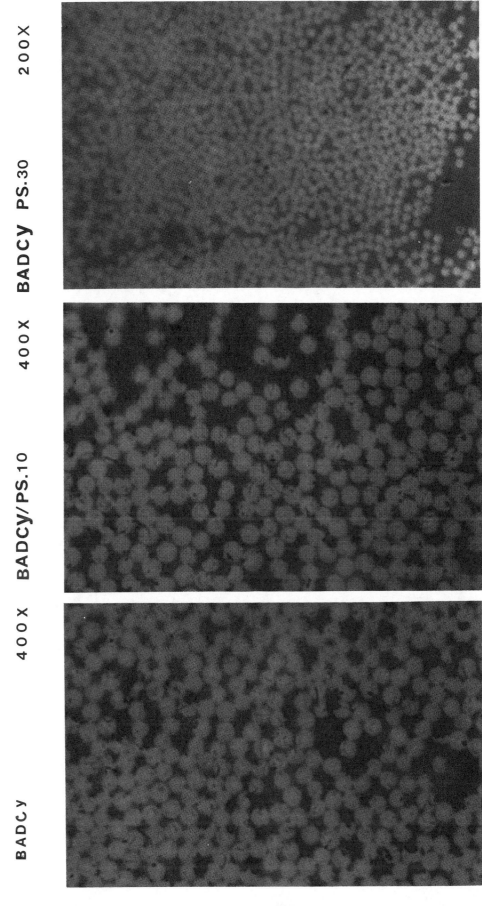

Figure 1: Optical Micrographs showing the cross sections of laminates.

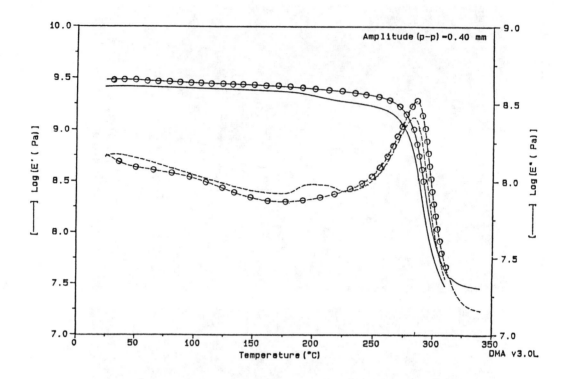

Figure 2: Dynamic mechanical properties of BADCy (-o-) and BADCy/PS.10 (---).

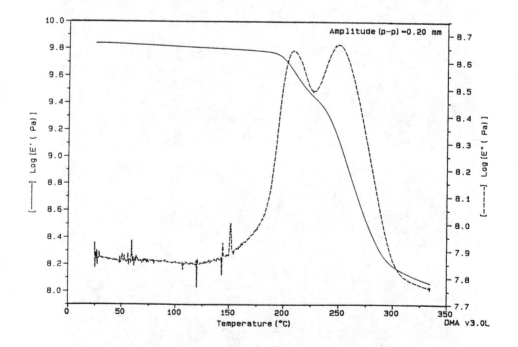

Figure 3: Dynamic mechanical properties of BADCy/PS.25.

BADCy/PS.10   400X

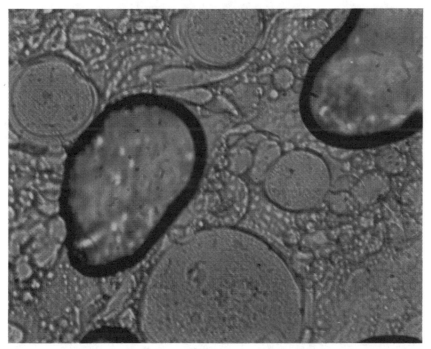

OPTICAL   400X

BADCy/PS.10/15 PHR CTBN x8

Figure 4: Optical micrographs showing the morphology of
            BADCy/PS.10 (top);
            and BADCy/PS.10 / 15phr CTBN x 8 (Bottom).

BADCy/PS 10

BADCY/PS.10/15 PHR CTBNx8

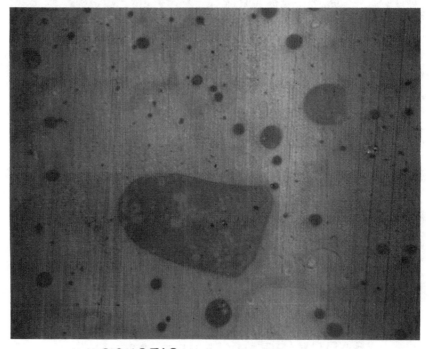

ACOUSTIC   400X

Figure 5: Acoustic micrographs of BADCy/PS.10 (top);
and BADCy/PS.10 / 15phr CTBN x 8 (Bottom).

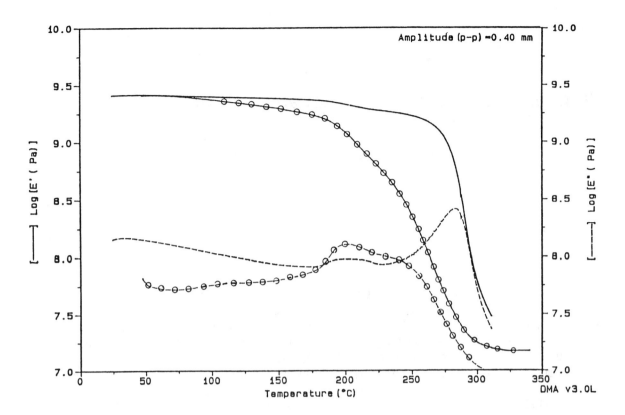

Figure 6: Dynamic mechanical properties of
BADCy/PS.10 (——); and
BADCy/PS.10 / 15phr CTBN x 8 (-o-).

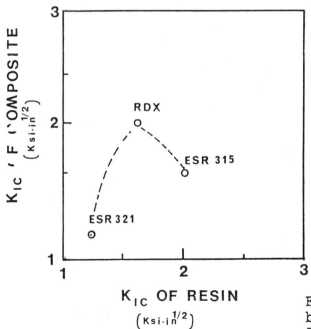

Figure 7: K_{IC} of dycyanate blend neat resins vs laminates.

Composite Materials in Orthopaedic Applications

RUSSELL D. JAMISON

ABSTRACT

Basic issues related to design of orthopaedic implants are explained and the physical environment and anatomical constraints to which these devices must be accommodated are described. Using the design, fabrication, and testing of a prototype composite material hip endoprosthesis as an example, the important factors which must be considered in using composite materials for such an application are identified and discussed.

INTRODUCTION

The use of metallic endoprosthetic devices in total hip and knee replacement surgeries has become an accepted and successful procedure for relief of pain and restoration of mobility in patients suffering from degenerative bone diseases such as osteoarthritis. Designs for these devices have become increasingly refined as biomedical research and clinical experience have advanced our understanding of the response of bone to implant under the complex anatomical loading which both must share. The issue of load-sharing between bone and implanted device is in fact central to long-term viability of the implant. The tendency of bone to resorb when it is stress-shielded by a rigid implant device is a major cause of failure of these procedures, requiring revision surgery in some cases.

The engineering challenge is to design devices which have elastic moduli closer to that of bone so that proper load-sharing can be achieved. Existing metallic materials suitable for implantation such as cobalt-chrome steel and titanium alloys with elastic moduli ranging from 100 GPa to 250 GPa are substantially different from cortical bone which has an elastic modulus of approximately 20 GPa. Composite materials on the other hand offer the prospect of a widened range of achievable moduli, approaching that of bone on the lower end without unacceptable compromise of strength or biocompatibility.

R. D. Jamison, Composites Research Manager, Richards Medical Company, 1450 Brooks Road, Memphis, TN 38116

BACKGROUND

Bone resorption is particularly problematic in reconstructive procedures of the hip or knee in which the stability of the implant device depends in large part upon the quality and vitality of bone remodelling around the implant postoperatively. If the implanted device carries too much of the joint load, the adjacent unloaded bone will resorb over time resulting in loosening of the implant [1,2]. Loosening of the implant is associated with ambulatory pain, and in isolated cases mechanical failure of the implant itself. In cases where an implant has been cemented into the bone to provide initial stability, resorption of the bone can lead to fragmentation of the cement mantle (commonly polymethylmethacrylate) and thereby cause pain and adverse biological response. In cases of implant loosening, revision surgery may be required. The incidence of implant loosening for total hip replacements has been estimated to be 20 percent after five years and 30 percent after ten years [3].

Recognizing that more than 200,000 total hip and knee replacement procedures are performed each year with that number increasing as the society ages, this matter is viewed with considerable interest and concern within the orthopaedic community and has stimulated the search for a for a more "bone-like" implant material. The bone itself is both anisotropic and inhomogeneous as reflected by its properties (Table 1).

TABLE 1.   Human Femur Mechanical Properties [4,5]

Elastic Modulus (GPa)	7 - 30
Tensile Strength (MPa)	90 - 122
Strain-to-Failure (%)	0.5 - 3.0
G_c (kJ/m^2)	0.6 - 5.0

TABLE 2.   Properties of Conventional Metallic Materials and Composites for Orthopaedic Applications [6]

PROPERTY	MATERIAL			
	Co-Cr-Mo	Stainless	Titanium	Carbon/Polysulfone (Quasisotropic)
Young's Modulus (GPa)	210	200	120	58
Shear Modulus (GPa)	82	75	44	24
Poisson's Ratio	0.30	0.33	0.36	0.30
Tensile Strength (MPa)	700	600	930	590

Bone is in fact composed of hydroxyapatite ceramic reinforced collagen organized into oriented osteons. It is perhaps not surprising then that composite materials have been viewed for a number of years as attractive candidates in orthopaedic applications. Compared to existing materials now being used (Table 2) they alone offer the prospect of lower modulus with adequate strength.

Beginning in the early 1970's work with carbon reinforced composites for bone plates and hip components [5,7-17] has been reported. Some of the early work was limited by the materials available at the time. Results of implant studies [5,11,14,18-21], principally in animals, have yielded generally encouraging results regarding potential biomechanical compatibility of composite implants. However, the range of issues which must be addressed and resolved before clinical implantation into humans can be considered is formidable. If one considers any one application, the hip prosthesis for example, the problem can be posed as one of constraints and requirements. Constraints on the geometry and requirements on the material define the necessary elements of a development program for composite materials in such an application.

DISCUSSION

A typical total hip replacement is shown in Figure 1. Whether the stem and acetabular cup are made of metal or composite, they must fit within the respective bone cavities. Moreover, because there is variability in human anatomy, there must be a range of sizes available. And although there is a general correlation between femur medullary canal dimensions and body weight, there are circumstances in which large patients require small stems. Hence strength is a primary requirement.

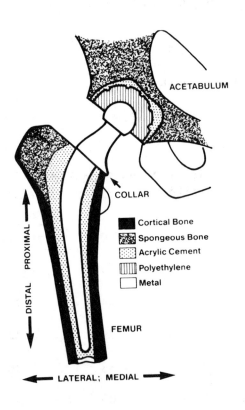

FIGURE 1.   Total hip replacement [16].

Determining the actual joint loads for normal physical activities like walking or ascending/descending stairs remains a topic of discussion and research [16,22,23]. One representation of hip loads due to normal walking is shown in Figure 2. Load spectra such as this one are typically computed from gait or force plate measurements, although efforts to measure and transmit loads <u>in vivo</u> by data telemetry have been reported recently [24]. It has been estimated that one year in the life of an active patient will be 2 x 10^6 cycles of such a spectrum [8].

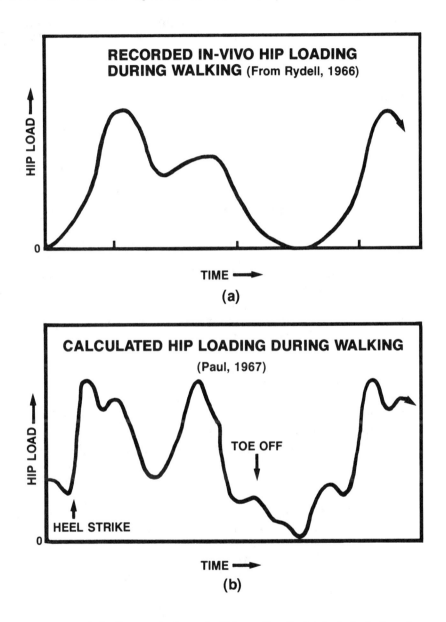

FIGURE 2. (a) Recorded and (b) calculated joint loads during walking [23,36].

The components of the compressive joint force applied at the spherical head of the prosthesis produce a combination of bending/compression (the medial-lateral component) and bending/torsion (the anterior-posterior component). The relative magnitudes of these components have been estimated to be approximately 5 to 1, the M-L component larger and acting approximately in-phase for most activities [22]. Table 3 shows the range of joint forces which are believed to occur for typical physical activities. The direction of the force components varies during any activity. For level walking the M-L component is approximately 20° from the stem axis and the A-P component is less than 20° from the stem axis.

TABLE 3.   Peak Joint Force Resultants for
                    Typical Physical Activities [16]

Activity	Peak Joint Force Resultant (Force - Body Weight)
Level Walking	4.9 - 7.6
Ascending/Descending Stairs	7.2/7.1
Ascending/Descending Ramp	5.9/5.1

The joint forces are not the only forces acting on the stem. The muscle groups attached to the femur and pelvis also produce time-varying forces during activity. The magnitude of these muscle forces has been the subject of study [25] but to the author's knowledge a complete treatment does not yet exist.

In addition to the anatomical constraints imposed, there is also the issue of biocompatibility. Current Co-Cr-Mo and titanium alloy metal implants are well-tolerated by the body. The biocompatibility of carbon makes carbon fibers a logical and popular candidate for composite implants. Carbon fiber has already seen limited commercial application in artificial ligaments and as a reinforcement for polyethylene. Polyethylene, which has demonstrated biocompatibility but poor interfacial bond strength, is mechanically unsuitable for the hip stem application. In fact, the number of polymers which meet the mechanical criteria and are known to be biocompatible is small. Obvious choices like the carbon/epoxies may not have the environmental stability required [26] and may produce a toxic response if unreacted monomer is present [27,28]. Carbon/carbon composites appear to be nonreactive [29] and hip stem prototypes have been produced by carbon vapor infiltration of a braided preform [25]. However, cost and fatigue strength are concerns. For some choices like polysulfone, a thermoplastic, there are prior biomedical applications and biocompatibility evidence [18]. For other choices which may ultimately represent a superior matrix material, biocompatibility must be demonstrated. In the United States, evidence of biocompatibility of a polymer must be accepted by the Food and Drug Administration before that polymer can be used clinically. This evidence is provided by the results of controlled studies in laboratory animals. Among the tests which are normally performed on a candidate polymer are those which determine its cell toxicity, mutagenicity, blood compatibility, and carcinogenicity as well as tissue response to wear particulate.

Individual tests can last from one day to two years and represents a significant investment of time and money. Biocompatibility testing of the polymer and carbon fiber separately does not necessary establish the biocompatibility of the composite since other constituents (fiber sizing, mold release, machining coolant, solvents) may be introduced in processing. A prudent course is to establish biocompatibility of the product as fabricated although fixing the fabrication technique for the several years required for testing may not always be possible.

DESIGN AND FABRICATION REQUIREMENTS

The mechanics problem presented by a hip stem prosthesis is complex in terms of geometry, boundary conditions, loading, and environment. Confident predictions of postoperative mechanical behavior require refined finite element analysis and extensive in-vitro testing. When the material itself has a complex structure like that of composites, the task is particularly challenging. Regardless of the approach to fabrication, the goal is to achieve lower modulus* without unacceptable compromise of strength. This can be achieved by varying the available parameters of constituent properties, fiber fraction, and fiber orientation. Given the range of available choices, the use of an optimization scheme is an obvious and necessary expedient.

Once an optimized design has been developed, the fabrication challenge presents itself. The composite stem is a thick part of varying cross-section. The fabrication of such a part taxes the current technology. There are a number of approaches to be taken including braiding, filament winding, machining from a block, net shape molding, and pultruding using combinations of woven fabric, carbon felt, random fiber compound, and unidirectional tape. Regardless of the approach the challenge is to produce stems of assured high quality, to very close tolerances, in a range of sizes, in quantity, at a cost which is not prohibitive.

The requirement of assured quality includes not only material integrity but also material purity. Inasmuch as the composite hip stem may remain in the body for many years, contaminants in even trace amounts may have a deleterious effect. Consequently, opportunity for contamination must be eliminated at every step beginning with the fabrication of the fiber and polymer. Among the issues which must be addressed in specifying "medical grade" composite prepreg are fiber sizing, polymer purity and assay, residual solvents, contamination during manufacture or handling, even the powder commonly found on rubber gloves. A similar set of issues is associated with fabrication of the stem itself as well as any post-fabrication machining or handling.

All implant devices are of course sterilized before implantation using one of three principal methods: gamma irradiation, ethylene oxide gas, or steam autoclave. While it has been established that none of these methods has a deleterious effect on metal implants, the same cannot be said for composites. Although there has been some study of the effect of radiation on epoxy matrix composites in space applications [38] this is not yet a settled issue for other methods and other composites. Steam autoclave sterilization which may occur multiple times for a given prosthesis before implantation, may be especially problematic for polymers susceptible to moisture uptake.

It may be necessary to encapsulate the composite stem with a continuous polymer coating. Such a coating would serve two purposes. One purpose would be to protect the body from the stem, particularly carbon fiber fragments which may be present on the surface. Although carbon is biocompatible, filamentary fragments may nevertheless stimulate the body's foreign body response and produce an adverse effect.

---

*More precisely, the issue is one of section modulus, the product of elastic modulus and second moment of inertia of the stem [7,30]. In the present discussion a solid, fully canal-filling stem is assumed and the only variable is elastic modulus.

The second purpose of a coating would be to protect the stem from the body environment. The liquid medium in which the composite prosthesis must function, the synovial fluid, is a mixture of fatty acids, cholesterol, phospolipids, oxidants, and glucose [31]. The long-term effect of these constituents on candidate materials is not well-characterized and providing a barrier for the stem material might increase longevity.

Polymer coatings can be applied in a number of ways including thermal deposition, solvent casting, and secondary molding. With each of these methods there are concerns regarding the affect of the coating application method on the substrate, the strength and integrity of the coating-to-substrate interface, and the wear behavior of the coating against bone in press-fit applications.

The issue of wear is one not only of material loss rates but also wear particle morphometry and morphology. Experience with wear debris from ultra-high-molecular weight polyethylene articulating surfaces in the acetabular cup (Figure 1) has shown that even a biocompatible material such as this can produce bone lysis [1] and soft tissue reaction [32] for critical wear particle characteristics. The effect of wear particles in joints can be studied by injecting laboratory-generated debris from pin-on-disk wear simulation into the joints of mice or rabbits [33,37].

If the stem is not coated, the anisotropic wear behavior exhibited by composite laminates must be considered [34]. In wear test of Kevlar/epoxy against silicon carbide abrasive paper for example, wear rates have been found to vary by almost one order of magnitude for the same conditions depending upon the orientation of the fibers relative to the surface and wear direction [35]. To further complicate this issue for the case of a composite hip stem, much of the wear debris is retained between the bone and stem creating a three-body wear problem.

The surgical procedure for uncemented total hip replacement requires that the prosthesis be impacted into the medullary canal of the femur to establish a tight press-fit and proper alignment. The magnitude of this impact loading can be substantial and is a potential source of damage initiation in a composite stem, in its coating, or both. Although this may be the only impact to which the stem is ever exposed, the prospect of diminished fatigue lifetime can not be discounted. It is necessary therefore to characterize this intraoperative impact and incorporate it as a preconditioning step into a fatigue testing protocol.

CLOSURE

It is in the formulation and execution of in vitro testing of composite hip stem prototypes that the greatest care is required. The conjunction of so many potential effects in such a critical application requires that testing be comprehensive and exhaustive. The prospect for composite applications in orthopaedic implants is bright despite the magnitude of the undertaking and the strain placed upon existing technology. It is a unique application with great potential for benefiting mankind. For that reason, the need for a careful, thorough, thoughtful approach cannot be overstated for those who undertake this challenge.

REFERENCES

1. Harris, W. H. et. al., "Extensive Localized Bone Resorption in the Femur Following Total Hip Replacement," J. Bone and Joint Surgery, Vol. 58A, 1976, pp. 612.

2. Djerf, K. and Gillquist, "Calcar Unloading After Hip Replacement: A Cadaver Model of Femoral Stem Designs," Acta Orthopaedia Scandinavia, Vol. 58, 1987, pp. 97-103.

3. Engh, C. A. et. al., "Symposium: Methods of Femoral Implant Fixation," Contemporary Orthopaedics, Vol. 15, No. 4, October, 1987, pp. 63-94.

4. Kennaway, A., "Medical Uses of Polymers," Plastics and Rubber International, June, 1986, pp. 16-19.

5. Musikant, S., "Quartz and Graphite Filament Reinforced Polymer Composites for Orthopaedic Surgical Applications," J. Biomedical Material Research Symposium, Vol. 1, 1971, pp. 225-235.

6. Dingman, Carol A., "An Analytical Investigation Demonstrating Improved Load Transfer to Bone Using Composite Hip Components," Master's Thesis, Ohio State University, 1987.

7. Hastings, G. W., "Carbon Fiber Composites for Orthopaedic Implants," Composites, July, 1978, pp. 193-197.

8. Bradley, J. S., et. al., "Carbon Fiber Reinforced Epoxy as a High Strength, Low Modulus Material for Internal Fixation Plates," Biomaterials, January, 1980, pp. 38-40.

9. Hunt, M. S., "Fiber Reinforced Composites for Orthopaedic Surgical Implants," The South African Mechanical Engineer, January, 1981, pp. 1-5.

10. Bradley, J. S. and Hastings, G. W., "Carbon Fibre-Reinforced Plastics for Orthopaedic Implants, Mechanical Properties of Biomaterials, Edited by G. W. Hastings and D. F. Williams, John Wiley & Sons Ltd., London, 1980, pp. 379-386.

11. Tayton, Keith, "The Use of Semi-Rigid Carbon Fiber Reinforced Plastic Plates for Fixation of Human Fractures," J. Bone and Joint Surgery, Vol. 64-B, No. 1, 1982, pp. 105-111.

12. Cook, William E. et. al., "Ceramic Metal Composites for Surgical Implants," J. Biomedical Materials Research Symposium, No. 2 (Part 2), pp. 443-466.

13. Fitzer, E., et. al., "Torsional Strength of Carbon Fiber Reinforced Composites for the Application as Internal Bone Plates," Carbon, Vol. 18, 1980, pp. 383-387.

14. Mendes, D. G., et. al., "A Composite Hip Implant," Orthopaedic Review, Vol. XVII, No. 4, April, 1988, pp. 402-407.

15. McKenna, Gregory B., et. al., "The Development of Composite Materials for Orthopaedic Implant Devices," Proceedings of 21st SAMPE Symposium and Exhibition, Los Angeles, April, 1976.

16. Davidson, J. A., "The Challenge and Opportunity for Composites in Structural Orthopaedic Applications," Journal of Composites Technology and Research, Winter, 1987, pp. 151-161.

17.   Soltesz, U. and Richter, H., "Investigation of Mechanical Behavior of Fiber-Reinforced Materials of Endoprosthetic Devices," Biomaterials 1980, Edited by G. D. Winter, John Wiley and Sons Ltd., London, 1982, pp. 33-38.

18.   Behling, C. A. and Spector, M., "Quantitative Characterization of Cells at the Interface of Long-Term Implants of Selected Polymers," J. Biomedical Materials Research, V. 20, 1986, pp. 653.

19.   Christel, P. et. al, "Development of a Carbon-Carbon Hip Prosthesis," J. Biomedical Materials Research: Applied Biomaterials, Vol. 21, No. A2, 1987, pp. 191-218.

20.   Hunt, M. S., "Development of Carbon Fibre/Polysulfone Orthopaedic Implants," Materials and Design, Vol. 8, No. 2, March/April 1987, pp. 113-119.

21.   Skirving, A. P., et. al., "Carbon Fiber Reinforced Plastic (CFRP) Plates in the Treatment of Fractures of the Tibiae in Dogs," Clinical Orthopaedics and Related Research, No. 224, November, 1987, pp. 117-124.

22.   Davidson, J. A., et. al., Biaxial Fatigue of Cold-Worked 316L Stainless Steel," Proceedings of the Sixth Southern Biomedical Engineering Conference, Dallas, October, 1987.

23.   Paul, J. P., "Forces Transmitted by Joints in the Human Body," Proceedings of the Institute of Mechanical Engineers, Part 3J., Vol. 181, 1966, pp. 8-15.

24.   Bergmann, et. al., "Multichannel Strain Gage Telemetry for Orthopaedic Implants," J. Biomechanics, Vol. 21, No. 2, 1988, pp. 169-176.

25.   Rohlman, A. et. al., "Finite Element Analysis and Experimental Investigation in a Femur with Hip Prosthesis," J. Biomechanics, Vol. 16, No. 9, 1983, pp. 727-742.

26.   Katz, J. L., "Present and Potential Biomedical Applications of Composite Materials Technology," Contemporary Biomaterials, Edited by John W. Borelos and Murray Eden, Noyes Publications, Park Ridge, NJ, 1984, pp. 453-476.

27.   Thorgeirsson, A. and Fregert, S., "Allergenicity of Epoxy Resins in the Guinea Pig," Acta Dermatovener, Vol. 57, 1987, pp. 253-256.

28.   Thorgeirsson, A. et. al., "Sensitization Capacity of Epoxy Resin Oligomers in the Guinea Pig," Acta Dermatovenor, Vol. 58, 1978, pp. 17-21.

29.   More, N., et. al., "Biocompatibility Studies of [14]C Labelled Carbon-Carbon Composites, Biological and Biomechanical Performance of Biomaterials, Edited by P. Christel, Elsevier, Amsterdam, 1986, pp. 343-348.

30.   Espiritu, E.T., et. al., "A Method for Quantifying Stiffness Parameters of Titanium Alloy Femoral Stems and Cobalt-Chrome Alloy Prostheses," Titanium Alloys in Surgical Implants, ASTM STP 796, Edited by H. A. Luckey and Fred Kubli, Jr., ASTM, Philadelphia, 1983, pp. 74-87.

31. Davidson, J. A. and Georgette, F. S., "State-of-the-Art Materials for Orthopaedic Prosthetic Devices," Proceedings of Manufacturing Engineers: Implant Manufacturing and Material Technology, Itasca, IL, December, 1986.

32. Skinner, H. B. and Makey, M. F., "Soft-Tissue Response to Total Hip Surface Replacement," J. Biomedical Materials Research, Vol. 21, 1987, pp. 589.

33. Rushton, Neil and Rae, Trevor, "The Intra-Articular Particulate Carbon Fiber Reinforced High Density Polyethylene and Its Constituents: An Experimental Study in Mice," Biomaterials, Vol. 5, November, 1984, pp. 352-356.

34. McGee, A. C. et. al., "Abrasive Wear of Graphite Fiber Reinforced Polymer Composite Materials," Wear, Vol. 114, 1987, pp. 97.

35. Cirino, M. et. al., "The Abrasion Wear Behavior of Continuous Fiber Polymer Composites," J. Materials Science, 33, 1987, pp. 2481-2492.

36. Rydell, N., "Forces in the Hip Joint II, Intravital Studies," Biomechanics and Related Bioengineering Topics, Pergamon, Oxford, 1966.

37. Neugebauer, R., et. al., "The Body Reaction to Carbon Fiber Particles Implanted into the Medullary Space of Rabbits," Biomaterials, Vol. 2, July 1981, pp. 182-184.

38. Memory, J. D. et. al., "Radiation Effects on Graphite Fiber Reinforced Composites," J. Reinforced Plastics and Composites, Vol. 7, January, 1988, pp. 33-65.

SYMPOSIUM VI

Interfaces in Composites

The Interdependence Between Fiber-Matrix Adhesion and Composite Properties

LAWRENCE T. DRZAL, MICHAEL J. RICH AND SHEKHAR SUBRAMONEY

ABSTRACT

Increasing attention has been given to adhesion between fiber and matrix as being not only a necessary condition for achieving acceptable composite mechanical properties, but also a variable by which composite properties may be strongly affected. As the variety of matrices and reinforcements increases the engineer's choices for composite constituent selection, an "a priori" understanding of the role of fiber-matrix adhesion is necessary for rational composite design. While an analytical model of the fiber-matrix interphase and its role on composite properties is not available, significant qualitative relationships have been established.

This paper will present results from interfacial studies with carbon fibers in thermoset matrices. These studies have quantified the level of adhesion attainable between carbon fibers and epoxy matrices using single fiber testing techniques. The effect of fiber surface treatment and the application of a fiber size or finish have been shown to make significant changes in not only the level of interfacial shear strength but also in the failure mode. Composite specimens fabricated from the same materials and processed in the same manner have been evaluated for mechanical properties. Results from these tests show that composite properties are altered by a factor of two with changes in fiber-matrix adhesion. Both the level of adhesion and the failure mode are important considerations in predicting composite behavior.

BACKGROUND

A series of studies have been conducted by the author directed at developing a molecular level understanding of the physics and chemistry of adhesion between carbon fibers and epoxy matrices. As part of this effort, commercial fiber surface treatments (oxidation) and the application of fiber sizings (finishes) were investigated as to their effect of fiber-matrix adhesion.

In the first study [1] intermediate modulus PAN based carbon fibers designated A-1 and supplied by Hercules, Inc. were characterized as to

L.T. Drzal, Professor of Chemical Engineering, M.J. Rich, Research Specialist, Composite Materials and Structures Center, Michigan State University, East Lansing, MI, 48824-1326, S. Subramoney, Research Associate, Dept. of Materials Science and Engineering, University of Arizona, Tucson, AZ 85721

their surface area, surface chemical and surface energetic changes that
occurred with surface treatment (electrolytic oxidation). This study
showed that the surface area was unaffected by surface treatment, but that
the surface chemistry and surface energetics were changed. The surface
oxygen content increased by a factor of two and the surface free energy
increased. Most of the change was in the polar component of the surface
free energy. An embedded single fiber test was used to measure the change
in the interfacial shear strength with each of these fibers. A
corresponding factor of two increase in the interfacial shear strength was
measured. Subsequent sectioning of the samples and electron microscopic
observation of the interface documented two distinct failure modes
attributable to the surface conditions. For the untreated fiber (AU)
failure occurred within the outer few nanometers of the fiber surface along
the interface. After surface treatment however, the surface treated fiber
(AS) attains a factor of two increase in the interfacial shear strength
with an interfacial crack growth failure mode with the locus following the
fiber-matrix interface. The surface treatment had removed the incoherent
fiber surface layer and left behind a structurally sound surface able to
withstand higher shear loadings without failure.

TABLE 1 - EFFECT OF SURFACE CONDITION ON INTERFACIAL SHEAR STRENGTH AND FAILURE MODE

FIBER	SURFACE TREATMENT	INTERF. SHEAR STRENGTH	FAILURE MODE
AU	Untreated	37.8 MPa (5.4 ksi)	Interfacial
AS	Surface Treated	62.3 MPa (8.9 ksi)	Interfacial
ASC	Surface Treated & Sized	82.8 MPa (11.8 ksi)	Matrix

In the second study [2] the effect of a fiber finish (size) on
adhesion was investigated. A thin 100 nanometer layer of epoxy without any
curing agent was deposited on surface treated fibers as a model sizing
agent. After curing two significant changes were noted. The level of
interfacial shear strength as measured by the single fiber interfacial
shear strength test was increased by thirty percent and the failure mode in
the specimen changed from interfacial to matrix fracture perpendicular to
the fiber axis at the point of fiber breakage. (Table 1)
Additional experimentation showed that the finish layer had created an
amine deficient layer around the fiber. This layer had properties that
included a higher tensile modulus than the bulk stoichiometric resin and a
lower fracture toughness and tensile strength. Overall this finish layer
acted as a brittle interphase increasing the level of interfacial shear
strength because of the better stress transfer properties, but at the same
time causing matrix failure in the specimen at the points where the fiber
breaks.
Although the relationship between interphase structure and adhesion
between fiber and matrix using single fiber techniques has been shown, the
question remained as to the suitability of extrapolating single fiber
results to composite specimens. This study was undertaken to investigate
the relationship between single fiber results and composite performance.

EXPERIMENTAL

Identical material used in the single fiber studies was combined to make approximately sixty volume percent composites. The fibers were A-4 fibers from Hercules in three different surface conditions. AU-4 was an untreated fiber. The AS-4 fiber was the AU-4 fiber that had been surface treated with Hercules' oxidative surface treatment. The AS-4C fiber was surface treated and then sized with diglydiyl ether of Bisphenol-A epoxy resin from solution to a thickness of 100 nanometers. All fibers regardless of surface condition have the same mechanical properties.

The matrix used was the same diglycidyl ether of Bisphenol-A prepared with the stoichiometric amount of meta-phenylenediamine as curing agent. The composites were prepared with the aid of a Research Tool hot-melt prepregging unit. Fiber tows were impregnated with the epoxy mixture at 52°C and wound on a drum to prepare the prepreg tape. Conditions were fixed so that all composites produced a nominal 60 fiber volume percent composite. Unidirectional plates were laid up, vacuum bagged and cured in a laminating press at 75°C for two hours followed by a 125°C post cure. The plates were slowly cooled to room temperature. Sample coupons were cut from the panel according to standard procedures and tested following ASTM protocols.

RESULTS and DISCUSSION

Table 2 lists the values for the interfacial shear strength measured by the single fiber test for each of the three fibers in this epoxy matrix as well as the values obtained from experimental measurements on composites made with these same three fibers and the same epoxy matrix. Average values for all tests are reported.

TABLE 2 - COMPOSITE MECHANICAL PROPERTY RESULTS

FIBER	AU-4	AS-4	AS-4C
INTERFACIAL SHEAR STRENGTH	37.8 MPa	62.3 MPa	82.6 MPa
3 PT. SHEAR STRENGTH	47.6 MPa	79.1 MPa	72.8 MPa
4 PT. SHEAR STRENGTH	33.6 MPa	67.2 MPa	59.5 MPa
90° TENSILE STRENGTH	22.4 MPa	31.5 MPa	21.0 MPa
0° TENSILE STRENGTH EFFICIENCY	1.38 GPa 59%	1.37 GPa 65%	1.26 GPa 57%

Since the single fiber research efforts centered around using a measurement of the interfacial shear strength as a mechanical property measurement characteristic of the fiber-matrix adhesion, interface sensitive composite tests should likewise be most sensitive to variations in fiber-matrix adhesion.

Three composite tests which primarily stress the fiber-matrix interface are the three point shear test (also known as the short beam shear test), the four point shear test and the 90 degree tensile test. Both the three and four point shear tests provide a measure of interlaminar

shear properties.   Shortcomings with the three point shear test or short
beam shear test have been pointed out by Browning at el. [3].   They propose
the four point shear test as the proper test methodology.   Both tests were
carried out.

Shear Strength.   As the data in Table 2 shows, increases are noted in
the composite shear strength with both surface treatment as well as with
the application of a surface finish compared to that obtained with the
untreated fibers.   The percentage increases are slightly higher for the
four-point shear test (99% for AS-4 and 76% for AS-4C) than for the three-
point test (66% for AS-4 and 52% for AS-4C) in all cases.   The percentage
increases compare well with the increase measured in the single fiber test
for the surface treated fiber over the untreated fiber (65%).   The
correspondence between single fiber and composite tests results ends when
data for the AS-4C fiber is compared.   The single fiber tests predict an
additional 32% increase with the application of the surface finish but the
composite test results indicate a slight decrease in both three and four-
point shear values for the finished fiber.

Observation of the composite failure mode and recollection of the
failure locus associated with the AS-4C fiber provides an explanation for
this behavior.   While the AU-4 and AS-4 composites fail in shear, the AS-4C
composites fail in a combined tension/shear mode.   Figure 1 contains
photomicrographs of the tensile face of the three composite under load in
the four point shear test near the loading points.   The AU-4 and AS-4
composites show occasional fiber fractures in the tensile skin of the
composite under load but the AS-4C composite specimen shows a very large
concentration of fiber fractures and coalescence of these cracks.   The
presence of the brittle interphase around each AS-4C fiber promotes matrix
cracking perpendicular to the fiber which results in matrix dominated
failure of the specimen.   No increase in shear properties for the AS-4C
coated fibers is detected over the other fibers because of this change in
failure mode.

90 Degree Tensile Strength.   Transverse strength depends on the
response to stress in a direction normal to the fiber-matrix interface.
The measurement of the 90 degree tensile strength also shows significant
changes with fiber surface treatment.   There is a 41% increase in
transverse strength in the AS-4 composite over the untreated AU-4 sample.
The AS-4C composite where the fibers are coated with the brittle interphase
shows no change in transverse strength compared to the AU-4 fiber and
exhibits a large drop in transverse strength compared to the AS-4 surface
treated fiber.   Yet the AS-4C fiber is also surface treated but in addition
contains the brittle interphase layer of 100 nanometers in thickness.

Figure 2 is a photomicrograph of the fracture surface of the 90 degree
tension composite coupons after fracture.   Pronounced differences in the
fracture surface are readily apparent.   The AU-4 sample shows a large
number of bare fibers pulled out from the surrounding composite.   These
fibers had the lowest level of adhesion as measured in shear. The AS-4
composite displays a surface in which all of the fibers are well aligned
and apparently wet by the matrix.   Some fibers have separated from the rest
but represent a small fraction of the total.   The AS-4C fibers however,
display almost a planar surface with the fracture path appearing very
smooth.   Examination of these fibers at high magnification shows that each
is covered by a thin layer of epoxy with the appearance of a brittle thin
coating.   Apparently the brittle interphase has acted as the locus of
failure for the AS-4C fibers under these loading conditions leading to a
lower than expected value for the 90 degree tensile strength.

0 Degree Tensile Strength. The values of 0 degree tensile strength measured for these composites reflect the slightly different volume fractions of the composites. A true comparison requires that the values measured should be compared to the theoretical ultimate tensile strength (i.e. the tensile strength predicted from fiber volume fraction calculations alone). That value is tabulated in the last row of Table 2. From these numbers it is again obvious that the AS-4 fiber composite has the greatest efficiency (65%) for translating fiber strength into composite strength. Both the AU-4 (59%) and the AS-4C (57%) composite are significantly less efficient.

All of the composites failed by a longitudinal splitting mechanism as recently discussed by Bader [4]. The AU-4 which has very poor adhesion to this matrix gives a brush-type failure. The photomicrograph of Figure 3 illustrates this failure mode. The micrographs are taken perpendicular to the fiber axis on the surfaces creating by the longitudinal splitting during testing. Many bare fibers are shown in the figure. Both the AS-4 and AS-4C composites also failed with longitudinal splitting but did not exhibit brush-type failure. The AS-4C composite gave qualitatively less axial splitting and more planar fracture perpendicular to the fiber axis than the AS-4. This intermediate behavior of the AS-4 was the most favorable mode since its efficiency at translating fiber strength into composite strength was the highest of any group tested.

CONCLUSIONS

1. Composite properties were measured on specimens prepared from the same fiber and matrix under identical processing conditions in which only the degree of adhesion between fiber and matrix was changed. Significant differences in composite mechanical properties were detected and attributable to the differences in adhesion.

2. There was general agreement between the relative changes in adhesion as measured by the single fiber interfacial shear strength test and composite tests as long as the failure mode remained interfacial. Divergence occurred when the failure mode changed to matrix dominated.

3. There appears to be an optimum level of adhesion for maximum composite properties. This level is the highest adhesion possible while maintaining an interfacial mode of failure.

REFERENCES

[1] Drzal, L. T. , Rich, M. J., and Lloyd, P. F., "Adhesion of Graphite Fibers to Epoxy Matrices: I. The Role of Fiber Surface Treatment", <u>Journal of Adhesion</u>, Vol. 16, 1983, pp. 1-16.

[2] Drzal, L. T. , Rich, M. J., Koenig, M., and Lloyd, P. F., "Adhesion of Graphite Fibers to Epoxy Matrices: II. The Effect of Fiber Finish", <u>Journal of Adhesion</u>, Vol. 16, 1983, pp. 133-152.

[3] Browning, C. E., Abrams, F. L., and Whitney, J. M., <u>ASTM STP 797</u>, C. E. Browning (ed.). American Society for Testing Materials, Philadelphia, PA (1983)

[4] Bader, M. G., "Tensile Strength of Uniaxial Composites", <u>Science and Engineering of Composite Materials</u>, Vol. 1, 1988, pp. 1-12.

AU4   **AS4**   **AS4C**

Figure 1.   Four Point Shear Specimen Tensile Surfaces under Load at 1000X

AU4   **AS4**   **AS4C**

Figure 2.   90 Degree Tensile Specimen Fracture Surfaces at 30X

AU4   **AS4**   **AS4C**

Figure 3.   0 Degree Tensile Specimen Axial Split Fracture Surfaces at 150X

Interphase Tailoring in Graphite-Epoxy Composites

R. V. SUBRAMANIAN, A. R. SANADI AND A. S. CRASTO

ABSTRACT

An electrodeposition technique has been used for the modification of the fiber-matrix interphase in graphite fiber-epoxy composites. A coating of poly (styrene-co-maleic anhydride) (SMA) polymer was electrodeposited from an aqueous solution on AU graphite fibers used as electrodes in an electrolytic cell. Different electrocoating parameters were initially used to establish optimum conditions to achieve thin uniform coatings suitable for functioning as interphases in composites reinforced by the coated fibers. Scanning electron microscopy was used to select coating conditions that resulted in coatings of the polymer that were uniform in thickness. The interfacial shear strength (IFSS), evaluated by a single fiber composite technique, showed that the SMA coating resulted in an improvement of about 50 % in IFSS compared to the commercially treated fibers (AS). This was achieved without sacrificing impact strength, which for the SMA coated fiber composites, was actually about 6 % higher than the AS fiber composites. It is suggested that good penetration of the epoxy into the SMA interphase results in a tough fiber-matrix interfacial region giving rise to high energy absorbing mechanisms that, in turn, yield a tough composite. Evidence of good epoxy penetration into the coating was determined by the use of electron microprobe line scans for bromine across the diameter of a filament in a single fiber composite. The bromine was introduced by using a brominated epoxy resin. Improvements in both impact and shear strengths are thus simultaneously achieved with fibers electrocoated by SMA.

INTRODUCTION

The performance of fiber reinforced composites is dependent on the fiber-matrix interface (1). The interlaminar shear strength (ILSS) is directly related to the efficiency of the fiber-matrix interfacial bond: the better the fiber-matrix adhesion the higher is the ILSS. A great deal of work has been done to change the surface properties of graphite fibers to obtain better coupling between the fiber and the matrix (2,3). For example, surface oxidation results in the improvement of ILSS which is due to the mechanical keying effect caused by surface roughening, and the formation of functional groups created on the surface that bond with the matrix resin.

---

* R.V. Subramanian, A.R. Sanadi, A.S. Crasto, Department of Mechanical and Materials Engineering, Washington State University, Pullman, WA 99164-2920

In general, however, this improvement in fiber-matrix interfacial bond strength (IFSS) results in the reduction in impact strength of the composites (1). The interface affects the toughness of the composite by the various energy absorbing mechanisms associated with the interface; these being, debonding, fiber stress relaxation and fiber pull-out. In the presence of a weak interfacial bond, debonding takes place ahead of the crack tip according to the Cook and Gordon mechanism (4,5). The energy absorbed due to this debonding depends on the debonded length. If the fiber-matrix bond is very strong the crack would preferentially penetrate the fiber rather than being blunted by the interface– this would minimize the contribution of the debonding energy. A weak interfacial bond is thus expected to provide high fracture toughness in composites; this has been observed by many workers (1).

A more efficient approach is to modify the interface by coating the fibers with various polymers which themselves contribute to the toughness of the composite and therefore improve both shear strengths and fracture toughness (6). Subramanian et.al. (7,8,9) coated various polymers onto the fibers by electrodeposition or electropolymerization: the advantage of the deposition is the ability to carefully control the thickness of the interlayer polymer, and also to obtain a very compact interphase. Such an interlayer absorbs energy through deformation and also reduces residual stresses around the interface arising from cure shrinkage of the resin; both of these can significantly affect the fracture toughness of the fiber composite (8-11). The conductive nature of the graphite fibers has been utilized in these studies to electrodeposit polymers or to electropolymerize monomers onto the fiber surface. In the present study the influence of an electrodeposited layer of poly(styrene-co-maleic anhydride) on the fiber-matrix interfacial strength and on the composite toughness has been investigated, as well as the influence of different electrodeposition parameters on the coating. The penetration of the matrix into the interlayer polymer is necessary to form a good matrix-interlayer bond (10) and an electron microprobe scan was therefore necessary. Izod impact strengths were conducted using an instrumented impact testing machine. The IFSS has been evaluated through a single fiber composite technique that takes into account changes in the fiber strength as a result of the electrodeposition process. A theory developed in our laboratory (12) uses the distribution of fiber strengths and the aspect ratio of fragment lengths through the single fiber composite test to obtain a distribution of the IFSS.

EXPERIMENTAL METHOD

Poly(styrene-co-maleic anhydride), SMA 1000 and SMA 2000 (ARCO Chemical Co.) used in electrodeposition had respectively a 1:1 and 1:2 ratio of styrene and maleic anhydride. Epon 828, (DGEBA, Shell Chemicals) was cured with 14.5 phr m-phenylene diamine (MPDA) (Aldrich Chemical Co.). A 33% solution (w/w) of a brominated epoxy, Epon 1120-A-80, in Epon 828 was used as the matrix in the single carbon monofilament composite used for electron microprobe analysis. Fortafil 5 (Grat Lakes) PAN-based graphite fibers, dog-bone shaped in cross-section and without any surface treatments were used to study the influence of various electrodeposition parameters. Continuous tows of circular cross-section Hercules high modulus untreated (HMAU) were electrocoated with SMA for composite preparation and comparison with HMAS fiber composites.

The general experimental approach was to electrocoat untreated graphite fibers, and use the graphite coated thus for epoxy composites. Single filament tensile specimens were used for interfacial strength measurements. The experimental procedures for electrodeposition, fiber strength, IFSS, and impact strength measurement, and electron microprobe analysis are described in detail in earlier publications from this laboratory (10,11).

RESULTS AND DISCUSSION

Electrodeposition

The electrodeposition of polymer follows Faraday's laws. The mass of the polymer deposited (M) is related to the equivalent weight of the polymer (R), the Faraday's constant and the couloumbs of charge (Q) used in the electrodeposition through the relationship, $M = (R/F)Q$. The equivalent weight of the polymer is related to the number of ionized carboxyl groups on the molecule and an increase in the latter lowers its equivalent weight.

The effect of pH on electrodeposition is shown in table 1. The addition of base to increase the pH also increases the number of ionized carboxylic groups on the polymer. This causes the equivalent weight of the polymer to decrease and hence reduces its couloumbic efficiency. The solubility of the SMA in water also increases with increased pH, which further decreases the amount of deposit. An increase in pH also results in a high rate of gas evolution that could cause the coating to be less compactly held.

In experiments where the solid content of the SMA-1000 solutions varied up to 5% (w/w), the amount of the the coating increased with the solid content of the bath. SEM studies indicated that the coatings from solutions of 5% SMA were uneven but an intermediate concentration of 2.5% SMA resulted in a fairly slow deposit rate that reduced the variations of coatings between specimens, under identical conditions. The weight of the deposit was found to increase with time, as expected from Faraday's law. No indication of the levelling of coating thickness was observed, probably because of the limited deposition times.

As the voltage was increased, both at 75 and 100 coulombs there was an increase in the evolution of oxygen at the anode and hydrogen at the cathode due to the electrolysis of water. This resulted in an increase in the amount of current carried by the discharge of OH^- and H^+ ions. This phenomenon causes a fall in the number of carboxylate ions discharged at the anode, and a decrease in the amount of the polymer deposited. Similarly at short deposition times (3 min), the deposit increased with current, as was expected from Faraday's laws; but at longer deposition times there was a slight decrease in the amount of polymer on the surface. This behavior was traced to the evolution of oxygen at the anode and the lack of fresh SMA solution around the center of the bundle of fiber. Similar behavior has been observed by Subramanian and Crasto (10) on their work on the deposition of poly(butadiene-co maleic anhydride) on graphite fibers. For this reason a flat band of fibers were preferred for the deposition of polymer.

When the coated fibers were washed in distilled water, it was observed that the amount of SMA on the fiber remained constant, after a washing time of of 15 seconds. SEM micrographs reveal that the deposit was compact near the fiber surface, and flaky or loosely adhering at the outer surface of thicker coatings. This would suggest that only the outer surface is being washed away during the distilled water rinse. Based on these observations, electrocoating of fibers for composite fabrication was conducted under the following conditions: bath concentration - 2.5%; 25° C, 9.4 volts, 0.4 amps, deposition time - 6 minutes, pH - 2.7 and wash time - 15 seconds.

Fiber Tensile Strength

The fiber tensile strengths distributions have been modeled with both lognormal and Weibull distributions (Table 2). The observed lowering of the strength of the AU/SMA fibers could be due to flaws induced on the fiber surface during electrodeposition or while separating a single fiber from the tow for testing. Previous studies (10) on the effect of electrodeposition voltage on the fiber strength indicate greater losses in strengths at higher voltages. It is interesting to note that the increase in the Weibull α parameter indicates a narrowing of the scatter in data, which is also clearly seen in the narrowing of the distribution curve in figure 1. The elimination of the high strength population in the distribution would indicate that weaker flaws are made more severe during electrodeposition. It is therefore important to optimize the electrodeposition parameters so as to achieve quality coatings.

Interfacial Shear Strength

Interfacial shear strength measurements through the single fiber composite technique results in a realistic estimator of the trends in the fiber-matrix bond strength: it takes into account that the fiber also changes in strength during electrodeposition for the AU/SMA fibers. The IFSS, τ, is related to the fiber strength, σ_f, the fiber diameter, d, and the fiber critical length, l_c, by the relationship:

$$l_c/d = \sigma_f/2\tau. \tag{1}$$

Since the strength and the critical aspect ratio are all experimentally determinable, it was possible to obtain values for IFSS (τ). Since the details of this evaluation are given in earlier publications (10,11), only a brief outline is included here. If the fiber strength is distributed lognormally with a lognormal (LN) mean μ_1, and a LN variance σ_1^2, and the critical aspect ratio (CAR) is also distributed lognormally, with it LN mean μ_2 and its LN variance σ_2^2 then an analytic solution is possible (12). In such a case the IFSS is also distributed lognormally with a LN mean μ_3 and a LN variance σ_3^2, defined by:

$$\mu_3 = \mu_1 - \mu_2 - ln(2) \tag{2}$$

$$\sigma_3^2 = \sigma_1^2 + \sigma_2^2 \tag{3}$$

If the interfacial shear strength follows a lognormal distribution, the mean and variance of the IFSS is given by:

$$Mean = exp(\mu + \sigma^2/2) \tag{4}$$

$$variance = exp(2\mu + \sigma^2/2)[exp(\sigma^2) - 1] \qquad (5)$$

The aspect ratios, the ratio of the length to the diameter (l/d) of all the surface conditions of AU, AS, AU/SMA-1000 and AU/SMA-2000 have lognormal distributions and the lognormal parameters are given in the table 3. There is a dramatic improvement in the IFSS of the electrocoated fibers, thus making the stress transfer more efficient: this is clearly observed in figure 1 and table 4. SMA copolymer is electrically compacted on the fiber surface and is also compatible with the matrix. The latter arises from facile diffusion of the epoxy into the SMA coating and results in the creation of a strong *interphase*. Electron microprobe analysis of Br-tagged epoxy confirms good interpenetration of SMA and epoxy matrix. The presence of carboxylic groups in the SMA could result in chemical bond formation between the SMA and epoxy. The greater number of carboxylic groups in SMA-1000 as compared to SMA-2000 (due to a greater number of anhydride groups in the former) may explain the observed difference in the IFSS in the two cases.

Impact Strength

Due to the large energy absorbing capabilities of a weaker interface through fiber pull-out and interface debonding, it is not surprising that impact energy was much greater for the AU composites as compared to the others (table 5). Therefore one would also expect a higher impact toughness of the AS composites one would expect a higher impact toughness of the AS composites compared to the SMA coated fiber composites due to the lower IFSS of the AS fiber-matrix interface. However, the SMA coated fibers show a slightly higher impact toughness over the AS composites even though the strength of SMA coated fibers is significantly lower than the AS fibers. This would then suggest that the interface is itself absorbing a great deal of energy, even though the critical aspect ratio of the SMA fibers is less. It is thus possible that the SMA interface also helps reduce the residual stresses around the fiber during thermal and cure shrinkage, which would affect the composite characteristics (13).

The load/energy versus time traces of the SMA composites during the impact test show a large absorption of energy subsequent to crack initiation, supporting this earlier evidence on the role of the interface in increasing fracture toughness (fig.2). SEM of the fracture surfaces of a SMA coated fiber composites, areas of fiber-interphase and interphase-matrix failure are clearly apparent (fig.3). This is in contrast to the high amount of matrix-fiber failure evident in the commercially treated fiber composite.

CONCLUSIONS

Modification of the fiber-matrix interphase by the electrodeposition of poly(styrene-co-maleic anhydride) results in considerable improvement (about 50%) in the fiber-matrix bond strength and a small improvement (about 6%) in the impact strength, as compared to the commercially treated AS fiber composites. The single fiber technique along with the statistical theory for deriving a log-normal distribution of the IFSS provides an effective method to estimate trends in the fiber-matrix bond strength. Electrodeposition parameters must be optimized to minimize any reduction in fiber strength during the electrodeposition process; any reduction in fiber strength can be taken into account in applying the theory. The difference in IFSS between the SMA-1000 and SMA-2000 coating reveals the sensitivity of IFSS even to minor changes in chemical structure of the interlayer. The improvement in impact strength is seen to arise from the absorption of energy through deformation of the interphase, accompanied by, crack tip blunting and diversion of crack path. The feasibility of modifying interfaces and composite properties through the manipulation of copolymer chemical structure provides a tool to tailor properties for different applications.

ACKNOWLEDGEMENTS

The authors would like to acknowledge Chris Davitt, Joyce Davis and J. H. Larsen for their help in scanning electron microscopy and Scott Cornelius and Diane Johnson for their assistance in electron microprobe analysis. This research was partially supported by a sub-contract from Battelle, Columbus Laboratories, Advanced Materials Center for the Commercial Development of Space (NASA).

REFERENCES

1. P. E. McMahon and L. Ying, "Effects of fiber/matrix interactions on the properties of graphite/epoxy composites," *NASA Contractor Report 3607*, (1982).
2. L. T. Drzal, M. J. Rich and P. F. Floyd, *J. Adhesion*, **16**, 1 (1982).
3. D. W. Mckee and V. J. Mimeault, *Chem. Phys. Carbon*, **8**, 151, (1973).
4. B. Harris, P. W. R. Beaumont and E. M. deFerran, *J. Mater. Sci.*, **6**, (1971).
5. J. Cook and J. E. Gordon, *Proc. Roy. Soc. Lond.*, **A282**, 508, (1964).
6. T. U. Marston, A. G. Atkins and D. K. Felback, *J. Mater. Sci.*, **9**, 447, (1974).
7. R. V. Subramanain, V. Sundaram and A. K. Patel, *33rd Ann. Tech. Conf., SPI, RP/C Institute, section 20-F, Washington DC*, (1978).
8. R. V. Subramanian and J. Jakubouski, *Polm. Eng. Sci.*, **18**, 590, (1978).
9. A. Crasto, Shi-Hau Own and R. V. Subramanian, in *Composite Interfaces*, H. Ishida and J. L. Koenig eds., (North-Holland, N.Y., 1986) p133.
10. R. V. Subramanian and A. Crasto, *Polym. Comp.*, **7(4)**, 201, (1986).
11. A. S. Crasto, S. H. Own and R. V. Subramanian, *Polm. Comp.*, **9(1)**, 78 (1988).
12. S. H. Own, R. V. Subramanian and S. C. Saunders, *J. Mater. Sci.*, **21**, 3912 (1986).
13. J. T. Lim, W. Bailey and M. R. Piggott, *SAMPE Quarterly*, **15**, 25, (1984).

Table 1
Effect of pH on Polymer Deposited

pH	Coulombs	Voltage (V)	Weight Deposited (%)
2.73	100	10	5.62
3.78	100	10	4.59
4.82	100	10	1.55
6.08	100	10	.522

Table 2
Fiber Strength and Strength Distribution Parameters

Fiber Type	Fibers Tested number	Mean Strength (MPa)	Std. Dev. (MPa)	Log-normal Mean	Log-normal Dev.	Weibull α	Weibull β
HM AU	99	5283.01	902.01	8.557	.1755	6.605	5665.1
HM AS	99	4763.67	1024.81	8.446	.2144	5.023	5183.5
AU/SMA	99	4245.71	434.20	8.345	.1031	10.271	4443.7

Table 3
Aspect Ratios with Distribution Parameters

Fiber Type	Fragments counted	Mean l/d	Std. Dev.	Log-normal	Log-normal
HM AU	227	87.496	27.877	4.423	.3093
HM AS	311	57.694	19.632	4.000	.3302
SMA-1000	365	32.524	8.517	3.449	.2544
SMA-2000	181	37.65	8.067	3.606	.2140

Table 4
Interfacial Shear Strength and Lognormal Parameters

Fiber Type	Mean IFSS (MPa)	Std. Dev. (MPa)	Log-normal mean	Log-normal dev.	percent increase in IFSS
HM AU	33.26	12.21	3.441	.3556	-27.9
HM AS	46.10	18.87	3.753	.3944	0
SMA-1000	69.58	19.53	4.207	.2671	+50.9
SMA-2000	59.05	14.19	4.050	.2370	+28.1

Table 5
Effect of Surface Condition on Impact Strength (IS)

Fiber Type	V_f	Mean IS (kJ/m^2)	Std. Dev.	percent change in IS
HM AU	64.0	249.71	48.30	+52.8
HM AS	62.5	163.43	14.71	0
SMA-1000	63.5	173.85	26.68	+6.38

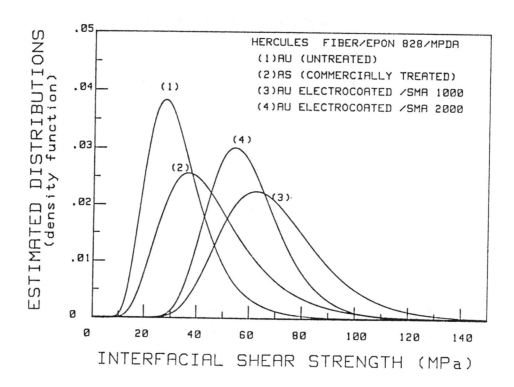

Fig.1 Lognormal distribution of interfacial shear strength.

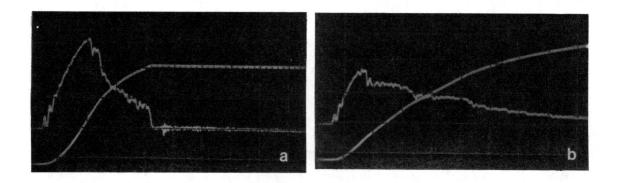

Fig.2 Load / Energy versus time for (a) AS and (b) SMA coated AU fiber-epoxy composites, under impact loading.

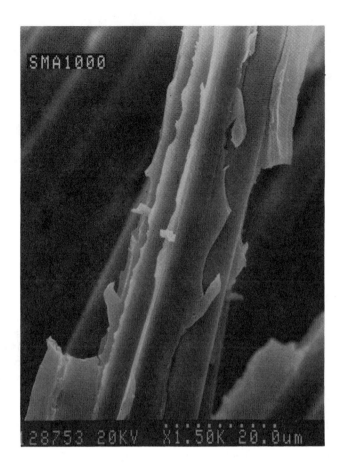

Fig.3 SEM photomicrograph of SMA coated AU fiber-epoxy composite showing regions of fiber-interphase failure and also interphase-matrix failure.

Effects of Plasma Treatment on Performance of Composite Flywheels

JOHN BORDER

ABSTRACT

By using plasma treatments, PDA has been able to promote specific surface chemistry that significantly increases the adhesion between fibers and matrices. In unidirectional composites, the transverse properties are improved. These property enhancements translate into increases in performance of the composite that is dependent on the application. This paper examines the impact of plasma treatments on the performance of high speed composite flywheels.

Three materials are examined using data measured at PDA. Failure modes considered include delamination, hoop rupture, and fatigue. Failure speeds are compared for each material and failure mode. Operating speeds are defined for each material. The impact of plasma treatments on the failure of composite flywheels is discussed.

1.0 INTRODUCTION

Fiber reinforced composites are by nature nonisotropic. In many applications, the performance of the composite is limited by the adhesion of the fiber to the matrix. PDA Engineering has been engaged in developing optimized plasma treatments for reinforcing fibers to increase the adhesion of fibers to matrices. This plasma chemistry approach has been utilized for a wide variety of fibers and matrices and has significantly increased the transverse tensile strengths of unidirectional composites.

To assess the impact these plasma treatments have on the performance of composites, an analytical study was undertaken to determine the change in failure modes produced by increasing the transverse strength. The application chosen for examination was a high speed composite flywheel, a schematic of which is shown in Figure 1.

J. N. Border, Development Engineer, PDA Engineering, 3754 Hawkins NE, Albuquerque, NM 87109

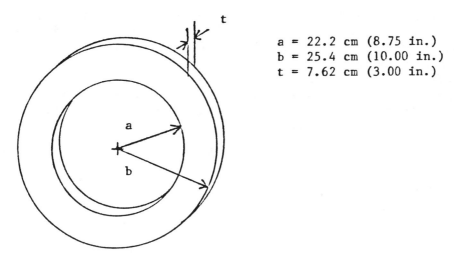

$$
\begin{aligned}
a &= 22.2 \text{ cm } (8.75 \text{ in.}) \\
b &= 25.4 \text{ cm } (10.00 \text{ in.}) \\
t &= 7.62 \text{ cm } (3.00 \text{ in.})
\end{aligned}
$$

Figure 1. Flywheel Geometry Used in Analysis.

The materials examined include Spectra 1000/epoxy, Kevlar 49/epoxy, and IM7 graphite/bismaleimide. The failure modes investigated include delamination, hoop rupture, creep distortion and fatigue rupture. The effect of stress concentrations was also examined. The purpose of the analytical investigation was to identify the limiting failure mode and to determine the effect of PDA's plasma treatments on the operating speed of the flywheel.

2.0 FLYWHEEL ANALYSIS

The equations for the stress state in a composite flywheel were developed by Morganthaler and Bonk [1]. The equations for radial stress (σ_{rr}) and hoop stress ($\sigma_{\theta\theta}$) are shown in Equations I and II.

$$
\sigma_{rr} = P_b \left(\frac{b}{a}\right)^2 \left[\frac{\left(\frac{r}{a}\right)^{-N-1} - \left(\frac{r}{a}\right)^{N-1}}{\left(\frac{a}{b}\right)^{-N-1} - \left(\frac{a}{b}\right)^{N-1}}\right] - P_a \left[\frac{\left(\frac{r}{b}\right)^{-N-1} - \left(\frac{r}{b}\right)^{N-1}}{\left(\frac{a}{b}\right)^{-N-1} - \left(\frac{a}{b}\right)^{N-1}}\right] \tag{I}
$$

$$
- \frac{(3 + \nu_{r\theta})\rho\omega^2 b^2}{(9 - N^2)} \left[\left(\frac{r}{b}\right)^2 - \frac{1 - \left(\frac{a}{b}\right)^{N+3}}{1 - \left(\frac{a}{b}\right)^{2N}} \left(\frac{r}{b}\right)^{N-1}\right.
$$

$$
\left. - \frac{1 - \left(\frac{a}{b}\right)^{-N+3}}{1 - \left(\frac{a}{b}\right)^{-2N}} \left(\frac{r}{b}\right)^{-N-1}\right]
$$

P_a = pressure at inner surface (psi)

P_b = pressure at outer surface (psi)

a = inner radius (inches)

b = outer radius (inches)

r = sample radius (inches)

ρ = density (lb in^2/sec^4)

ω = rotational speed (rad/sec)

$\nu_{r\theta}$ = Poisson's ratio of contraction in the r direc- due to extension in the θ direction

$$
\sigma_\theta = N P_a \left[\frac{\left(\frac{r}{b}\right)^{-N-1} + \left(\frac{r}{b}\right)^{N-1}}{\left(\frac{a}{b}\right)^{-N-1} - \left(\frac{a}{b}\right)^{N-1}}\right] - N P_b \left(\frac{b}{a}\right)^2 \left[\frac{\left(\frac{r}{a}\right)^{-N-1} + \left(\frac{r}{a}\right)^{N-1}}{\left(\frac{a}{b}\right)^{-N-1} - \left(\frac{a}{b}\right)^{N-1}}\right] \tag{II}
$$

$$
- \frac{(3 + \nu_{r\theta})\rho\omega^2 b^2}{(9 - N^2)} \left\{\left(\frac{r}{b}\right)^2 \frac{N^2 + 3\nu_{r\theta}}{3 + \nu_{r\theta}} - N \left[\frac{1 - \left(\frac{a}{b}\right)^{N+3}}{1 - \left(\frac{a}{b}\right)^{2N}}\right] \left(\frac{r}{b}\right)^{N-1}\right.
$$

$$
\left. + N \left[\frac{1 - \left(\frac{a}{b}\right)^{-N+3}}{1 - \left(\frac{a}{b}\right)^{-2N}}\right] \left(\frac{r}{b}\right)^{-N-1}\right\}
$$

Equations I and II can be used to examine stress distributions within flywheels as a function of the material properties. The stresses within flywheels made of two different materials are shown in Figures 2 and 3.

As Figures 2 and 3 show, the anisotropic nature of the filament wound flywheel produces a very anisotropic stress state. Because the fibers are oriented to accept the maximum load, filament windings are well suited to the stress state developed in a spinning flywheel. The factors that determine the stress distribution are the ratio of the inner and outer diameters and the value of $N = (E_\theta/E_r)^{1/2}$. Both radial and hoop stresses vary considerably with radial position in the flywheel.

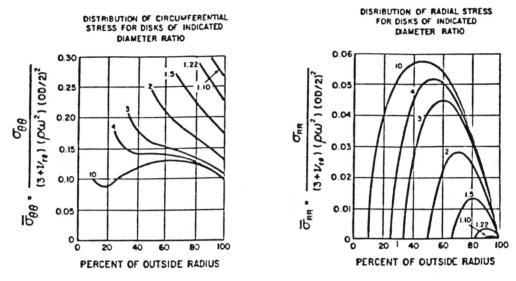

Figure 2.   Radial and Tangential Stress Distributions for Various Diameter Ratios for Orthotropic Disks with N Equal to 1.75 [1]

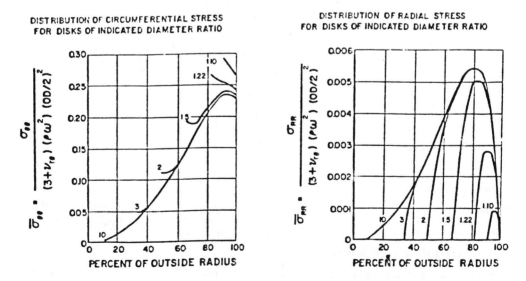

Figure 3.   Radial and Tangential Stress Distributions for Various Diameter Ratios for Orthotropic Disks with N Equal to 10 [1]

The material properties used in this analysis for the different material systems examined are shown in Table I. The data for the materials has been compiled from several sources and includes results from materials testing at PDA. As Table I shows, the transverse tensile strengths of materials treated at PDA are 32-93% higher than those of the untreated materials.

TABLE I

MATERIAL PROPERTIES USED IN THE ANALYSIS

Material	Density g/cc(lb/in^3)	Longitudinal Strength MPa (Ksi) Modulus GPa (Msi)	Transverse Strength MPa (Ksi) Modulus GPa (Msi)	Poisson Ratio
S 1000/Epoxy Untreated	1.1 (0.040) [9]	1200 (174) [9] 50.3 (7.3) [9]	3.98 (0.58)*[5] 3.51 (0.51) [9]	0.32
S 1000/Epoxy Treated	1.1 (0.040) [9]	1104 (160)* 50.3 (7.3) [9]	7.72 (1.12)*[5] 3.51 (0.51) [9]	0.32
Kev 49/Epoxy Untreated	1.38 (0.50) [2]	1380 (200) [2] 76 (11) [2]	13.5 (1.96)*[5] 4.5 (0.65) [2]	0.34
Kev 49/Epoxy Treated	1.38 (0.50) [2]	1380 (200) [2] 76 (11) [2]	22.5 (3.26)*[5] 4.5 (0.65) [2]	0.34
IM7/Bismal Untreated	1.63 (0.059) [13]	2922 (424) [13] 175 (25) [13]	44.4 (6.44)*[16] 8.9 (1.29) [2]	0.27
IM7/Bismal Treated	1.63 (0.059) [13]	2922 (424) [13] 175 (25) [13]	58.6 (8.51)*[16] 8.9 (1.29) [2]	0.27

*Taken from PDA measured values.

3.0 FAILURE ANALYSIS

The materials were evaluated by comparing the rotational speed needed to produce each of the different types of failure. In each case, both treated and untreated materials were evaluated. As a result of the analytical study, design limits and relative performance of the different materials were calculated. The collected results of this analysis are presented in Table II, and the individual results are discussed in Sections 3.1-5.

3.1 Delamination

For a filament wound flywheel, delamination occurs when the radial stress exceeds the strength between fibers. Delamination has been the limiting failure mode exhibited in most composite flywheel programs [1,2,3,4]. By examining Figures 2 and 3, it can be seen that the location of maximum radial stress depends on the diameter ratio and the value of N for the material. For the flywheel geometry shown in Figure 1, the diameter ratio is 1.2. For the materials being examined, the values of N are close to 4. Figures 2 and 3 show maximum radial stress occuring 45% through the radial width of the flywheel. For the flywheel shown in Figure 1, maximum radial stress occurs at r = 23.1 cm (9.09 in.).

Equation I was solved for rotational speed to produce a given radial stress. The rotational speeds to produce delamination failure are shown in Table II. As can be seen in Table II, the plasma treatments significantly increase the speed at which delamination failure occurs.

TABLE II

EFFECT OF PLASMA TREATMENTS ON ROTATIONAL SPEEDS
AT FAILURE FOR DIFFERENT FAILURE MODES

Material	Delam. (rpm)	Hoop Rupture (rpm)	Fatigue Limit (rpm)	Operating Speed for No Fatigue (rpm)	Design Stress W/ Stress Concen.	Stress Concen. Delam. Speed (rpm)
Spectra 1000/ Epoxy						
Treated	31110	38700	No Data		1.6 MPa (232 psi)	18490
Untreated	22330	40400			1.3 MPa (189 psi)	13270
Kevlar 49/ Epoxy						
Treated	46090	38200	64% UTS [11]	30560	7.6 MPa (1103 psi)	27400
Untreated	35730	same	(R = .1)		4.6 MPa (668 psi)	21240
IM7 Graphite/ Bismal.						
Treated	70400	51600	60% UTS [13]	39970	19.9 MPa (2890 psi)	41850
Untreated	61330	same	(R = .1)		15.1 MPa (2190 psi)	36460

3.2 Hoop Rupture

Hoop rupture occurs when the circumferential stress in the flywheel exceeds the longitudinal tensile strength of the material. In Figures 2 and 3 at a diameter ratio of 1.2, maximum hoop stress is reached at the inner diameter.

Values for longitudinal strength have been compiled from various sources. Tensile tests on single fibers that have been plasma treated have shown that only Spectra 1000 exhibits a noticeable change in strength following plasma treatment. Spectra 1000 showed a 7.8% decrease in tensile strength [5]. In all the cases except Spectra 1000, the longitudinal tensile strength was assumed to be unchanged by plasma treatment. In calculating the speed at which hoop rupture occurs, the value for the tensile strength of the Spectra 1000 composite was decreased by 7.8% to reflect the decrease measured in the single fiber tests.

Equation II was solved for rotational speed to produce a given stress. Hoop failure speeds for the different materials are shown in Table II. In the case of hoop rupture, failure is dependent on the strength of fibers alone. Consequently, plasma treatments of the fibers have little effect on hoop rupture.

3.3 Fatigue

A further consideration in determining the design strength for a rotating mechanism such as a flywheel is the fatigue resistance of the material. An imbalance in the flywheel will lead to a cyclic stress state within the flywheel. The large number of rotations experienced by a high speed flywheel during its lifetime dictates that fatigue be considered.

The limiting operating speed from the previous analyses was used to calculate the total number of rotations experienced during the flywheel lifetime as $1\text{-}2 \times 10^8$ cycles. This information was used to determine the stress level necessary to prevent fatigue failure based on the hoop strength and the fatigue limit. Due to a lack of information on graphite/bismaleimide materials, information on IM7 graphite/epoxy is presented as representative of fatigue in graphite based composites. Fatigue data for Kevlar 49 and IM7 graphite are presented in Table II.

From the data presented in Table II, it can be seen that fatigue can be the limiting factor in flywheel failure. However, it is very important to take into account the stress conditions in evaluating fatigue data. For the fatigue data presented, unidirectional samples were tested in cyclic tension, and the ratio of the minimum to maximum stress was 0.1. This stress ratio is much higher than the stress ratio that would most likely be encountered in a flywheel, although resonant vibrations in the flywheel may increase the effect on the flywheel itself [3]. Another factor to consider is that longitudinal tension fatigue data does not include the effects of radial tensile fatigue which would also occur in the flywheel. The onset of fatigue cracking is dependent on both the cyclic stress being applied and the nature of the stress concentrations present in the composite.

3.4 Stress Concentrations

Defects within a material can alter the stress distribution to produce a local area of higher stress called a stress concentration. The degree to which the stress is concentrated depends on the geometry of the material and the defect. The effect of stress concentrations can be further complicated by fatigue. Stress concentrations are expressed as a multiplier to the nominal stress.

The major source of stress concentrations in filament wound composites comes from a relatively high void content. They typically have a void content of 3-5%, while composites produced by other processes have < 1% voids [6]. Due to the high concentration of voids in filament wound parts, it is reasonable to assume that stress concentrations are always present, and that failure of the composite will initiate in the regions of high stress surrounding the voids [7].

For a unidirectional composite that is loaded perpendicular to the direction of the fibers, the stress concentration produced by a spherical hole is 2.9 [7]. For the case of an elongated prolate spheroid cavity oriented perpendicular to the stress in an isotropic material, the stress concentration is 2.95 [8]. The close agreement of the stress concentration factors for these two cases indicates that, for the stress state of tension perpendicular to unidirectional fibers, the composite can be approximated as provisionally isotropic. Because the filament winding action will produce voids that are best approximated as drawn out prolate spheroids, a stress concentration factor of 2.95 was used in the analysis.

Given that the transverse tensile strengths listed in Table II are for specimens prepared in a hot press, this data can be assumed to be for materials without stress concentrations. When the reduced failure strength due to the stress concentrations is considered, a new set of lower operating speeds is defined. The operating speeds, which predict radial tensile failure initiated by stress concentrations, are shown in Table II. Again, plasma treatments allow the operating speeds to be increased substantially without failure.

3.5 Creep Deformation and Rupture

Creep in a material results in a continuous increase in strain under a constant load. The degree that different materials will creep varies dramatically. If the material creeps enough, under a tensile load, it will rupture. As the material in a flywheel creeps, the diameters of the flywheel increase. This, in turn, increases the stress within the flywheel and accelerates the creep of the flywheel.

The materials were examined to define the maximum flywheel operating speed without creep rupture. As this analysis was originally undertaken for the application of energy storage for an electromagnetic launcher, the minimum design lifetime was 85 hours [5]. The results of the analysis are shown in Table III. Data on the creep of IM7 graphite/epoxy is presented as representative of the creep resistance of graphite based composites. Like hoop rupture, creep failure is dependent on the creep properties of the fibers, and plasma treatments have little effect.

TABLE III
CREEP ANALYSIS RESULTS

Material	Applied Load	Lifetime	Result	Equiv. Speed
S1000/Epoxy [9]	352 MPa (51 ksi)	85 Hrs.	5% Creep	22130 rpm
Kev 49/Epoxy [12,13]	965 MPa (140 ksi)	150 Hrs.	Rupture	31960 rpm
IM7 Gr/Epoxy [13]	2340 MPa (340 ksi)	1000 Hrs.	Intact	46152 rpm

4.0 CONCLUSIONS

From the discussions above it can be easily seen that failure in an operating flywheel can manifest itself in several distinctly different ways. In considering the different failure modes it is important to realize that some failure modes, such as hoop rupture and creep distortion, are dependent on the strength of the fibers. Other failure modes, including delamination, fatigue, and stress concentrations, are dependent on the fiber/matrix bond. The analytical study has shown that the failure modes dependent on fiber/matrix bonding are predominant for high speed flywheels.

PDA's plasma treatments effectively increased the predicted flywheel performance for all three of the materials considered. The maximum operating speed varied considerably depending on the materials. The limiting failure mode predicted in all cases was delamination induced by stress concentrations within the material, although the mechanism of the failure could be fatigue related or simple failure depending on how well the flywheel was balanced (see IM7 graphite/bis.). This failure mode manifests itself as crack propagation along the weak fiber-resin interface [14]. PDA's plasma treatments, which produce strong chemical covalent bonding of fibers and resin, increase the transverse strength of the composite by 32-93%. Shear strength, which is important for the case of a flywheel that is decelerated rapidly, is also increased. In addition, ultrasonic inspection has shown a significant reduction in the void and microvoid content in materials treated at PDA [15], so it is likely that the impact of the plasma treatments on fatigue may be even greater than studies so far have indicated.

The end result of the analysis is that the plasma treatments increased the maximum operating speeds for the three materials from 15-39%. Table IV shows the maximum operating speeds for the treated and untreated materials. From the operating speed, the energy stored in the flywheels was calculated using Equation III. As can be seen from the data in Table IV, PDA's plasma treatments increased the maximum energy stored in the flywheels by 32-93%.

$$E = \pi \omega^2 t \rho (b^4 - a^4)/4 \tag{III}$$

TABLE IV

MAXIMUM OPERATING SPEEDS AND ENERGY STORED
IN FLYWHEELS OF DIFFERENT MATERIALS

Material	Max. Operating Speed	Max. Energy Stored
S1000/Epoxy		
Treated	18490 rpm	0.149 KWh (508 Btu)
Untreated	13270	0.077 (262)
Kevlar 49/Epoxy		
Treated	27400	0.41 (1400)
Untreated	21240	0.24 (818)
IM7 Gr./Bismal.		
Treated	41850 (Balanced)	1.13 (3800)
Treated	39970 (Fatigued)	1.03 (3500)
Untreated	36460	0.86 (2900)

5.0 ACKNOWLEDGEMENTS

The work presented in this paper was undertaken as part of an SBIR Phase I program psonsored by the U.S. Army Armament Research, Development & Engineering Center. The title of the program was "Improved Composite Flywheel Storage Devices for Electromagnetic Launchers," Contract No. DAAA21-87-C-0115.

6.0 REFERENCES

1. G. F. Morganthaler and S. P. Bonk, "Composite Flywheel Stress Analysis and Materials Study," 12th National SAMPE Symposium, 1967, P. D-5.
2. E. D. Reedy Jr., "A Composite-Rim Flywheel Design," SAMPE Quarterly, Vol. 9, No. 3, April 1978.
3. G. A. Cuccuru, F. Ginesu, B. Picasso and P. Priolo, "Characterization of Composite Materials for Filament Wound Flywheels," J. Composite Materials, Vol. 14, Jan. 1980, p. 31.
4. A. P. Coppa, "Composite Ring-Disk Flywheel Design, Fabrication and Testing," General Electric Document No. 83SDS4248, August 1983, Prepared for Lawrence Livermore National Laboratory, Subcontract No. 6624409.

5. R. E. Allred, J. N. Border, and C. D. Bowland, "Improved Composite Flywheel Storage Devices for Electromagnetic Launchers," SBIR Phase I Final Report, submitted to Army Armament, Munitions and Chemical Command,

6. B. E. Spencer, "Application of the Filament Winding Process," Advanced Composites: The Latest Developments, Proceedings of the Second Conference on Advanced Composites, Nov. 1986, Published by ASM International, p. 107.

7. G. Kardos, "Stress Concentrations in Composites," 31st International SAMPE Symposium, April 1986, P. 396.

8. F. B. Seely and J. O. Smith, <u>Advanced Mechanics of Materials</u>, John Wiley & Sons, Inc., New York, 1952.

9. Spectra Technical Literature, Allied Fibers Technical Center, Petersburg, VA.

10. R. H. Ericksen, "Room Temperature Creep of Kevlar 49/Epoxy Composites," Composites, July 1976, p. 189.

11. G. Lubin (Ed.) <u>Handbook of Composites</u>, Van Nostrand Reinhold Company, 1982, New York.

12. J. A. Rinde, "LLL Materials Program for Fiber-Composite Flywheels," Proceedings of the 1979 Mechanical and Magnetic Energy Storage Contractors' Review Meeting, Washington, DC, August 19-22, 1979.

13. Graphite Fiber Technical Literature, Hercules Inc., Wilmington, DE.

14. A. K. Munjal, "Use of Fiber Reinforced Composites in Rocket Motor Industry," SAMPE Quarterly, Volume 17, No. 2, Jan. 1986, p. 1.

15. H. M. Stoller, N. J. Delollis, and C. Rodacy, "Plasma Treatment Optimization of Polyaramid Filaments to Improve Kevlar/Epoxy Composites," SBIR Phase I Final Report, submitted to Army Material Technology Laboratory, May 1986.

16. PDA Unpublished Work.

Strength of Glass-Epoxy Laminates Based on Interply Resin Failure

JOHN C. FISH
*McDonnell Douglas
Helicopter Co.
Mesa, AZ 85205*

SUNG W. LEE
*Center for Rotorcraft Education and Research
Department of Aerospace Engineering
College Park, MD 20742*

ABSTRACT

The delamination onset strength of sixteen ply quasi-isotropic glass-epoxy laminates is studied. An alternate method of strength prediction based on failure of the interply resin layer between the delaminating plies is investigated. A strength-of-materials approach, utilizing the average stress concept and a delamination failure criterion, is applied to interlaminar stress distributions obtained from quasi-three-dimensional finite element models. Good correlation between experimental and numerical results is obtained by basing strength on interply resin failure and using an interlaminar stress averaging distance of one-half of a ply thickness.

INTRODUCTION

Delamination is among the most critical failure modes in composite structures, limiting their strength and useful life. In particular, the free-edge delamination problem has received a great deal of attention [1-8]. Interlaminar stresses may arise near the free edges of composite laminates in order to mainatin stress equilibrium within the structure. Delamination will occur if these stresses are large enough.

Methods of predicting delamination onset are desirable for designing efficient composite structures. One approach to the problem is to assume that failure (delamination onset) occurs when the stresses within the structure are large enough to exceed the strength of the composite material. This strength-of-materials approach has been used by several investigators [5, 7, 8] to predict laminate strength based on ply failure (composite properties) at the critical (delamination prone) ply interface within a laminate. Reasonable correlation between analytical and experimental results was shown for the graphite-epoxy laminates studied. However, much of this work only considered layups for which one of the interlaminar stress components dominates the others.

When composite laminates are cured, resin layers are formed between the plies that make up the laminate. These interply resin layers represent a different material from the plies. Since resin layers exist at the interface between plies and

242

delamination is the out-of-plane separation of plies in a laminate, failure of the interply resin should be related to delamination in composite structures.

This study examines delamination onset in sixteen ply quasi-isotropic glass-epoxy laminates under tensile loading. Three layups, which develop significantly different free-edge stress states, are studied. In addition to strength predictions based on ply failure, an alternate method of delamination onset prediction based on failure of the interply resin layer between the delaminating plies is investigated.

EXPERIMENTS

The test specimens were made from SP-250-S29 glass-epoxy uni-directional prepreg tape manufactured by 3M. The quasi-isotropic layups investigated are: $(\pm 45/0_2/90_2/\pm 45)_s$, $(+45/0_2/-45/90_2/\pm 45)_s$, and $(0_2/\pm 45/90_2/\pm 45)_s$. These layups will be referred to as laminates A, B and C, respectively. The coupon specimens were 10 inches in length and 1.5 inches in width. Fiberglass tabs, approximately 2 inches in length were adhesively bonded to the ends of all test specimens to prevent grip damage during the tension tests. Finally, to aid in visual observation of the damage development, black ink was swabbed on to the edges of the test specimens.

The tension tests were performed with a 24 kip screw-driven test machine. Load and strain were continuously monitored using an x-y plotter with the crosshead rate set to 0.03 in./min. Visual inspection for damage was conducted by viewing the blackened edges while a lamp was used to pass light through the translucent material.

ANALYSIS

FINITE ELEMENT MODELS

A quasi-three-dimensional elasticity formulation, which assumes that all laminate cross-sections are under a state of uniform axial extension [9], and the finite element method are used to determine the free-edge stress state in the composite laminates. For symmetric laminates, only one-quarter of the laminate cross-section needs to be modeled, as shown in Figure 1. The X, Y and Z coordinates represent the laminate length, width and thickness directions, respectively.

Four-node quadrilateral elements are used to model the composite laminates. Each node has three translational degrees-of-freedom (DOF) for an element total of twelve DOF. The elements are based on an assumed stress hybrid formulation [10] which provides more accurate stresses than standard displacement based elements.

Three different finite element models are used to determine the ply stresses at the critical ply interface where delamination occurs. The difference in the models is the degree of mesh refinement through the thickness of the two plies adjacent to the critical ply interface. The three models are represented by Figure 2, where the diagonally shaded area denotes the two plies that may be modeled by more than one element through each ply thickness. The thickness of each of the two plies adjacent to the critical ply interface is modeled with either one, three or five elements. Each of the remaining six plies is modeled by one element through the thickness. All of the finite element models use thirty elements across the laminate half-width (b), twenty of which are located near the free-edge. The elements near the free-edge have a length in the Y-direction of one-quarter of a ply thickness (0.25H). Thus, the three models used to determine ply stresses have 837 DOF (240 elements), 1209 DOF (360 elements) and 1581 DOF (480 elements) when the thickness of the plies adjacent to the critical ply interface is modeled with one, three and five elements, respectively.

A fourth finite element model is used to determine the interply resin stresses in the critical ply interface for the laminates studied. This model is shown in Figure 3, where the shaded area represents the interply resin between the plies at the critical interface. An interply resin layer thickness (R) of one-tenth of a ply thickness has been assumed by other authors [11, 12] and is used for this study as well. The mesh refinement in the width direction (Y) is identical to the previously mentioned models. Each of the plies and the interply resin layer is modeled with one element through the thickness. Therefore, this interply resin model contains a total of 930 DOF (270 elements).

Each ply is assumed to be homogeneous and orthotropic. The material properties for the uni-directional prepreg tape, as specified by the manufacturer, are presented in Table 1. The out-of-plane elastic modulus is assumed to be equal to the transverse elastic modulus. Furthermore, the out-of-plane shear moduli are assumed to be equal to the in-plane shear modulus.

The material properties for the interply resin are presented in Table 2.

FAILURE CRITERION

Since delamination is an out-of-plane failure mode, out-of-plane stresses should be the primary contributors to any delamination strength criteria. Therefore, a modification of the Tsai-Wu failure criterion [13] will be developed as a candidate to predict the delamination strength of composite structures.

The basic assumption of the Tsai-Wu strength criterion is that there exists a surface in stress-space which can be expressed in scalar form as

$$F_i \, \sigma_i + F_{ij} \, \sigma_i \, \sigma_j = 1 \qquad\qquad (1)$$

where F_i and F_{ij} (i, j = 1, 2, 3, ..., 6) are strength tensors and σ_i and σ_j are stress tensors equivalent to σ_{xx}, σ_{yy}, σ_{zz}, ..., σ_{xy}, respectively. If delamination is considered an out-of-plane failure mode, independent of the inplane stress state, the inplane stresses can be uncoupled from the out-of-plane stresses, reducing the failure criteria to

$$F_3 \, \sigma_3 + F_4 \, \sigma_4 + F_5 \, \sigma_5$$

$$+ F_{33} \, \sigma_3^2 + 2 \, F_{34} \, \sigma_3 \, \sigma_4 + 2 \, F_{35} \, \sigma_3 \, \sigma_5$$

$$+ F_{44} \, \sigma_4^2 + 2 \, F_{45} \, \sigma_4 \, \sigma_5$$

$$+ F_{55} \, \sigma_5^2 = 1 \tag{2}$$

[handwritten marginal notes: t of interply is 1/10 of t ply. width of element in interply or ply is 1/4 t ply. Bottom line is stresses in interply elements are pretty poor as far as any peaking goes]

Furthermore, if strength is considered to be independent of the sign of the shear stress, all components containing a linear shear stress term may be excluded, leaving

$$F_3 \, \sigma_3 + F_{33} \, \sigma_3^2 + F_{44} \, \sigma_4^2 + F_{55} \, \sigma_5^2 = 1 \tag{3}$$

Finally, if the two out-of-plane shear strengths are assumed to be equal ($F_{55} = F_{44}$), the failure criterion can be expressed as

$$F_3 \, \sigma_3 + F_{33} \, \sigma_3^2 + F_{44} \left[\sigma_4^2 + \sigma_5^2 \right] = 1 \tag{4}$$

The strength parameters can be obtained in terms of the out-of-plane strengths, using experimental tests in tension, compression and shear. Thus, switching to the X, Y, Z coordinate system, the modified Tsai-Wu failure criterion for predicting delamination strength can be represented by

$$\frac{ZC - ZT}{ZC \; ZT} \, \sigma_{ZZ} + \frac{1}{ZC \; ZT} \, \sigma_{ZZ}^2 + \frac{1}{S} \left[\sigma_{YZ}^2 + \sigma_{XZ}^2 \right] = 1 \tag{5}$$

where ZT, ZC and S are the through-the-thickness tensile, compressive and shear strengths, respectively.

For out-of-plane ply failure, the through-the-thickness tensile and compressive strengths are assumed to be equal to the transverse tensile and compressive strengths, respectively. In addition, the through-the-thickness shear strength is assumed to be equivalent to the in-plane shear strength.

For interply resin failure, the resin shear strength is assumed to be equal to the resin tensile strength. Furthermore, since the interply resin layer is very thin, the resin compressive strength is assumed to be equal to the transverse compressive strength of the composite.

The composite and resin strengths are given in Tables 1 and 2, respectively.

AVERAGE STRESS CONCEPT

Failure is assumed to occur when the average stresses over some distance, d_o, force the failure criterion above unity. This stress averaging concept [5, 7, 8, 14, 15] is applied to the out-of-plane stresses obtained from the finite element model. For example, the average interlaminar normal stress near the free edge may be expressesed as

$$\bar{\sigma}_{ZZ} = \frac{1}{d_o} \int_{b-d_o}^{b} \sigma_{ZZ}(Y) \, dY \qquad (6)$$

The smallest averaging distance considered is one-quarter of a ply thickness ($d_o = 0.25H$), which is the dimension (in the Y-direction), of the smallest finite elements. Stress averaging distances of up to 1.5H (in increments of 0.25H) are investigated.

RESULTS AND DISCUSSION

EXPERIMENTAL

Delamination onset was observed as a loss in laminate stiffness that was characterized by a tearing noise upon growth into the width of the laminate. Most of the delamination was evident along the θ/90 interface, although some delamination was observed in the 90/θ interface as well.

Table 3 shows the experimental means, standard deviations (SD) and coefficients of variation (CV) for the strengths of laminates A, B and C. The laminate strengths are given in terms of the average axial stress upon delamination onset. The mean strengths of laminates A, B and C are 47.1 ksi, 48.6 ksi and 49.1 ksi, respectively.

ANALYTICAL

The interlaminar stresses obtained using the three successively refined finite element models (Figure 2) provided the same general stress distributions for each of the laminates, however the magnitudes of the stresses increased as more elements were used to model the thickness of the plies adjacent to the critical interface (interface plies).

The interlaminar stress distributions using five elements to model the thickness of the interface plies are shown in Figures 4, 5 and 6, for laminates A, B and C, respectively. All of the stress distributions shown correspond to an average axial stress of one ksi. Figure 4 shows that laminate A is dominated by the interlaminar normal stress, σ_{zz}. Laminates B and C, however have significant contributions from the interlaminar shear stresses, σ_{yz} and σ_{xz}, with less involvement of σ_{zz} (Figures 5 and 6).

The interply resin stress distributions obtained using the finite element model shown in Figure 3 are given in Figures 7, 8 and 9, for laminates A, B and C, respectively. Once again, the stress state of laminate A (Figure 7) is dominated by the interlaminar normal stress, while the stress states of laminates B and C (Figures 8 and 9) show that large interlaminar shear stresses exist.

Each of the interlaminar stress distributions are averaged near the free edge and the modified Tsai-Wu failure criterion is applied to predict laminate strength for various interlaminar stress averaging distances. The laminate A strength predictions based on the stress distributions from the four finite element models are shown in Figure 10 for stress averaging distances of $0.25H$ to $1.5H$. The experimental mean for laminate A is also shown in the figure. The strength predictions based on ply interface failure are nearly identical whether the thickness of the interface plies is modeled with one, three or five elements. The strength of laminate A based on ply failure can be reasonably predicted using an interlaminar stress averaging distance of one ply thickness ($d_o/H=1$). (This is in agreement with the ply stress averaging distance successfully used in reference 5.) However, if strength is based on interply resin failure, a stress averaging distance of about one-half of a ply thickness ($d_o/H=0.5$) is appropriate.

The laminate B strength predictions are shown in Figure 11. Differences in the ply failure strength predictions are evident as more elements are used to model the interface plies. The best predictions for the strength of laminate B based on ply failure are obtained with the smallest stress averaging distance considered ($d_o/H=0.25$). This is substantially different from the best stress averaging distance for laminate A strength prediction based on ply failure ($d_o/H=0.25$ vs. $d_o/H=1$). On the other hand, adequate strength predictions based on interply resin failure can once again be attained using a stress averaging distance of one-half of a ply thickness.

Figure 12 shows the strength predictions for laminate C. The composite failure strength predictions are quite sensitive to the degree of through-the-thickness mesh refinement. The strength of laminate C based on ply failure is severely overestimated even with the finest mesh and the smallest stress averaging distance considered. However, a stress averaging distance of one-half of a ply thickness continues to provide reasonable predictions when strength is based on interply resin failure.

[handwritten margin note:] Does the interply d_o of ½ ply stay constant because of poor stress resolution in the interply caused by using only one element thru the interply thickness

[handwritten note at bottom:] since interply elements are one thru thickness (1/10 t ply) × ¼ t ply wide, average would be taken over 2 of these to get over ½ t ply

Comparisons of the interlaminar stress distributions from the three finite element models used to predict ply failure in laminate C are shown in Figures 13, 14 and 15. Figure 13 shows that the interlaminar normal stress is fairly insensitive to mesh refinement. However, Figures 14 and 15 show that the interlaminar shear stresses are very sensitive to mesh refinement. Thus, since laminate B, and particularly laminate C, have large interlaminar shear stress contributions, the ply failure strength predictions in Figures 11 and 12 should be expected to be sensitive to mesh refinement.

SUMMARY AND CONCLUSIONS

The delamination strength of sixteen ply quasi-istoropic glass-epoxy laminates with significantly different free-edge stress states was studied. A strength-of-materials approach, utilizing the average stress concept and a delamination failure criterion, was applied to interlaminar stress distributions obtained from quasi-three-dimensional finite element models. The interlaminar normal stress distributions obtained from the stresses in the plies adjacent to the delaminating interface were insensitive to through-the-thickness mesh refinement. However, the interlaminar shear stress distributions were sensitive to mesh refinement. Basing laminate strength on ply failure at the critical ply interface resulted in poor agreement in the appropriate stress averaging distances for each of the laminates studied. However, basing laminate strength on failure of the interply resin layer between the delaminating plies provided good results for all laminates when an interlaminar stress averaging distance of one-half of a ply thickness was used.

ACKNOWLEDGEMENT

This work was sponsored by the U.S. Army Research Office (Contract No. DAAG-29-83K002). The program monitor was Dr. Robert Singleton.

REFERENCES

1.   Pipes, R. Byron and Pagano, N.J., "Interlaminar Stresses in Composite Laminates Under Uniform Axial Extension," Journal of Composite Materials, Vol. 4, October 1970, pp. 538-548.

2.   Wang, A.S.D. and Crossman, F.W., "Some New Results on Edge Effect in Symmetric Composite Laminates," Journal of Composite Materials, Vol. 11, January 1977, pp. 92-106.

3.   O'Brien, T.K., "Characterization of Delamination Onset and Growth in a Composite Laminate," Damage in Composite Materials, ASTM STP 775, American Society for Testing and Materials, 1982, pp. 140-167.

4.   Wang, S.S., "Edge Delamination in Angle-Ply Composite Laminates," AIAA Journal, Vol. 22, No. 2, February 1984, pp. 256-264.

5.   Kim, R.Y. and Soni, S.R., "Experimental and Analytical Studies on the Onset of Delamination in Laminated Composites," Journal of Composite Materials, Vol. 18, January 1984, pp. 70-80.

6.   O'Brien, T.K., "Mixed-Mode Strain Energy Release Rate Effects on Edge Delamination of Composites," Effects of Defects in Composite Materials, ASTM STP 836, American Society for Testing and Materials, 1984, pp. 125-142.

7.   Soni, Som R. and Kim, Ran Y., "Delamination of Composite Laminates Stimulated by Interlamiar Shear," Composite Materials: Testing and Design (Seventh Conference), ASTM STP 893, American Society for Testing and Materials, 1986, pp. 286-307.

8.   Lagace, Paul A., "Delamination in Composites: Is Toughness the Key?," SAMPE Journal, Society for the Advancement of Material and Process Engineering, November/December 1986, pp. 53-60.

9.   Lekhnitskii, S.G., Theory of Elasticity of an Anisotropic Body, MIR Publishers, Moscow, 1981, English translation of the revised 1977 Russian edition.

10. Lee, S.W., Rhiu, J.J. and Wong, S.C., Hybrid Finite Element Analysis of Free-Edge Effect in Symmetric Composite Laminates, Technical Report, Department of Aerospace Engineering, University of Maryland, June 1983.

11. Wang, S.S., "An Analysis of Delamination in Angle-Ply Fiber-Reinforced Composites," Journal of Applied Mechanics, Vol. 47, March 1980, pp. 64-70.

12. Kemp, B.L. and Johnson, E.R., "Response and Failure Analysis of a Graphite/Epoxy Laminate Containing Terminating Internal Plies," Proceedings of the AIAA/ASME/ASCE/AHS 26th Structures, Structural Dynamics and Materials Conference, Part 1, Orlando, Florida, April 15-17, 1985, pp. 13-24, AIAA Paper No. 85-0608.

13. Tsai, S.W. and Wu, E.M., "A General Theory of Strength for Anisotropic Materials," Journal of Composite Materials, Vol. 5, 1971, pp. 58-80.

14. Whitney, J.M. and Nuismer, R.J., "Stress Fracture Criteria for Laminated Composites Containing Stress Concentrations," Journal of Composite Materials, Vol. 8, July 1974, pp. 253-265.

15. Nuismer, R.J. and Whitney, J.M., "Uniaxial Failure of Composite Laminates Containing Stress Concentrations," Fracture Mechanics of Composites, ASTM STP 593, American Society for Testing and Materials, 1975, pp. 117-142.

TABLE 1 - SP-250-S29 GLASS-EPOXY UNI-DIRECTIONAL TAPE

ELASTIC MODULI	E_L = 7.0 x 10⁶ psi	
	E_T = 2.1 x 10⁶ psi	
SHEAR MODULUS	G_{LT} = 0.8 x 10⁶ psi	
POISSON'S RATIO	ν_{LT} = 0.26	
LONGITUDINAL TENSILE STRENGTH	260 ksi	
LONGITUDINAL COMPRESSIVE STRENGTH	145 ksi	
TRANSVERSE TENSILE STRENGTH	6.2 ksi	
TRANSVERSE COMPRESSIVE STRENGTH	29 ksi	
INPLANE SHEAR STRENGTH	14 ksi	

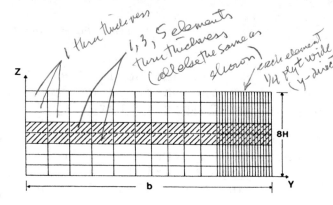

Figure 2 - Finite element model for ply interface stresses

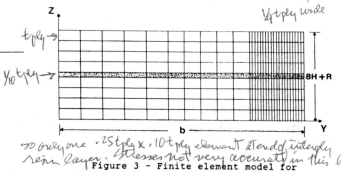

Figure 3 - Finite element model for interply resin stresses

TABLE 2 - RESIN PROPERTIES

ELASTIC MODULUS	0.57 x 10⁶ psi
SHEAR MODULUS	0.15 x 10⁶ psi
POISSON'S RATIO	0.37
TENSILE STRENGTH	9.0 ksi

TABLE 3 - EXPERIMENTAL LAMINATE STRENGTHS (KSI)

LAMINATE	SPECIMENS	MEAN	SD	CV(%)
A	5	47.1	1.83	3.89
B	5	48.6	0.78	1.60
C	4	49.1	3.63	7.38

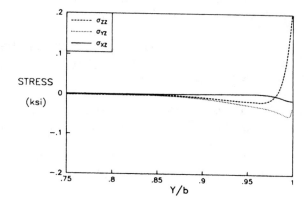

Figure 4 - Laminate A ply interface stresses

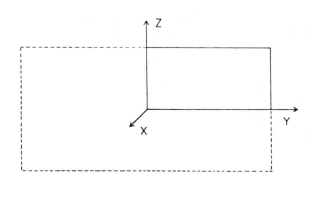

Figure 1 - Modeled section of symmetric laminate

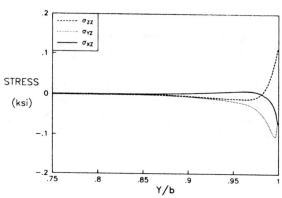

Figure 5 - Laminate B ply interface stresses

250

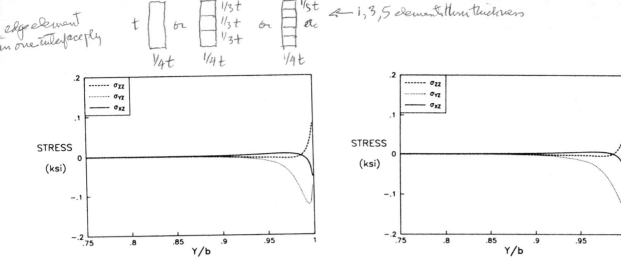

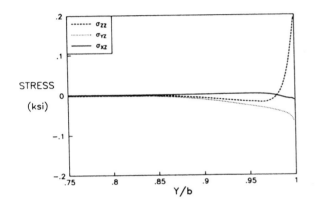

Figure 6 - Laminate C ply interface stresses

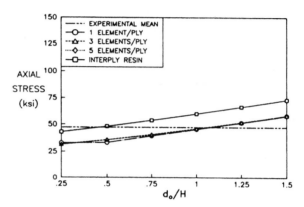

Figure 9 - Laminate C interply resin stresses

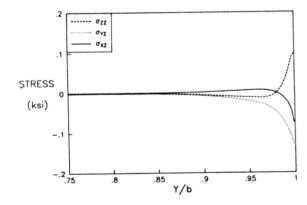

Figure 7 - Laminate A interply resin stresses

Figure 10 - Laminate A strength predictions

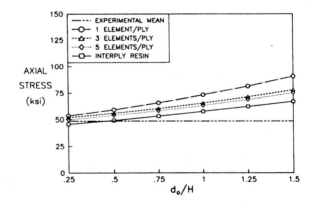

Figure 8 - Laminate B interply resin stresses

Figure 11 - Laminate B strength predictions

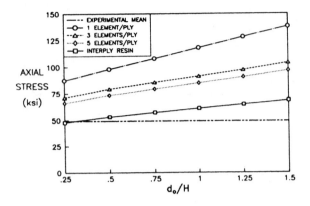

Figure 12 - Laminate C strength predictions

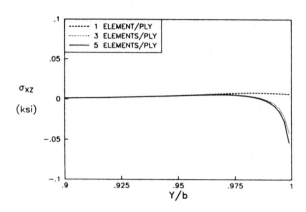

Figure 15 - Comparison of σ_{XZ} distributions

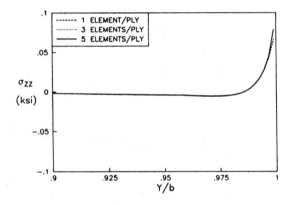

Figure 13 - Comparison of σ_{ZZ} distributions

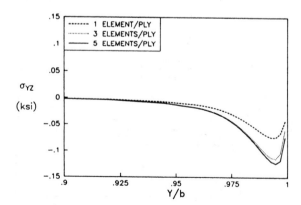

Figure 14 - Comparison of σ_{YZ} distributions

Advances in Materials and Process Engineering

Application of Sliding Mode Control to the Cure of a Thermoset Composite

A. S. TAM AND T. G. GUTOWSKI

Abstract

Present cure cycles for thermoset composites are determined empirically; they are probably inefficient and suboptimal in most cases and may not be able to handle possible material, part, or process variations. A control strategy is presented, based on the sliding mode control concept, which monitors and alters each cure cycle in-process. Performance is simulated for the compression molding of a Hercules AS4/3501-6 flat laminate. The system reaches a specified degree of cure and fiber volume fraction within given time limits. The controller displays good robustness to parameter changes and model uncertainties. Results indicate that relevant parameters are the size of the part and its initial moisture content. With this control method, parts can be made which cannot be successfully cured by using the standard cure cycle.

Introduction

The cure cycle for processing a thermoset composite part is typically one determined empirically by the material supplier to ensure complete cure of the matrix. Since it usually does not take into account such aspects as part geometry, material variations, or process changes, the recommended cure cycle is necessarily conservative and is sub-optimal for many cases. Also, it requires little or no feedback measurement during processing of the part, so no guarantee exists that the part has cured properly. Because each part usually has a large inherent value (from both labor and material), an improperly cured part incurs a large cost. An overly conservative cure cycle, though, decreases throughput and makes inefficient use of the machinery.

This research addresses the above problems by exploring methods to optimize the cure cycle *for each individual part*. The approach is to use *closed loop control*, shown schematically in Figure 1. At this point, we would like to differentiate this form of control from those common in industry:

Albert S. Tam, Research Assistant, and Timothy G. Gutowski, Associate Professor of Mechanical Engineering and Director of the MIT-Industry Composites and Polymer Processing Program, Laboratory for Manufacturing and Productivity, Massachusetts Institute of Technology, Cambridge, MA 02139.

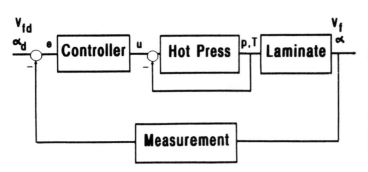

Figure 1.   Schematic representation of a closed loop control system.

(numerical) machine control and statistical process control. "Machine control" means only that the processing machinery is made to follow input commands to the best of its ability. This corresponds to the inner loop in Figure 1, which ensures that the hot press is supplying the pressure and temperature specified by a preset cure cycle. With "statistical process control," a nominal cure cycle is first used and the finished parts are then randomly tested downstream. The whole process is adjusted only if a part is not within specification. It is in fact a type of closed loop control, with changes made to the cure cycle after-the-fact. However, it does not work well if only a small number of each part is to be made, and for large batch runs, may incur large costs in bad parts and wasted processing time.

Closed loop control means that measurements of critical parameters are obtained during the process and compared with desired values. A controller (typically a microprocessor) then uses the discrepancy between the measured and desired values to change the cure cycle *during each individual process*. What remains is to choose a proper controller formulation to guarantee good performance and stability. To determine how to alter the cure cycle to drive the system to the desired state, the controller requires a mathematical model of the process physics. Several major problems exist. The process dynamics are highly nonlinear and are not well handled by classical controller formulations. Moreover, a certain amount of error is inherent in the models which is aggravated by possible material and process variations. The sliding mode formulation was chosen for its particular ability to handle these concerns.

Model Development

The process to be studied is the compression molding of a Hercules AS4-3501-6 flat laminate. This process contains many aspects of the general cure problem, but is simple enough that a good understanding of the underlying physics has been achieved and is available in the literature. Let us first identify the following state variables x_i and control efforts u_i:

$$
\begin{aligned}
x_1 &\equiv V_f, & u_1 &\equiv p, \\
x_2 &\equiv \alpha, & u_2 &\equiv T_{in}, \\
x_3 &\equiv T_{part},
\end{aligned}
\tag{1}
$$

where V_f is the fiber volume fraction of the part, α is the degree of cure of the resin, T_{part} is the temperature of the part, p is the applied pressure, and T_{in} is the applied temperature. The state variables chosen are sufficient to completely describe the system status during the process.

The system dynamics may be described in standard form as:

$$\dot{\underline{x}} = \underline{f}(\underline{x}) + \underline{g}(\underline{x})\,\underline{u},$$
$$\underline{y} = \underline{h}(\underline{x}), \tag{2}$$

where $\underline{x}$ is the state vector, $\underline{y}$ is the output vector, and the dot indicates differentiation with respect to the time, t.

For this process, the elements of the $\underline{f}$ and $\underline{g}$ matrices are:

$$\underline{f}(\underline{x}) = \begin{bmatrix} - (3S_{xx}\sigma x_1)/(\mu a^2) \\ f_2(x_2,x_3) \\ - x_3/\tau \end{bmatrix}, \tag{3a}$$

$$f_2 = \begin{cases} (K_1+K_2 x_2)(1-x_2)(B-x_2) & 0 < x_2 \leq 0.3 \\ K_3(1-x_2) & 0.3 < x_2 \leq 0.5, \end{cases} \tag{3b}$$

$$K_i = A_i \exp(\, -\Delta E_i/(Rx_3)\,) \qquad , i=1..3, \tag{3c}$$

$$\underline{g}(\underline{x}) = \begin{bmatrix} (3S_{xx}\sigma x_1)/(\mu a^2) & 0 \\ 0 & 0 \\ 0 & 1/\tau \end{bmatrix}, \tag{4}$$

$$\underline{u} = \begin{bmatrix} u_1 \\ u_2 \end{bmatrix}, \tag{5}$$

where S_{xx} is the permeability of the fiber bed, a is half the part length, σ is the fiber stress, μ is the viscosity of the resin, τ is the heat transfer lag, A_i are the pre-exponential factors, ΔE_i are the activation energies, and R is the universal gas constant. Both S_{xx} and σ are non-linear functions of V_f.

The outputs of interest are the fiber volume fraction and degree of cure, so the output vector is:

$$\underline{y}(\underline{x}) = \underline{h}(\underline{x}) = \begin{bmatrix} x_1 \\ x_2 \end{bmatrix}. \tag{6}$$

The dynamics of the consolidation is obtained from Gutowski's model [1] and the cure kinetics for this particular resin system from Springer's model [2]. In addition, Springer provides the following relationship between the degree of cure, part temperature, and viscosity:

$$\mu(\alpha,T_{part}) = \mu_\infty \exp(\, (U/(RT_{part})) + K\alpha\,), \tag{7}$$

where U is the activation energy and μ_∞ and K are empirical constants. The viscosity couples all the state variables and introduces a further non-linearity into the system.

We have two available control inputs, p and T_{in}, and two main variables of interest, V_f and α. The system is controllable in the sense that we can affect the desired outputs with the given inputs. However, not included in the dynamics is the fact that the process is irreversible. Once the resin is squeezed out of the part, it cannot be sucked back in, so V_f must be monotonically increasing. Moreover, for a thermosetting epoxy, the degree of cure cannot be reversed and must also be monotonically increasing.

There is also a third quantity of interest, the mean resin pressure p_r given by:

$$p_r = p - \sigma(V_f). \tag{8}$$

This quantity cannot be controlled independently, but it is of prime importance since it provides an indication of the possibility of void growth in the resin. Kardos [3] has found that to suppress the predominant water-based voids in a graphite/epoxy system, we need to maintain a minimum pressure in the resin given by:

$$p_v > 4.962 \times 10^3 \exp(-4892/T_{part}) \, (RH)_0, \tag{9}$$

where p_v is the void suppression pressure in atmospheres (absolute), and $(RH)_0$ is the initial equilibrium relative humidity of the part. For a part with a certain moisture content, this criterion states that we must apply a certain pressure for a given part temperature. This minimum pressure level tends to squeeze the resin out of smaller parts much faster than larger parts. Resin loss in turn leads to a *lower* resin pressure since more load is supported by the fibers, and we are left with a potentially unstable situation.

MIMO Sliding Mode Control

The full mathematics of multi-input-multi-output (MIMO) sliding mode controller design and proofs of its stability and performance are beyond the scope of this paper. Utkin [4], Slotine [5], and Fernandez and Hedrick [6] will provide an in-depth treatment of this controller formulation for the interested reader. Here, we present the underlying equations from which application of standard sliding mode design follows.

We first define the errors e_i between the measured outputs y_i and the desired output values y_{id}:

$$e_i = y_i - y_{id}(t) \quad , \ i=1,2. \tag{10}$$

We next chose "sliding surfaces" S_i in state variable space. These surfaces simply describe where we would like the states to be. The controller then calculates the efforts required to drive the system to these surfaces and keep it there once the surface is reached. We define each surface in terms of the errors so that being on the surface means that the system is tracking the desired trajectory:

$$S_1 = e_1 + \lambda_1 \int_0^t e_1 \, d\tau,$$

$$(11)$$

$$S_2 = \dot{e}_2 + \lambda_2 e_2,$$

where λ_i are chosen to give desired response dynamics on the surface.

The control efforts are now calculated to ensure that the surfaces are "attractive". This means that whenever the system state is not on the surface, the control will tend to move the states toward it:

$$u_1 = - [\eta_1 \mathrm{sgn}(S_1) + f_1 + \lambda_1 e_1 - \dot{y}_{1d}]/g_{11}(\underline{x}),$$

$$(12)$$

$$u_2 = - \left[\eta_2 \mathrm{sgn}(S_2) + \frac{\partial f_2}{\partial x_2} f_2 - \frac{\partial f_2}{\partial x_3} \frac{x_3}{\tau} + \lambda_2 (f_2 - \dot{y}_{2d}) - \ddot{y}_{2d} \right] \frac{\tau}{\dfrac{\partial f_2}{\partial x_3}},$$

$$(13)$$

where η_i are non-negative control gains chosen to ensure the "attractiveness" condition.

The benefit of using a sliding mode controller in this application is its robustness to modeling errors. As long as we have an idea where the modeling errors occur, and we can place bounds on the error magnitudes, we can account for it in the controller. The primary modeling error occurs in the empirical expression for the viscosity, which in turn greatly affects the compaction dynamics. The value of K in Eq. 7 is accurate to within 10%, resulting in uncertainty in both $f_1(\underline{x})$ and the first control gain, $g_{11}(\underline{x})$. With this information on the error, we can calculate a new (time-varying) η_1 to ensure that the sliding surface is still "attractive".

We next design the command paths to the controller to maintain some desired resin pressure and enhance void suppression ability. We examine the cure process only up until the resin gels ($y_2=0.5$), after which no more flow is possible. A power-law path is selected for y_2, allowing a moderate temperature build up. The path for y_1 is chosen to apply pressure up until gelation, but not squeeze so hard that the desired compaction is reached too early and the applied pressure drops:

$$y_{1d}(t) = \begin{cases} V_0 & t \le t_s \\ V_0 + (V_{fd}-V_0)(1-\exp(-(t-t_s)/\tau_c)) & t > t_s, \end{cases}$$

$$(14)$$

$$y_{2d}(t) = 0.5(t/t_{gel})^n, \quad n \ge 1,$$

$$(15)$$

where V_{fd} and V_0 are the desired and initial fiber volume fraction, respectively, t_{gel} is the desired time to reach gelation, t_s is the time to start compaction, and τ_c is a suitably chosen time constant.

Simulations and Results

The system equations and controller were simulated using the nonlinear package SIMNON to determine performance. The specifications were to achieve gelation and a certain degree of compaction within a reasonable time. If possible, the resin pressure at gelation was to be kept above the necessary void suppression pressure at that temperature. The system parameters used for the simulation were obtained from the models cited above.

A time limit of one hour to gelation was given to the controller. A parabolic trajectory for the degree of cure gave satisfactory results. The compaction was set to begin after approximately 12.5 minutes. The command for fiber volume fraction was made to be a first order response to a step with four time constants elapsing until gelation time.

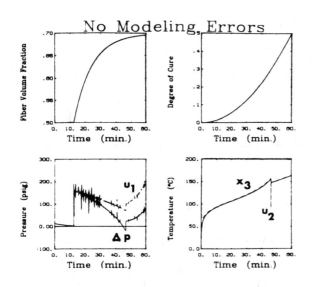

Figure 2. System response assuming no modeling errors for a = 6 in.

Figure 2 shows the results for a=6 in. (12 in. long part) and V_{fd}=0.70 with no modeling errors. The sliding mode gains η_i and λ_i were all set at 1. Both the fiber volume fraction and the degree of cure tracked the desired trajectory perfectly. In addition, the pressure plot shows the applied pressure and the void suppression pressure difference, Δp, given by

$$\Delta p = p_r - p_v \qquad (16)$$

where p_v is the minimum pressure needed to suppress voids at the current part temperature, assuming the worst possible initial moisture content with RH_0=100%. For voids to be suppressed, Δp must be positive, ideally throughout the cure.

The required applied pressure was certainly within the capabilities of a typical hot press. The initial temperature rise may be a bit steep, but should still be achievable; also, the temperature did not exceed the typical hold temperature for this resin of 177°C. Though Δp was negative for a short time, it was well above zero at gelation and void growth should be suppressed.

When the system was simulated with the smaller part size of a=3 in., the command trajectories were still followed, but the void suppression criterion was not met. To remedy the problem, a saturation limit was introduced into the control laws: we placed an upper bound on the applied temperature to make sure the available resin pressure was greater than the needed void suppression pressure. The results of the simulation with this restriction imposed and a=3 in. are shown in Figure 3. Both the compaction and void suppression were satisfactory, but the resin still had not reached gelation after 80 min. The temperature ceiling imposed caused the resin to cure slower, and though the final degree of compaction had been reached,

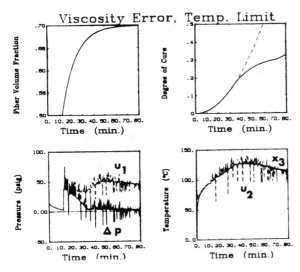

Figure 3. System response with actual viscosity lower than assumed for a = 3 in.

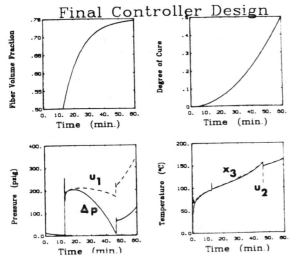

Figure 4. Final controller design response for worst case scenario: a = 3 in.; 100% RH; viscosity lower than assumed.

the resin had not gelled and could still flow.

We next investigated the effect of the modeling errors. The possible error in viscosity means $g_{11}(\underline{x})$ and $f_1(\underline{x})$ may vary up to 80%. The controller was modified to ensure the sliding surfaces were still "attractive". No problem existed if the actual viscosity was higher than predicted, since the controller could simply apply more pressure. However, if the actual viscosity was lower, then squeezing harder caused more resin to flow, in turn causing a large drop in resin pressure. For this case, the problem was solved by commanding the part to a higher V_{fd}. In addition, a first order filter was introduced between the calculated control pressure and the actual applied pressure to account for the mechanical response of the hot press. Figure 4 shows the simulation result for the worst case scenario—small part (a=3 in.), 100% RH, and viscosity lower than expected. Both compaction and gelation were achieved within 60 min. and the void suppression criterion was met. The part temperatures in both cases were reasonable. The required applied pressure was higher than before, but should still be within the capabilities of a typical hot press.

Concluding Remarks

The sliding mode controller design provides excellent control over the cure cycle necessary to cure a part to a certain fiber volume fraction. In addition, the tailored command trajectories enhance the ability to suppress voids. Control effort restrictions imposed do not seriously compromise system performance and introduce no instabilities. From the simulations, it is apparent that the small, wet parts pose the greatest problems. However, the controller can overcome these problems by either compacting to a higher fiber volume fraction or by taking longer to cure the part. In this manner, the control scheme can make parts not possible with the standard cure cycle.

In terms of implementation, the controller assumes full state feedback. The fiber volume fraction is obtainable by measuring the part height. The degree of cure and the part temperature are both obtainable by using a microdieletric sensor available from Micromet Instruments, Inc. The viscosity and thence degree of cure may also be reconstructed by measuring the rate of change of the part height over time. Though these measurements may be noisy, the dynamics of our system is slow enough that most of the noise can be filtered out, either with suitable signal processing or a Kalman filter in the controller. Likewise, the sliding mode controller requires a significant amount of calculations, but again, with currently available microprocessors, the computation time should be insignificant with respect to the required system bandwidth.

References

1.   Gutowski, T.G., Cai, Z., et.al., "Consolidation Experiments for Laminate Composites," Journal of Composite Materials, Vol. 21, June 1987, p. 650.

2.   Lee, I.W., Loos, A.C., and Springer, G.S., "Heat of Reaction, Degree of Cure, and Viscosity of Hercules 3501-6 Resin," Journal of Composite Materials, Vol. 16, Nov. 1982, p. 510.

3.   Dave, R., Kardos, J.L., and Dudukovic, M.P., "Process Modeling of Thermosetting Matrix Composites: A Guide for Autoclave Cure Cycle Selection," Proceedings of ASFC, First Tech. Conf., Oct 7-9, 1986, p. 137.

4.   Utkin, V.I., "Variable Structure Systems with Sliding Modes", IEEE Transaction on Automatic Control, AC-22, April 1977, p. 212.

5.   Slotine, J., "Sliding Controller Design for Nonlinear Systems", International Journal of Control, Vol. 40, 1984, p 421.

6.   Fernandez, B., and Hedrick, J.K., "Control of Multivariable Non-Linear Systems by the Sliding Mode Method," International Journal of Control, Vol. 46, Sept. 1987, p. 1019.

Rheological and Chemical Characterization
of a Glass Reinforced Polyester Prepreg Material

LORETTA A. PETERS

ABSTRACT

Although extensive characterization efforts have been carried out by the aerospace companies on the advanced graphite epoxy prepreg systems, very little data exist on the rheological and chemical characterization of the glass reinforced polyester prepreg materials used in non-aerospace applications like the thick laminate ballistic composites used on military combat vehicles. This information is important because chemistry and rheology have a large impact on part performance. This study will document the methods and results of film-fiber fixture rheology, reverse phase liquid chromatography, and size exclusion chromatography measurements.

INTRODUCTION

In the aerospace industry, it is accepted that no allowables program, component test program, or in-service evaluation of epoxy graphite prepregs can achieve its full potential without the assurance that the initial material is of high quality, uniformity, and continuing consistency [1]. Since the expected increase in ballistic threats will result in heavier military combat vehicles unless lighter composite armor materials are used, a significant effort is now underway on the use of composites in the Army's land vehicles. Specifically, the S-2 glass/polyester composite panels have demonstrated particularly good ballistic armor performance against fragment simulation projectiles [2]. Further, due to their non-metallic composition, glass reinforced plastic (GRP) composites have shown potential for reducing spallation from shaped charges. In order to successfully apply composite materials in ground vehicles, improved quality assurance methods for the initial polyester-glass prepreg materials are of utmost importance, just like in the aerospace industries.

In the present paper, a preliminary quality assurance methodology, consisting of dynamic mechanical analysis (DMA), reverse phase liquid chromatography (RPLC) and size exclusion chromatography (SEC) will be applied to two versions of a polyester resin matrix in order to show their utility in assuring quality, uniformity, and consistency.

L. A. Peters, Member of Technical Staff, Polymer Technology, FMC-Central Engineering Laboratories, 1205 Coleman Avenue, Santa Clara, CA 95052

LIQUID CHROMATOGRAPHY: SIZE EXCLUSION (SEC) AND REVERSE PHASE (RPLC)

The advent of high performance liquid chromatographic (HPLC) methods have greatly aided the analysis of oligomeric molecules or polymers of modest molecular weight [3]. The chromatograms in Figure 1 (obtained by size exclusion on a 500 Angstrom crosslinked styrene-divinyl benzene column) show four identifiable components with elution times at 5.4, 6.5, 9.2 and 10.4 minutes for an earlier version A of the polyester resin matrix prepreg and only three identifiable components with elution times at 5.4, 6.5, and 9.2 minutes for the latter version B of supposedly the same polyester resin matrix prepreg. By injection of standards, the peaks in both version A and version B at 5.4, 6.5, and 9.2 minutes were identified to be prepolymer diallyl phthalate (DAP), chlorinated polyester, and DAP-based polyester monomer, respectively. In other words, these peaks are associated with the low molecular weight components and the unsaturated polyester base resin used in the formulation. The extra peak at 10.4 minutes in version A was identified to be a .1% solution of quinhydrone, or roughly 10 ppm (Figure 2). Substituted hydroquinones are of low molecular weight (100 to 150) and are strong chromophores which are used to retard resin cures. The SEC chromatographic conditions are as follows:

Use approximately 0.30 gram of the as-received material in 10 ml Tetrahydrofuran (THF). Shake for 10 minutes. Filter the solution through a 0.5 micron Fluoropore filter before sample injection. Observe the following instrument conditions:

Column:	500 $\overset{o}{A}$ Waters Ultrastyragel, 7.8 mm ID x 30 cm
Solvent:	THF
Flow Rate:	1.0 ml/min
Sample size:	10 microliter
Detector:	UV 254 nm

Previous work has documented, using reverse-phase gradient liquid chromatography (RPLC) with a fluorescence detector, the utility of RPLC in obtaining quantitative information on the low molecular weight components (material eluting after 9 minutes in the SEC) [4]. This in conjunction with the SEC's capacity of detecting changes in the base resin can be used to completely chemically characterize the resin.

RHEOLOGICAL CHARACTERIZATION USING FILM FIBER FIXTURE

The tensile storage modulus (E'), the tensile loss modulus (E"), and the tan delta (δ) of the two versions of the polyester prepreg were measured using a Rheometrics Mechanical Spectrometer, Model RMS-800. Using the film fiber fixture (Figure 3), the motion of the motor is transferred to linear motion or elongation, so that the film of prepreg is stretched rather than sheared. The prepreg film is initially stretched before the application of the sinusoidal strain in order to prevent the film from buckling from the oscillatory motion. A frequency sweep from .1 to 100 radians at .02% strain and 21°C nominal was performed. These frequency curves are overlaid in Figure 4 and show the version B prepreg to be stiffer than the version A prepreg across the frequency range.

SUMMARY

The objective of this paper was to show that quality assurance methods adapted from the aerospace industries can be used on polyester glass prepreg. Specifically, it was shown that the combined liquid chromatographic methods of SEC and RPLC are very sensitive to resin formulation variations. Dynamic mechanical properties of the prepreg in tension will be determined more by the fabric reinforcement than the resin matrix. Visually, prepreg version B had a tighter weave and was measured to have a higher elastic modulus, E', than prepreg version A. In order to prove the efficacy of these quality assurance methods in predicting the performance of the thick laminate composites, these methods will have to be correlated to processing parameters and mechanical properties.

REFERENCES

1.   Chen, J.S., and Hunter, A.B., <u>Development of Quality Assurance Methods for Epoxy Graphite Prepreg</u>, NASA Contractor Report 3531, Prepared for Langley Research Center under Contract NAS1-15222, March 1982, page 2.

2.   Vasudev, A., and Mehlman, M. J., "A Comparative Study of the Ballistic Performance of Glass Reinforced Plastic Materials," <u>SAMPE Quarterly</u>, Vol. 18, No. 4, July 1987, pp. 43-48.

3.   Mazur, S., "Ester Interchange Reactions, The Determination of Sequence and Block Size Distributions in Copolyesters," American Chemical Society National Meeting, Las Vegas, Nevada, 1980.

4.   Dunn, D., Army Materials Technology Laboratory, Watertown, MA, Personal Communication, March 1988.

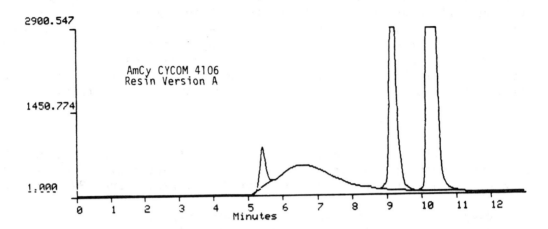

Peak Name	Ret time	Area	Height	Type	Amount	RF
Unknown peak	5.42	7765449	691130	BB	0.000	0.0000e+00
Unknown peak	9.23	46934972	2829539	BB	0.000	0.0000e+00
Unknown peak	10.47	71049224	2845716	BB	0.000	0.0000e+00

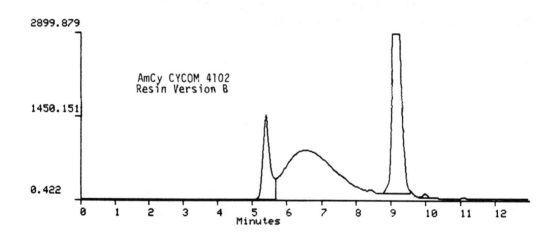

Peak Name	Ret time	Area	Height	Type	Amount	RF
Unknown peak	5.40	20738972	1473461	**	0.000	0.0000e+00
Unknown peak	9.08	57172736	2783506	**	0.000	0.0000e+00
Unknown peak	9.98	484975	57484	**	0.000	0.0000e+00

Fig. 1.   Size-exclusion chromatogram of
          prepreg version A (top) and prepreg
          version B (bottom)

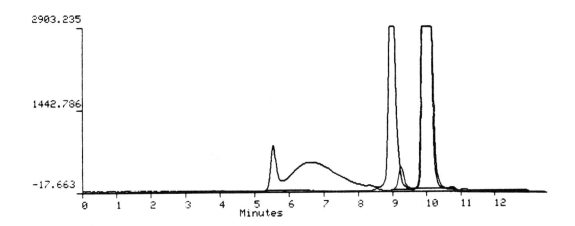

Peak Name	Ret time	Area	Height	Type	Amount	RF
Unknown peak	9.25	4919912	395124	**	0.000	0.0000e+00
Unknown peak	10.13	66688304	2873989	**	0.000	0.0000e+00

Fig. 2. Size-exclusion chromatogram of a
.1% solution of quinhydrone
overlaid on prepreg version A resin
matrix

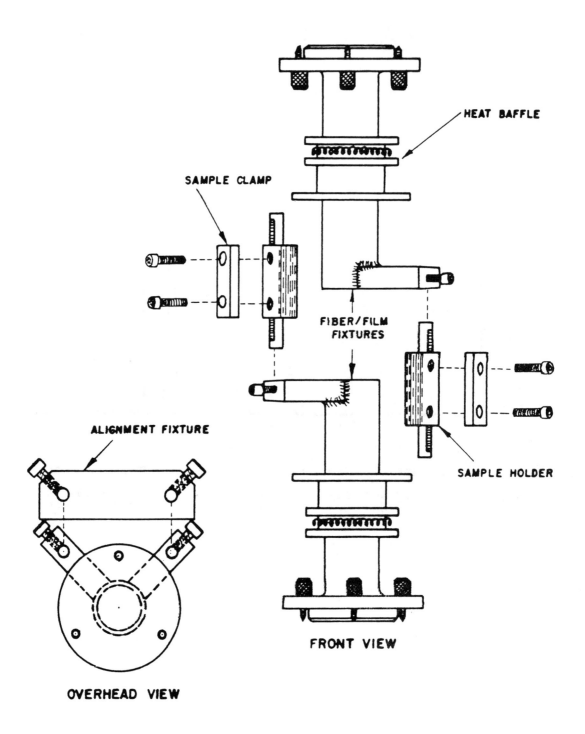

HEAT BAFFLE

SAMPLE CLAMP

FIBER/FILM
FIXTURES

SAMPLE HOLDER

ALIGNMENT FIXTURE

OVERHEAD VIEW

FRONT VIEW

Fig. 3. View of the film-fiber fixture used
in the RMS-800

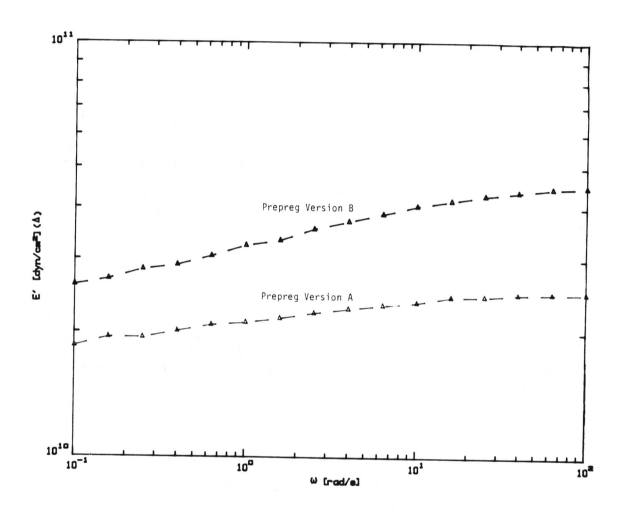

Fig. 4. Comparison of the rate sweep curves
between prepreg version B (top) and
prepreg version A (bottom)

Toughening of a Thermoplastic Matrix Composite: Processing Issues

D. S. PARKER* AND A. F. YEE*

ABSTRACT

The use of a rubber modified thermoplastic resin has been investigated as a method to improve the impact damage and the Mode I interlaminar fracture toughness of a unidirectional continuous carbon fiber composite. Test results show that the improvement in the fracture toughness is less than expected due to rubber particle agglomeration, poor fiber/matrix adhesion and the induced crystallization of the matrix. These restrictions must be eliminated in order to evaluate the maximum potential of a rubber toughened thermoplastic matrix on a composite's fracture toughness.

INTRODUCTION

High performance continuous fiber composites offer stiffness and strength which are superior to metals on a per weight basis [1]. This has lead to an increase in the use of composites in the aerospace and automotive industries. The most commonly used polymeric matrices and fibers are epoxies and glass and carbon fibers. However, a major problem with these composites is that they are sensitive to out-of-plane impact which, in turn, reduces the subsequent compressive strength and interlaminar fracture toughness [2,3]. The compressive strength after an impact event can be reduced as much as 50% [3]. Composite structures must be tolerant of both manufacturing defects and damage caused by impacts during service.

During impact of a composite, cracks may develop due to fiber failures, processing defects or high stresses developing between the plies. If the structure is subsequently loaded in compression, crack propagation may occur (Figure 1). The weakest fracture mode in composites is crack propagation between plies (delamination) or within plies parallel to the fibers (splitting) [4,5]. Crack propagation is influenced by three primary

---

* Department of Materials Science and Engineering, University of Michigan. Ann Arbor, MI 48105

variables: fiber/matrix adhesion, matrix toughness and fiber volume fraction. In high performance composites, the fiber volume fraction is maximized. Assuming good fiber/matrix adhesion, crack propagation will be controlled by the properties of the matrix [3,6-8]. Therefore, to improve the damage tolerance of a composite the majority of research efforts are focused on producing tough or damage tolerant high performance composites through modification of the matrix.

Hunston and co-workers have recently supplied a survey of literature fracture data comparing a resin's fracture toughness with that of a composite utilizing the same resin as a matrix material [9,10]. Of course it is impossible to compare each test directly due to differences in the fiber volume fractions, testing rates and conditions, etc., however, the trends are very interesting. For brittle epoxy resins with a strain energy release rate value < 200 J/M^2, the corresponding composite's G_{1C} value can be as high as three times that of the matrix resin's bulk value. However, for tough matrix resins such as rubber modified epoxies and thermoplastic resins, the composite's G_{1C} value is about one third that of the matrix resin. Reasons cited for these shortcomings are: 1) the plastic zone size is constrained by the fibers, 2) the rubber particles are too large in diameter, 3) poor fiber/matrix adhesion, and 4) residual stresses due to processing [2,8,9,11-15].

In order to develop toughened composites utilizing tough matrix resins, we must provide answers to the following questions. What are the primary deformation mechanisms that are operative within the highly constrained interfibril regions of a composite? Are the toughening techniques used for neat resins applicable in enhancing these deformation mechanisms? And finally, what variables are important to allow the toughening method chosen to operate within a composite structure? Our objective then is to determine these deformation mechanisms and develop an appropriate toughening technique which will bring about an increase in interlaminar fracture toughness.

Recently we have investigated using a toughened thermoplastic resin as a matrix material for a continuous fiber composite. Our approach for improving the interlaminar fracture toughness will be briefly described. A tough polycarbonate (PC) matrix can be toughened using core shell rubber particles having a diameter ($< 1\mu$m) much smaller than the interlaminar spacing. These core shell rubber particles have an advantage over liquid rubber tougheners or block co-polymers in that the particle size is independent of the processing conditions. In PC, which deforms by shear yielding, the rubber particles will enhance the formation of shear bands which results in an increase in fracture toughness. This toughening technique is described in the included references [16-19]. The toughened resin is in turn used as a matrix material for a continuous fiber composite. During loading, the rubber particles will reduce the constraint developed between the fibers by cavitating and will reduce the tendency for brittle fracture. The stress concentration at the cavitation sites will promote shear yielding, and by stable shear band growth the deformation zone will expand beyond the interlaminar spacing (Figure 2). Since the volume of material contained within the deformation zone is now larger than that in the unmodified matrix, the interlaminar fracture toughness would be expected to increase. Although our work will involve a thermoplastic matrix material, this will not preclude the possibility of transferring the insight gained here to other polymer matrix composites.

Through our work we have discovered a number of processing factors which produce a weak fiber/matrix interface in a toughened thermoplastic matrix composite and in turn limits the composite's fracture toughness which utilizes these tough resins. These limiting factors are the focus of our discussion. Although some of these variables are specific to our approach to toughening a thermoplastic matrix composite, others are independent of the toughening technique.

EXPERIMENTAL

Polycarbonate (General Electric Lexan® 141 resin) was blended with 10 Wt% rubber particles. The types of core shell rubber particles used were Acryloid ® KM330 and KM653 which are products of Rhom & Haas. Composite pre-pregs were made using the unmodified and modified PC resins and unsized Hercules AS4 continuous carbon fibers. These were produced by a solution pre-pregging method. The pre-preg solution consisted of a 50:50 mixture of methylene chloride and chloroform with 17 Wt% solids. Twenty-four ply unidirectional laminated composite plates 152.4 x 152.4 x 3.3 mm (approximately 56 Wt% fibers) were produced by compression molding.

The resistance to delamination growth of the composite was determined by evaluating the Mode I interlaminar fracture toughness strain energy release rate, G_{1C} using a double cantilever beam specimen (DCB). The DCB specimen is generally used to model the delamination process for the following reason. Once a delamination has occurred due to an impact, subsequent loading in compression causes material on either side of the delamination to buckle outward (Figure 1) causing further delamination by crack propagation in Mode I [20]. We used a hinged double cantilever beam test (HDCB) and determined a G_{1C} value by the area integration method described in NASA publication 1092 [21]. The dimensions of the test specimens are provided in Figure 3. Analysis of the HDCB fracture surface was performed utilizing a scanning electron microscope (SEM).

The rubber distribution within the composite was investigated by using a sodium hydroxide etch. The etching solution was prepared by mixing a saturated solution of methanol and sodium hydroxide. The etchant was applied to the fracture surface for 2 minutes followed by a rinse using distilled water. The surface was inspected using an SEM.

RESULTS

The HDCB results are provided in Table I. These results indicate that the fracture toughness of the unmodified matrix composite is greater than the modified matrices. SEM photomicrographs of the surface indicate that the adherence of the control resin to the carbon fibers is better than that of the rubber toughened systems (Figure 4-5). By utilizing a sodium hydroxide etch, it was determined that the rubber particles had agglomerated and preferentially lie along the fiber/matrix interface (Figure 6). By simulating the solution pre-pregging process in small glass vials, we discovered that the rubber particles rise to the surface of the solvent during the impregnation process. This provides a mechanism for preferentially coating the fiber surfaces and contributes to a weak fiber/matrix interface.

By using a differential scanning calorimeter, we found that solvents normally used for producing pre-pregs cause the polycarbonate to crystallize once the solvents are removed. Thin sections of the fractured DCB specimens indicated signs of crystalline PC along the fiber/matrix interface (Figure 7). The presence of crystalline PC could be a result of residual crystalline material which remains after molding the pre-pregs into a composite, and/or upon cooling, the carbon fibers act as nucleation sites. Crystalline PC properties are very poor (brittle) and would also result in a poor fiber/matrix interface.

CONCLUSION

The fiber/matrix interface of a polycarbonate matrix composite can be weakened due to the presence of crystalline material. The crystalline structure may be produced by solvents used during pre-pregging or by the carbon fibers acting as nucleation sites. In either case, a weakened fiber/matrix interface is created. Results specific to a thermoplastic matrix composite toughened by rubber particles indicate that solvents used for pre-pregging must not allow separation of rubber particles to occur. If a separation does occur, the fiber surface may be preferentially coated with rubber particles and the distribution of rubber particles throughout the matrix will not be homogeneous. The combination of the rubber particles along the fiber/matrix interface, agglomeration of the rubber particles and the presence of crystalline material all lead to a reduced interlaminar fracture toughness for the rubber modified composite. The potential of this toughening approach cannot be evaluated at this time because the interface failure prevents the toughening mechanisms from being fully operable.

This work demonstrates that there are several factors which influence the interfacial strength of a thermoplastic composite. If the factors observed here are also occurring in the production of other thermoplastic matrix composites, then certainly the poor transfer of toughness to the composite is understandable. These restrictions must be eliminated in order to properly evaluate the potential use of a toughened thermoplastic resin as a matrix material for a high performance fiber composite.

ACKNOWLEDGEMENT

We would like to thank Dr. K. Riew (B. F. Goodrich) and Dr.   Flexman (Dupont) for producing our modified matrix materials. We also wish to thank NASA and Dr. N. Johnston for supporting this work (NASA grant NAG-1-607) and producing our composite samples.

REFERENCES

1. Morley, J. G., <u>"High Performance Fibre Composites,"</u> London: Academic Press, 1987.

2. Manders, P. W., and Harris, W. C., "A Parametric Study Of Composite Performance In Compression- After- Impact Testing," <u>SAMPE Journal,</u> Nov./Dec. 1986,pp. 47-51.

3. Lang, R. W., Heym. M.; Tesch, H., and Stutz, H. "Influence of Constituent Properties On Interlaminar Crack Growth In Composites," <u>High Tech- The Way Into the Eighties, European SAMPE</u>, Edited by K. Brunsch, H. -D. Golden and C. -M. Herkert, Elsevier Science Publishers, Amsterdam, 1986, pp. 261-272.

4. Lagace, P. A., "Delamination In Composites: Is Toughness The Key?," <u>SAMPE Journal </u>, Nov./Dec. 1986, pp. 53-60.

5. Crick, R. A., Leach, D. C., and Moore, D. R. "Interpretation Of Toughness In Aromatic Polymer Composites Using A Fracture Mechanics Approach," <u>SAMPE Journal</u>, Nov./Dec. 1986, pp. 30-36.

6. Lee, S. M., and DiSalvo, G. D. M. "Resin Properties/Laminate Fracture Toughness Correlation," <u>30th National SAMPE Symposium</u>. March 19-21, 1985

7. Devitt, D. F., Schapery, R. A., and Bradley, W. L., "A Method For Determining The Mode 1 Delamination Fracture Toughness Of Elastic And Viscoelastic Composite Materials," <u>Journal Of Composite Materials,</u> Vol. 14, 1980, pp. 270-285.

8. Hibbs, M. F., Tse, M. K., and Bradley, W. L., "Interlaminar Fracture Toughness and Real-Time Fracture Mechanism of Some Toughened Graphite/Epoxy Composites," <u>Toughened Composites,</u> ASTM STP 937, edited by N. J. Johnston, American Society for Testing and Materials, Philadelphia, 1987, pp. 115-130.

9. Hunston, D. L., "Composite Interlaminar Fracture: Effect of Matrix Fracture Energy", <u>Composites Technology Review</u>, Vol. 6, no. 4, Winter 1984, pp. 176-180.

10. Hunston, D. L., Moulton, R. J., Johnston, N. J. and Bascom, W. D. "Matrix Resin Effects in Composite Delamination: Mode I Fracture

Aspects," <u>Toughened Composites,</u> ASTM STP 937, edited by N. J. Johnston, American Society for Testing and Materials, Philadelphia, 1987, pp. 74-94.

11. Schultz, J., Lavielle, L., and Martin C., "The Role of The Interface In Carbon Fibre-Epoxy Composites," <u>Journal of Adhesion,</u> Vol. 23, 1987, pp. 45-60.

12. McGarry, F. J., Mandell, J. F., and Wang, S. S., "Fracture Of Fiber Reinforced Composites," <u>Polymer Engineering and Science</u>, Vol. 16, no. 9, Sept. 1976, pp. 609-614.

13. Chakachery E. A., and Bradley W. L., "A Comparison of The Crack Tip Damage Zone For Fracture of Hexcel F185 Neat Resin and T6T145/F185 Composite," <u>Polymer Preprints,</u> Vol. 26, no. 2, Sept. 1985, pp. 117-118.

14. Jordan, W. M., and Bradley, W. L. "Micromechanisms of Fracture in Toughened Graphite-Epoxy Laminates," <u>Toughened Composites,</u> ASTM STP 937, edited by N. J. Johnston, American Society for Testing and Materials, Philadelphia, 1987,(pp. 95-114).

15. Lee, S. M., "Correlation Between Resin Material Variables and Transverse Cracking in Composites," <u>Journal of Materials Science,</u> Vol. 19, 1984, pp. 2278-2288.

16. Kinloch, A. J., Shaw, S. J., and Hunston, D. L., "Deformation and Fracture Behavior of a Rubber-Toughened Epoxy: 2. Failure Criteria," <u>Polymer,</u> Vol. 24, 1983, pp. 1355-1363.

17. Yee, A. F., and Pearson, R. A., "Toughening Mechanisms In Elastomer-Modified Epoxies Part 1 Mechanical Studies," <u>Journal of Materials Science,</u> Vol. 21, 1986, pp. 2462-2474.

18. Pearson, R. A.; Yee, A. F.,"Toughening Mechanisms In Elastomer-Modified Epoxies Part 2 Microscopy Studies," Journal Of Materials Science, Vol. 21, 1986, pp. 2475-2488.

19. Bascom, W. D., Ting, R. Y., Moulton, R. J., Riew, C. K., and Siebert, A. R.,"The Fracture of an Epoxy Polymer Containing Elastomeric Modifiers," <u>Journal of Materials Science,</u> Vol. 16, 1981, pp. 2657-2664.

20. Manders, P. W., and Harris, W. C.,"A Parametric Study Of Composite Performance In Compression- After- Impact Testing," <u>SAMPE Journal,</u> Nov./Dec. 1986, pp. 47-51.

21. "Standard Test for Toughened Resin Composites," <u>NASA Reference Publication 1092,</u> National Aeronautics and Space Administration, Langley Research Center, Hampton, Virginia, July 1983.

Table 1: HDCB TEST RESULTS

I.D.	% Resin	Material	G_{IC} KJ/M^2 Area Method
JS120-1	37.1	PC + AS4 fibers (Control)	1.23
JS104-1	34.7	PC + 10 wt% KM330 + AS4 fibers	0.83
JS93-1	31.3	PC + 10 wt% KM653 + AS4 fibers	0.69

A displacement rate of 0.5 in./min. (12.7 mm/min.) was used for all tests. The values shown are based on the average value of four tests.

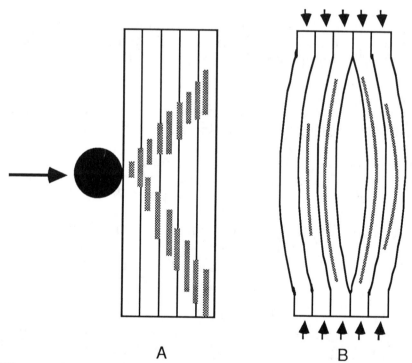

A   B

Fig. 1. Schematic of impact initiated compression failure. A) On impact a damage zone is created. B) Subsequent loading in compression causes delamination and buckling.

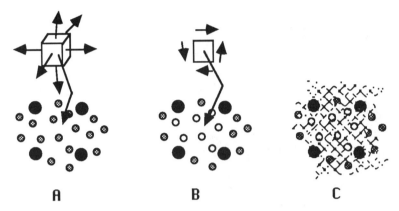

Fig. 2. Schematic of the development of the deformation zone within a rubber toughened composite. A) During loading the matrix between the fibers is highly constrained (hydrostatic tension). B) Cavitation of the rubber particles reduces the hydrostatic tension and the deviatoric stresses are enhanced. C) Formation and growth of stable shear bands extend the plastic zone beyond the interfibril spacing.

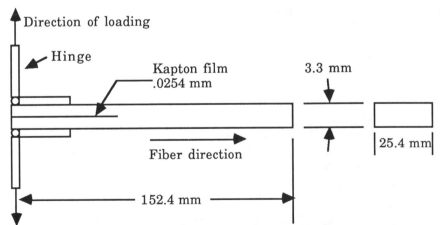

Fig. 3. Hinged double cantilever beam (HDCB) specimen. (Hercules AS4 fiber, 24 ply unidirectional)

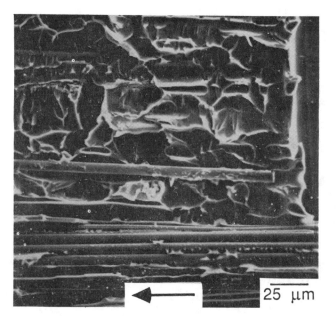

Fig. 4. SEM Photomicrograph of the fracture surface of the DCB specimen JS120-1 (control). Good adhesion of the resin to the fibers is apparent. Arrow indicates the direction of crack propagation.

25 μm

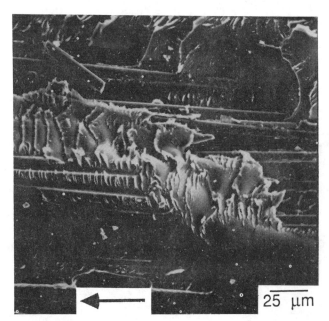

Fig. 5. SEM photomicrograph of the fracture surface of the DCB specimen JS93-1. The fibers along the fracture plane have pulled free from the matrix during testing with relatively little matrix deformation. Arrow indicates the direction of crack propagation.

25 μm

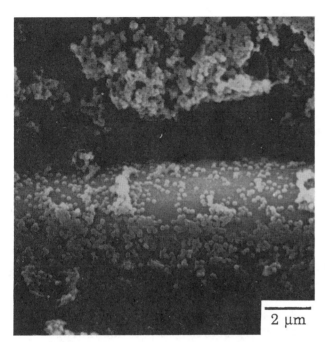

2 µm

Fig. 6. Sodium Hydroxide etch of the fracture surface of DCB specimen JS93-1 reveals that the rubber particles have agglomerated and lie along the fiber surface.

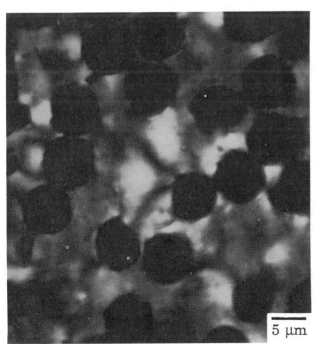

5 µm

Fig. 7. A thin section of DCB specimen JS104-1. Using optical microscopy and cross-polarized light, crystalline polycarbonate at the matrix/fiber interface is observed.

Rubber-Toughened Polycarbonate

C. KEITH RIEW, ROGER E. MORRIS AND MARTIN E. WOODS

The BFGoodrich Company
Research and Development Center
9921 Brecksville Road
Brecksville, OH 44141

ABSTRACT

Recently, fiber reinforced engineering thermoplastics became commercially important composites. The best known members of the class are reinforced nylon and polycarbonate. We attempted to toughen unfilled polycarbonate matrix resin with preformed elastomeric particles.

Circular compact tension specimens were prepared by injection molding and used to determine the fracture toughness, K_{IC}, according to ASTM E399 procedure. We have made various efforts in order to meet validity of the test procedure, i.e., to assure all fractures to be plane strain mode. When the rubber-toughened polycarbonates were subjected to Mode I loading in the compact tension geometry, many of them failed in plane-stress mode even at high displacement rates, at low temperatures down to -40°C, and/or use of thick specimen. This indicates qualitatively the preformed elastomeric particles did significant toughening.

We studied Izod impact strength at various toughener levels and test temperatures to understand efficiency of the rubbery tougheners. We hoped that we could correlate this test to the K_{IC} test. We found that the notched Izod test has the similar problem of plane stress to plane strain failure modes as in the K_{IC} determination.

Therefore, for quantitative comparison purposes, we recommend the J-integral determination, the energy per unit area of crack extension, J_{IC}, which should be a better technique to quantify the toughness.

Fractographs from the Izod impact test specimens by transmission and scanning electron microscopy were analyzed to elucidate rubber-toughening mechanisms. Polycarbonates deform through shear yielding with or without discrete rubbery second phase. With the rubbery domains, however, toughness is enhanced by a simultaneous cavity formation (a dilatational process). The cavities are formed in rubber domains and/or in matrix surrounding the rubber domains within the plastic zone ahead of crack-tips.

Graft Interpenetrating Polymer Networks of Poly(propylene oxide) Glycol-Based Polyurethane and Epoxy

KUO-HUANG HSIEH AND JIN-LIN HAN

ABSTRACT

A fraction of isocynate group in an excessive amount of polyurethane prepolymer (PU) reacted first with pendant secondary hydroxyl group in the epoxy resin, then both simultaneous bulk polymerized individually to form polyurethane/epoxy graft interpenetrating polymer networks (Graft-IPNs). The mechanical properties of the Graft-IPNs were investigated.

The microphotographs of the Graft-IPNs exhibited the sedimentation of PU particles existed in the microscoped structure of the PU/Epoxy Graft-IPNs. A maximum tensile strength is noted at the PU/Epoxy ratio of approx. 25/75. In addition, a maximum value in the impact property also emerges at the same composition. Meanwhile, it has been found that PU/Epoxy Graft-IPNs with PU based on Poly (propylene oxide) glycol, M.W. = 400 (PPG-400) demonstrated better impact property than does those with PU based on PPG-1000. The dynamic mechanical analysis showed that PU/Epoxy Graft-IPNs with PU based on PPG-400 shifted the glass transition temperature (Tg) to a lower temperature, whereas those with PU based on PPG-1000 increased the intensity of soft segment phase without changing Tg of the resultant Graft-IPNS.

INTRODUCTION

The epoxy resin (DGEBA) is inherent with high glass transition temperature, (Tg> 100°C) brittle and hard properties. [16,17] It is successful in commercial to improve the toughness of the epoxy resin by using carboxyl-terminated butadiene acrylonitrile copolymer (CTBN)[1,2,11, 12, 17, 25] In addition, the toughened epoxy resin by using reactive liquid polymer (RLP) has begun as early as 1965, and with reasonable achievements.[11, 20, 24-28] Gillham believed that the toughening of the epoxy resin lies in the so-called two-phase morophology.[18-21] In 1975, Rowe et al. found that the toughening effect would be achieved by incorporating bisphenol A in the CTBN/Epoxy system; they also reasoned on the basis of the particle size in the two-phase morophology. It is noted that the presence of small rubber particles is to strengthen the shear band, whereas the existence of large rubber particles is to increase the crazes in the system. As both large and small particles existed together, both the strengthened shear band and the more crazes will coexist in the system whereby optimum tonghening effect will be created because of two toughening mechanisms contributed at the same time.

Department of Chemical Engineering, National Taiwan University, Taipei, Taiwan, 10764, R.O.C.

In 1960, Sperling and Friedmen[22] proposed the simultaneous interpenetrating network (SIN) of epoxy resin (DGEBA) and n-butyl acrylate monomer by means of bulk polymerization to yield a two-phase morophology. Later, Scartio and Sperling[23] have utilized such a simultaneous bulk polymerization of DGEBA and n-butyl acrylate monomer to improve the impact property of epoxy resin, and they indeed have come out with substantive amelioration.

The present paper relates to the preparation of PU/Epoxy Graft-IPNs with the same simultaneous bulk polymerization.[3-8] The PU prepolymer based on two different molecular weight of Poly (propylene oxide) glycol (PPG) were employed to prepare Graft-IPNs with epoxy resin. The mechanical property, impact property and morophology with varying amount of PU incorporated into the Graft-IPNs will be discussed.

EXPERIMENTAL

The materials required for this experiment are listed in Table 1. Poly (propylene oxide) glycol with M.W. = 400 (PPG-400) and 1000 (PPG-1000), crosslinking agents (TMP), chain extenders (1,4 − BD) and epoxy resin (DGEBA) were moderately heated to 60°C in the flask, and they were stirred, and continuously drafted in vacuum overnight before use.

To prepare PU prepolymers, two equivalent weight of MDI were fed to a reaction kettle and heated to a melting state, whereupon one equivalent weight of PPG was added and stirred. The temperature is maintained at approx. 68°C and the reaction takes place under the dried nitrogen atmosphere. As the −NCO content, which is determined by using n-butylamine titration method,[9] reaches the theoretical value, the reaction is stopped.

The preparation of PU/Epoxy Graft-IPNs was made in two steps; the first step involves feeding of excessive amount of PU prepolymers and epoxy resins into the reaction kettle. The temperature was maintained at 68−70°C. Let reaction take place in circumstances of dried nitrogen. Sample will be taken on every hour for determination of the reaction in IR spectra. The ratio of the absorption peak of −NCO group (2270cm^{-1}) relative to that of the epoxy resin group (920 cm^{-1})[10] will decrease with the reaction time, and eventually come to a contant ratio. This indicate that the pendant secondary hydroxyl group has completely been reacted with the − Nco group of the PU prepolymer.[13-15] Because of excessive amount of the PU prepolymer used, there is still present the −NCO group in the mixture. Next, the above mixture will be added catalyst (TDMP), 2 phr (based on DGEBA) and 1,4-BD/TMP mixture ratio of 4/1. The mixture was agitated mechanically for about five seconds, then drafted for about half a minute, poured into the aluminum mold at 80°C and pressed for two hours. It was further heated to 120°C for an additional six hours. Let the pressure of the hot press be maintained at 140 kgw/cm² throuhout the process. Upon removal from the mold, the samples will be stored in a container of which the relative humidity is maintained at 50% for three days before they are tested on their mechanical properties.

TESTING METHODS

The stress-strain property test is made using Instron TM-SM universal test unit. The test procedure is followed and listed in ASTM-D638 with crosshead speed at 1 cm/min. The scanning electron microscopy (SEM) was made for the morphology study on the fractured surface of the sample.

Infrared (IR) analysis was made using HITACHI 270-30. Infrared Spectrophotometer. The sample is directly applied by dabbing onto KBr pellet to varify the reaction of the −NCO group with the pendant secondary hydroxyl group in the epoxy resins. The dynamic mechanical analysis was made using the DMA unit (Du Pont 9900) with the operation temperature ranging between −100°C through 250°C, the frequency set at 110 Hz, the sample size used is approximate 6 cm x 1 cm x 0.2 cm.

RESULTS AND DISCUSSION

Stress-Strain Properties

The stress-strain properties of the PU/Epoxy Graft-IPNs are shown in Fig. 1. It is seen that both the tensile strength and elongation of the resultant PU/Epoxy Graft-IPNs are appreciably improved for the PU content less than 27 wt.%, this is owing to the grafted reaction of the −NCO group in the PU prepolymer to the pendant secondary hydroxyl group in the epoxy resin to form crosslinking structure and interpenetrating networks between the epoxy chains.

In the PU/Epoxy Graft-IPNs with PU based on different molecular weight of PPG, it is noted that an increase of PU content is always accompanied by an increase in its tensile strength. Furthermore, a maximum value emerges as the PU content reaches 22wt.% and 27wt.% for the PU based on PPG-400 and PPG-1000, respectively. The more pronounced increase in tensile strength of the Graft-IPNs with the PU based on PPG-400 is possible due to higher effectiveness for crosslinking by the shorter chain length of PPG-400 than PPG-1000. If the PU content is increased beyond the critical point where the maximum value taken place, serious phase separation is observed as a result of poor compatibility and the tensile strength comes down consequently.

Izod Impact Property

It has been made on the subject of the toughening theory of epoxy resin modified with liquid rubber.[11,12] A generalized review of such researches gives that the rubber particle size and the quantity in use are also the key factor to create optimum toughening effect in addition to requisties of rubber sedimentation and chemical bonds serving to link two phase.

Listed on Table 2 are the Izod impact property of the PU/Epoxy Graft-IPNs. It indicates that evident improvement has been achieved in impact property for the introduction of PU in epoxy resin. The Izod impact property is promoted along with an increase of the PU rubber content below 27wt.% in which the size of the rubber sedimentation is about 2000Å to 3000Å. The drastic drop in impact property, as the PU content exceeds 27 wt.%, results from the macro-phase separation taken place as shown in Fig. 2 and 3.

A further point to note is that impact property of PU/Epoxy Graft-IPNs with PU based on PPG-400 of the shorter soft chain turned out to be better than those based on PPG-1000 of the longer soft segment. This would be due to more toughening effect to the rigid epoxy matrix by introduction of shorter chain than that by introduction of longer chain.

Dynamic Mechanical Property

From Table 3, it can be seen that the glass transition temperature at high temperature (Tgh) for the Graft-IPNs with the PU based on PPG-400 obviously shift to a low temperature as the PU introduced in the system, whereas there is almost no change in Tgh for the case of the Graft-IPNs with the PU based on PPG-1000. This imply that the shorter chain of PU which based on PPG-400 is more compatible with epoxy resin than the longer chain of PU which based on PPG-1000.

CONCLUSIONS

With an increase of the PU content in PU/Epoxy Graft-IPNs, the tensile strength and impact property increase to a maximum value. As the PU content increases consequently, macrophase separation occurs, and thereby results in decrease of the tensile strength and impact property. The toughening PU/Epoxy Graft-IPNs are found through sedimentation of effectual PU rubber particles sized approx. 2000Å through 3000Å.

ACKNOWLEDGEMENT

The authors wish to express their gratitude to the National Science Council, Taipei, Taiwan, R.O.C. for the financial support through Grant No. NSC77-0405-E002-10.

REFERENCES

1. L.H. Sperling, Ed. "Interpenetrating polymer networks and related materials" Planum Press, New York, 99 (1981).

2. C.B. Bucknall, "Toughness Plastics" Applied science, London 33, (1977).

3. E.F. Lassidy, H.X. Xiao, K.C. Frisch, H.L. Frisch, J. Polym. Sci., Polym. Chem. Ed. 22(8), 1839 (1984).

4. D. Klempner, L. Berkowski, K.C. Frisch, K.H. Hsieh, R. Ting, Rubber World, Vol. 192, No. 6, 16, (1985).

5. E.F. Cassidy, H.X. Xiao, K.C. Frisch, H.L. Frisch, J. Polym. Sci. Polym. Chem. Ed., 22(8) 1851 (1984).

6. L.H. Sperling Ed. "Interpenetrating polymer networks and related materials" Plenum Press, New York, 88 (1981).

7. R. Pernice, K.C. Frisch, R. Navare, J. Cell. Plast, 18(2), 121 (1982).

8. E.F. Cassidy, H.X. Xiao, K.C. Frisch, H.L. Frisch, J. Polym. Sci. Polym. Chem. Ed. 22(10), 2667 (1984).

9. C. Hepburn Ed., "Polyurethane Elastomer", 1982, Applied Science Publishers, p. 280.

10. T.I. Kadurina, V.A. Prokopenko and S.I. Omelchenko, Eur. Polym. J. Vol. 22, No. 11, 1865 (1986).

11. Clive B. Bucknall and Toshiya Yoshi, The British Polymer Journal, Vol. 10, 53-59 (1978).

12. J.N. Sulton and F.J. McGarry, Polym. Eng. and Sci. 13(1), 29(1973).

13. H. Lee and L. Neville, "Epoxy resin," McGraw-Hill, New York (1957).

14. K.C. Frisch and L.P. Rumao, J. Macromol. Sci-Revs. Macromol. Chem. C5, (1), 103, (1970).

15. K.C. Frisch, D. Klempner, and S.K. Mukherjee, and H.L. Frisch, J. of Appl. Polym. Sci. Vol. 18, 689, (1974).

16. H. Lee, K. Neville, "Handbook of Epoxy resin", McGraw-Hill, New York (1972).

17. R.J. Perez, in "Epoxy Resin Technology", P.S. Bruins, Ed., Interscience, New York, 45,

(1968).

18. J.A. Manson; L.H. Sperling, "Polymer blends and Composites", Plenum Press. New York, (1976).

19. F.J. McGarry; A.M. Willner, "Toughening of Epoxy resin by an Elastomeric Second Phase", R68-8, March, MIT (1968).

20. L.T. Manzione; J.K. Gillham, ACS, Org. Coat. and Plast. Chem., 364 (1979).

21. L.T. Manzione, J.K. Gillman, C.A. Mcpherson, J. Polym. Sci., 26, 889, (1981).

22. L.H. Sperling: D.W. Friedmen, J. Polym. Sci., Part A-2, 7, 425 (1969).

23. P.R. Scarito, L.H. Sperling, Poly. Eng. and Sci., 19, 297 (1970).

24. E.H. Rowe, 26th Annu. Tech. Conf., Reinforced Plastics/Composites Div. SPI, Sec. 12-E, 1 (1971).

25. C.K. Riew, E.H. Rowe, A.R. Siebert, in "Toughness and Brittleness of Plastics", ADVANCES. IN CHEM. SERIES No. 154, ACS, Washington D.C., 326 (1976).

26. E.H. Rowe, 24th Annue. Tech. Conf. Reinforces Plastics/Composoties Div. SPI, Sec. 11-A, 1 (1969).

27. E.H. Rowe, A.R. Siebert, R.S. Drake, Modern Plastics, 47, 110 (1972).

28. R. Drake, A. Siebert, SAMPE Quarterly, July (1975).

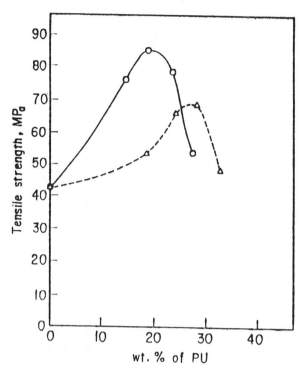

Fig. 1　Effect of PU Contents on tensile strength of PU/Epoxy Graft-IPNs, ——○— : PU based on PPG 400. ——△— : PU based on PPG 1000.

Table 1　Materials

Designation	Description	Source
MDI	4,4'diphenyl methane diisocyanate	Bayer Chemical Co.
PPG-400	Poly(propylene oxide)glycol M.W. = 400	Chiun Glong Co.
PPG-1000	Poly(propylene oxide)glycol M.W. = 1000	Chiun Glong Co.
1,4-BD	1,4-Butanediol	Hayashi Pure Chem.
TMP	Trimethylol propane	Hayashi Pure Chem.
DGEBA	Dighycldyl ether of Bisphenol A, EEW. = 186	Dow Chemical Co.
TDMP	2,4,6-Tris(dimethylamino- -methylphenol)	Merck Chemical Co.

Table 2 Impact Value and PU Particle Size in PU/Epoxy Graft-IPNs

Type of PU In. PU/Epoxy Graft-IPNs	Excessive PU content (phr)	PU content PU (wt %)	Izod Impact (J/m)*	Particle Diameter (Å)
PPG400-based	–	0	6.57	0
"	5	15.7	8.01	5886
"	10	19.1	8.54	3167
"	15	22.3	9.28	3006
"	25	27.8	5.39	–
PPG1000-based	–	0	6.57	0
"	5	21.7	7.21	2700
"	10	24.6	7.31	2160
"	15	27.4	7.53	2250
"	20	30	6.72	–

* 1(J/m) = 0.6 (lb-ft/in)

Table 3 Transition Temperatures of PU/Epoxy Graft-IPNs

Type of PU in PU/Epxoy Graft-IPNs	Ecessive PU content (phr)	PU content (wt%)	T_{gh}*(℃)	T_{gl}*(℃)
PPG400-based	–	0	106.6	-72.51
"	5	15.7	89.40	-78.40
"	10	19.1	82.92	-71.19
"	15	22.3	80.76	-66.15
"	25	27.8	68.52	-69.03
PPG1000-based	–	0	106.6	-72.51
"	5	21.7	98.77	-87.04
"	10	24.6	109.6	-72.25
"	15	27.4	111.7	-74.07
"	20	30	112.4	-64.71

* T_{gh} : Transition Temperatures at high temperature

* T_{gl} : Transition Temperatures at low temperature

288

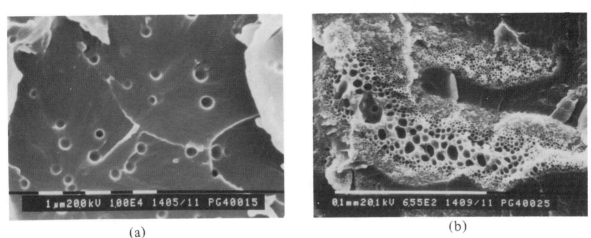

(a)   (b)

Fig. 2 Photomicrographs of SEM for PU/Epoxy Graft-IPNs with PU based on PPG 400 at various PU/Epoxy ratio of (a) 22.3/77.7 (b) 27.8/72.2

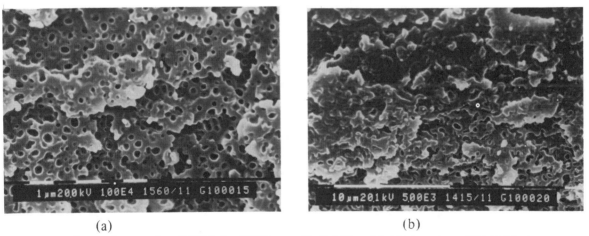

(a)   (b)

Fig. 3 Photomicrographs of SEM for PU/Epoxy Graft-IPNs with PU based on PPG 1000 at various PU/Epoxy ratio of (a) 27.4/72.6 (b) 30/70.

Advances in Materials and Process Engineering

Pultrusion Impregnated Thermoplastic Tape

MILTON F. CUSTER AND KENNETH E. VICKNAIR

ABSTRACT

The properties and potential production cost reductions of thermoplastic polymer composites make them extremely interesting candidates for both military and commercial applications. Unfortunately, the high molecular weight inherent to these polymers create viscosities much higher than those achievable with thermosetting resins, and thus complete saturation or wet out of unidirectional fiber reinforcements is extremely difficult. A process of impregnation using proprietary pultrusion technology has been developed, allowing substrates to be produced, in either tape or tow configuration, with fully encapsulated filaments. This process is applicable to fibers, up to intermediate modulus, and all melt fusible thermoplastic polymers. The mechanical properties of tape made by this process technology show an excellent translation of stress to the fiber, indicating a high degree of fiber wet out. For the purpose of this study, a (poly)arylene sulfide matrix and fiberglass substrate were used. The study suggests that an impregnation process which fully encapsulates the reinforcing filaments with high viscosity thermoplastic polymer is essential if optimum composite properties are to be achieved.

INTRODUCTION

Thermoplastics bring a new dimension to polymer composites as damage tolerant corrosion resistant matrices which can be formed to net shape and be manufactured at a low cost. However, despite the ultimate benefits of thermoplastics, there are numerous difficulties involved in processing final parts with optimum mechanical properties. The viscosity of molten thermoplastics, under conditions of low shear, can range up to 100,000,000 centipoise. Consequently, during typical

M.F. Custer, Group Leader, Advanced Materials Development; K.E. Vicknair, Process Development Specialist; Hexcel Corporation; 11711 Dublin Blvd.; Dublin, CA 94568-0705; (415) 828-4200

consolidation cycles, very little flow can be expected from the majority of thermoplastics; this situation becomes worse as the glass transition temperature of the thermoplastic matrix increases.

Given these high viscosities, it is extremely difficult to fill voids in the prepreg during consolidation. Therefore, if dry fibers are to be avoided in the final part, it is essential that as much air as possible be removed from the prepreg prior to consolidation. More importantly, what flow does occur during consolidation will be in the direction of least resistance which is parallel to the axis of the fiber not perpendicular to it. Such flow does little to wet out dry areas of the fiber bundles and tends to result in polymer migrating out of the part. Furthermore, the pressure applied for consolidation tends to force the fibers together, making complete encapsulation of filaments by the polymer even more difficult. If total encapsulation of the filaments is not achieved, the resulting flaws, whether voids or touching fibers, will lower the mechanical properties.

EXPERIMENTAL

The matrix of choice for this particular study was Fortron [1], a semi-crystalline (poly)arylene sulfide polymer. The polymer was impregnated in an E-glass substrate. Panels were press cured using a matched metal, positive pressure mold. Mechanical properties were determined on 8-ply and 16-ply (SBS only) unidirectional panels.

Flexural strength and modulus were determined by ASTM D-790; a transducer was used to measure the deflection at the beam mid-point. Short Beam Shear (SBS) strength was determined using ASTM D-2344. Density of the panels was determined using ASTM method D-792.

RESULTS AND DISCUSSION

Mechanical property data is shown in Table 1. Failure occurred on the tensile side of the flexural test specimens. SBS specimens demonstrated standard shear failures.

At this time, mechanical properties are far from comprehensive, however, the initial data indicates that this pultrusion process yields impregnated materials which can be fabricated into high strength panels. At present, variations in both consolidation cycles and methods are being studied for optimization. We feel that further refinement of the consolidation and impregnation processes will lead to even higher properties.

Developmental capability of the Hexcel pultrusion process for impregnating materials ranges from a single tow to three inch wide unidirectional tape. Successful impregnation has been demonstrated on carbon (1K to 12K fiber bundles), fiberglass and Kevlar [2]. Experimental production has been demonstrated for semi-crystalline, liquid crystalline and amorphous polymers, including non-endcapped thermoplastic polyimides. Cross-sectional photomicrographs and scanning electron micrographs indicate a high degree of encapsulation of the individual filaments. This impregnation process is applicable to all melt fusible thermoplastic polymers and fibers, up to intermediate modulus.

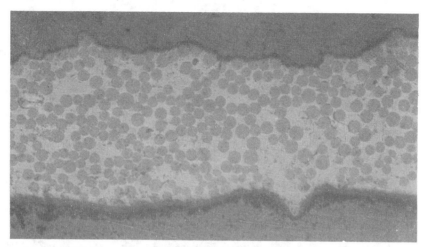

Figure 1. Fortron/E-Glass

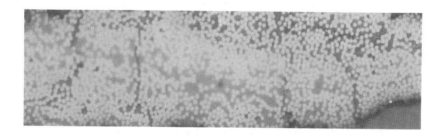

Figure 2. Xydar/IM6

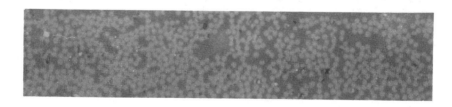

Figure 3. Udel/AS4

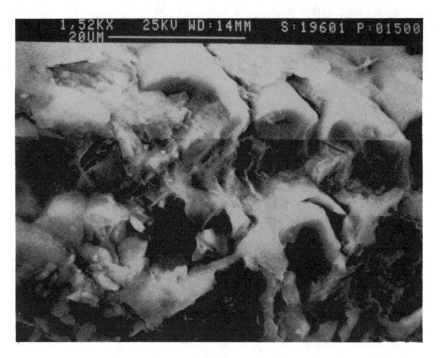

Figure 4.   Fortron/E-Glass

CONCLUSION

A proprietary pultrusion impregnation process has been developed which allows for the production of high quality prepreg. By this process, fully impregnated materials are produced from the melt. The mechanical properties of panels made from this tape verify that an effective transfer of stress to the fibers has been achieved, indicating a high degree of fiber encapsulation is present.

ACKNOWLEDGMENTS

The assistance of Rosanna Falabella, Bill Sherwood, Michelle Cakebread, Dave Doty and Mark Bowie in material preparation and testing is gratefully acknowledged.

REFERENCES

1. Kevlar is a registered trademark of E.I. DuPont De Nemours & Co. (Inc.).

2. Fortron is a registered trademark of Hoechst Celanese.

TABLE 1

Room Temperature Properties of Fortron®/E-Glass Laminates

Property	ASTM Test	Average	C.V., %
Fiber Vol, %	–	52	–
Density	D-792	2.2	–
SBS, ksi	D-2344, 5:1 span	9.7	1.4
Flexural Strength, ksi	D-790, 16:1 span 3-pt.	239.0	2.9
Flexural Modulus, msi	D-790, 16:1 span 3-pt.	7.5	2.0

Fabrication of Low Cost Missile Structure

C. H. SHEPPARD, B. A. KOETJE AND R. KRONE

Abstract

The use of fiber reinforced composites in airframe structure has been widely advocated for design versatility. However, their actual use has been somewhat limited because of manufacturing cost, and in the case of the epoxy thermosetting matrices, their environmental stability. A solution to both of the problems appears achievable by using recently developed high performance matrices know as thermoplastics. The preponderance of work to date has concentrated in the evaluation of their performance in primary aircraft structure. Most studies concerning manufacturing have shown their manufacturing costs to be higher than the epoxy reinforced composites. This paper discusses a program whose objective was to design and manufacture a major section of a missile structure, using a thermoplastic system as the low cost material system. The structure selected was the center body section of a generic air-to-ground tactical missile.

Introduction

It has been established that thermoplastic matrix composites possess mechanical properties sufficient to warrant usage in structural applications. While considerable work is still needed to establish design data bases for these materials, manufacturing studies using these systems must be accelerated. The most recent work in the aerospace industry has shown that the cost of converting the thermoplastic broadgoods to structure (in most cases complex structure) is often higher than currently available thermosetting systems. There are numberous reasons contributing to this fact, among them being the initial design philosophy, the selection process for the material systems, and the limited types of processing facilities in the aerospace industry. The current processing equipment and

C. H. Sheppard, Materials Research Engineer, B. A. Koetje, Materials Research Engineer Boeing Aerospace, Seattle, Washington 98124 and J. R. Krone, Manufacturing Research Engineer, Phillips Chemical Company, Bartlesville, Ok 74004

facilities have been designed for making structures using thermosetting matrices but are not suitable for efficiently making structures using the processing characteristics of high performance thermoplastic system. The objective of this program was to demonstrate that a low cost missile system could be designed and manufactured, if all these restraints were removed. In the program, the overall "cost" of the system remained paramount. This paper will describe only the manufacturing portion of the effort.

Materials Selection

A detailed survey of material systems was conducted with emphasis being the trade-off of structural properties and the processing methods used to obtain those properties. The most attractive candidates, from the processing view, were the high performance thermoplastic materials, using fiberglass as the reinforcement. These materials could be procured and handled similiar to metals (sheetstock). The conversion of the sheetstock to structure was projected to be simple (heating and shaping) in commercially available thermoforming equipment. The next step was to select a resin matrix possessing the best "set" of characteristics, using cost, availability of fiberglass sheetstock, structure properties, enviromental stability and processing conditions as the key criteria. This resulted in the selection of E glass reinforced polyphenylene sulfide (PPS) as the candidate material system.

Part Design and Manufacture

The initial design trade study selected a stiffened skin construction for the demonstration structure. This resulted in the need for three material forms, unidirectional prepreg, fabric prepreg, and glass swirl mat. These forms would be used in fabricating the upper and lower skins, channels and stiffeners and the rails (longerons) (Figure 1). The designers were very cognizant of the requirement to minimize the laminate (sheetstock) stacking sequence. Their diligence resulted in only one configuration for the top skin, bottom skin and the longerons. The stiffeners used the glass mat swirl as the material form. The laminate configuration was 60 percent unidirectional glass tape at zero degrees and 40 percent glass fabric at +/-45 degrees and 0.225 inches thick. The physical size of the upper skin was determined to be approximately 54 inches x 72 inches, while the lower skins were made in three separate sections of approximately 30 inches x 54 inches (the 0 degree direction being the 72 inches and the 30 inches respectively). The reinforcing stiffeners were approximately 6 inches x 54 inches x 0.100 inches thick and horseshoe in shape. The longeron was approximately 72 inches x 6 inches x 0.225 inches thick, a "C" section and contained a complex joggle.

The manufacturing methods selected included thermoforming for the skins and compression molding for the reinforcing stiffeners. Due to the cost of the program, most all of the tooling was made from wood

(Figures 2 and 3). Boeing didn't have the processing equipment necessary to form the various parts and so cooperated with Phillips Chemical Company to accomplish the forming. All processes were accomplished in less than 15 minutes for the most complicated part. Also due to the high projected cost of tooling (Figure 4 for metal tool)and inadequate thermoforming facilities, the longeron was fabricated using conventional autoclaving processes (It should be noted that in a production hardware environment, the longeron could be redesigned so that it would also be thermoformed.)

CONCLUSIONS

The work conducted in this program gave positive indication that a thermoplastic composite airframe structure could be designed and fabricated within the projected cost of thermoplastic structure (i.e. 50 percent the cost of metal and conventional glass reinforced composite). The manufacturing processes used demonstrated that the thermoforming of thick (0.225 inch) composite could be done (Figures 5 thru 7). The program was valuable in other ways, such as validation of manufacturing costs, processing techniques and methods of delivery of sheetstock into the thermoforming machine.

ACKNOWLEDGEMENT

The authors would like to express their appreciation to the many people at Boeing Aerospace and Phillips Chemical Company that made this program a success. At Boeing Aerospace, thanks to Verla Monroe, Eric Freitas, Frank Horan and Arlene Brown in the Parts Materials and Processes Group. From Structures Technology Group, Victor Lindsay and Dave Carbery contributed the structural design for the demonstration hardware. A very special thanks to Phillips Chemical Company for the generous use of their thermoforming equipment, for their contribution of the PPS glass laminates, their processing techniques and the forming of the various parts used in the demonstration hardware.

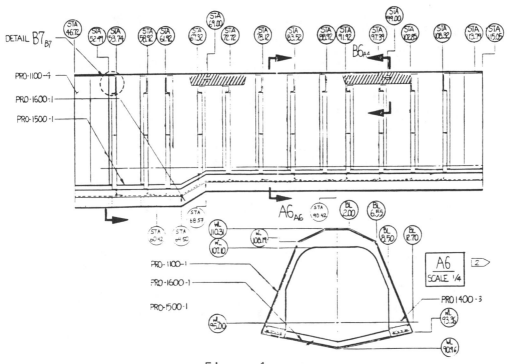

Figure 1
Engineering Drawing of Low Cost Missile Section

Figure 2
Picture of the Wood Tooling for Upper Skin

Figure 3
Picture of the Bottom Skin Wood Tooling

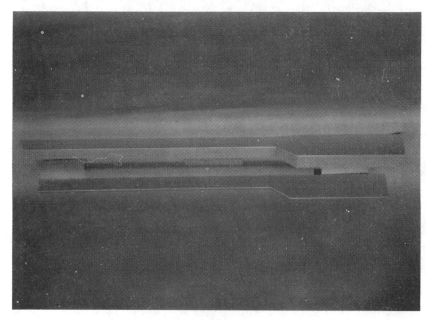

Figure 4
Picture of the Longeron Metal Tooling

Figure 5
Picture of the thermoformed upper skin

Figure 6
Picture of the Thermoformed Lower Skins and Compression Molded Stiffeners

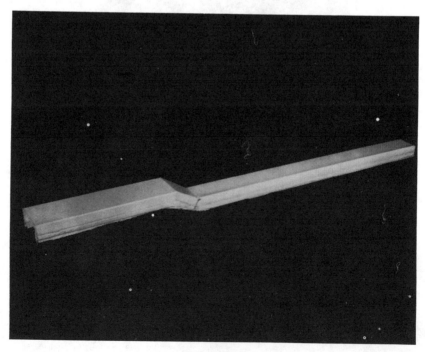

Figure 7
Picture of the Autoclave Formed Longeron

Microwave and Thermal Heating of Polymer Composites Containing Thermally Conductive or Non-Conductive Reinforcement

JINDER JOW, MARTIN C. HAWLEY AND JOHN D. DeLONG

ABSTRACT

Composite materials of relatively high thermal conductivity, epoxy/carbon, and of relatively low thermal conductivity, epoxy/glass, have been heated using a microwave cavity. In the case of epoxy/carbon, very small differences in temperature between the center and boundary of the sample have been measured. The small temperature gradient is due to the high thermal conductivity of the carbon. In the case of epoxy/glass, significant temperature difference has been noted between the center and boundary temperatures of the sample during heating, due to the low thermal conductivity of the material and cooling at the boundary. The temperature gradient has been reduced by heating the epoxy sample in a combined thermal/microwave process.

INTRODUCTION

Polymer composite materials are conventionally cured under pressure in a variety of thermal processes. The use of thermal curing cycles results in high quality material in the case of thin composites (less than about 1/4 inch in thickness). Thermal curing of thick-section composites (greater than 1/4 inch) often results in the formation of significant thermal stresses due to temperature gradients experienced in the polymer during cure. These stresses can lead to premature failure in a thick-section composite structure.

The desire to fabricate thick-section composites with low residual stresses has led to research into the development of microwave curing as an alternative to thermal curing. Microwave curing offers several advantages over thermal curing, including 1) selective and controlled heating of the material due to absorption of microwave energy by polar functional groups of the polymer, 2) decreased thermal degradation due to rapid uniform bulk heating, and 3) increased control of material temperature-time profile and cure cycle. These advantages may cause microwave cured materials to have superior mechanical characteristics when compared to conventionally cured materials. The ability to heat

J. Jow, Research Assistant, J. D. DeLong, Research Associate, and Martin C. Hawley, Professor, Department of Chemical Engineering, Michigan State University, East Lansing, Michigan 48824-1226

rapidly and controllably offers the potential to virtually eliminate temperature gradients during cure and thus reduce thermal stresses. In addition, microwave curing has been shown to improve adhesion at the interface in carbon fiber/ epoxy composites by 15 percent [1] and increase the glass transition temperature of an epoxy by nearly 100 °C [2] when compared to thermally cured materials.

Typical temperature profiles observed in a cylindrically shaped epoxy/amine resin during thermal curing and during continuous microwave curing without temperature control are shown in Figures 1 and 2, respectively. This data has been reported previously [2]. In both cases, significant temperature excursions are observed due to the exothermic reaction of the epoxy with the amine curing agent. Due to the slow conductive heat transfer, an effective control system is not possible to maintain constant temperature during thermal cure. Temperature control at a given position in the polymer is possible with microwave curing, however, by rapidly pulsing the power in response to a temperature-sensing feedback control loop. The result of such an experiment is shown in Figure 3 [2]. As shown in Figure 3, by controlling the power input the cure temperature is held constant with time. There remains, however, a temperature gradient between the center and boundary of the epoxy specimen due to heat loss from the boundary to the cooler ambient air in the cavity.

This paper reports the results of a study of the efficacy of microwave, thermal, and combined microwave/thermal heating on reduction of temperature profiles in epoxy/glass or epoxy/carbon mixtures.

EXPERIMENTAL METHODS

Microwave heating studies were carried out in a tunable, single-mode, 2.45 GHz microwave applicator designed at Michigan State University that has been described in detail previously [3]. A schematic diagram of the applicator system is shown in Figure 4. This system can be used for material processing as well as for dielectric diagnosis [4]. The single-mode applicator offers a distinct advantage over multimode cavities in that changes in the dielectric properties of the material during processing can be compensated for by tuning the applicator, i. e., by changing the axial dimension of the cylindrical cavity. A resonant electromagnetic field is thus maintained and the applied power is focussed into the material, leading to a more energy efficient and controllable process than is possible in a multimode cavity.

Epoxy specimens consisted entirely of diglycidyl ether of bisphenol A (DGEBA, DER 332). Composite specimens were made by adding approximately 50 percent by weight of finely ground AS4 carbon fiber or glass fiber to the epoxy. No curing agent was added to the epoxy. The liquid samples were poured into cylindrical teflon holders 9.5 cm in diameter and 5 cm in length. The teflon holders were then suspended concentrically in the microwave applicator, and, after initial tuning of the cavity, a pre-programmed processing cycle was carried out. The center and boundary temperatures of the heated material were monitored continuously using fluoroptic sensors inserted in the mixture.

The system shown in Figure 4 was modified for combined microwave/thermal heating by forcing hot air of the desired temperature into the microwave cavity.

RESULTS AND DISCUSSION

The results of heating a 50 percent by weight ground glass in epoxy sample using rapidly pulsed microwave power is shown in Figure 5. Data from a similar experiment using a plain epoxy sample are also shown. In both cases, a nearly identical temperature gradient is established between the center of the sample and the boundary. The addition of the glass has had no effect on the temperature gradient.

Figure 6 shows the results of heating a 50 percent by weight ground carbon fiber in epoxy using rapidly pulsed microwave power, with the temperature profiles for the plain epoxy heating experiment included for comparison. In the case of the epoxy/carbon sample, the temperature gradient fluctuates from a value approximately equal to that for the plain epoxy, to a point where the temperature difference is nearly zero. The points of minimum gradient correspond to the times when the power is turned on (with a slight lag time) and the maxima correspond to the times when the power is turned off during the pulsed power cycle. The presence of carbon in the heated material has tended to reduce the temperature gradient between the center and boundary.

The reduction in temperature gradient in the epoxy/carbon material as compared to the epoxy/glass is due to the greater thermal conductivity of the carbon. In the case of the epoxy/glass material, the components are of similar thermal conductivity, 1.04 watts/m °C for glass [1] and .21 watts/m °C for epoxy [5]. The thermal conductivity for carbon fiber is considerably higher, with a value of 105 watts/m °C [1]. Heat generated in the interior of the sample by molecular relaxation after microwave absorption is transported to the boundary more rapidly by the epoxy/carbon material than in the epoxy/glass. A greater boundary temperature is consequently observed for the epoxy/carbon material.

As can be seen in Figure 6, there is an extreme fluctuation in the boundary temperature for the epoxy/carbon material. This is due to the pulsing of the microwave power. The microwave applicator system is currently equipped with an on/off switch (Figure 4) for the microwave power input to the cavity. The boundary temperature rapidly approaches the set point temperature when the power is turned on and rapidly falls when the power is turned off. The cycle repeats for each power pulse. Reduction of this fluctuation should be possible through use of a power attenuator as part of an improved temperature control loop.

The temperature gradient observed during microwave heating of lower thermal conductivity material such as epoxy can be reduced through the use of a combined thermal/microwave process. Figure 7 shows the temperature profiles measured for two epoxy samples, one heated thermally and the other using microwave energy. The greater temperature gradient at steady state for the microwave heated sample is due to the much cooler boundary, which is exposed to unheated ambient air. The temperature profiles for an epoxy sample heated in a combined thermal/microwave process are shown in Figure 8. As expected, the center and boundary temperatures measured at steady state are identical. This type of hybrid thermal/microwave process maintains the rapid heating feature of the microwave method and but eliminates the unwanted temperature gradients.

REFERENCES

1.   Agrawal, R.K., and L.T. Drzal, "The Characterization of the
     Composite Fiber-matrix Interface Formed under Microwave Curing",
     presented at Conference on Emerging Technologies in Materials,
     August 18, 1987, Minneapolis, Minnesota.
2.   Jow, J., M. Hawley, S. Singer, C. Schwalm, and J. DeLong,
     "Microwave Curing of Epoxy Resins," presented at the Polymer
     Processing Society International Meeting, May 8-11, 1988, Orlando,
     FL.
3.   Asmussen, J., H. Lin, B. Manring, and R. Fritz, "Single Mode or
     Controlled Multimode Microwave Cavity Applicators for Precision
     Materials Processing", <u>Rev. Sci. Instrum.</u>, 58 (1987), 1477.
4.   Jow, J., M. Hawley, M. Finzel, J. Asmussen, H. Lin, and B. Manring,
     "Microwave Processing and Diagnosis of Chemically Reacting
      Materials in a Single-mode Cavity Applicator", <u>IEEE Trans.</u>,
     <u>Microwave Theory and Techniques</u>, MTT-35 (1987), 1435.
5.   Lee, H. and Neville, K., <u>Handbook of Epoxy Resins</u>, McGraw-Hill, New
     York, 1972.

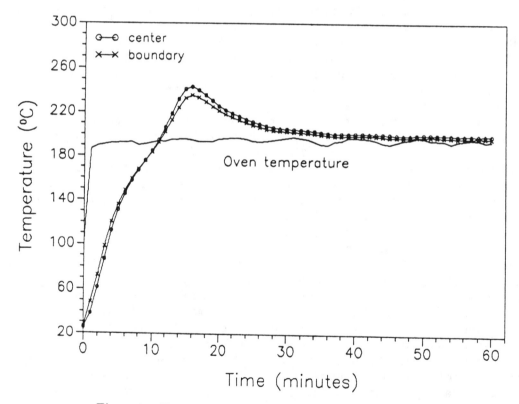

Fig. 1. Thermal curing of epoxy/amine resin

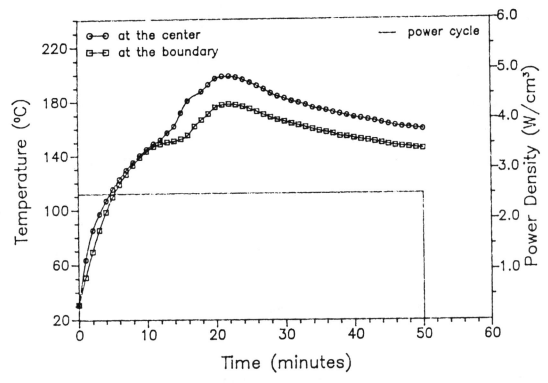

Fig. 2. Continuous microwave curing of epoxy/amine

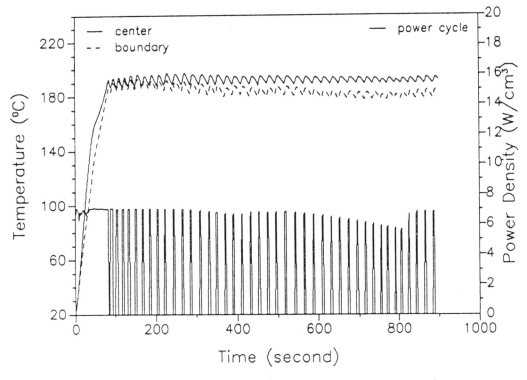

Fig. 3. Pulsed microwave curing of epoxy/amine

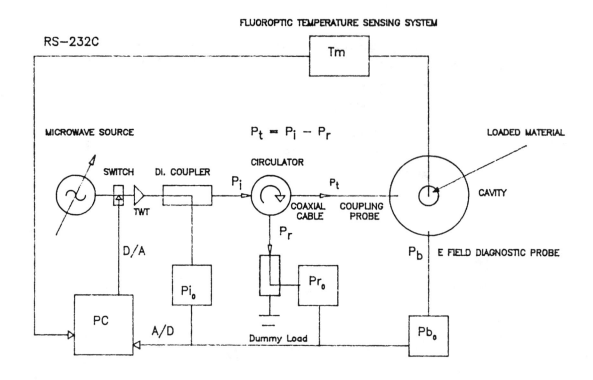

Fig. 4. Computer—aided microwave processing system

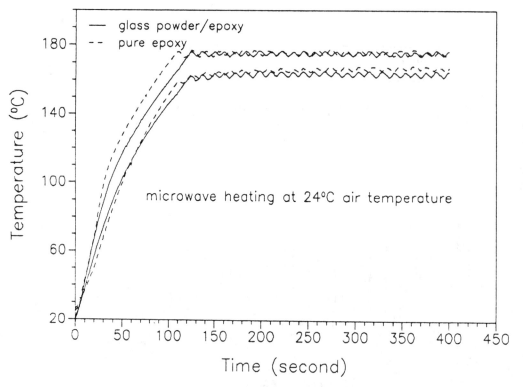

Fig. 5. Microwave heating of epoxy and glass/epoxy

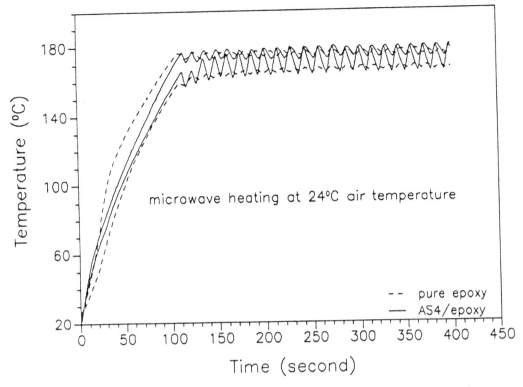

Fig. 6. Microwave heating of epoxy and AS4/epoxy

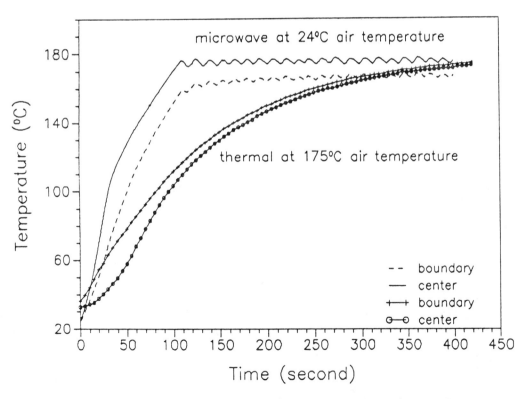

Fig. 7. Thermal and microwave heating of epoxy

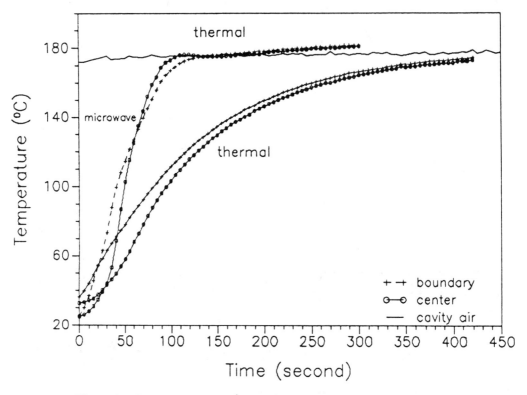

Fig. 8. Thermal w/o microwave heating of epoxy

Monitoring Cure of Composite Resins Using Frequency Dependent Electromagnetic Sensing Techniques

**D. E. KRANBUEHL, M. HOFF, D. A. EICHINGER,
A. C. LOOS AND W. T. FREEMAN, JR.**

ABSTRACT

A nondestructive in-situ measurement technique has been developed for monitoring and measuring the cure processing properties of composite resins. Frequency dependent electromagnetic sensors (FDEMS) were used to directly measure resin viscosity during cure. The effects of the cure cycle and resin aging on the viscosity during cure were investigated using the sensor. Viscosity measurements obtained using the sensor are compared with the viscosities calculated by the Loos-Springer cure process model. Good overall agreement was obtained except for the aged resin samples.

INTRODUCTION

Composite structures constructed from continuous fiber, thermosetting resin matrix prepreg materials are fabricated by laminating multiple plies into the desired shape and then curing the material in an autoclave by simultaneous application of heat and pressure. Elevated temperature applied during cure causes resin polymerization and crosslinking while applied pressure consolidates the individual prepreg plies by squeezing out excess resin. The magnitude and duration of cure cycle temperatures and pressures significantly influence the final physical and mechanical properties of the composite. A composite that is processed using an optimum cure cycle will result in a void free structure that is uniformly cured to the desired resin content and glass transition temperature in the shortest amount of time.

D. E. Kranbuehl, M. S. Hoff, and D. A. Eichinger, Professor, Research Associate, and Graduate Student, respectively, Department of Chemistry, College of William and Mary, Williamsburg, VA 23185.

A. C. Loos, Associate Professor, Department of Engineering Science and Mechanics, Virginia Polytechnic Institute and State University, Blacksburg, VA 24061.

W. T. Freeman, Jr., Aerospace Technologist, Applied Materials Branch, NASA-Langley Research Center, Hampton, VA 23665.

The numerous material properties and processing parameters that must be specified and controlled during cure of a composite laminate make trial-and-error procedures to determine the cure cycle extremely inefficient. Analytical process simulation models are clearly a far superior alternative for determination of optimum cure cycles. In-situ sensors which can measure critical processing properties i.e., viscosity and reaction advancement, are essential for verifying the predictions of the analytical model and continuous monitoring of a part throughout cure. Ultimately both analytical models and sensing technology are required for a closed-loop intelligent production control system.

Frequency dependent electromagnetic sensing techniques (FDEMS) are currently being developed in our laboratories as an in-situ measuring technique for use in automated composite processing and quality control [1-4]. FDEMS sensors are being used to monitor the curing process of autoclave cured composites, and the sensor output has been used to control the cure cycle to produce a more uniformly cured laminate [5,6]. The FDEMS technique was recently applied to measure resin viscosity at different ply positions during cure of a thick 192-ply graphite-epoxy laminate. These measurements were used to verify the viscosity calculated using the Loos-Springer cure process model. The FDEMS measured viscosity agreed remarkably well with the viscosity calculated using the model.

In this paper we examine the use of FDEMS sensors and the Loos-Springer cure process model to monitor and predict the advancement of composite resins as a function of out-time, and to assess the effects of resin age on viscosity during cure. Absolute viscosity measurements are obtained directly from FDEMS sensor output and compared with the viscosities calculated using the Loos-Springer cure process model. Also examined were the effects of increasing the cure cycle heating rate and the intermediate hold temperature on the processing properties.

EXPERIMENTAL

Dynamic dielectric measurements were made using a Hewlett-Packard 4192A LF Impedance Analyzer controlled by a 9836 Hewlett-Packard computer. Measurements at frequencies from 50 to 5×10^6 Hz were taken at regular intervals during the cure cycle and converted to the complex permittivity, $\varepsilon^* = \varepsilon' - i\varepsilon''$. Measurements were made with the geometry independent Dek Dyne FDEMS Sensor which was embedded in the resin.

Viscosity measurements were made using a Rheometrics System IV dynamic mechanical spectrometer. All measurements were made with Hercules 3501-6 catalyzed epoxy resin. At the beginning of the test program simultaneous FDEMS and viscosity measurements were made on fresh resin samples. Additional samples were aged at room temperature and 75% relative humidity for 30 days. Only FDEMS measurements were made on aged samples.

ELECTROMAGNETIC SENSING THEORY

Measurements of capacitance, C, and conductance, G, were used to calculate the complex permittivity, $\varepsilon^* = \varepsilon' - i\varepsilon''$ where

$$\varepsilon' = \frac{C \text{ material}}{C_o}$$

and   (1)

$$\varepsilon'' = \frac{G \text{ material}}{C_o 2\pi f}$$

This calculation is possible when using the Dek Dyne probe whose geometry independent capacitance, C_o, is invariant over all measurement conditions. Both the real and the imaginary parts of ε^* have an ionic and dipolar component. The dipolar component arises from diffusion of bound charge or molecular dipole moments. The dipolar term is generally the major component of the dielectric signal at high frequencies and in highly viscous media. The ionic component, ε_i^*, dominates ε^* at low frequencies, low viscosities and/or higher temperatures.

Analysis of the frequency dependence of ε^* in the Hz to MHz range is, in general, optimum for determining both the ionic mobility-conductivity, σ, and a mean dipolar relaxation time, τ. In turn, the ionic parameter, σ, and the dipolar parameter, τ, are directly related on a molecular level to the rate of ionic translational diffusion and dipolar rotational mobility and thereby to changes in the molecular structure of the resin which reflect degradation, changes in viscosity and degree of cure.

CURE PROCESS MODEL

A cure process model has been developed which can be used to simulate the curing and consolidation processes of composites constructed from continuous fiber-reinforced prepreg materials. A detailed description of the model has been previously reported in references 7-9. Only a brief summary of the information that can be generated by the model will be presented here.

The model relates the cure cycle and prepreg properties to the thermal, chemical, and rheological processes occurring in the composite during cure. For a flat plate composite, the model can be used to calculate the following parameters for a specified process cycle:

    a)   The temperature distribution inside the composite;
    b)   the temperature drop across the tooling, bleeders, and bagging material;
    c)   the degree of cure of the resin as a function of position and time;
    d)   the resin viscosity as a function of position and time;
    e)   the resin flow from the composite and the fiber/resin distribution as a function of time;
    f)   the number of prepreg plies that are fully compacted as a function of time;
    g)   the void sizes, temperature, and pressures inside the voids as functions of void location and time;
    h)   the residual stress distribution inside the composite after cure; and
    i)   the thickness and mass of the composite as a function of time.

Different aspects of the model have been verified experimentally in previous investigations [10,11]. Results of these studies indicate that the model describes adequately the temperature distribution and resin flow of flat plate graphite-epoxy laminates.

RESULTS AND DISCUSSION

Fig. 1 is a plot of the loss factor ε'' of an epoxy resin scaled by the frequency during a multiple ramp-hold cure cycle. Measurements were made over a 50 Hz to 1 MHz frequency range. Plots of ε'' times frequency (ω) are convenient because overlapping lines indicate the frequencies and time-temperature periods during cure where ε'' is dominated by ionic diffusion. Non-overlapping lines which exhibit a systematic series of peaks with frequency and time-temperature can be used to determine a characteristic dipolar relaxation time. Fig. 1 shows the value of ε'' is dominated by ionic contributions throughout the first hold and at the low frequencies in the second hold. The low frequency values of ε'' monitor the ionic mobility and thus reciprocally monitor changes in viscosity. Fig. 1 shows the resin goes through an ionic mobility maximum, i.e. a viscosity minima, at the beginning of the 1st hold and half way through the ramp to the final cure temperature.

The magnitude of the ionic mobility as measured by the conductivity σ is determined from the low frequency overlapping ($\varepsilon'' \cdot \omega$) values using $\sigma(\text{ohm}^{-1}\text{cm}^{-1}) = \omega\varepsilon''(\omega)/\varepsilon_o$ where $\varepsilon_o = 8.854 \times 10^{-14} \text{J}^{-1}\text{C}^{-1}/\text{cm}^{-1}$. The reciprocal relationship of ionic mobility as measured by σ to viscosity has been discussed previously [1-4]. Calibration curves relating σ to viscosity were generated for representative time-temperature cure cycles by making simultaneous σ and η measurements in the rheometer in which one plate was replaced by a sensor.

Fig. 2 is a plot of the viscosity determined by the sensor for unaged 3501-6 resin. The specimen was cured using the manufacturer's recommended cure cycle consisting of a 60 minute hold at 116°C and a 120 minute hold at 177°C. Also shown on the figure is the resin viscosity calculated using the Loos-Springer model which incorporated a chemorheology model for 3501-6 resin from reference 9. The agreement between sensor measured and model predicted viscosity is very good.

Shown in Fig. 3 is the sensor measured and model calculated viscosities when the cure cycle intermediate hold temperature is increased from 116°C to 137°C and the ramp speed is increased from 2°C/min to 2.4°C/min so that the hold is reached at the same time. Agreement between the trends in the measured and calculated viscosities is good. Compared to the original cure cycle, the modified cycle results in a slight decrease in the magnitude of the viscosity at the first minima; but a significant increase in the magnitude of the viscosity at the second minima. This result was observed in both the sensor measurements and model calculations.

The effects of the cure cycle on the resin flow process during autoclave cure of graphite-epoxy composites were examined using the Loos-Springer model. The properties of AS-4 graphite fiber and 3501-6 resin were used in the calculations [7]. The resin flow shown in Fig. 4 is computed for both the standard thermal cure cycle (Fig. 2) and the modified cure cycle (Fig. 3) for a 32-ply composite. The resin flow in Fig. 4 represents the total resin mass loss with respect to the initial mass of the composite. The lower resin viscosity due to the higher initial heating rate used in the modified cure cycle results in a higher rate of resin loss from the composite. Also, with the modified cure cycle, the time required to complete the resin flow process and fully compact the laminate is reduced by 10 minutes.

Resin viscosity determined by the FDEMS sensor for 3501-6 resin aged for 30 days at room temperature is shown in Fig. 5. The magnitude of viscosity of the aged resin is considerably higher than the viscosity of unaged resin throughout the cure (Fig. 2). The increase in resin viscosity would most likely decrease the rate of resin loss and increase the time required to consolidate a laminate during composite processing.

The viscosity of aged resin during cure was calculated using the Loos-Springer model and is plotted in Fig. 5. As can be seen, the model does not predict the viscosity of aged resin as accurately as the viscosity of the fresh resin. This is expected since the chemorheology model for 3501-6 used in the Loos-Springer model was developed using elevated temperature kinetic and rheological data from fresh resin samples. Hence, the model is unable to accurately predict the extent of reaction that occurs during room temperature aging of the resin.

CONCLUSIONS

Cure monitoring sensors and analytical cure processing models are useful for determining the effects of processing parameters on the composite curing process. Frequency dependent electromagnetic sensors (FDEMS) were used to monitor the effects of changes in the cure cycle on the curing process of composite resins. The FDEMS sensor output gives a direct measure of resin viscosity. Agreement between the sensor measured viscosity and the predicted viscosity of the Loos-Springer Model was good.

FDEMS sensors were also used to monitor the advancement of composite resins as a function of out-time to assess the effect of resin advancement on the curing process. Agreement between the sensor measured viscosity of the aged resin during cure and the viscosity computed using the Loos-Springer model was not very good. The difference is most likely due to the fact that the model is unable to accurately predict resin advancement during room temperature aging.

REFERENCES

1. Kranbuehl, D. E., "Electrical Methods for Cure Monitoring," *Developments in Reinforced Plastics - 5*, Elsevier Appl. Science Publishers Ltd., 1986, pp. 181-204.

2. Kranbuehl, D. E., Delos, S. E., and Jue, P. K., "Dielectric Properties of the Polymerization of an Aromatic Polyimide," *Polymer*, Vol. 27, 1986, pp. 11-18.

3. Kranbuehl, D. E., Delos, S. E., Hoff, M. S., and Weller, L., "Dynamic Dielectric Analysis: A Means for Process Control," *National SAMPE Symposium Series*, Vol. 31, 1986, pp. 1087-1094.

4.   Kranbuehl, D. E., Delos, S. E., Hoff, M. S., Weller, L., Haverty, P., and Seeley, J., "Dynamic Dielectric Analysis: Monitoring the Chemistry and Rheology During Cure of Thermosets," ACS Polymer Materials Science and Engineering, Vol. 56, 1987, pp. 163-168.

5.   Loos, A. C., Kranbuehl, D. E., and Freeman, W. T., "Modelling and Measuring the Curing and Consolidation Processes of Fiber-Reinforced Composites," Intelligent Processing of Materials and Advanced Sensors, H. Wadley, P. Parrish, B. Rath, and S. Wolf, Eds., The Metallurgical Society, Inc., Warrendale, PA, 1987, pp. 197-211.

6.   Kranbuehl, D. E., Hoff, M. S., Haverty, P., Loos, A. C., Freeman, W. T., "Insitu Measurement and Control of Processing Properties of Composite Resins in a Production Tool," 33rd International SAMPE Symposium and Exhibition, Vol. 33, 1988, pp. 1276-1284.

7.   Loos, A. C., and Springer, G. S., "Curing of Epoxy Matrix Composites," Journal of Composite Materials, Vol. 17, 1983, pp. 135-169.

8.   Loos, A. C., and Springer, G. S., "Calculation of Cure Process Variables During Cure of Graphite-Epoxy Composites," Composite Materials: Quality Assurance and Processing, ASTM STP 797, C. E. Browning, Ed., American Society for Testing and Materials, Philadelphia, 1983, pp. 119-118.

9.   Lee, W. I., Loos, A. C., and Springer, G. S., "Heat of Reaction, Degree of Cure, and Viscosity of Hercules 3501-6 Resin," Journal of Composite Materials, Vol 16, 1982, pp. 510-520.

10. Loos, A. C., and Freeman, W. T., "Resin Flow During Autoclave Cure of Graphite-Epoxy Composites," High Modulus Fiber Composites in Ground Transportation and High Volume Applications, ASTM STP 873, D. Wilson, Ed., American Society for Testing and Materials, Philadelphia, 1985, pp. 119-130.

11. Loos, A. C., "Modelling the Curing Process of Thermosetting Resin Matrix Composites," Review of Progress in Quantitative Nondestructive Evaluation, Vol. 5B, ed. D. O. Thompson and D. E. Chimenti, Plenum Press, New York, 1986, pp. 1001-1012.

ACKNOWLEDGEMENTS

    This work was made possible through the support of the National Aeronautics and Space Administration - Langley Research Center grant no. NAG1-237 with the College of William and Mary and NAG1-343 with Virginia Tech.

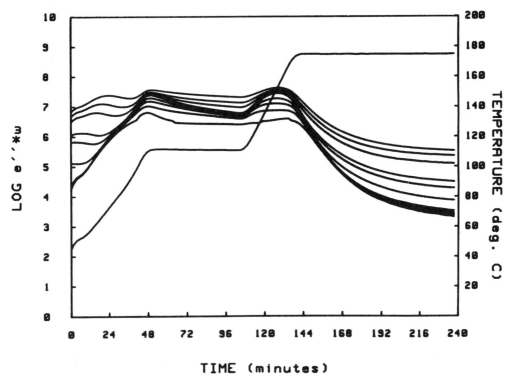

Fig. 1 Log (ε" · ω) vs time during a multiple ramp-hold cure cycle of the amine cured epoxy, 3501-6.

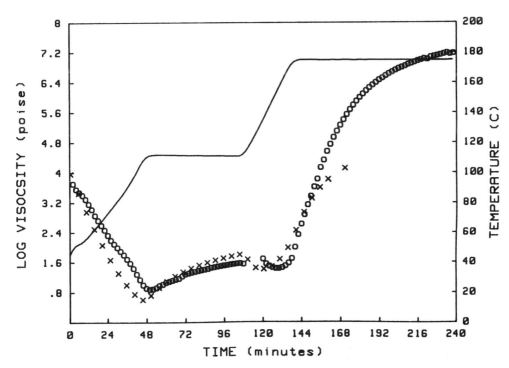

Fig. 2 The viscosity as determined by the FDEMS sensor of an unaged 3501-6 resin during the manufacturer's recommended cure cycle (O). Also shown is the resin viscosity calculated using the Loos-Springer model (x).

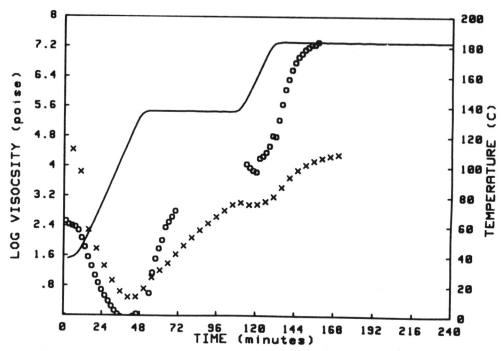

Fig. 3 Sensor measured (0) and model calculated (x) viscosities using an alternate cure cycle on unaged 3501-6 resin.

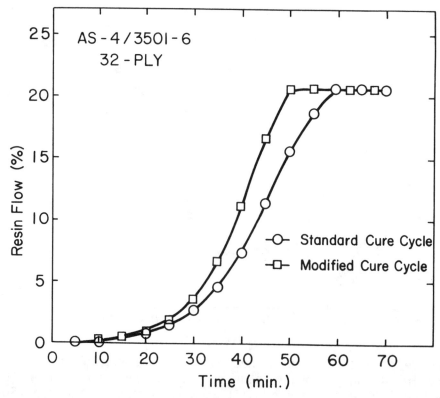

Fig. 4 Resin flow during cure calculated using the Loos-Springer model

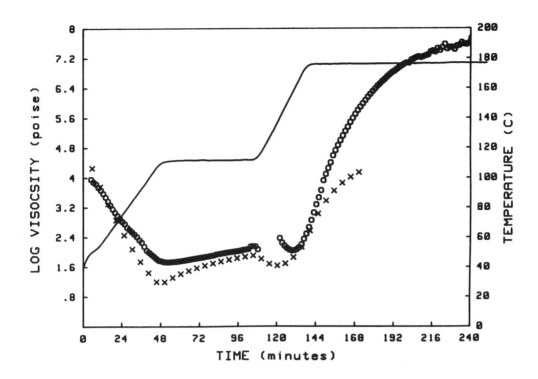

Fig. 5 Sensor measured (O) and model calculated (x) viscosities for 3501-6
aged 30 days at room temperature.

SYMPOSIUM VIII

Impact and Damage Tolerance

Impact-Induced Damage Initiation Analysis: An Experimental Study

S. P. JOSHI

ABSTRACT

Experimental results indicate that most of the damage is produced in the distal layers. The purpose of this paper is to experimentally investigate the crack initiation in these layers. The first segment of the experiment is to bracket the velocity range in which the crack initiates in the distal ply. In the second segment, strain history is recorded under the impact zone on the distal side of the specimen. Results indicate that the crack initiates at a critical maximum average transverse strain (9000 μstrain) in all the laminates.

INTRODUCTION

Impact-induced damage in advanced laminated composites has been a subject of research interest in recent years. Such interest stems from the fact that fiber-reinforced composites have very low tensile strength transverse to fiber direction, which makes them susceptible to damage due to impact by a foreign object. The stresses in the vicinity of impact can be high enough to initiate matrix cracks, even at a very low impact velocity (energy). However, a low-velocity impact does not produce a catastrophic failure, but it makes the structure susceptible to an overall failure. A laminated composite structure subjected to a low-velocity impact by a foreign object does not have visible incipient damage, which makes preventive maintenance difficult. However, the negligible incipient damage may propagate under various structural loadings at a later time if not detected. It is apparent from the above arguments that the understanding of low-velocity impact-induced damage is crucial to safer and reliable use of laminated composites.

A qualitative explanation of damage initiation in laminated composites has been presented by Joshi and Sun [1, 2]. Their discussions are based on impact experiments and a plane strain analysis using a finite element formulation. The stress field obtained using two-dimensional plane strain analysis provided an adequate approximation to the local stress field in the impact region and was used to identify regions of stress concentration [3]. The qualitative comparison implies that the skew cracks in the proximal layers are primarily due to transverse shear stress. Skew cracks in the middle layers are also due to transverse shear stress. Vertical cracks in the distal layer are primarily due to flexural stress transverse to the fiber orientation of the layer. The transverse normal stress does not play an important role in damage initiation. Figure 1 identifies layers and cracks.

S. P. Joshi, Assistant Professor, Department of Aerospace and Mechanical Engineering, University of Arizona, Tucson, AZ 85721

Fig. 1. A typical transverse cross section of an impacted cross-ply laminate.

The experimental results indicated that most of the damage is produced in the distal layers. Extensive delamination on the distal side is initiated because of vertical cracks in the distal layer and transverse matrix cracks in the adjacent layer. The distal side in most of the structures is difficult to inspect and, therefore, it is very important to understand crack initiation in this layer. The purpose of this paper is to experimentally investigate the crack initiation in the distal layer.

EXPERIMENTAL METHODS

The experimental quantification of transverse matrix crack initiation in the distal layer requires careful and controlled experiments. Experiments were devised to extract the common mechanisms causing crack initiation in the distal ply of various laminated plates. Two stacking sequences for the cross-ply laminate and three stacking sequences for the quasi-isotropic laminate were impact tested to obtain quantitative information. The three quasi-isotropic laminates were cut from the same panel with different cut orientations in order to determine the effect of geometric boundary conditions. Quasi-isotropic and cross-ply laminates differed both in stacking sequence and layer thickness.

Fabrication

Hercules AS4/3501-6 graphite-epoxy prepreg was laid up and cured at the composites materials laboratory autoclave facility at Purdue University. A curing cycle similar to the 3501-5A cure cycle recommended by the manufacturer was followed. Square panels 30 cm × 30 cm were fabricated and the first few were C-scanned to assure the quality of specimens. One-inch-wide (25.4 mm) strips were cut from the panels.

Impact Testing Apparatus

The experimental apparatus is shown schematically in Fig. 2. A pressurized air gun was used to shoot the hard steel ball. The impactor ball travels 1.3 m inside the barrel, whose position can be adjusted. The ball comes out of the barrel 6 inches (15 cm) away from the mounted specimen. The impactor velocity is obtained by using light-emitting diode-photovoltaic cell pairs to measure the time interval between the interception of the impactor with the transverse light beams, which are a fixed distance apart. The impactor velocity so obtained is corrected using an experimentally obtained calibration curve. The specimens are clamped 6 inches apart and positioned with the help of a pointer so that the impactor hits the marked center point. Fast-drying spray paint is applied on the impacting side just before the impact to get an impression of the contact area. An impactor ball of 12.7 mm diameter was chosen.

Experimental Procedure

The first segment of the experiment is to bracket the velocity range in which the crack initiates in the distal ply. Starting with a very low value, the velocity was

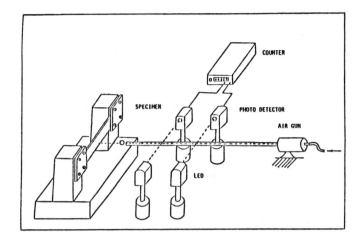

Fig. 2. Schematic diagram of impact testing apparatus.

gradually incremented. At this stage, impacted specimens were optically inspected to bracket the velocity range covering damage initiation. These rough estimates were further narrowed down by impacting more specimens. The initiation crack is not visible to the naked eye and may also go undetected by a low-magnification microscope. The specimens with fine initiation cracks are observed with the help of an X-ray radiograph.

Once the narrow range of velocity for crack initiation is established, the second set of experiments are done. In the second segment, strain history is recorded under the impact zone on the distal side of the specimen. This is where the crack initiates in the distal ply when impacted. Strain gages are mounted transverse to fiber orientation of the distal ply in each laminate. The strain gage records flexural strain transverse to fiber direction. All five lamination configurations are tested in the already established crack initiation velocity range.

The recording setup is shown schematically in Fig. 3. The pressurized air gun and impactor velocity measuring device are the same as described in the previous section. Additional instrumentation used for recording dynamic strain is described below. A wheatstone bridge circuit with a constant voltage source is used to obtain voltage proportional to strain. This voltage signal is amplified using a differential amplifier (Textronics AM 502). The amplified signal is digitally recorded on a Biomation instrument, and a hard copy is obtained using an analog plotter. All of the instruments have a frequency response adequate for recording strain signals (100 kHz). The strain gage and the instrument settings are:

Gage type	EA-13-125AC-350
Resistance in ohms	350 + 0.15%
Gage factor at 75°F	2.115 + 0.5%
Gage grid	3.14 mm × 3.14 mm
Voltage source	2.4 volts
Amplifier gain	100
Biomation sensitivity	2 volts full scale
Signal samples after triggering	1900
Sample spacing	0.2 μsec

EXPERIMENTAL OBSERVATION

Table 1 establishes the narrow range in which cracks appear in the distal ply of cross-ply laminates. Panel 1 is cured using the same procedure as used for curing the panels for specimens presented in Table 2. The curing procedure is different for panel 2. The results presented for panel 2 are therefore not considered for comparison purposes. Table 1 (only panel 1) suggests that the crack initiation impact velocity is in the range 10.5 to 11.5 m/s.

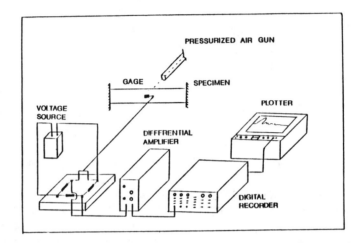

Fig. 3. Schematic diagram of strain response recording setup.

Table 1. Damage identification in impacted cross-ply laminates to bracket the threshold velocities.

Panel 1			Panel 2					
$[0_5/90_5/0_5]$			$[0_5/90_5/0_5]$			$[90_5/0_5/90_5]$		
Identi-fication Code	Impact Velocity (m/sec)	Damage	Identi-fication Code	Impact Velocity (m/sec)	Damage	Identi-fication Code	Impact Velocity (m/sec)	Damage
4	10.54	no crack	15	10.65	no crack	2	10.43	no crack
1	10.65	no crack	14	11.65	no crack	6	10.63	no crack
3	11.39	crack	11	11.73	no crack	7	11.03	no crack
6	11.44	crack	12	11.77	no crack	8	11.08	no crack
5	11.49	no crack	10	11.90	no crack	12	11.25	crack
10	11.61	crack	8	12.05	no crack	9	11.38	no crack
9	11.63	crack	9	12.53	crack	11	11.67	crack
8	11.85	crack	6	12.67	no crack	4	11.87	crack
7	12.09	crack	13	12.76	crack	10	12.33	no crack
2	14.03	crack	3	13.57	crack	5	13.04	no crack
			4	14.43	crack?	15	13.61	no crack?
			5	14.45	crack	14	14.45	no crack
			16	14.70	crack	3	18.60	crack
			7	15.14	crack	16	27.85	crack?
			1	53.15	crack?	13	--	no crack?
			2	57.15	crack?	1	--	crack

Table 2. Damage identification in impacted quasi-isotropic laminates to bracket the threshold velocities.

$[0_2/90_2/45_2/-45_2]_s$			$[-45_2/45_2/0_2/90_2]_s$			$[90_2/0_2/-45_2/45_2]_s$		
Identification Code	Impact Velocity (m/sec)	Damage	Identification Code	Impact Velocity (m/sec)	Damage	Identification Code	Impact Velocity (m/sec)	Damage
3	14.89	crack	42	14.16	no crack	4	10.96	no crack
7	15.51	crack	51	14.52	no crack	1	11.08	no crack
6	15.76	crack	72	14.59	crack	2	11.64	crack
8	15.90	crack	41	14.90	crack	3	12.89	crack
1	15.94	no crack	102	15.22	crack	6	13.31	crack
5	16.28	crack	71	15.25	crack	5	13.39	crack
2	17.42	crack	91	15.27	no crack	8	14.93	crack
4	18.20	crack	92	15.27	crack			
			101	15.28	no crack			
			82	15.31	no crack			
			11	15.34	crack			
			31	15.52	crack			
			12	15.58	crack			
			22	15.78	crack			
			21	15.85	crack			
			81	15.90	crack			
			32	16.24	crack			

The impact velocity at which the quasi-isotropic laminates are examined for crack initiation are given in Table 2. The $[0_2/90_2/45_2/-45_2]_s$ laminate appears to crack in the 15 to 16 m/s impact velocity range. An impact velocity range of 14.5 to 15.3 m/s initiates cracking in the $[-45_2/45_2/0_2/90_2]_s$ laminate. Cracking initiates in the impact velocity range from 11 to 13 m/s in the case of the $[90_2/0_2/-45_2/45_2]_s$ laminate.

Figure 4 shows dynamic strain response from the strain gage mounted transverse to the fiber orientation of the distal ply for the $[0_5/90_5/0_5]$ laminate. The maximum strain is 9000 μstrain at an impact velocity of 11.45 m/s. Figure 5 shows dynamic responses for the cross-ply $[90_5/0_5/90_5]$ laminate at 8.84 and 10.39 m/s. The maximum strain at the impact velocity of 8.84 m/s is higher than at 10.39 m/s. Except for the initial peak, the strains are higher for higher velocity. The discrepancy in peak strain is due to off-centered impact. The strain gradients are high within the portion covered by the strain gage and, therefore, the peak strain is lower when impact is not at the center. The response afterwards due to reflected energy from boundaries is not affected by a slightly off-centered impact. Transverse strain in the distal ply of the quasi-isotropic laminate $[0_2/90_2/45_2/-45_2]_s$ under the impact center is plotted in Fig. 6. Similar plots for the strain history under the impact center in $[-45_2/45_2/0_2/90_2]_s$ and $[90_2/0_2/-45_2/45_2]_s$ are plotted in Figs. 7 and 8, respectively.

ANALYSIS OF RESULTS

The distal ply has a matrix crack parallel to the fibers in the ply. The crack direction is known a priori because of the nature of the crack. The inclination of the crack in the ply provides information about the stresses which possibly initiate that crack. It is experimentally observed that the crack is vertical (perpendicular to the interface), which suggests that the crack is produced by a tensile flexural stress.

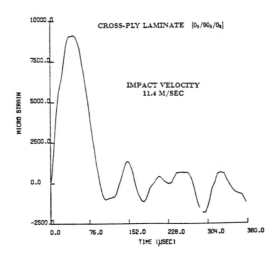

Fig. 4. Transverse strain in the distal ply of the $[0_5/90_5/0_5]$ laminate under the impact center.

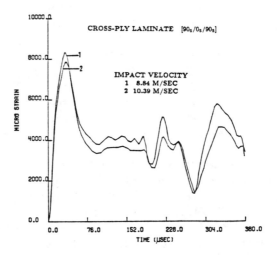

Fig. 5. Transverse strain in the distal ply of the $[90_5/0_5/90_5]$ laminate under the impact center.

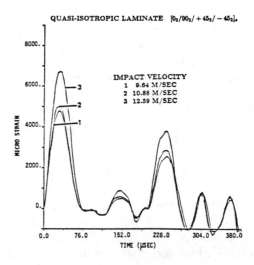

Fig. 6. Transverse strain in the distal ply of the $[0_2/90_2/45_2/-45_2]_s$ laminate under the impact center.

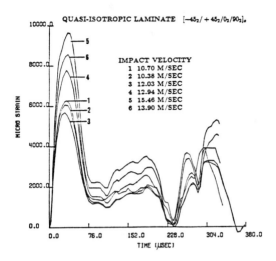

Fig. 7.   Transverse strain in the distal ply of the $[-45_2/45_2/0_2/90_2]_s$ laminate under the impact center.

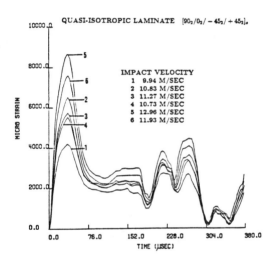

Fig. 8.   Transverse strain in the distal ply of the $[90_2/0_2/-45_2/45_2]_s$ laminate under the impact center.

The experimental results are rearranged in the following graphs to bring out the salient features of the analysis already presented. The strain responses of three different quasi-isotropic laminates at about the same impact velocity are compared in Fig. 9. Studying these strain responses in view of the crack initiation impact velocities obtained experimentally shows that the crack initiates at a critical maximum transverse strain in all three laminates. The strain responses for the cross-ply and the quasi-isotropic laminates at the crack initiation impact velocities are shown in Fig. 10. These experimental results imply that the matrix crack in distal ply occurs at about 9000 μstrain. The quasi-isotropic laminate with 90° fiber orientation of the outer ply is 30% lower than the expected value of 9000 μstrain at crack initiation. A careful examination of the data used to bracket the crack initiation impact velocity in Table 2 suggests that 13 m/s could be the threshold velocity. The experimentally measured strain response at 12.96 m/s for this lay-up (Fig. 8) reaches a maximum strain of 9000 μstrain. The limited experimental results establish that the vertical crack in the distal ply is produced when the transverse flexural strain (E_{22}) in the ply reaches a critical value (9000 μstrain) in the impact zone.

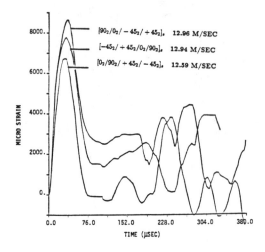

Fig. 9. Strain response for the three-quasi-isotropic laminates.

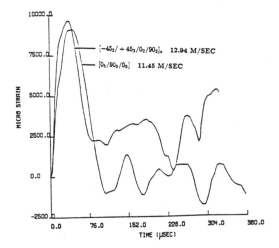

Fig. 10.   Strain response for the cross-ply and quasi-isotropic laminates at threshold velocities.

It is known that the in-situ transverse lamina strength in graphite-epoxy laminates depends strongly on ply thickness and fiber orientation of adjacent plies in the case of in-plane loading [4]. The dependence of in-situ strength is not established in the case of an impact-induced stress field. The strength of the material calculated on the basis of the maximum values of the theoretical stresses is not applicable in the case of high stress gradients which occur during an impact. The strength calculated by averaging stresses over a small area defines damage initiation in a better way. The present study takes an average strain near the impact center (3 mm × 3 mm) as the crack initiation strain.

The ultimate transverse strain in the case of uniform static in-plane loading is 0.6% for unidirectional laminates. The present study shows that the ultimate transverse strain in the distal ply in the impact zone is of the order of 0.9%. The mean strength under uniform tension is always lower than the mean strength under a non-uniform state of stress [5]. The ultimate transverse strain in the distal ply in the case of static cylindrical bending is 0.9%, which is the same as the average crack initiation strain in the impact region.

REFERENCES

1. Joshi, S. P. and Sun, C. T., "Impact Induced Fracture in Laminated Composites," *Journal of Composite Materials*, Vol. 19, No. 1, 1985, pp. 51-66.

2. Joshi, S. P. and Sun, C. T., "Impact Induced Fracture in a Quasi-Isotropic Laminate," *Journal of Composites Technology and Research*, Vol. 9, No. 2, 1987, pp. 40-46.

3. Joshi, S. P. and Sun, C. T., "Impact Induced Fracture Initiation and Detailed Dynamic Stress Field in the Vicinity of the Impact," Proceedings of the American Society for Composites, Second Technical Conference, September 23-25, 1987, pp. 177-185.

4. Flaggs, D. L. and Kural, M. H., "Experimental Determination of the In-Situ Transverse Lamina Strength in Graphite/Epoxy Lamiantes, *Journal of Composite Materials*, Vol. 16, 1982, pp. 103-116.

5. Bolotin, V. V., *Statistical Methods in Structural Mechanics*, translated by S. Aroni, Holden-Day, Inc., Oakland, Calif., 1969.

Simulated Impact Damage in a Thick Graphite/Epoxy Laminate Using Spherical Indenters

C. C. POE, JR.

ABSTRACT

The extent of fiber damage due to low-velocity impacts was determined for very thick graphite/epoxy laminates. The impacts were simulated by pressing spherical indenters against the laminates. After the forces were applied, the laminate was cut into smaller pieces so that each piece contained a test site. Then the pieces were deplied and the individual plies were examined to determine the extent of fiber damage. Broken fibers were found in the outer layers directly beneath the contact site. The locus of broken fibers resembled cracks. The cracks were more or less oriented in the direction of the fibers in the contiguous layers. The maximum length and depth of the cracks increased with increasing contact pressure and indenter diameter. The length and depth of the cracks were also predicted using maximum compression and shear stress criteria. The internal stresses were calculated using Hertz's law and Love's solution for pressure applied on part of the boundary of a semi-infinite body. The predictions and measurements were in good agreement.

INTRODUCTION

NASA is developing light-weight filament-wound cases (FWC) for the solid rocket motors of the Space Shuttle. They are made of graphite/epoxy using a wet filament-winding process. The FWC's are 3.6-cm (1.4-in.) thick except very near the ends where they are thicker to withstand concentrated pin loads.

Tests [1-4] revealed that impacts by blunt objects with low velocity could reduce the uniaxial tension strength of the FWC laminate by as much as 37 percent without making visible surface damage. Sharp impacters caused visible surface damage but not much more strength loss than the blunt impacters [1]. Radiographs and conventional ultrasonic attenuation maps did not even reveal the internal damage caused by the blunt impacters.

In reference [5], numerous specimens with simulated impact damage were deplied to measure the extent of fiber damage in terms of contact pressure and indenter shape. Plates that were cut from an actual FWC were used. Impacts were simulated under quasi-static conditions by pressing hemispherically shaped indenters against the laminate at different locations. For a thick laminate and a given maximum force during impact, the damage caused by

C. C. Poe, Jr., Senior Research Engineer, NASA Langley Research Center, Hampton VA 23665-5225

low-velocity impacts and these simulated impacts should not be significantly different. After the contact forces were applied, the laminate was cut into small squares, each containing a contact site, and deplied. The size of damage was also predicted using maximum compression and shear stress criteria. The internal stresses were calculated using Hertz's law and Love's solution for pressure applied on part of the boundary of a semi-infinite body. The measured and predicted sizes of damage were in good agreement. Some of the force measurements in reference [5] were discovered to be questionable, and the tests were repeated. This report summarizes reference [5] with the questionable results replaced and also includes observations of matrix cracks in two specimens that were sectioned rather than deplied.

MATERIAL

In the previous impact investigations [1-4], a 0.76-m-diameter (30-in.) full-thickness cylinder was used to represent the region away from the ends. No material from the cylinder remained to be used in this investigation. Instead, two 30.5-by-30.5-cm (12-by-12-in.) plates from an actual FWC were used. The materials of the cylinder and the plates were as nearly the same as manufacturing considerations would allow. Both were wound by Hercules Inc. using a wet process and AS4W-12K graphite fiber and HBRF-55A epoxy resin except for the hoop layers of the cylinder, which were hand laid using prepreg tape. The layup of the cylinder and plates were essentially identical. The plates were 3.68 cm (1.45 in.) thick.

From outside to inside, the laminate orientation of plate #1 was

$$\{(\pm\phi)_2/0°/[(\pm\pmb{\phi})_2/0°]_3/[(\pm\phi)_2/0°]_7/[\pm\phi/(0°)_2]_2/\pm\phi/(0°)_4/\pm\phi/(0°)_2/(\pm\phi)_2/\text{cloth}\}$$

and that of plate #2 was

$$\{\pm\pmb{\phi}/0°/[(\pm\pmb{\phi})_2/0°]_3/[(\pm\phi)_2/0°]_7/[\pm\phi/(0°)_2]_2/\pm\phi/(0°)_4/\pm\phi/(0°)_2/(\pm\phi)_2/\text{cloth}\}.$$

The 0° layers are the hoops and the $\pm\phi$ layers are the helicals, where $\phi = 56.5°$. The underlined and emboldened helical layers have about 1.6 times as many tows per in. of width as the other helical layers and are thus thicker in the same proportion. The cloth layer at the inner surface has an equal number of fibers in the warp and weave directions. The plates are balanced (equal numbers of $+\phi$ and $-\phi$ layers) but not symmetrical about the midplane. Most of the hoop layers are closer to the inner surface than the outer surface. Only the outside helical layers for the two plates are different.

TEST APPARATUS AND PROCEDURE

The contact forces were applied with a 500-kN closed-loop, servo-controlled, hydraulic testing machine operating in a load-control mode. The load was increased very slowly by turning a potentiometer. After reaching the desired maximum load, the load was decreased slowly to zero in the same manner. Load and displacement were recorded during each test. Each plate lay on a 36x46x5-cm (14x18x2-in.) aluminum platen, which had one flat surface and one curved surface that mated with the plates. See Fig. 1. The aluminum platen was fastened to a flat 25x64x10-cm (10x25x4-in.) steel platen that was fastened to the hydraulic actuator. The indenters were screwed into an aluminum rod, which was held firmly by steel "L" grips. The indenter was centered on the fixed aluminum platen, and each plate was moved around to align the indenter with the center of a square. The load vector was always normal to the curved surface of each plate. The hemispherical indenters were made of a hardened steel and had diameters of 1.27, 2.54, and 5.08 cm (0.50, 1.00, and 2.00 in.).

For the 2.54-cm-diameter (1.00-in.) indenter, it was found [2-4] that the impact-force threshold for nonvisible damage was about 75.2 kN (16.9 kips), which corresponded to an average contact pressure of 640 MPa (93 ksi). Contact forces were chosen for each of the three indenter diameters to give a range of average contact pressure from 64 to 116 percent of 640 MPa (93 ksi), focusing attention on the threshold for nonvisible damage. Plate #1 was used for all tests with the 5.08-cm-dia. (2.00-in.) indenter and for those with the 2.54-cm-dia. (1.00-in.) indenter that had pressures greater than 514 MPa (85.4 ksi). Plate #2 was used for all tests with the 1.27-cm-dia. (0.50-in.) indenter and for those with the 2.54-cm-dia. (1.00-in.) indenter that had pressures less than or equal to 514 MPa (85.4 ksi).

After the contact forces were applied, the plates were cut into squares. One square for each combination of indenter diameter and contact pressure were deplied. In the deply process, the squares were heated to 400°C (752°F) for 60 to 90 minutes to partially pyrolyze or burn away the epoxy matrix. Following pyrolysis, the laminate was mostly a loose stack of graphite layers. The top layer of the stack was the side that was contacted by the indenter. The length of cracks was measured using an optical microscope. For specimens from plate #1, crack length was measured only on the bottom side of the layers, but for specimens from plate #2, crack length was measured on the top and bottom sides of the layers.

RESULTS AND ANALYSIS

Hertz Law

An excellent treatise on the contact problem for composite materials and its application to impact was given by Greszczuk [6]. The following development is taken from Greszczuk. For a semi-infinite body that is homogeneous and transversely isotropic, the local displacement or indentation is given by

$$u = R_i^{-1/3} (\frac{P}{n_o})^{2/3} \tag{1}$$

where R_i is the radius of the sphere (indenter) and n_o depends on the elastic constants of the indenter and semi-infinite body (target).

The corresponding contact radius is given by

$$r_c = (\frac{PR_i}{n_o})^{1/3} \tag{2}$$

and the pressure distribution on the surface is given by

$$p(\rho) = \frac{3}{2} P_c (1 - \rho^2)^{1/2} \tag{3}$$

where $\rho = r/r_c$ and

$$P_c = \frac{P}{\pi r_c^2} \tag{4}$$

is the average contact pressure. The pressure varies from a maximum of $1.5 p_c$ at $\rho = 0$ to zero at $\rho = 1$.

The load-displacement curves for loading were used to determine n_o. A value of n_o was calculated with equation (1) for each test using the maximum load and displacement or the load and displacement at 80 percent of the estimated impact-force threshold for nonvisible damage, whichever was smaller. The data was restricted because equation (1) does not account for damage. The average of the n_o values was 4.52 GPa (656 ksi) and the coefficient of variation was 0.0984. Values of contact radius and pressure were calculated for each test using equations (2) and (4).

Internal Stresses

Love's solution for stresses in a semi-infinite body produced by pressure on part of the boundary [7] was used to calculate the internal stresses in the FWC plates. The pressure distribution is given by equation (4). The solution is for a homogeneous isotropic body. Even though the laminate is made of orthotropic layers, the results should at least be qualitative.

Contours of maximum shear stress τ_{max} and maximum compression stress σ_{max} are plotted in Fig. 4 for various values of average contact pressure. The maximum compression stress is in the plane of the composite layers, but the shear stress is not. The depth from the surface and distance from the center of contact are normalized by the contact radius r_c. The stresses are axisymmetric about the center of the contact region. Each contour represents the extent of damage according to a maximum stress criterion ($\tau_{max} = \tau_u$ or $\sigma_{max} = \sigma_u$). Since damage is not included in Love's solution, the contours only approximate the size of the damage region. Actually, the stresses from Love's solution are exact only for predicting the onset of damage, which occurs on the axis of symmetry for $p_c/\tau_{max} = 2.15$ and $p_c/\sigma_{max} = 0.834$. Notice that damage from the shear stress initiates below the surface at a normalized depth of 0.482, whereas damage from the compression stress initiates at the surface. It can be shown with equations (2) and (3) that the contact radius r_c increases in proportion to indenter radius R_i and average contact pressure p_c. Thus, the size of the damage contours in Fig. 4 increase in proportion to R_i.

Fiber Damage

Broken fibers were visible in the outer layers of the deplied specimens. Photomicrographs with high magnification of a layer with broken fibers are shown in Fig. 2. The locus of fiber breaks resembles a crack. Broken fibers were found in all the layers above this layer, but not below. This specimen, which was taken from the cylinder [1-4] and not the FWC plates, was impacted using the 2.54-cm-diameter (1.00-in.) indenter, producing a contact force of 54.3 kN (12.2 kips). The fiber breaks for actual impacts and for simulated impacts having equal contact forces were very similar, indicating the equivalence of the tests. The photograph in Fig. 2(a) shows the entire crack, and the photograph in Fig. 2(b) shows a small portion of the crack at an even higher magnification. These cracks were parallel to the direction of fibers in the contiguous layers. By coincidence, the cracks appear to be nearly normal to the fibers. When the fibers in the contiguous layers were not parallel to each other, the direction of the cracks wandered between the direction of the fibers in the contiguous layers.

Two additional specimens with simulated impacts, one using the 2.54- and one using the 5.08-cm-diameter (1.00- and 2.00-in.) indenter, were not deplied but were sectioned through the thickness and examined using a scanning electron microscope. The sections, which were oriented to reveal the breaks (crack) in the hoop fibers, revealed matrix cracks in the contiguous helical layers. The matrix cracks were located on planes of maximum shear stress. It appeared that the matrix shear cracks caused the fiber breaks. Fiber kinking, which is associated with in-plane compression failures, was also observed in a layer near the surface of the specimen tested with the 5.08-cm-diameter (2.00-in.) indenter.

The number and length of cracks were measured in each layer for 20 of the deplied specimens with simulated impacts. The half-length of cracks in each layer is plotted in Fig. 3 for the three indenters and six values of contact pressure. For specimens from plate #1, crack length was assumed to be constant through the thickness of a layer. When layers had more than one crack, the length of the longest crack was plotted. In general, the length and depth of the cracks increase with contact pressure and indenter diameter. Cracks were not found in any layers for the smallest contact pressure of 408 MPa (59.2 ksi) nor for 514 MPa (74.6 ksi) and the two smallest indenters. At a pressure of 648 MPa (94.0 ksi), the reported threshold for nonvisible damage [2-4], the fiber damage is extensive.

For an average contact pressure of 742 MPa (108 ksi), tests were duplicated for the 2.54- and 5.08-cm-diameter (1.00- and 2.00-in.) indenters. The results for the duplicate tests, which are not presented here for the sake of brevity, were in good agreement.

The maximum depths of fiber breaks for the deplied specimens are plotted in Fig. 5. The damage depth was normalized by the contact radius. The filled symbols indicate visible surface damage, and the open symbols indicate no visible surface damage. The 5.08-cm-diameter (2.0-in.) indenter made no visible surface damage for average contact pressures below 590 MPa (86 ksi). The pressures to cause visible damage were a little larger for the smaller indenters. Considering the wide range of r_c values, the normalized damage depths coalesce fairly well for the various indenter diameters, indicating that the damage depth is proportional to the contact radius as predicted.

The maximum depths of the contours in Fig. 4 are plotted in Fig. 5 using values of τ_u = 269 MPa (39 ksi) and σ_u = 587 MPa (85 ksi). These values were chosen to fit the deply data. The maximum depths of damage correspond to the maximum shear stress criterion, and the damage threshold corresponds to the maximum compression stress criterion. Notice that the damage caused by the in-plane compression stress is predicted to be very shallow.

Although the allowables τ_u = 269 MPa (39 ksi) and σ_u = 587 MPa (85 ksi) were chosen to make the analysis and deply data agree, they are also realistic. The average compression strength of disks cut from the cylinder were 620 MPa (90 ksi) [2-4], which corresponds to a maximum shear of 310 MPa (45 ksi). The ultimate compression strain for unidirectional AS4/3501-6 is about 0.0175 [8], which corresponds to an in-plane compression stress of approximately 610 MPa (88 ksi) for the FWC plates. These values are in good agreement with the values of τ_u and σ_u used in Fig. 4, especially considering that Love's solution is for a homogeneous isotropic body and does not account for damage.

The maximum width of the damage contours in Fig. 4 are plotted against the average contact pressure in Fig. 6 for values of τ_u = 269 MPa (39 ksi) and σ_u = 587 MPa (85 ksi). The maximum half-crack lengths for the deplied specimens are plotted for comparison. When a layer contained more than one crack, the length of the longest was plotted. Both predictions and measurements are normalized by contact radius as before. In this case, the curve for the maximum shear stress criteria is a little below the mean of the deply data, but still the agreement is good. Notice that similar crack lengths are predicted using the maximum compression and shear stress criteria, whereas the predicted crack depths in Fig. 5 were quite different. This result is also apparent from the stress contours in Fig.4.

According to Love's solution, indenter diameter and contact pressure are the principal parameters of the contact problem. However, one cannot judge the effect of indenter diameter on damage using these results alone since the impact threat is usually defined in terms of kinetic energy and mass, not contact pressure. In reference [1], residual strengths were slightly lower for sharper indenters (smaller diameter) and a given kinetic energy and mass, indicating that the extent of damage was also slightly larger for sharper indenters. It follows from the stress analysis in Figs. 4-6 that contact pressures must have been much larger for a smaller indenter diameter than for a larger indenter diameter. On the other hand, the visibility of impact damage increased with decreasing indenter radius. Therefore, impacts with sharper indenters are more critical with regard to residual strength but less critical with regard to visibility of damage. Since visibility and damage size were found to vary with contact pressure, the analysis presented here can be used to predict the effect of indenter shape on damage size and visibility in terms of impact force.

CONCLUSIONS

The extent of fiber damage due to low-velocity impacts was determined for very thick graphite/epoxy laminates. The impacts were simulated by pressing spherical indenters against the laminates. The specimens were deplied, and broken fibers were found in the outer layers directly beneath the contact site. The locus of broken fibers resembled cracks, which were more or less oriented in the direction of the fibers in the contiguous layers. The maximum length and depth of the cracks increased with increasing contact pressure and indenter diameter. The cracks initiated at a critical value of contact pressure less than that required to cause visible surface damage. Several pieces were not deplied but were sectioned and examined in a scanning electron microscope. Shear type cracks were found in the matrix next to the broken fibers, indicating that matrix cracking due to shear precipitated the fiber failures. Some shear kinking type of fiber failures that are usually associated with in-plane compression failures were also found near the surface.

The internal stresses in the laminate were calculated using Hertz's law and Love's solution for pressure applied on part of the boundary of a semi-infinite body. The maximum length and depth of the cracks were predicted using maximum compression and shear stress criteria. The predictions and measurements were in good agreement with the deply results.

ACKNOWLEDGMENTS

Mr. Mickey R. Gardner, a Langley Research Center engineering technician, performed the laboratory work. He was assisted in deplying the specimens by Ann E. Davis, a participant in the Summer High School Apprenticeship Research Program.

REFERENCES

1. Poe, Jr., C. C.; Illg, W.; and Garber, D. P.: A Program to Determine the Effect of Low-velocity Impacts on the Strength of the Filament-wound Rocket Motor Case for the Space Shuttle. NASA TM-87588, September 1985.

2. Poe, Jr., C. C.; Illg, W.; and Garber, D. P.: Tension Strength of a
      Thick Graphite/epoxy Laminate after Impact by a 1/2-In.-Radius Impact-
      er. NASA TM-87771, July 1986.

3. Poe, Jr., C. C.; Illg, W.; and Garber, D. P.: Strength of a Thick Gra-
      phite/epoxy Laminate after Impact by a Blunt Object. NASA TM-89099,
      February 1987.

4. Poe, Jr., C. C.; and Illg, W.: Tensile Strength of a Thick Graphite/E-
      poxy Rocket Motor Case After Impact by a Blunt Object. <u>1987 JANNAF
      Composite Motor Case Subcommittee Meeting</u>, CPIA Publication 460, Feb.
      1987, pp. 179-202.

5. Poe, Jr., C. C.: Simulated Impact Damage in a Thick Graphite/epoxy Lam-
      inate using Spherical Indenters. NASA TM-100539, Jan. 1988.

6. Greszczuk, Longin B.: Damage in Composite Materials due to Low Velocity
      Impact. <u>Impact Dynamics</u>, John Wiley & Sons, Inc., 1982, pp. 55-94.

7. Love, A. E. H.: The Stress Produced in a Semi-infinite Solid by Pres-
      sure on Part of the Boundary. <u>Phil. Trans. Roy. Soc. Lond. Series A</u>,
      vol. 228, 1929, pp. 377-420.

8. DOD/NASA Advanced Composites Design Guide. Volume IV-A - Material. First
      Edition. Contract No. F33615-78-C-3203, Flight Dyn. Lab., U.S. Air
      Force, July 1983.

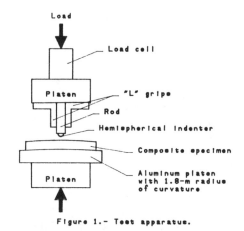

Figure 1.- Test apparatus.

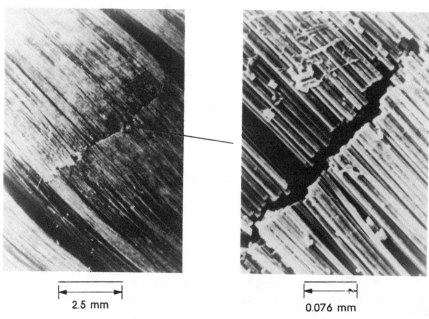

Figure 2. Photomicrographs of fibers broken by impact (layer 7, the deepest with damage.)

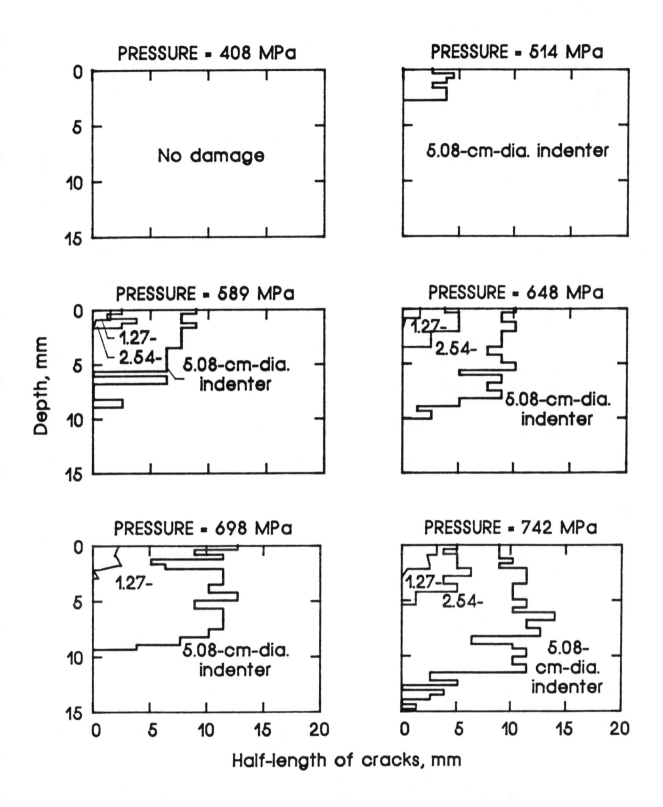

Figure 3. Fiber damage contours from deplied specimens for various average contact pressures.

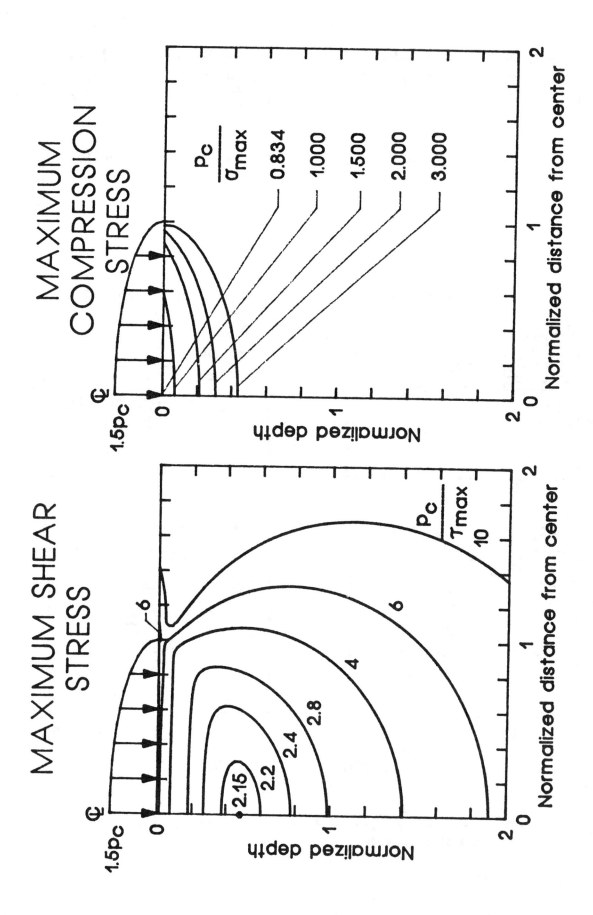

Figure 4.- Maximum shear and compression stress contours according to Love's solution (Poisson's ratio = 0.3).

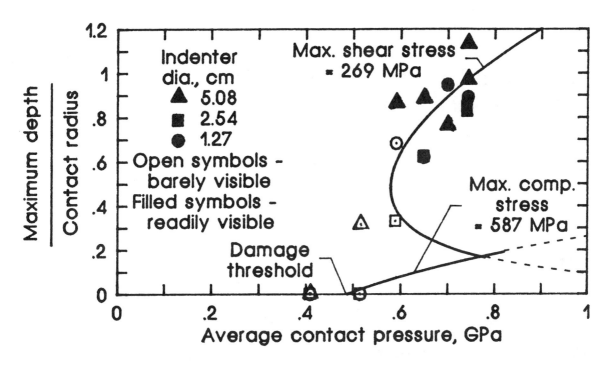

Figure 5. Comparison of predicted and measured damage depth.

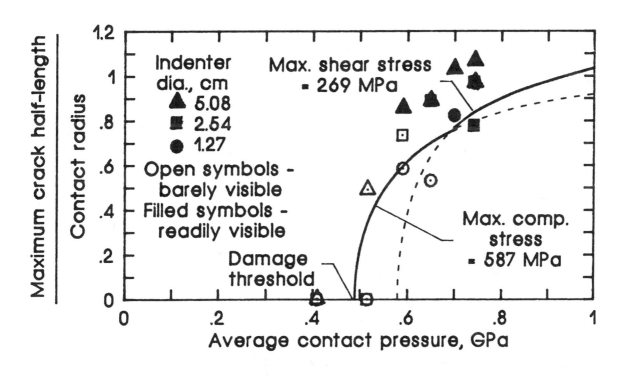

Figure 6. Comparison of predicted and measured crack length.

Impact Damage Characterization of Graphite/Epoxy Laminates

J. H. GOSSE AND P. B. Y. MORI

ABSTRACT

A method is presented for determining the spatial positions (depths and planar orientations) of delaminations present in laminated composite plates subject to transverse impact. The K-rule is used to identify the delamination positions and depths and is described here in detail. A qualitiative free-body analysis is used to explain how the K-rule works. Fractography and ultrasonics are used to verify the K-rule for a quasi-isotropic layup of T-300/934.

INTRODUCTION

One obstacle to the effective utilization of laminated composite structures is a general lack of knowledge as to their behavior with respect to transverse impact. Laminated composite plates, for example, can undergo severe internal damage resulting from low velocity impacts (i.e. 4-16 feet/second). Evidence of this damage may or may not be visibly apparent. Even so, any indication of surface damage will most likely be minor compared to the complete damage state which exists within the laminate.

It is well known that the presence of laminate impact damage in composite structure can result in a significant reduction in the load bearing capacity of the structure, particularly with respect to post-impact compression. It is therefore imperative that a practical understanding of the damage mechanisms

Materials and Processes, Boeing Aerospace, P. O. Box 3999 M/S 2E-01
Seattle, WA. 98124-2499

resulting from the transverse impact be established so that the characteristic damage state (CDS) may be constructed for any general laminate subjected to a general impact loading state. Once the CDS is developed, structural analysis of the global structure with respect to the CDS may be conducted more effectively.

Complete characterization of the CDS involves three separate phases. The first phase is application of a relationship between strain and material organization to locate spatial geometry of the delaminations and the transverse cracks which define their boundaries throughout the impacted laminate. This relationship is the K-rule. Verification of the K-rule is the primary topic of this paper. The second phase entails analytical verification of locations of all transverse cracks which "couple", and therefore bound, delaminations from phase I [1]. Finally, using phases I and II, phase III involves calculation of the lateral extent of the delaminations identified in phase I (as a function of a given impact loading condition). Phase III employs an analytical model which is heavily dependent on the K-rule employed in phase I [2].

Attempts to more completely describe the CDS can be found in the literature [3], [4]. It is desirable to establish a method by which the CDS determined matches evidence of the actual damage state established by time-of-flight pulse-echo ultrasonics and fractography. The method should also be employed as a design aid and therefore be a practical analytical tool. This paper will describe, in detail, a method to relate the post-impact damage state to the laminate material organization (tape layup). The lateral (or radial) extent of these delaminations are not determined by the K-rule.

APPROACH

The approach used to investigate and verify the K-rule was interdisciplinary involving relevant aspects of nondestructive evaluation (NDE), materials science, and qualitative aspects of solid mechanics. The main goal was to firmly establish that the K-rule could be used to predict spatial geometry of the delaminations and coupling transverse cracks resulting from an impact on a quasi-isotropic laminated composite plate (tape layup). The impacted part was scanned using a time-of-flight pulse-echo system [5] and subsequently cut and polished at specific locations to verify the K-rule. Once established on the quasi-isotropic layup, the K-rule was investigated for other layup configurations. Complete descriptions of the layup used and its fabrication can be found in the references [5]. The material used in this paper is T-300/934. Other layups investigated but not reported here were $(\pm26/(90)_2/-45/0/45/(90)_2)/0/(90)_2)_s$ and $(-45/(90/45)_3/(0/-45)_2/0)_s$.

The technical approach specifically involved the following:

1) Composite coupon fabrication [5]
2) Impact loading of the composite coupon [5]
3) Ultrasonic scanning of the impacted region [5]
4) Cutting and polishing of the impacted part at specific locations
5) Verify the K-rule predictions with fractographs and pulse-echo C-scans
6) Repeat steps 1-5 for other layup configurations for general verification of the K-rule (work to be reported at a later time).

RESULTS

The K-rule relates transverse crack initiation to subsequent delamination development and propagation. The present discussion involves low velocity impact in 0.125 to 0.25 inches thick composite plates (4.0 inches x 6.0 inches) only. Damage is dominated by transverse matrix cracks and delamination for such geometry, as opposed to fiber breakage. To understand the proposed mechanism of delamination initiation, a qualitative stress analysis (free body analysis) will be employed to see how resolved stresses and strains create damage. If an appropriate stress analysis of the impacted plate is conducted, stress and strain distribution in the loaded laminate can be used to describe resulting internal damage [1].

From a free body diagram approach and using figure 1a, it can be shown that the stress or strain state anywhere in the laminate can be reduced to a set of principal stresses or strains with respect to a given element. For purposes of discussion this element will have dimensions on the order of the thickness of the plies and our analysis will be restricted to strain space. The solid arrows represent the strain components resulting from an impact stress analysis. The dashed arrows represent the resolved tensile principal strains and planes on which they act. Fiber orientation of the element in figure 1a is such that fibers are normal to the plane of the page. The tensile principal strains are assumed to initiate the subsequent transverse crack. An analysis of this event involving the three dimensional state of strain has confirmed results of this free body analysis [1].

If the element in our analysis is placed within a general laminate we can see how displacements of triangular separations lead to delaminations with respect to upper and lower bounding plies (Figure 1b). Here ξ, Φ and Ψ are arbitrary ply orientations such that $\xi < \Phi < \Psi$, (e.g. $\xi = 0^0$, $\Phi = 45^0$, and $\Psi = 90^0$). When resolved principal tensile strain exceeds the transverse ultimate tensile strain of the ply material, a transverse tensile crack is initiated. This crack is then established as the "coupling" between two developed delaminations. It therefore becomes the boundary between these two delaminations. This mechanism of transverse crack initiation exists in every ply. The result is a series of coupling transverse cracks and subsequent delaminations. This is graphically shown in Figure 1c and shown in actual photomicrographs of Figure 2. The polished section of an impacted composite laminate (pulse-echo C-scan shown in Figure 3) was cut through the center of an impact in the 0^0 direction. The crack pattern in Figure 2 is identical to that discussed by the free-body analysis in figure 1, as indicated by the arrow .

The ply by ply description of transverse cracks and delamination can be presented by extending free body analysis, through the laminate thickness (starting at the surface ply from the impact side). Figure 4 illustrates the series of events at each ply and couples these events into a reproduction of the pulse-echo (time-of-flight) C-scan of an impacted composite plate (Figure 3). Ultrasonics cannot reveal delaminations beyond the first delamination encountered [6] (i.e., pulse reflects off the first delamination, leaving all other delaminations behind it undetected). As a result, this NDE alone cannot be used to map the entire CDS. Instead, the K-rule is needed to analyze NDE information and map the entire CDS. Complete delamination geometries are shown,using the K-rule, in Figure 5.

Fractographic evidence for further verification of the K-rule is shown in Figures 2 and 6. Figures 2, 5, and 6 indicate that the resulting delamination patterns are accurately reproduced using the K-rule.

The triangular delaminations marked 1, 2, and 3 in Figure 3 are represented in Figure 6 as shown. Other delamination positions can also be identified using Figures 2, 5, and 6. Note that delaminations not visible by the C-scan in Figure 3 are exposed in Figure 6 as predicted by the K-rule (Figure 5). The symmetric layup of the laminate result in a reversal of transverse crack orientation at the mid-plane. This is observed using the K-rule (Figure 5) and is validated in Figure 2, 5, and 6. An analysis of delamination size (lateral or radial extent) with respect to a given loading condition is presented elsewhere [2].

Other layups of T-300/934 and additional materials were investigated to further verify the K-rule. The layups were $(\pm26/(90)_2/-45/0/45/(90)_2/0/(90)_2)_s$ and $(-45/90/45)_3/0/-45)_2/0)_s$. Again good comparison between K-rule predictions and actual measurements of the CDS were obtained. Additionally, laminates of IM7/8551-7 with a layup of $(-45/0/45/90)_{3s}$ were investigated, also indicating the validity of the K-rule as applied to tougher tape laminate systems. A detailed account of these additional studies will be given at a later date.

CONCLUSIONS

The K-rule has been presented as a relationship between transverse impact loading events, material organization (tape layup), and subsequent damage patterns throughout the resulting damage zone. Analytical models designed to quantitatively calculate damage size (delamination extent and transverse crack through-the-thickness orientation) are discussed elsewhere [1], [2]. Evidence has shown that K-rule predictions of internal damage patterns of impacted composite laminates agrees well with the actual damaged state of impacted laminates (as determined by pulse-echo ultrasonics and fractography). Once the CDS is defined, appropriate structural analysis may be conducted on the global laminate (including the CDS) to determine the laminate damage tolerance. The results of this analysis are reported elsewhere [7].

REFERENCES

1. Gosse, J. H., Sorcher, S. M., "An Integrated Impact Damage Analysis of Laminated Composites Plates", to be published in <u>Proceedings of the Review of Progress in Quantitative Nondestructive Evaluation</u>, 1988.

2. Gosse, J. H., Mori, P. B. Y., and Avery, W. B., "The Relationship Between Impact-Induced Stress States and Damage Initiation and Growth in Composite Plates", to be published in the <u>Proceedings of the 20th International SAMPE Technical Conference</u>, September, 1988.

3. Joshi, A. P., Sun, C. T., "Impact-Induced Fracture in a Quasi-Isotropic Laminate", 1987 <u>American Society for Testing and Materials.</u>

4. Wu, H. T., Springer, G. S., "Impact Induced Stresses, Strains, and Delaminations in Composite Plates", to appear in the <u>Journal of Composite Materials.</u>

5. Gosse, J. H., Hause, L. R., "A Quantitative Nondestructive Evaluation Technique for Assessing other Compression-After-Impact Strength of Composite Plates", Proceedings of <u>Progress in Quantiative Nondestructive Evaluation</u>, Volume 7B, 1987.

6. Kolsky, H., "Stress Waves in Solids", Dover Publications, dated 1963.

7. Ilcewicz, L. B., Dost, E. F., and Gosse, J. H. "Sublaminate Stability Based Modeling of Impact-Damaged Composite Laminates", to be published in <u>Proc. American Society for Composites</u>, 3rd Technical Conference on Composite Materials, September 1988.

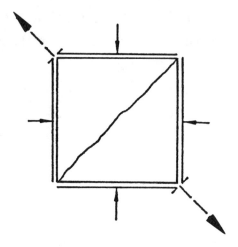

a   Principal stress resolution leading to transverse crack initiation within ply

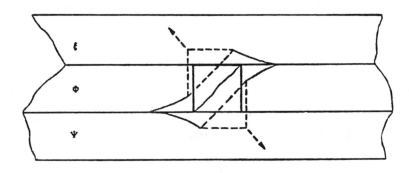

$\xi < \Phi < \Psi$ (ply angle orientation)

b   Transverse crack initiation and subsequent delamination propagation within the laminate

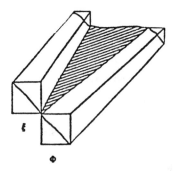

c   Propagated delamination bounded by transverse cracks in the indicated plies

Figure 1 a,b,c   Free body analysis of impact damage initiation and propagation

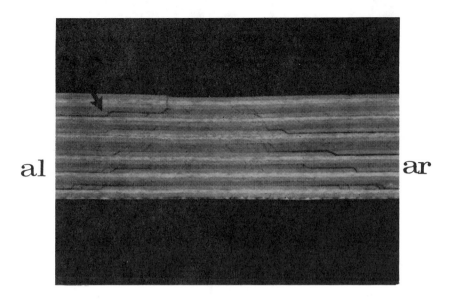

al   ar

Figure 2   Polished section through a-a (see Figure 3)

bl al

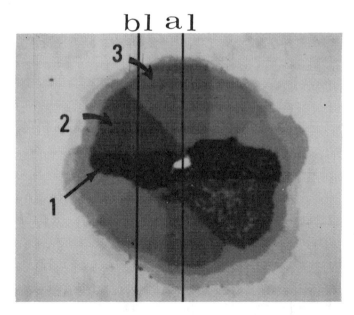

br ar

Figure 3   Pulse-echo C-scan (time of flight) showing depths of the
first-echo delaminations of a quasi-isotropic laminate
$(-45/0/45/90)_{3s}$

351

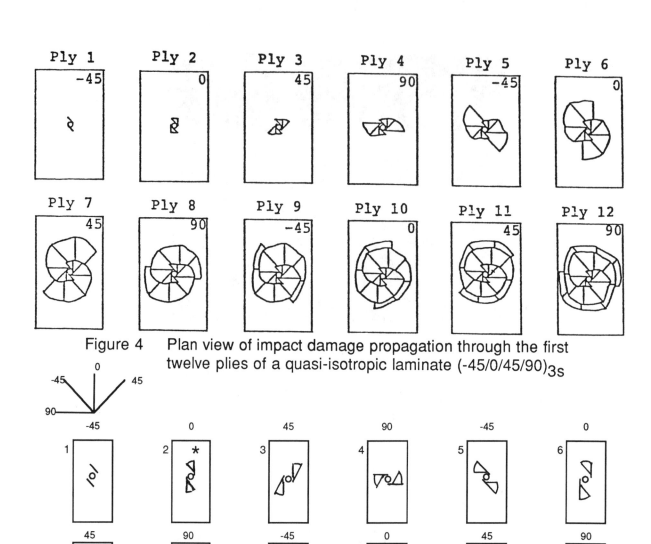

Figure 4   Plan view of impact damage propagation through the first
twelve plies of a quasi-isotropic laminate (-45/0/45/90)$_{3s}$

* Delaminations shown exist between the present ply and the previous ply

Figure 5   Graphical representation of the K-rule for a quasi-isotropic
layup of T-300/934 (-45/0/45/90)$_{3s}$

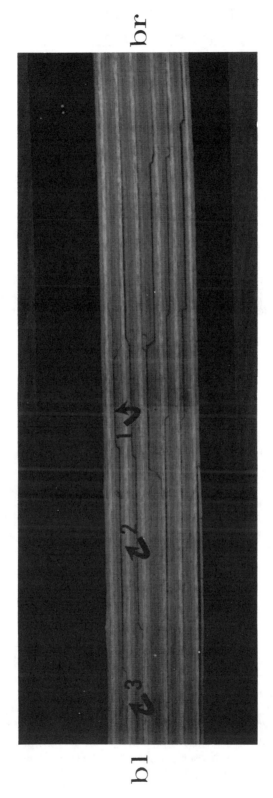

Figure 6   Polished section through b-b (see Figure 3)

Sublaminate Stability Based Modeling of Impact-Damaged Composite Laminates

ERNEST F. DOST, LARRY B. ILCEWICZ AND J. H. GOSSE

ABSTRACT

A sublaminate stability based approach to predicting post-impact compressive strength in composite laminates is presented and applied to a special case. The approach requires an accurate description of the impact damage state. Impact damage in a particular class of quasi-isotropic laminates was found to consist of a spiral array of transverse cracks and delaminations which form 4-ply thick circular sublaminates. Post-impact compressive strength tests with coupons indicated good comparisons between theory and experiment. Finite width effects were modeled and found to be an important consideration when interpretting results from impacted coupon tests.

INTRODUCTION

Composite materials, such as those consisting of an epoxy resin matrix reinforced with carbon fibers, are used on structures in commercial and military aircraft. One important design consideration for composite laminates is low velocity impact by foreign objects (e.g., hail, runway debris, tool drop). Such impact can create nonvisable damage that significantly degrades composite performance. The ability to predict this degradation is essential to the efficient design of composite primary structure [1].

In order to simulate the effects of impact on composite performance, the characteristics of the damage state must be understood. A typical damage zone caused by impact to primary structure consists of matrix cracks and delaminations through the laminate thickness, while impact damage in thinner-gage structure may also involve fiber failure. Matrix damage extends radially from the point of impact, restructuring the laminate into sublaminates, while fiber failure is more localized. The loss of stability of sublaminates and fiber failure are both critical to stress redistribution that generally lowers compressive residual strength. A model to predict the former effect is covered in the current paper.

Commonly used nondestructive test methods, such as through-thickness ultrasonic C-Scans, yield limited information on the damage state (e.g., an inplane view of the area extent of the damage). Without detailed information on the damage state, empirical models have been developed. For example, the relationships between impact damage and equivalent open holes or implanted delaminations have been studied [e.g., 2-5]. Although empirical models are useful tools for interpolation within a data set, they become limited when applied to optimizing damage tolerant composite structure.

The current study reviews the development of a sublaminate stability based model for predicting post-impact compressive strength in composite laminates. In order to allow general application to large structural configurations, the analysis simulates impact damage rather than modeling it discretely when possible.

Authors current address: The Boeing Company, P.O. Box 3707, MS 79-60, Seattle, WA 98124-2207

Empirical factors were not used in any of the models; however, an accurate description of the impact damage state is required to perform the analysis. Experimental verification was obtained using quasi-isotropic laminates. The nondestructive methods described in [6] were used to help characterize the damage states in these laminates.

CHARACTERISTIC DAMAGE STATE

Current advances in ultrasonic inspection methods combined with a strength of materials based K-Rule have lead to the identification of a characteristic damage state (CDS) for several materials and laminate stacking sequences [6]. In a typical CDS, transverse matrix cracks were found to connect delaminations, forming circular sublaminates that repeat with ply stacking sequence through the laminate thickness. For a given stacking sequence the CDS spanned an area that extended radially from the point of impact. This area was found to depend on material toughness; however, the connection and location of damage through the laminate thickness was effectively independent of toughness for the materials studied.

Quasi-isotropic laminates were used to verify the analysis developed in the current study because they represent layups currently used for design and material qualification tests. These layups also have known CDS as described in detail by [6] for laminates with thicknesses on the order of 0.125 to 0.25 in. A similar CDS was described in [7], but with subtle differences. The CDS for repeating stacking sequences of four ply orientations with each adjacent ply or ply group oriented by a difference in angle of 45 or -45 degrees (e.g., $(45,90,-45,0)_{nS}$ or $(45,0,-45,90)_{nS}$), consists of a spiral array of interconnected transverse cracks and delaminations which separate a laminate into sublaminates (see Figure 1). A brief review of the critical features of this CDS as related to the structure of resulting sublaminates will be given below.

The CDS is best envisioned by linking damage ply by ply, through the thickness of a laminate. Consider a circular region in the plane of the panel, centered at the point of impact and spanning the damage zone. This circular region can be broken into wedge-shaped octants (divided at $\pi/4$ intervals) which are numbered one to eight in a clockwise sense from the 0-degree orientation. The octant geometry can be used to map matrix damage and sublaminate structure. It is important to remember that matrix damage is spiraling continuously from octant to octant, through the laminate thickness. This implies that a continuous link between sublaminates also exists within the damage zone.

Starting from the surface opposite that of impact, the outer-most sublaminate (sublaminate 1 in Figure 1) has variable thickness (i.e., the delaminated surface has a spiraled appearance). It is created by the connections between matrix cracks and wedge-shaped delaminations that occur in each octant. The outer-most delamination encountered in a $(45,0,-45,90)_{nS}$ occurs at the 0/-45 interface in octants one and five. The sublaminate layup is (45/0) in these octants. Transverse cracks in the next ply (-45) connect this damage with delamination at the -45/90 interface in octants two and six. The sublaminate layup is (45,0,-45) in this octant. Transverse cracks in the next ply (90) connect this damage with delamination at the 90/45 interface in octants three and seven. The sublaminate layup is (45,0,-45,90) in this octant. Transverse cracks in the next ply (45) connect this damage with delamination at the 45/0 interface in octants four and eight. The sublaminate is (45,0,-45,90,45) in this octant. This completes the description of the outer-most circular sublaminate which has an average thickness of 3.5 plies.

The matrix damage pattern described above repeats through the thickness of the laminate, creating additional sublaminates. Each octant of sublaminate 2 in Figure 1 has a constant thickness of 4 plies. A spiral pattern of damage now occurs on both faces of the sublaminate. As a result, each pair of octants has a unique stacking sequence: (-45,90,45,0) for octants one and five, (90,45,0,-45) for octants two and six, (45,0,-45,90) for octants three and seven, and (0,-45,90,45) for octants four and eight. Sublaminate 2 represents the basic unit that repeats through most of the laminate thickness. The exceptions are at the plane of symmetry where a reflection occurs and at the front surface where the sublaminate has structure similar to that of sublaminate 1 shown in Figure 1. Sublaminates approaching the front surface are also somewhat smaller in diameter.

Note; the sublaminates in Figure 1 are not completely separated by matrix damage, instead a link of three continuous plies exist between neighboring sublaminates at two locations. For example, octants 4 and 8 of sublaminate 1 link to 1 and 5 of sublaminate 2, respectively. In other words, the sublaminates all connect, forming a spiral array.

ANALYTICAL APPROACH

The general analysis approach to the problem of predicting post-impact compressive strength involves several steps. First, the CDS must be idealized. Next, the compressive stability of impact damage is predicted. Following loss of stability, load redistribution is calculated. Finally, a suitable failure criteria is applied. The remainder of this section describes this approach as applied to the CDS defined in the previous section. Note that certain features of this CDS allow simplifications in the analysis which would not be accurate for other layups.

Hypothetically, stability of the spiral array of sublaminates described above can be estimated by modeling individual circular sublaminates. The predicted loss of stability for the repeating unit should lead to a good estimate of the loss of stability for the entire damage zone because sublaminates are all close to the same size and connect only after complete circular delaminations have been formed. The length dimension characterizing damage for this case is the sublaminate diameter, D.

An analytic series solution [8] is used to predict the stability of the sublaminates. This solution assumes a thin symmetric sublaminate on a thick, quasi-isotropic base laminate. These assumptions result in boundary conditions applied to the sublaminate which depend on base laminate deformations. A modification to the method of [8] was made to account for reduced bending stiffnesses [RBS] due to bending-extensional coupling in unsymmetric sublaminates. The bending moduli [D] are reduced by the term $[B][A]^{-1}[B]$ as described in [9], where [B] are coupling moduli and [A] are inplane moduli. As expected, sublaminate buckling loads calculated when accounting for reduced bending stiffness are appreciably lower than the buckling loads calculated without.

As discussed earlier, the basic 4-ply thick repeating sublaminate has four different stacking sequences, each spanning two octants. The buckling strains calculated depended slightly on whether or not the 0-degree ply is a surface ply. Since a single stacking sequence must be assumed for the sublaminate when using the analytic solution, the buckling loads for all possible stacking sequences that constitute the basic sublaminate were calculated and averaged.

A limited number of STAGSC-1 [10] finite element analyses were used to evaluate the reduced bending stiffness approach to buckling prediction for the unsymmetric sublaminate layups studied in this paper. The entire base laminate and the first major sublaminate were modeled using two dimensional plates. The sublaminate was seperated from the base laminate using gap elements, a variation on the non-linear mount element, to prevent element crossover.

Various methods of accounting for sublaminate stiffness properties were evaluated with the different finite element input options. Sublaminate properties were input in three ways; [A] and [D] matricies, [A] and [RBS] matricies, and unsymmetric laminated plate input in which [A], [B] and [D] matricies are automatically generated (i.e., ply by ply properties are entered). The results from [A] and [RBS] input compared well with that of unsymmetric laminated plate input. This helped justify the use of a reduced bending stiffness approach to unsymmetric sublaminate buckling for the layups under study. As expected, finite element analysis with [A] and [D] matrix input yielded significantly higher sublaminate buckling predictions.

Sublaminate buckling predictions from finite element and analytic solutions compared well for sublaminates with diameters greater than 1.5 in. Finite element predictions for diameters less than 1.5 in tended to be higher than those of the analytic solution. For example, finite element predictions using unsymmetric laminated plate and RBS input were approximately 50% greater than that of the RBS analytic solution for a diameter of 0.8 in. This will be discussed later.

Initially, matrix cracks and the decoupled unsymmetric ply groups that comprised the structure of sublaminates were thought to lead to a significant local stiffness reduction prior to buckling, as is the case with some through-width delaminations (e.g., [11]). This would promote a local inplane strain concentration which would lower sublaminate buckling and failure strength predictions. Later it was determined that this behavior occurs only when plies are isolated (i.e., matrix crack and delaminations combine to remove plies from the load path). Without ply isolation, local stiffness reduction is minimized by enforcing strain compatibility with the base laminate. Past experimental studies on the effects of local delamination on tensile strength [12] and finite element analysis using STAGSC-1 also confirmed that the initial strain concentration due to general matrix cracking and ply decoupling is small. Since this effect was much less than that associated with buckled sublaminates, it was not considered in the current predictions.

Load redistribution around the impact damaged region in compressively loaded structure is a function of local sublaminate stabilities. After sublaminates buckle, the damaged region becomes relatively soft with respect to the undamaged material (i.e., as load increases, buckled sublaminates effectively carry a lower percentage of the total load).

In the current analysis, individual sublaminates are assumed to follow a linear load-deflection path until they buckle. At this point, the load carried by the sublaminate becomes constant, as in Euler buckling. A geometrically nonlinear STAGSC-1 analysis was used to justify this assumption. The analysis performed used the discrete model for a repeating sublaminate described earlier. This analysis indicated that the Euler buckling assumption provided a good approximation.

Following buckling, the damaged region is simulated as a reduced stiffness inclusion whose modulus depends on sublaminate stability. This is done to simplify calculations of load redistribution after buckling. Analysis of a reduced stiffness anisotropic inclusion in an infinite width anisotropic plate was given in [13].

The effective reduced modulus of the soft inclusion is shown schematically in Fig. 2. As illustrated, both simulated damage and actual sublaminates have the same end shortening and carry the same load. Prior to sublaminate buckling, the reduced modulus is the same as that of undamaged material. After buckling, the reduced modulus is a decreasing function of laminate stress. Although the damaged region response changes with increasing load, the most important reduced modulus is that occuring at specimen failure. A maximum strain criteria is used to predict post-impact strength, with failure occuring when the undamaged laminate strength is reached. The reduced modulus at failure can be determined by ensuring strain compatibility at the boundary (i.e., effective strain in the damage equals undamaged failure strain of the laminate). The reduced modulus defined by this point is sufficient to calculate load redistribution and predict laminate failure.

The reduced modulus, E^*, of the simulated damage is defined at failure as the ratio,

$$E^* = \sigma_{LB} / \varepsilon_0 \qquad [1]$$

where σ_{LB} is the laminate stress at local buckling of sublaminates and ε_0 is the undamaged laminate failure strain. A modulus retention ratio, MRR, can be defined as

$$MRR = (E^* / E) \qquad [2]$$

where E is the axial modulus of the undamaged laminate. The symbol $MRR(\%)$ will be used when MRR is expressed as a percentage (i.e., MRR x 100). The remaining moduli (transverse and shear) for the soft inclusion are assumed to be reduced by the same ratio, while Poisson ratio remains the same as that of undamaged material. Note that the reduced moduli approximation given above will only be accurate for the quasi-isotropic case considered; however, the concept should be applicable to more complex damage states and layups with some modification to account for anisotropy.

Finite element models were used to calculate the strain concentration occuring next to the damage. Finite elements were used instead of the analytic solution [13] in order to consider the effects of finite specimen width, W. Damage was simulated as a soft inclusion whose stiffnesses are calculated as discussed above. Models of one quarter of the specimen were used due to symmetry. Stiffness of the inclusion was varied from a $MRR(\%)$ of 100% to 0.1%, while damage size was varied from a D/W of 0.1 to 0.85.

The finite width correction factor (FWC) is defined as the ratio of strain concentration in a finite width plate (K_t) to that of an infinite width plate (K_t^∞). Results from finite element analyses defined a surface for K_t as a function of MRR and D/W. As illustrated in Fig. 3, the FWC factor for an open hole is not applicable to a soft inclusion except in the limit as $MRR(\%)$ approaches 0%. The FWC equation for an open hole [14] was modified with an exponent that is a function of MRR. The resulting equation is

$$FWC = \{ (2 + (1 - D/W)^3)/(3(1 - D/W)) \}^{(1-MRR^N)\,M} \qquad [3]$$

This form for the exponent was chosen because it reverts to open hole and undamaged cases as $MRR(\%)$ approaches 0% and 100%, respectively. The constants N and M were determined by a nonlinear regression analysis of finite element results, with N = 0.722 and M = 2.56 giving the best curve fit. This equation was found to represent all finite element results to within ±5%.

EXPERIMENTAL METHODS

Two different graphite fiber material types (fiber volumes of 0.57), one with a relatively brittle matrix (Hercules IM6/3501-6) and one with a toughened matrix (Hercules IM7/8551-7), were used for experimental verification of the approach presented above. In both cases, the nominal ply thickness was 0.0074 in. Two quasi-isotropic layups, (45,0,-45,90)$_{3S}$ and (45,90,-45,0)$_{3S}$, were fabricated at The Boeing Company using 350° F cure cycles. Large panels were machined into 4.0 in by 6.0 in specimens with the 0-degree fibers oriented in the long direction.

Impact specimens were placed in an impacting jig with a 3.0 in by 5.0 in opening and impacted by dropping a 12.0 lb weight from a specified height. The impactor had a 0.625 in diameter, rounded nose. A measure of incipient impact energy was defined as the weight of the impactor times the drop height. Incipient impact energies in the range from 75.0 to 1000.0 in-lb were used. The impact damage was quantified by pulse echo ultrasonic scans as described in [6]. A limited number of specimens were sectioned to verify the damage state. The rest were compressively failed in a fixture that had edge supports designed to prevent global coupon buckling [4]. All post-impact strength tests were performed in a laboratory environment (≈70° F/50% RH).

RESULTS and DISCUSSION

In the past, post-impact strength has been plotted versus impact energy to indicate impact damage tolerance in composite laminates. Such plots can be misleading when attempting to identify physical relationships with material and laminate properties. A distinction between damage resistance and damage tolerance for a material should be made to help clarify this. Throughout the following discussion, damage resistance will refer to a material's ability to resist damage during an impact event. Damage tolerance will denote a material's tolerance to a given impact damage (e.g., post-impact compressive strength versus damage diameter). The IM7/8551-7 material required much higher impact energies to obtain the same damage size as IM6/3501-6; therefore, it has significantly higher damage resistance. This is believed to relate to higher material toughness.

The CDS and properties of the various materials and layups studied resulted in post-impact strength predictions which were the same in all cases. For a given stacking sequence, the CDS for both IM6/3501-6 and IM7/8551-7 material types were found to be essentially the same. The only difference between (45,0,-45,90)$_{3S}$ and (45,90,-45,0)$_{3S}$ stacking sequences was in the direction of damage spiraling (clockwise versus counter-clockwise). Although IM6/3501-6 and IM7/8551-7 have significantly different toughnesses, the stiffness characteristics of both are similar. The stability of sublaminates are a function of their layup, stiffness, size, and thickness, but not of toughness; hence, the buckling predictions versus damage diameter for these materials are identical. The failure prediction depends on both the buckling and undamaged failure strains. The undamaged failure strains of the quasi-isotropic laminates studied were 0.009 in/in for IM6/3501-6 and IM7/8551-7. This leads to post-impact strength versus damage diameter predictions that were also the same for both materials.

As shown in Fig. 3, strain concentration was found to be a nonlinear function of *MRR* and *FWC*. Small sublaminate diameters had relatively high buckling loads which lead to large *MRR* and correspondingly low strain concentration. Smaller damage also has a relatively small *FWC* for the 4.0 in wide coupons. Large sublaminate diameters had very low buckling loads which lead to small *MRR* and strain concentrations approaching that of an open hole. Damage approaching a 3 in diameter has a very high *FWC*.

Fig. 4 shows buckling and failure predictions for the layups and materials studied. Experimental data for the failure strains are also superposed. Experimental results confirm theoretical predictions. Note that both materials have nearly the same damage tolerance. This result indicates that toughness is not important to the failure prediction and implies little or no damage growth prior to failure in the layups studied. Following failure, some specimens were cross-sectioned to evaluate the failure mechanism. Fig. 5 shows distinct sublaminate groups in a buckled configuration. This basically agrees with the failure mechanism assumed in the analysis.

Fig. 4 shows that predictions for small damage diameters are somewhat conservative. This may be due to the limits of the analytic prediction of buckling strain. As discussed earlier, the buckling prediction for a damage diameter of 0.8 in from finite element analysis was 50% higher than that of the analytic solution. As shown in Fig. 4, this results in a prediction more in line with experiments for small damage. Failure predictions that assumed buckling was not affected by the reduced bending stiffness of unsymmetric sublaminates were unconservative, forming an upper bound on experimental data.

The choice of failure criteria may be another reason why conservative failure predictions were made for small damage. Modifying the maximum strain criteria with one of many existing characteristic dimension approaches [15] would increase the failure predictions for small damage, while not affecting calculations with bigger damage. Such approaches would require significant modifications in the analysis described earlier. For example, *MRR* would have to be determined by an iterative scheme. Considering the range of damage sizes that have a significant effect on post-impact strength for matrix damage dominated CDS, efforts to incorporate the characteristic dimension into failure predictions may not be warranted. The same could not be said of CDS dominated by fiber failure, for which small damage has very high K_ts.

Fig. 6 shows predictions of failure strain versus damage diameter for 4 in, 5 in, 7 in and infinite width coupons. The effect of coupon width for large damage sizes (i.e., greater than 1.5 inches) is seen to be significant. Experiments with a range of coupon widths have confirmed these predictions. These results will be presented at a later date. An important point to remember when judging a material's damage tolerance is that the lowest point on the failure strain versus damage diameter curve for infinte width panels corresponds to the undamaged failure strain divided by the K_t^∞ of an open hole (i.e., 3.0 for quasi-isotropic layups).

Without analysis of finite width effects, the use of coupon level testing to determine a materials damage tolerance can lead to erroneous judgements on the behavior of materials in large structure. This is particularly true when damage resistance is not considered separately from damage tolerance. For example, some material selection tests judge post-impact strength for coupons that were subjected to constant impact energy. If the damage resistance of materials differ, then different damage diameters will result. Each damage diameter has its own *FWC* which, if not properly accounted for, can artificially magnify the difference between the post-impact strength of tough and brittle materials. For instance, 270 in-lb of impact energy caused a damage diamiter of 2.5 in IM6/3501-6, while IM7/8551-7 sustained only 1.0 in damage diameter. The difference in post-impact strength for these materials would be on the order of 2.4 for 4.0 in wide coupons, while the difference in large panels is only 1.6.

CONCLUSIONS

An analysis approach to predict the residual compressive strength of impact damaged composite laminates was developed. The characteristic damage state in a quasi-isotropic layup was shown to cause sublaminates that repeat through the laminate thickness. Analysis steps to predict the effects of this damage included a model of sublaminate stability, an equivalent reduced stiffness calculation for the buckled damage zone, finite element analysis of stress redistribution and application of failure criteria. Finite element analysis was needed to accurately model finite width effects, yielding a modification to the equation commonly used for open hole specimens.

Experimental results had excellent correlation with post-impact strength predictions. A distinction between damage resistance and damage tolerance was made to classify different aspects of impact analysis (i.e., the event or the aftermath). The sublaminate properties affecting damage tolerance included reduced bending stiffness, thickness, and diameter. The critical material properties affecting damage tolerance included moduli and undamaged compressive strength. Finite width effects were found to be significant in coupon tests commonly used to evaluate composite impact performance, with the deficiency of more brittle matricies being amplified.

ACKNOWLEDMENTS

The authors wish to acknowledge A.N. Baumgarten, L.R. Hause, R.E. Horton, J.T. Quinlivan and K.H. Schreiber of The Boeing Company for technical support and review of the manuscript. In addition, comments by D.S. Cairns of The Hercules Aerospace Company on the analysis approach were found to be helpful. Encouragement from our families during preparation of the manuscript was also greatly appreciated.

REFERENCES

1) Baker, A. A., Jones R., and Callinan, R. J., "Damage Tolerance of Graphite/Epoxy Composites," Composite Structures, 4, 1985, pp 15-44.

2) Walter, R. W., Johnson, R. W., June, R. R., and McCarty, J. E., "Designing for Integrity in Long Life Composite Aircraft Structures," Fatigue of Filamentary Composite Materials, ASTM STP 636, edited by K.L. Reifsnider and K.N. Lauraitis, American Society for Testing and Materials, Philadelphia, 1977, pp 228-247.

3) Starnes, J. H., Rhodes, M. D., and Williams, J. G., "Effect of Impact Damage and Circular Holes on Compression Strength of a Graphite-Epoxy Laminate," NASA TM-78796, 1978.

4) Byers, B. A., "Behavior of Damaged Graphite/Epoxy Laminates Under Compression Loading," NASA CR-159293, August 1980.

5.) Horton, R., Whitehead, R., et al, "Damage Tolerance of Composites, Final Report," AFWAL-TR-87-3030, Vol. 3, May 1988.

6) Gosse, J. H. and Mori, P. B. Y., "Impact Damage Characterization of Graphite/Epoxy Laminates," Submitted for publication in Proc. American Society for Composites, 3rd Technical Conference on Composite Materials, September 1988.

7) Guynn, E. G. and O'Brien, T. K., "The Influence of Lay-up and Thickness on Composite Impact Damage and Compression Strength," in <u>Proc. of AIAA/ASME/ASCE/AHS 26th Structures, Structural Dynamics and Materials Conf.</u>, AIAA-85-0646, April 1985, pp 187-196.

8) Shivakumar, K. N. and Whitcomb, J. D., "Buckling of a Sublaminate in a Quasi-Isotropic Composite Laminate," <u>Journal of Composite Materials</u>, Vol. 19, Jan. 1985, pp 2-18.

9.) Ashton, J. E., "Approximate Solutions for Unsymmetrically Laminated Plates," <u>Journal of Composite Materials,</u> Vol. 3, Jan. 1969, pp 189-191.

10.) Almroth, B. O., Brogan, F. A., and Stanley, G. M., "Structural Analysis of General Shells," <u>Vol. 2 User Instructions for STAGSC-1</u>, Lockheed Palo Alto Research Laboratory, Palo Alto, CA, Jan. 1983.

11.) O'Brien, T. K., "Analysis of Local Delaminations and Their Influence on Composite Laminate Behavior," <u>Delamination and Debonding of Materials</u>, ASTM STP 876, edited by W.S. Johnson, American Society for Testing and Materials, Philadelphia, 1985, pp 282-297.

12.) Lagace, P. A. and Cairns, D. S., "Tensile Response of Laminates to Implanted Delaminations," <u>32nd International SAMPE Symposium</u>, April 6-9, 1987, pp 720-729.

13.) Lekhnitskii, S. G., <u>Anisotropic Plates</u>, Gordon and Breach Science Publ., 1968.

14.) Nuismer, R. J. and Whitney, J. M., "Uniaxial Failure of Composite Laminates Containing Stress Concentrations," <u>Fracture Mechanics of Composites</u>, ASTM STP 593, 1975, pp 117-142.

15.) Awerbuch, J. and Madhukar, M. S., "Notched Strength of Composite Laminates: Predictions and Experiments-A Review," <u>Journal of Reinforced Plastics and Composites</u>, Vol. 4, 1985, pp 1-159.

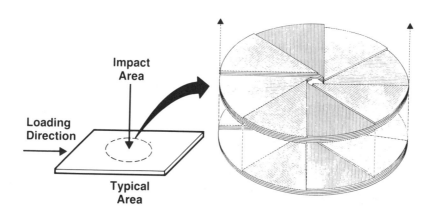

Figure 1: Sublaminate Geometries for $(45,0,-45,90)_{nS}$

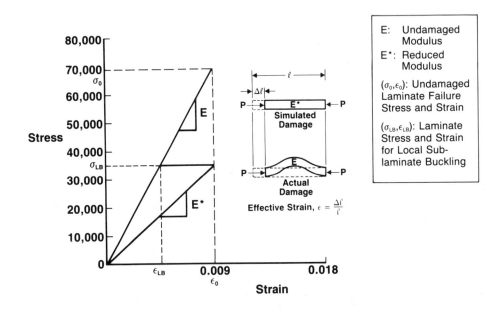

Figure 2: Reduced Modulus Calculation

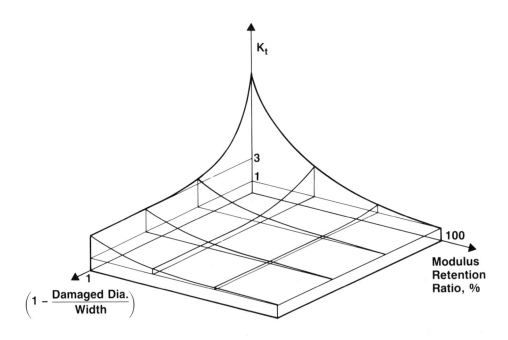

Figure 3: Schematic Diagram of Strain Concentration vs *MRR* and *D/W*

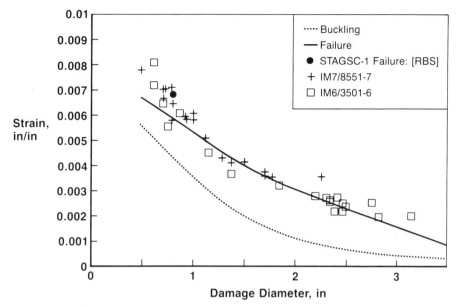

Figure 4:   Prediction versus Experimental Data

Figure 5: Photomicrograph of Section Through Failed Part

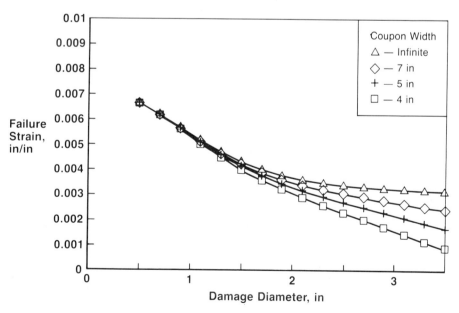

Figure 6: Finite Width Effects on Post-Impact Strength

Smooth Indentation of a Composite Laminate

B. V. SANKAR

ABSTRACT

Finite difference method is used to analyze the problem of smooth contact between a rigid indenter and a laminated circular plate clamped at the edges. The plate consists of transversely isotropic layers. The interlaminar shear stresses in the contact region are found to be much different from the laminate plate theory solutions. It has been found that low-modulus interlayers cause significant reduction in the maximum interlaminar shear stresses.

INTRODUCTION

During the past several years considerable amount of research has been done in understanding low-energy impact response and damage in composite structures. Current efforts in this area focus on developing simple design procedures so that the impact damage can also be taken into account in the design of composite structures. In this regard a through understanding of the detailed stress field in various types of laminates under various impact conditions has become crucial. Recent studies [1] have confirmed that static indentation tests may provide useful information about the failure mechanisms and failure loads for impacts by a large mass (1-10 kg) at low velocities (1-3 m/s). Further static indentation tests are inexpensive and well suited for small samples of material systems that are in development stage. It has also been found that interleaving the laminates with PEEK layers will reduce the extent of delamination due to impact [2]. But, a systematic and scientific study is needed to estimate the optimum location and extent of the PEEK interleaves that will be effective in arresting delamination propagation.

B. V. Sankar, Department of Aerospace Engineering, Mechanics, and Engineering Science, University of Florida, Gainesville, Florida 32611.

Laminate theories are not adequate in describing the stresses due to indentation or for that matter impact. Figure 1 shows the load-deflection diagram of two different 32 layer quasi-isotropic graphite/epoxy laminated plates indented by a 6.25 mm diameter steel indenter. The dimensions of the plate are 125x125 mm and 70x70 mm respectively. According to the laminate theory the failure load of plate 2 should be about 1.8 times that of Plate 1. But, it was found [1] that the failure loads strongly depend on the indenter diameters rather than on the extent of the laminate. This suggests that three-dimensional analysis is essential in the understanding of stresses due to indentation. The present study will focus on the interlaminar shear stresses in a composite laminate indented by a rigid spherical indenter (see Figure 2). The plate is assumed to be circular, made up of transversely isotropic layers, and clamped at the boundary.

FINITE DIFFERENCE FORMULATION OF THE CONTACT PROBLEM

Aboudi [3] has solved several 2-D and 3-D dynamic indentation problems using finite difference method. The equilibrium equations for axisymmetric problems are

$$C_{11}(u,_{rr}+r^{-1}u,_{r}-r^{-2}u)+C_{44}u,_{zz}+(C_{13}+C_{44})w,_{rz}=0, \qquad (1)$$

and

$$(C_{13}+C_{44})(u,_{rz}+r^{-1}u,z)+C_{44}(w,_{rr}+r^{-1}w,_{r})+C_{33}w,_{zz}=0, \qquad (2)$$

where u and w are displacements in the r and z directions respectively. The elastic constants C_{ij}'s are defined in [4]. The finite difference approximation to equations (1) and (2) are obtained by replacing all the derivatives by central differences, and applying them to all the interior points in each layer. The points on the top and bottom surfaces, on an interface between two layers, and on the z-axis, all need special consideration.

The initial contact between the rigid sphere and the indenter is assumed to occur at the central node. As the indenter is moved down, successive nodes come into contact with the indenter, and their boundary conditions are changed. Gauss-Seidel iteration is used to solve the system of linear equations. At each stage the contact stress distribution and the interlaminar stresses are calculated.

In the numerical examples the plate diameter was assumed to be 10 mm and the thickness 2 mm. The indenter radius was 10 mm, and the computations were performed for a maximum contact radius of 1 mm.

RESULTS AND DISCUSSION

Two different problems were studied. The material properties and the laminate configurations for each problem are presented in Tables 1 and 2 respectively. The plate with isotropic layers (Problem 1) was chosen to bring out the important features of the laminate contact problem. The

materials in Problem 2 represent some realistic composite
material systems. The properties of Material 1 are
approximately equal to that of a four ply quasi-isotropic
laminate idealized as a homogeneous transversely isotropic
layer. Properties of Material 2 are obtained by similar
idealization of glass/epoxy layers. Material 3 is assumed as
an epoxy resin.

Results of Problem 1 are presented in Figures 3 through 6.
In these figures the legends Homogeneous, Top and Interleaf
represent the configurations shown in Table 1. From Figure 3
it may be seen that the contact stresses in the homogeneous
plate is Hertzian, whereas there is considerable deviation in
the other two plates. In the presence of a soft surface layer
on top, the contact stresses peak at the center. When the soft
layer is used as an interleaf the contact stress concentration
is reduced by 20%. The variation of contact force with the
contact radius is presented in Figure 4 for all the three
cases. The two extreme curves represent the homogenous plates
made up of materials 1 and 2. The interlaminar shear stress
distribution in the three plates for a given contact force is
shown in Figure 5. The stresses for each case is presented at
a radius where the maximum shear stress occurs. The effect of
surface layer and interlayer in reducing the interlaminar
shear is evident. The variation of maximum interlaminar shear
stress with respect to the contact force is presented in
Figure 6. Once again one can see that the interlayer and
surface layer have significant effect in the maximum shear
stresses.

The results of Problem 2 are presented in Figures 7
through 11. In these figure the legends G/E, Hybrid and
Interleaf represent the homogenous graphite/epoxy laminate,
plate with an interlayer of glass/epoxy, and plate with a
interlayer of epoxy resin respectively. The contact stresses
in all the three plates were close to the Hertzian
distribution, and they are not presented here. There were not
significant differences in the load contact radius relations
also. The reason may be that the top layers of all three
plates are the same. The interlaminar shear stress
distribution in the three laminates are shown in Figures 7, 8
and 9 respectively. In the presence of interleaf (Figure 9)
the maximum interlaminar shear still occurs in the top layer,
but it is considerably less than that in the homogeneous plate
(cf. Figure 7). The same trend can be observed in Figure 10
where the shear stresses due to a given contact force are
presented. Figure 11 shows the variation of maximum
interlaminar shear with the contact force, and the presence of
interlayer significantly reduces the maximum shear stress.

The present numerical study shows that the interlaminar
shear stresses in a composite laminate subjected to static
indentation can be considerably reduced by interleaving with a
layer of epoxy resin. One should expect similar results with
PEEK interleaf also. Experimental studies are underway to see
if this could reduce the extent of damage during static
indentation. The results can be extended, at least

qualitatively, to low-velocity impact situations.

ACKNOWLEDGEMENTS

This study was initiated when the author was working at the Wright-Patterson Air Force Base Materials Laboratory under the subcontract RI-51542X through the University of Dayton Research Institute. The author is grateful to Mr. T.M. Cordell and Dr. N.J. Pagano for many helpful discussions and encouragement. Partial support was provided by the NASA Langley Research Center Grant NAG-1-826. Mr. C.C. Poe, Jr. was the contract monitor. The support of Pittsburgh Supercomputer Center through the NSF Grant MSM 8708764 is also acknowledged.

REFERENCES

1. B.V. Sankar: Low-Velocity Impact Response of Graphite-Epoxy Laminates. Final Report. Department of Engineering Sciences, University of Florida. WPAFB Contract F33615-84-C-5070, September 1987.

2. P.O. Sjoblom, J.T. Hartness and T.M. Cordell: Private Communication.

3. Aboudi, J.: Comp. Methods App. Mech. Engineering. 13 (1978) 189-204.

4. B.V. Sankar: Contact Law for Transversely Isotropic Materials. Paper No. AIAA-85-0745. Presented at the AIAA SDM Conference, 1985.

TABLE 1

Laminate Configuration and Material Properties For Problem 1

Material 1:   ISOTROPIC, E=200 GPa, ν=0.25
Material 2:   ISOTROPIC, E=20 GPa, ν=0.25

Laminate 1A: HOMOGENEOUS

MATERIAL 1
THICKNESS=2 mm

Laminate 1B: LAYER ON TOP

MATERIAL 2
THICKNESS=0.25 mm
MATERIAL 1
THICKNESS=1.75 mm

Laminate 1C: INTERLEAF

MATERIAL 1
THICKNESS=0.25 mm
MATERIAL 2
THICKNESS=0.25 mm
MATERIAL 1
THICKNESS=1.25 mm

TABLE 2

Laminate Configuration and Material Properties for Problem 2

Material	E_r GPa	E_z GPa	G_{rz} GPa	ν_{rz}	$\nu_{r\theta}$
1	50	7	3.8	0.285	0.265
2	20	8	3.6	0.270	0.255
3	3	3	1.1	0.350	0.350

Laminate 2A: HOMOGENEOUS

> MATERIAL 1
> THICKNESS=2 mm

LAMINATE 2B: HYBRID

> MATERIAL 1
> THICKNESS=0.25 mm
> MATERIAL 2
> THICKNESS=0.25 mm
> MATERIAL 1
> THICKNESS=1.5 mm

LAMINATE 2C: INTERLEAF

> MATERIAL 1
> THICKNESS=0.25 mm
> MATERIAL 3
> THICKNESS=0.25 mm
> MATERIAL 1
> THICKNESS=1.5 mm

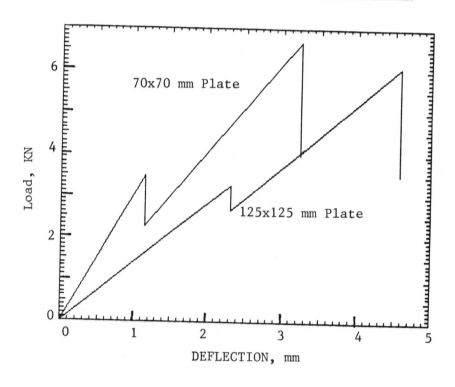

Figure 1   Load Deflection Diagram for Graphite/Epoxy
Laminated Square Plates

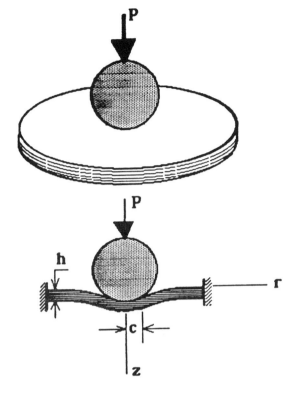

Figure 2   Indentation of a
            Transversely Isotropic
            Laminated Circular Plate

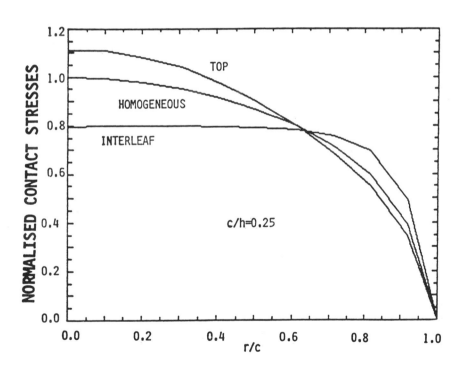

Figure 3   Normalized Contact Stresses (Problem 1)

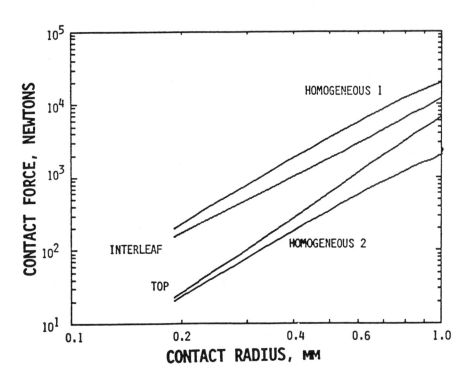

Figure 4   Contact Force–Contact Radius Relations (Problem 1)

Figure 5   Interlaminar Shear
Stresses for
P=6552 N (Problem 1)

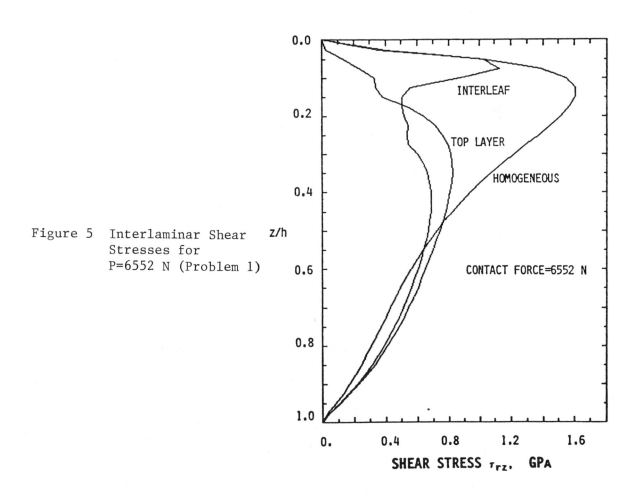

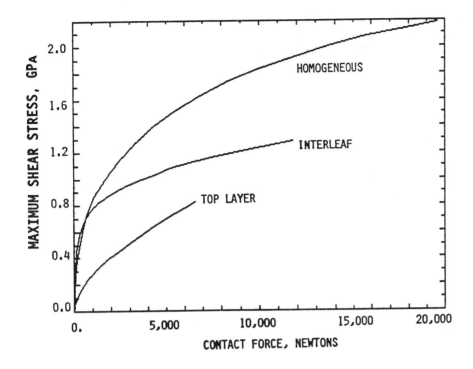

Figure 6   Maximum Interlaminar Shear-Contact Force Relations (Problem 1)

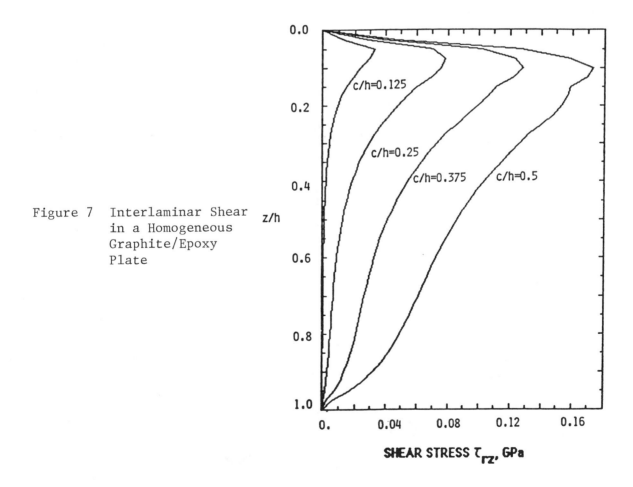

Figure 7   Interlaminar Shear
in a Homogeneous
Graphite/Epoxy
Plate

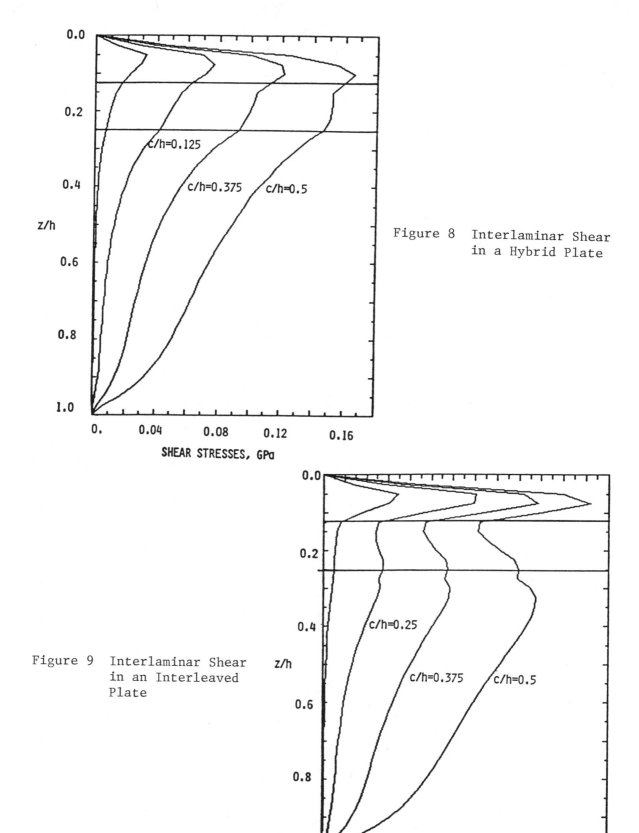

Figure 8 Interlaminar Shear
in a Hybrid Plate

Figure 9 Interlaminar Shear
in an Interleaved
Plate

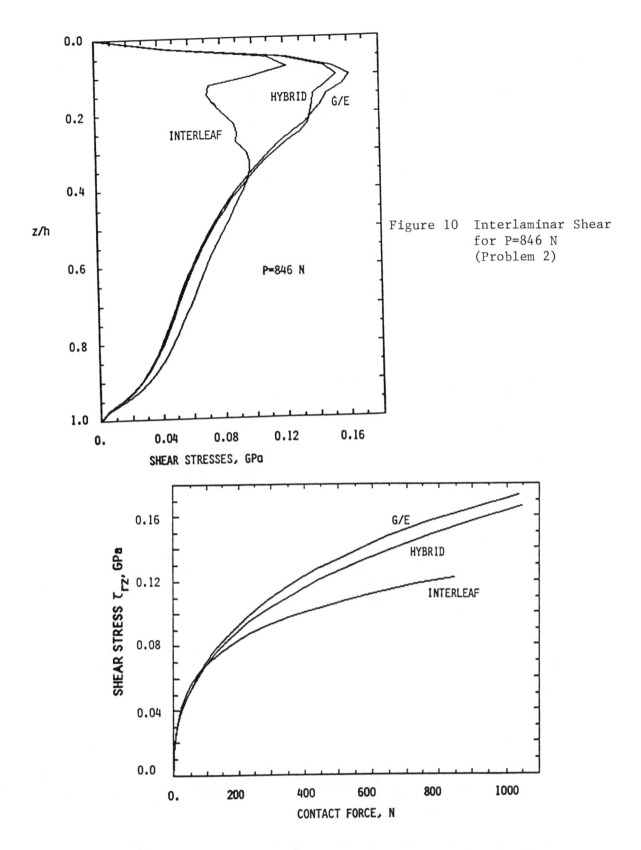

Figure 10 Interlaminar Shear
for P=846 N
(Problem 2)

Figure 11 Maximum Interlaminar Shear Stress-Contact Force
(Problem 2)

Parametric Studies of Impact Testing in Laminated and Woven Architectures

LEO E. TASKE, II AND AZAR P. MAJIDI

ABSTRACT

The impact behavior of cross-ply laminates, and two- and three-dimensionally woven graphite/PEEK composites have been studied. The parameters examined in these systems during impact testing included weave geometry, fiber volume fraction, and part thickness. The effect of thickness, in the range of 1.7 to 4.7mm, was studied on cross-ply laminates and two-dimensional (2-D) woven structures. The flexural moduli and the reduced stiffnesses of the respective system were used to determine the dependence of the absorbed impact energy on thickness. The data showed that, similar to the bending stiffness, the impact energy is linearly proportional to the cube of the thickness of the composite. This relationship was used to compare minor thickness variations between two 3-D woven commingled angle interlock structures. The residual strength of the impacted panels was evaluated by cutting the panel into a number of strips and testing in four point flexure. The 3-D woven materials contained the damage to a smaller area than the unidirectional cross-ply and the 2-D woven laminates at 110J of incident energy.

INTRODUCTION

Recently there has been a large influx of new engineering materials and manufacturing configurations available for composite part production. Depending on the design specification, parts can reach thicknesses of inches or even feet. Research has begun to look at the thickness effects on composite part performance. Increased part thicknesses can generate processing problems in thermosetting composites due to large exotherms. Other researchers have been looking at the mechanical property aspects of thick composites such as fracture toughness [1,2], impact damage and compression strength [2,3]. In out-of-plane testing of materials, whether isotropic or anisotropic, the thickness becomes a critical parameter.

This research originated by looking for architectural effects of three-dimensionally woven, commingled thermoplastic composites [4]. The processing of these commingled graphite/PEEK systems uncovered another parameter besides architecture which needed addressing. The thickness of the three-dimensionally woven, commingled systems changed during consolidation as a result of the melting of PEEK matrix fibers and the removal of the voids in the preform. The architecture of the three-dimensional weaves affects the packing efficiency and therefore the density of the unprocessed commingled preform. Various styles of

L. E. Taske, II, Graduate Student, and A. P. Majidi, Assistant Professor, Center for Composite Materials and Department of Mechanical Engineering, University of Delaware, Newark, DE 19716

woven preforms, though initially of the same thickness, can consolidate to different thicknesses due to the initial density of the preform.

This paper addresses the effect of thickness on the impact testing of cross-ply laminates, two-dimensional and three-dimensional woven systems. The cross-ply laminates and the 2-D woven specimens were fabricated in transversely isotropic form to resemble the two primary fiber orientations of the three-dimensionally woven materials. The impact energy level chosen was high to insure that no panel would absorb all the incident energy of the drop weight impact tup. Holding the incident energy level and the boundary conditions for the impact specimen constant, the effects of the thickness on the material systems were examined. From lamination theory, the D_{11} and D_{22} components of the stiffness matrix were calculated and plotted versus thickness for the cross-ply laminates and the 2-D 8HS woven system. This examination led to a thickness scaling factor which was used to normalize impact data accumulated for various thickness 3-D woven systems.

The composite thickness and fiber architecture affect not only the impact energy absorbed but also the mode and the extent of the impact damage. A study of the effects of these parameters on the residual strength after impact of the composites was therefore of particular interest to this investigation. However, the standard Boeing Compression After Impact Test [5] could not be used for this purpose because of the limited supply of material and the configuration of the specimens available. Instead, a residual strength test method was developed in which the impacted panel was sectioned into parallel-sided strips. Each strip was then tested in four point flexure testing such that the backsurface, where the impact damage was concentrated, was subjected to compressive loads. Because the impact specimen was sectioned across its width, the flexure testing gave detailed information about the extent of damage propagation from the point of impact and the residual strength in that damaged region.

MATERIALS

The unprocessed materials were consolidated into flat rectangular shaped plates in tooling designed for polymeric diaphragm forming technology [6,7]. Each plate contained three impact panels, three axial flexure specimens and three transverse flexure specimens. Two material systems, an APC-2/AS4 cross-ply laminate and a 2-D 8HS commingled woven laminate, were fabricated specifically to test the thickness effects during impact. All specimens were evaluated nondestructively using ultrasonic C-scanning.

The APC-2 $[0/90]_{ns}$ laminates had ply counts of 16, 20, 24, and 32 plies giving a final thickness range from 2.0 mm to 4.2 mm. The fiber volume fraction of the APC-2/AS4 material system was 0.61. The cross-ply fiber orientation mirrored the fiber direction of the 3-D woven materials.

The 8HS commingled woven laminates, $[0/90]_s$, were fabricated to have the same fiber volume fraction, 0.51, as the 3-D woven architectures. Initially the graphite fiber, Celion G30-500, was commingled with PEEK 150G fibers to make a 6K bundle with a fiber volume fraction of 61%. The commingled fiber bundles used in the 3-D woven architectures were plied (helically wrapped) with additional PEEK fibers which enabled the fiber tow to move more easily in the loom. This additional PEEK wrapping reduced the fiber volume fraction to 0.51. Because the 8HS commingled material did not have this additional plied PEEK fibers, additional PEEK matrix was added. The PEEK, in the form of a five mil thick film, was weighed and interspersed symmetrically throughout the laminate to obtain the same fiber volume fraction for the overall composite plaque as the 3-D woven materials.

The 3-D commingled angle interlock (AI) woven preforms were fabricated by Textile Technologies Inc. in two different weave geometries. In one geometry, AI #2, the fill yarns interlaced three warp yarns while in the other geometry, AI #3, two layers of warp yarns at a time were interlocked. Details of these two weave geometries have been given previously [7].

EXPERIMENTATION

Impact Tests

Drop-weight impact tests were conducted on a Dynatup Model 8200A Impact Tower equipped with a GRC Model 730-I Data Acquisition System. Specimens, with dimensions of 76mm by 76mm by thickness, were clamped on a raised support with a two inch diameter center opening and impacted with a five-eighth inch diameter steel tup. Both the dart energy and velocity were maintained constant during the tests, at 110J and 4.0 m/sec., respectively. The 110J incident energy was chosen in order to determine the maximum energy absorption for each panel configuration. By having excess incident energy, each panel will absorb the maximum for its configuration with additional energy remaining in the tup.

Four Point Flexure Tests

Four point flexure tests were conducted on both undamaged baseline composites and sectioned impacted specimens with dimensions of 76mm by 10mm. The four point flexure tests followed the ASTM D790-84 standard with the span to depth ratio of 16 to 1 for most specimens. The 2mm to 3mm specimens were not cut 25mm wide as the standard recommended. These specimens were kept 10mm wide to obtain more detailed information about the damage propagation in the impacted specimens and also to have a greater number of test specimens. Table I gives the geometrical measurements of the flexure specimens. The

TABLE I. Dimensions used in the Four Point Flexure Tests

Material System	Ply Code	Width (mm)	Depth (mm)	Span (mm)	Span to Depth
APC-2	16 Ply	10.0	2.0	40.0	20.0
	20 Ply	10.0	2.6	40.0	15.4
	24 Ply	10.0	3.1	48.0	15.5
	32 Ply	10.0	4.2	64.0	15.2
8 HS	6 Layer	10.0	1.8	40.0	22.2
	10 Layer	10.0	2.9	48.0	16.6
	12 Layer	10.0	3.4	64.0	18.8
	16 Layer	10.0	4.7	64.0	13.6
3D AI# 2	na	10.0	3.3	48.0	14.5
3D AI# 3	na	10.0	3.8	64.0	16.8

flexure test dimensions were held constant for each material system regardless of its damage status. The flexure testing provided a basis for evaluating the residual flexural strength of impacted specimens.

The flexure specimen was configured such that the backsurface from the impact test was oriented to the compressive or top side during the flexure testing. This puts the region with a high delamination density under the compressive bending load in attempt to simulate the Boeing Compression After Impact Test. Because the four point flexure test gives higher strength values than tensile testing, the four point flexure test should not be used to obtain design data [8].

THICKNESS DEPENDENCE OF THE BENDING STIFFNESSES

When structures are loaded in bending, the maximum stress is a function of the distance away from the neutral axis, thus, the thickness. The maximum stress seen in a four point flexure specimen is

$$S = \frac{3PL}{4bd^2}$$

where d is the thickness or depth of the specimen. Specimens impacted experience biaxial bending when supported on a circular clamping ring. With the bending seen in impact testing of thinner specimens, normalizing the absorbed impact energy by unit thickness loses information about the testing. Impact dynamics and lamination theory were examined to find a more realistic basis for the normalization of impact data.

The impact dynamics for isotropic materials have been extensively developed [9]. Additional factors leading to the impact dynamics development for anisotropic materials have been examined by the same researchers. The basic principle in impact testing falls on conservation of energy. The initial energy of the impact tup is transferred to the panel by energy absorbed due to plate bending effects and Hertzian contact forces. The plate bending effects dominate the mechanisms in the thin laminates and decrease with increasing thickness. In thicker laminates, the failure mode results from the contact forces which generate local, subsurface damage [9]. The plate bending mechanisms become insignificant at some thickness range and lamination theory should give clues when this transition in failure mechanisms occurs.

Using lamination theory, the resultant moments are computed using the B and D matrices. The reduced stiffness components, D_{11} and D_{22}, are the bending-curvature relations for the [0°] and [90°] directions, respectively. For the 8HS woven system, the D_{11} and D_{22} components of the reduced stiffness matrix, were calculated from one lamina using the bridging model for woven fabrics [10]. The effective lamina properties from the bridging model were used to calculate the bending stiffness components. The D_{11} and the D_{22} values were averaged to arrive at the value, D_{avg}, used for the comparison. In computing the D_{avg}, the importance of the ply thickness and location to the bending stiffness increases as the laminates get thicker.

For both the cross-ply laminate and the 8HS woven laminate, the D_{avg} points were related to the thickness from lamination theory by the equation, $D_{avg} = D_0 t^3$. The D_0 coefficient incorporates the fiber volume fraction and the effective lamina properties; D_0 increases with higher v_f and increased lamina properties. Figure 1 shows the plot of the D_{avg} versus the thickness cubed for the $[0/90]_{ns}$ cross-ply laminates and the 8HS woven laminates. Logarithmic curves were fitted through both laminate systems with similar relations between the bending stiffnesses and the composite thickness cubed.

It is shown in the following that the thickness dependence for the absorbed impact energies of the APC2 and 8HS laminates follows relationships similar to those shown above for the bending stiffnesses.

RESULTS

Table II gives a summary of the results of the impact tests.

The APC-2 cross-ply laminates underwent a change in the macroscopic fracture damage with thickness variation in the panel. The thinner laminates split and delaminate on the backside surface to a greater extent than the thicker APC-2 laminates. The 32 ply APC-2 laminate had the greatest resistance to penetration of any specimen configuration because the panel did not allow the impact tup to perforate and rebound off the impactor stops; the tup was wedged in the panel after puncturing the specimen. The outer plies did not fracture near the perimeter of the damage area because of the uncrimped continuous fibers. The C-scans showed delaminations throughout the two-inch diameter unclamped region of the APC-2 impact panels.

The 8HS woven laminates failed in a consistent manner for all of the panel thicknesses. All specimens failed with four dog-eared flaps broken away from the point of impact. The holes in the 8HS laminates were much larger than the five-eighth inch diameter tup because the damage was diamond-shaped. The plies sustained large amounts of interply shear in the damage region at the locations where the additional PEEK film was added. This was verified visually and by C-scanning. The 8HS woven materials absorbed less energy than the APC-2 material at comparable thicknesses. Two reasons for the smaller energy absorption in the 8HS woven laminates are less sustained delaminations and a lower fiber volume fraction. A high fiber volume fraction in the APC-2 system allowed higher maximum loads to be reached; due to the high stiffness of the APC-2 system, the energy also propagated further in the high fiber volume fraction specimen.

TABLE II. Average Impact Data for 110J of Incident Energy (E_I)

Material System	Specimen Code	No. of Tests	Panel Thick (mm)	Max. Load (kN)	Init. E (J)	Prop. E (J)	Total E (J)	% E_I
APC-2	16 Ply	3	2.0	4.14	8.2	26.2	34.4	32
	20 Ply	3	2.6	4.88	9.2	29.4	38.6	35
	24 Ply	3	3.1	6.57	14.2	34.9	49.1	45
	32 Ply	3	4.2	9.66	20.4	58.8	79.2	72
8 HS	6 Layer	3	1.8	2.48	10.4	10.4	20.7	19
	10 Layer	3	2.9	5.23	14.0	21.1	35.1	32
	12 Layer	3	3.4	6.66	16.5	28.1	44.6	40
	16 Layer	3	4.7	10.0	21.9	45.0	66.8	61
3D AI#2	n a	1	3.3	6.20	22.4	36.6	58.0	53
3D AI#3	n a	1	3.8	6.63	24.3	21.3	45.6	42

The 3-D woven structures contained the damage area to only the region impacted by the tup. A punctured hole was sustained in both the angle interlock #2 and #3 structures but with backside damage only located immediately surrounding the hole. The C-scans of the impacted panels showed no damage beyond these holes.

Using the relationships of the average bending stiffnesses, D_{avg}, to the cube of the laminate thickness, the amount of energy absorbed during impact was compared with the thickness of the material systems. In figure 2, the energy absorbed for both laminate systems were plotted against the the cube of the thickness. The graph indicates linear relationships of the energy absorption to the thickness cubed with good correlation (APC-2: R=1.00, 8HS: R=0.99). The linear fits of the points intercepted the E-axis near 25J rather than passing through the origin which deviates from the expectation. One possible reason for the non-zero intercept could be the use of static rather than dynamic bending stiffnesses. Figures 3 and 4 show the breakdown of absorbed energy versus the cube of the laminate thickness for the APC-2 laminates and the 8HS woven laminates, respectively.

An explanation has not yet been attempted for the linear relationship between the energy (initiation, propagation and total) and D_{avg}. Such a relationship is expected to exist when the composite's response to impact is fully elastic. However, in the present experiments, the material does not behave elastically; damage develops during the initiation portion of the load-deflection curve and propagates through the whole thickness of the composite during the propagation stage.

The normalization relationship determined from lamination theory, t^3, was used to compare the thickness variations between the angle interlock #2 and #3 structures. The raw impact data, shown in figure 5, was clarified by the thickness normalization shown in figure 6 as the thickness difference was scaled out. The unscaled data points at 57J in figure 5 show the thicker AI #3 absorbing more energy than the thinner AI #2. When the data was normalized by the composite thickness cubed, this trend was reversed. The thickness normalization allowed the architectural effects during the impact results to be examined directly.

The residual flexural strengths of the various impacted composite architectures were compared with the baseline strengths which were determined from undamaged specimens. In the residual strength tests the width and depth of the undamaged region was used to calculate the maximum stress. This residual strength calculation assumes the specimen is undamaged. If the specimen is damaged, the residual strength is proportional to the amount of damage incurred by the specimen at that location in the impact panel. Figure 7 shows plots of the ultimate flexural strength as a function of distance from the impact point along with the associated C-scan for an APC-2 cross-ply laminate, an 8HS woven laminate and the 3-D AI#2 woven material. The dashed lines in figure 7 signify the mean flexural strength for each material obtained from the undamaged flexural specimens.

The APC-2 and 8HS laminates have higher flexural strengths and more extensive damage propagation than the 3-D woven structures. The APC-2 cross-ply laminates have the highest flexural strength of all the specimens due to their higher fiber volume fraction, 0.61, and their continuous, uncrimped fibers. The 8HS laminates, which have the same fiber volume fraction (0.51) as the 3-D structures but less crimp in their axial fibers, have a higher flexural strength than the 3-D structures. The 3-D woven materials, though, have a higher damage tolerance than the laminates as shown by the localized damage in figure 7.

CONCLUSIONS

The effects of fiber architecture and composite thickness on the impact behavior have been studied using graphite/PEEK cross-ply laminates (APC-2), commingled 2-D 8 Harness woven laminates (8HS) and two commingled 3-D angle interlock woven composites. The following conclusions have been derived:

• The initiation, propagation, and total absorbed energies of both APC-2 and 8HS laminates showed a linear relationship with the cube of the composite thickness. This indicates that impact, within the range of thicknesses considered, is primarily controlled by the flexure of the composites.

• The above relationship was assumed to hold for the 3-D composites and was used to correct impact energies from two 3-D angle interlock structures for a minor thickness variation between the two structures. After the correction, the 3-D AI#2 structure showed a significantly higher absorbed energy than the 3-D AI#3 structure.

• At 110J of incident energy, the two 3-D woven structures were not better in terms of energy absorption than the APC-2 cross-ply laminates of comparable thickness but somewhat higher fiber volume fraction. The 3-D structures, however, did confine the damage incurred to a much smaller region than the APC-2 cross-ply laminate at this incident energy level.

• The four point flexure test qualitatively verified the nondestructive C-scan evaluations with the residual strength of the impacted composite panels. The four point flexure test reinforced the C-scan evaluations of the damage propagation due to drop-weight impact tests.

ACKNOWLEDGEMENTS

This research was supported by the Army Research Office-University Research Initiative, U.S. Army Center for Manufacturability, Maintenance and Repairability, and the Center for Composite Materials at the University of Delaware.

REFERENCES

1.   Harris, C. E., and Morris, D. H., "Effect of Laminate Thickness and Specimen Configuration on the Fracture of Laminated Composites," *Composite Materials: Testing and Design (Seventh Conference), ASTM STP 893*, J. M. Whitney, Ed., American Society for Testing and Materials, Philadelphia, Pa., 1986, pp. 177-195.

2.   Carlile, D. R., and Leach, D. C., "Damage and Notch Sensitivity of Graphite/PEEK Composite," 15th National SAMPE Technical Conference, Cincinnati, Ohio, 1983, pp. 82-93.

3.   Guynn, E. G., and O'Brien, T. K., "The Influence of Lay-up and Thickness on Composite Impact Damage and Compression Strength," AIAA/ASME 26th Technical Conference, Orlando, Florida, 1985, pp. 187-196.

4.   Taske, L. E., II, Master's Thesis under preparation, University of Delaware, 1988.

5. Boeing Specification Support Standard BSS 7260.

6. Mallon, P. J. and O' Brádaigh, C. M., "Development of a Pilot Autoclave for Diaphragm Forming of Continuous Fiber-Reinforced Thermoplastics," *Composites*, Vol. 19, No.1, 1988.

7. Rotermund, M. J., Majidi, A. P., and Taske, L. E., "Thermoplastic Preform Fabrication and Processing," *SAMPE Journal*, Volume 24, January/February 1988.

8. Whitney, J. M., Isaacs, I. M., and Pipes, R. B., *Experimental Mechanics of Fiber Reinforced Composite Materials*, rev. ed., Society for Experimental Mechanics, Englewood Cliffs, New Jersey: Prentice-Hall, 1984, pp. 165-169.

9. Zukas, J. A., Nicholas, T., Swift, H. F., Greszczuk, L. B., and Curran, D. R., *Impact Dynamics*, New York, New York: John Wiley & Sons, 1982, pp. 55-93.

10. Ishikawa, T. and Chou, T. W., "Strength and Stiffness Behavior of Woven Fabric Composites," *Journal of Material Science*, Volume 17, 1982, pp. 3211-3220.

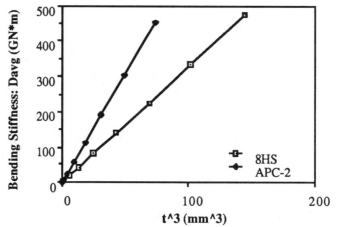

Figure 1. Bending stiffness relationships versus the cube of the composite thickness for the APC-2 laminate and the 8HS woven laminate. The APC-2 laminate is stiffer, as shown by the steeper slope, due to its higher fiber volume fraction and its uncrimped continuous fibers.

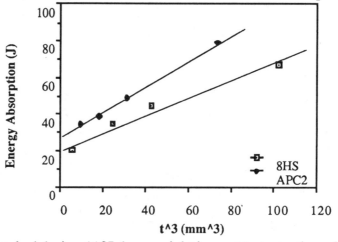

Figure 2. Energy absorbed during 110J drop-weight impact tests vs. the cube of the thickness.

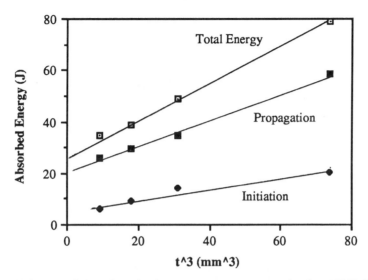

Figure 3. Breakdown of the absorbed energy components in the APC-2 cross-ply laminates during drop-weight impact testing at 110J energy and 4.0m/s.

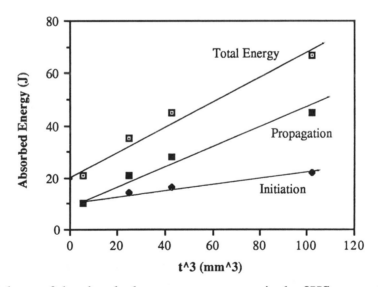

Figure 4. Breakdown of the absorbed energy components in the 8HS woven laminates during drop-weight impact testing with an incident energy of 110J and velocity of 4.0 m/s.

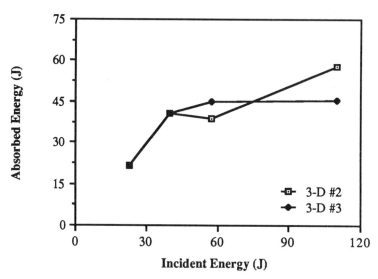

Figure 5. Impact energy absorption of the two 3-D angle interlock structures versus the incident energy level. The thickness difference of the two structures were ignored.

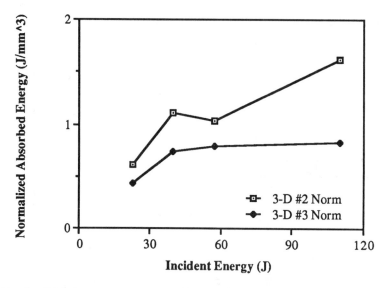

Figure 6. Absorbed impact energy normalized by the cube of the composite thickness versus the incident energy.

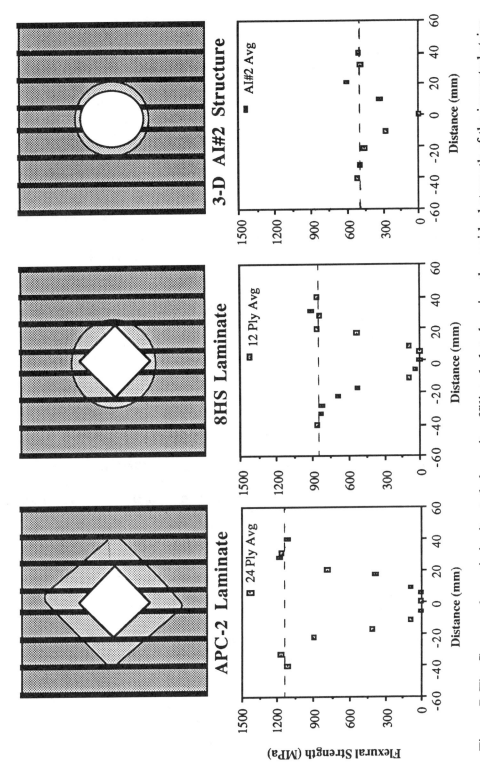

Figure 7. The C-scan schematic is situated above its affiliated plot showing the residual strength of the impacted strips versus the distance away from the point of impact (110J incident energy). The panel order is the APC-2 24 ply laminate, the 8HS 12 ply laminate, and the 3-D Al#2 woven structure, respectively. Note the limited damage in the 3-D system.

Experimental Investigation of Quasi-Isotropic Graphite/Epoxy Laminate with a Reinforced Hole

J. H. LEE AND S. MALL

ABSTRACT

The strength and failure mechanism of composite laminate (quasi-isotropic graphite/epoxy) with a reinforced hole were investigated experimentally. Two types of reinforcement boundary conditions were investigated; adhesive bonded reinforcement and snug-fit unbonded plug. For each case, four different sizes of hole diameter and three types of reinforcing material (aluminum, plexiglass, steel) were employed.

INTRODUCTION

The investigation of a composite laminate containing a circular hole has been undertaken by several investigators, e.g. see Reference 1. However, very few studies have been reported for the evaluation of strength for composite laminate containing circular hole with reinforcement. Kocher and Cross [2] investigated the reinforced cutouts in composite laminates. These reinforcements were cocured with the composite laminate so as to form an integral part of the structure. O'Neill [3], Pickett and Sullivan [4] studied the reinforcement of hole in very large composite plate (26" long and 10" wide). They employed the reinforcing technique where additional layers of same composite materials of circular shape were cocured on the sides of laminate with hole. Recently, Tan and Tsai [5] have developed closed form analysis which provides the stress distribution near a reinforced hole in the composite laminate. This analysis has shown that the strength of notched laminate can be drastically increased with the proper selection of reinforcement. However, no experimental verification of this analysis [5] has been reported so far.

J. H. Lee, ROKAF (former graduate student) and S. Mall, Professor, Department of Aeronautics and Astronautics, Air Force Institute of Technology, Wright-Patterson AFB OH 45433

The objective of present study was, therefore, to investigate experimentally the strength and failure mechanism of composite laminate with a reinforced hole.

EXPERIMENTS

The purpose of the test program was to investigate the effects of the following parameters on the strength of composite laminate with a reinforced hole.

1. Type of reinforcement - (a) adhesively bonded plug and (b) snug-fit (finger press fit) plug.

2. Material of reinforcement - (a) steel, (b) aluminum and (c) plexiglass.

3. Hole sizes - four different diameters - 0.1, 0.2, 0.4 and 0.6 inches.

4. Loading type - (a) tension and (b) compression

The material selected for the present study was AS4/3501-6 graphite/epoxy composite of quasi-isotropic lay-up of $[0/\pm45/90]_{2s}$. The nominal thickness was 0.08 inch. To investigate the effect of material of reinforcement, three materials having different stiffnesses were used. Steel has much higher stiffness than base laminate. On the other hand, the stiffness of aluminum is almost same as of quasi-isotropic graphite/epoxy composite and plexiglass has much lower stiffness than composite.

Gage length of tension specimen was selected as 6 inch to exclude edge effect, but for compression it was selected as 4 inch to prevent buckling of specimen. The end tabs were 1.5 inch long on both sides for both tension and compression specimens. The widths were 1 inch for 0.1 and 0.2 inches hole diameters and 1.5 inches for 0.4 and 0.6 inches hole diameters. The holes in specimens were drilled by a carbide tipped drill at the specimen center. Drilling the hole in each specimen was started by using a small drill, with additional aluminum plate attachment on both top and bottom surface of specimen to prevent burs. The hole was then carefully enlarged to its final dimensions.

In the case of bonding, 0.004 inch clearance between plug and hole was selected to obtain a good bond between plug and composite. But for unbonded case, the plug was made of almost same size (less than 0.001 inch clearance) as of hole to have a tight fit with hard finger pressure, and have the maximum contact with hole but not damage the inside wall of the hole. The structural adhesive EA 9302 (Hysol) was employed to bond the reinforcement in the hole. The bonding procedure involved the standard steps of surface preparation and curing at room temperature for at least five days. To make the maximum contact between plug and hole, and for convenience of handling (especially during bonding), the plug length was selected equal to 4 times the thickness of laminate for both tension and compression test specimens.

Both tension and compression tests were conducted in Instron test machine at cross-head speed of 0.02 inch per minute. A typical test involved the testing of specimen until complete failure occurred. However, several specimens were loaded incrementally up to failure at a regular interval, and specimens were taken out from the test machine after increasing each interval of load to examine the progression of failure mechanism by using X-ray technique. During compression tests, an anti-buckling fixture was used to prevent the buckling of the specimen.

RESULTS AND DISCUSSIONS

Initially tension and compression tests of unnotched and open hole specimens were conducted to develop the data base for comparison with the results of reinforced hole specimens. Thereafter, tests with unbonded and bonded reinforcement were conducted which are discussed separately in the following.

Unbonded Reinforcement

During tension loading of reinforced composite laminate, it was very important to maintain the maximum contact between plug and surrounding hole throughout the whole load history. To do this it is necessary to make the same size of plug as hole as possible, but not damage the inside of the hole when plugging in. By the experience of several preliminary tests, it turned out that snug-fit clearance showed generally the maximum increase of ultimate strength. Here snug-fit means the condition that the plug can be fitted with hard finger pressure and it can be moved inside the hole by finger pressure. The results of all these tension tests are shown in Figure 1 in terms of Strength Reduction Factor, SRF which is defined as:

$$SRF = \frac{\text{Notched strength with or without reinforcement}}{\text{Unnotched strength}}$$

For aluminum reinforced case, there was about 5 to 12% improvement in the strength from open hole case for the hole diameter greater than 0.2 in. But, for 0.1 in. hole there was no such improvement. This may be due to very small amount of interaction between plug and composite laminate during tension loading. The results of plexiglass reinforcement showed no increase. Because the stiffness of plexiglass is too low in comparison to laminate, it did not respond to the load interaction inside the hole. In most cases the plexiglass reinforcement fractured at same time when the laminate failed. In case of steel inclusion, the results for 0.1 and 0.2 inches showed similar response as in case of aluminum, but for the case of larger hole, it showed less increase than aluminum.

Since the stiffness of steel is much higher than that of composite, it did not follow the deformed shape of laminate hole with increase of the load.

In general, the aluminum reinforcement showed little improvement in ultimate strength relative to open hole. While, the steel and plexiglass inclusion showed no increase. This may be attributed to the difference in the moduli of laminate and reinforcement. As mentioned previously, aluminum has same Young's modulus as composite, plexiglass has far less than composite, and steel has greater than composite. Thus it can be concluded that the reinforcement material should have the same Young's modulus as that of base laminate for proper reinforcement.

The compression test results are shown in Figure 2. During compression test also, as in tension case, aluminum showed largest amount of improvement. For 0.1 in. diameter, as the reinforcing area was very small, the improvement was relatively smaller than other cases. But for the larger hole sizes, the increase was significantly high (e.g. 48% increase for 0.6 in. diameter). In case of plexiglass reinforcement the improvement was relatively small because of material property. However, it was higher than tension case because of the compressive load which caused more interaction between plug and hole. The results of steel inclusion also showed large amount of improvement as in aluminum case. This behavior is different than in tension case. This can be attributed again to increased interaction between plug and hole during compression.

Bonded Reinforcement

Several preliminary tests were conducted involving different adhesives and clearances between hole and plug. These tests showed that EA 9302 adhesive was better in comparison to other adhesive and clearance of 0.004 inch was the appropriate clearance from bonding consideration as well as in getting the maximum improvement in strength.

The results of bonded reinforcement case are plotted in Figure 3 for case of tension loading. These results show that there was no improvement due to bonded reinforcement for any holes size and for any plug material contrary to the expected increase in strength. To investigate the reason for this, the initiation and progression of damage in these tests were examined. Several specimens with bonded reinforcement were loaded and unloaded at increment of certain percentage of failure load. And at each interval, these specimens were inspected by microscope and X-ray technique. The initiation of crack in bondline was seen at approximately 90 to 95 percent of the failure load. In almost all the cases, failure initiated at the interface, either between adhesive and inclusion or between adhesive and composite laminate. Once the crack initiated, the reinforcement was no more useful. Thereafter, the catastrophic failure occurred immediately.

The compression test with bonded reinforcement was run with two types of reinforcement, i.e. steel and aluminum. These results are shown in Figure 4 for comparison. These clearly show the improvement in strength due to reinforcement in comparison to open or unreinforced hole which is very similar to its counterpart with unbonded reinforcement under comparison. In general, the increase in strength was comparatively more in case of bonded reinforcement than unbonded reinforcement for both aluminum and steel. However, bonded aluminum of large diameter significantly improved in strength in comparison to open hole (e.g. 68% increase for 0.6 inch diameter).

As mentioned previously, Tan and Tsai [5] have developed an analysis which provides the stress distribution near the reinforced hole in a composite laminate. Using this analysis, the strength of a laminate with an inclusion can be predicted which has shown that the maximum strength could be achieved by using inclusion that has similar elastic properties as that of base laminate. Hence aluminum inclusion should provide the maximum improvement in strength of the tested quasi-isotropic graphite/epoxy laminate, i.e. almost 100% recovery of strength (or SRF = 1) in comparison to steel or plexiglass. However, it should be mentioned that this analysis [5] is based on the assumption that there is perfect bonding between the base laminate and inclusion. The present experimental investigation is, in a general way, in agreement with the analysis of Tan and Tsai [5] if this assumption of perfect bonding between inclusion and base laminate could be realized.

SUMMARY

The tested composite laminate with a reinforced hole showed, in general, improvement of ultimate strength relative to the laminate with open hole. The largest improvement was obtained with the reinforcing material having same stiffness as that of laminate. This improvement was also related to the size of reinforcement. Further, improvement in strength was more pronounced in compression than tension due to larger interaction between hole and reinforcing plug during loading.

The composite laminate with bonded reinforcement under tension showed practically no increase in ultimate strength because of adhesive failure. Thus, if a strong bond between reinforcement and plug could be developed, the ultimate strength is expected to increase better than unbonded reinforcement.

ACKNOWLEDGEMENTS

The authors would like to express their thanks to Dr. S. W. Tsai who sponsored this project, and to Drs. S. C. Tan and R. Y. Kim for their valuable help and suggestions during the course of this investigation.

REFERENCES

1. Awerbuch, J. and Madhukar, M. S., "Notched Strength of Composite Laminates; Prediction and Experiments - A Review," <u>Journal of Reinforced Plastics and Composites</u>, Vol. 4, 1985.

2. Kocher, L. H. and Cross, S. L., "Reinforced Cutouts in Graphite Composite Structures," <u>Composite Materials (2nd Conference)</u>, ASTM STP 497, 1972, pp. 382-389.

3. O'Neill, S. S., <u>Asymmetric Reinforcement of a Quasi-Isotropic Gr/Ep Plates Containing a Circular Hole</u>. MS thesis, Naval Post Graduate School, Monterey, CA, 1982.

4. Pickett, D. H. and Sullivan, P. D., <u>Analysis of Symmetrical Reinforcement of Quasi-Isotropic Gr/Ep Plates with a Circular Cutout Under Uniaxial Tension Loading</u>. MS thesis, NPGS, Monterey, CA, 1983.

5. Tan, S. C. and Tsai, S. W., "Notched Strength," <u>Composites Design</u>, 3rd Edition, 1987.

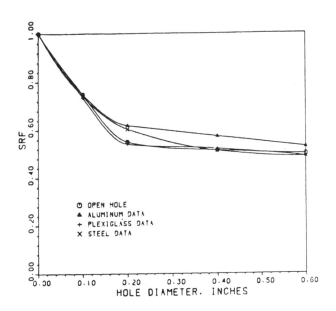

Fig. 1. Comparison of SRF among all unbonded reinforcements and open hole (tension).

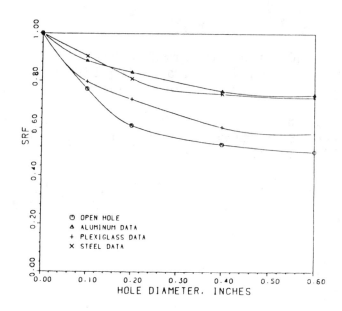

Fig. 2. Comparison of SRF among all
unbonded reinforcements and
open hole (compression).

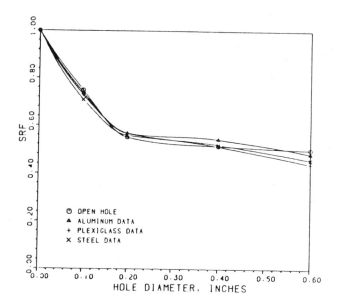

Fig. 3. Comparison of SRF among all
bonded reinforcements and
open hole (tension).

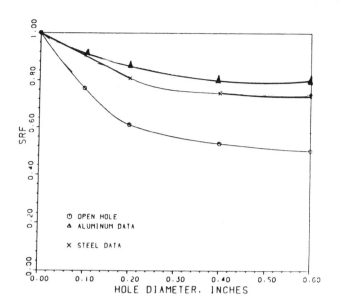

Fig. 4.   Comparison of SRF among all
bonded reinforcements and
open hole (compression).

SYMPOSIUM IX

Reinforcement Science and Engineering

Continuous Silicon Carbide Fiber Reinforced Metal Matrix Composites

MELVIN A. MITTNICK AND **JOHN McELMAN**

ABSTRACT

Continuous silicon carbide (SiC) fiber reinforced metals (FRM) have been successfully applied on numerous aerospace development programs fulfilling primary design objectives of high specific strength over baseline monolithic materials. This presentation will review the current state-of-the-art in silicon carbide fiber reinforced metals through discussion of their application. Discussion will include a review of mission requirements, program objectives and accomplishments to date employing these FRMs.

INTRODUCTION

The term composite usually signifies a combination of two or more constituent elements (in this case a high-performance, low-weight filler embedded in a metal) to form a bonded quasi-homogeneous structure that produces synergistic mechanical and physical property advantages over that of the base elements. Theoretically, there are three types of composites: (1) a particulate based material formed by the addition of small granular fillers into a binder that generally derives an increase in stiffness but not strength; (2) a whisker/flake filler that realizes a greater proportion of the filler strength due to its higher aspect ratio and hence a greater ability to transfer load; and (3) a continuous fiber system (i.e., Fiber Reinforced Metal, FRM) that, due to fiber continuity, derives the full properties of the high-performance fiber (strength and stiffness).

There are various metal matrix systems commercially available today that conform to the three types of composites discussed above, the mechanical properties of these various systems vary significantly. The major import is the clear distinction between continuous fiber systems and the discontinuous (particulate and whisker/flake) systems. In the latter, only greater stiffness is generally realized, whereas in the former, both the modulus and the strength of the continuous reinforcing fiber are fully translated into the composite. In practice, if all of the filler strength is to be translated into the composite, then (a) the filler must have sufficient aspect ratio to transfer the load from "filler" to

Melvin A. Mittnick, John McElman, Textron Specialty Materials, 2 Industrial Avenue, Lowell, Massachusetts 01851

"filler" through the matrix, (b) the filler particles must be in close proximity to each other to avoid large matrix distortions (high volume loading), and (c) a large number of "filler termination sites" placed in close proximity must be avoided to preclude stress concentrations that overload the filler and matrix. In practice it is very difficult to satisfy all of the requirements, and hence achieving the theoretical potential of the filler is close to impossible.

Although the mechanical properties of the continuous fiber composite system (FRM) are superior to those of the discontinuous systems, the ease of providing material isotrophy is a significant advantage, for there are many stiffness-controlled applications that benefit from discontinuous reinforced material. These applications generally involve complex geometries where it is difficult to position continuous fibers during the fabrication process. However, where there is sufficient volume of material to allow for orientation of fibers, the FRM systems will usually provide the most efficiency (weight savings).

SILICON CARBIDE FIBER

Since, the advent of high-strength, high modulus, low-density boron fiber, the role of fibers produced by chemical vapor deposition (CVD) in the field of high-performance composites has been well established. Although best known for its use as a reinforcement is resin-matrix composites,[1,2] boron fiber has also received considerable attention in the field of metal matrix composites.[3,4,5] Boron/aluminum was employed for tub-shaped truss members to reinforce the Space Shuttle orbiter structure, and has been investigated as a fan blade material for turbofan jet engines. There are drawbacks, however, rapid reaction of boron fiber with molten aluminum[6] and long-term degradation of the mechanical properties of diffusion-bonded boron/aluminum at temperatures greater than 480°C (900°F) preclude its use both for high-temperature applications and for potentially more economically feasible fabrication methods such as casting or low-pressure, high-temperature pressing. These drawbacks have led to the development of the silicon carbide (SiC) fiber.

SILICON CARBIDE FIBER PRODUCTION PROCESS

Continuous SiC filament is produced in a tubular glass reactor by CVD. The process occurs in two steps on a carbon monofilament substrate which is resistively heated. During the first step, pyrolytic graphite (PG) approxiamtely 1 μm thick is deposited to smooth the substrate and enhance electrical conductivity. In the second step, the PG coated substrate is exposed to silane and hydrogen gases. The former decomposes to form beta silicon carbide (βSiC) continuously on the substrate. The mechanical and physical properties of the SiC filament are:

    Tensile Strength = 3400 MPA (500 ksi)
    Tensile Modulus = 400 GPa (60 msi)
    Density = 3.045 g/cm^3 (0.11 lb/in^3)
    Coefficient of Thermal Expansion
       CTE = 1.5 x 10^{-6}/°F) (2.7 x 10^{-6}/°F)
    Diameter = 140 mm (0.0056 in.)

Various grades of fiber are produced, all of which are based on the standard βSiC deposition process described above where a crystalline structure is grown onto the carbon substrate. The βSiC is present as such across all of the fiber cross-section except for the last few microns at the surface. Here, by altering the gas flow in the bottom of the tubular reactor, the surface composition and structure of the fiber are modified by, first, an addition of amorphous carbon that heals the crystalline surface for improved surface strength, followed by a modification of the silicon-to-carbon ratio to provide improved bonding with the metal.

PROCESSING CONSIDERATIONS

As in any vapor deposition or vapor transport process, temperature control is of utmost importance in producing CVD SiC fiber. The Textron process calls for a peak deposition temperature of about 1300°C (2370°C).

Temperatures significantly above this temperature cause rapid deposition and subsequent grain growth, resulting in a weakening of tensile strength. Temperatures significantly below the optimum cause high internal stresses in the fiber, resulting in a degradation of metal matrix composite properties upon machining transverse to the fiber.[7]

Substrate quality is also an important consideration in SiC fiber quality. The carbon monofilament substrate, which is melt-spun coal tar pitch, has a very smooth surface with occassional surface anomalies. If severe enough, the surface anamoly can result in a localized area of irregular deposition of PG and SiC which is a stress-raising region and a strength-limiting flaw in the fiber. The carbon monofilament spinning process is controlled to minimize these local anamalies sufficiently to guarantee routine production of high-strength > 3450 MPa (> 500 ksi) SiC fiber.

Another strength-limiting flaw which can result from an insufficiently controlled CVD process is the PG flaw[8]. This flaw results from irregularities in the PG deposition. Two causes of PG flaws are: (1) disruption of the PG layer due to an anomaly in the carbon substrate surface and (2) mechanical damage to the PG layer prior to the SiC deposition. PG flaws often cause a localized irregularity in the SiC deposition, resulting in a bump on the surface. Poor alignment of the reactor glass can result in mechanical damage to the PG layer by abrasion. A series of PG flaws results in what is called a "string of beads" phenomenon at the surface of the fiber. The mechanical properties of such fiber are severely degraded. These flaws are minimized by careful control of the PG deposition parameters, proper reactor alignment, and the minimization of substrate surface anomalies.

The surface region of Textron's SiC fibers is typically carbon rich. This region is important in protecting the fiber from surface damage and subsequent strength degradation. An improper surface treatment or mishandling of the fiber (e.g., abrasion) can result in strength-limiting flaws at the surface. Surface flaws can be identified by an optical examination of the fiber fracture face. These flaws are minimized by proper process control and handling of the fiber (minimizing surface abrasion).

Typical mechanical properties of the Avco CVD Sic fiber consist of average tensile strength of 3790 to 4140 MPa (550 to 600 ksi) and elastic moduli of 400 to 415 GPa (58 to 60 msi). A typical tensile strength histogram shows an average tensile strength of 4000 MPa (580 ksi) with a coefficient of variation of 15%.

FIBER VARIATIONS

The surface region of the SiC fiber must be tailored to the matrix. SCS-2 has a 1 μm carbon-rich coating that increases in silicon content as the outer surface is approached. This fiber has been used to a large extent to reinforce aluminum. SCS-6 has a thicker (3 μm) carbon-rich coating in which the silicon content exhibits maxima at the outer surface and 1.5 μm from the outer surface. SCS-6 is primarily used to reinforce titanium.

SCS-8 has been developed as an improvement over SCS-2 to give better mechanical properties in aluminum composites transverse to the fiber direction. The SCS-8 fiber consists of 6 μm of very fine-grained SiC, a carbon-rich region of about 0.5 μm, and a less carbon-rich region of 0.5 μm.

COST FACTORS

From an economic standpoint, SiC is potentially less costly than boron for three reasons: (1) the carbon substrate used for SiC is lower cost than the tungsten used for the boron; (2) raw materials for SiC (chlorosilanes) are less expensive than boron trichloride, the raw material for boron; and (3) deposition rates for SiC are higher than those for boron, hence more product can be made per unit.

COMPOSITE PROCESSING

The ability to readily produce acceptable SiC fiber reinforced metals is attributed directly to the ability of the SiC fiber to (a) readily bond to the respective metals and (b) resist degradation of strength while being subjected to high-temperature processing. In the past, boron and Borsic^{TM} fibers have been evaluated in various aluminum alloys and, unless complex solid-state (low-temperature, high-pressure) diffusion bonding procedures were adopted, severe degradation of fiber strength has been observed. Likewise in titanium, unless fabrication times are severely curtailed, fiber/matrix interactions produce brittle intermetallic compounds that again drastically reduce composite strength.

In contrast, the SCS grade of fibers has surfaces that readily bond to the respective metals without the destructive reactions occurring. The result is the ability to consolidate the aluminum composites using less complicated high-temperature casting and low-pressure (hot) molding. Also for titanium composites, the SCS-6 filament has the ability to withstand long exposure at diffusion bonding temperatures without fiber degradation. As a result, complex shapes with selective composite reinforcement can be fabricated by the innovative superplastic forming/diffusion bonding (SPF/DB) and hot isostatic pressing (HIP) processes.

In the following discussion, further details of fabrication techniques will be discussed; however, first the production of intermediary products such as preforms and fabrics used in the component fabrication are described. These are required to simplfy the loading of fibers into a mold and to provide correct alignment and spacing of the fibers.

COMPOSITE PREFORMS AND FABRICS

"Green tape" is an old system consisting of a single layer of fibers that are collimated/spaced side by side across a layer, held together by a resin binder, and supported by a metal foil. This layer constitutes a prepared (in organic composite terms) that can be sequentially "laid up" into the mold or tool in required orientations to fabricate laminates. The laminate processing cycle is then controlled so as to remove the resin (by vacuum) as volatilization occurs. The method normally used to wind the fibers onto a foil-covered rotating drum, overspraying the fibers with the resin, followed by cutting the layer from the drum to provide a flat sheet of "prepreg".

"Plasma-sprayed aluminum tape" is a more advanced "prepreg" similar to "green tape" but replaces the resin binder with a plasma-sprayed matrix of aluminum. The advantages of this material are (a) the lack of possible contamination for resin residue and (b) faster material processing times because of the hold time required to ensure volatilization and removal of the resin binder is not required. As with the green tape system, the plasma-sprayed preforms are laid sequentially into the mold as required and pressed to the final shape.

"Woven fabric" is perhaps the most interesting of the preforms being produced since it is a universal preform concept that is suitable for a number of fabrication processes. The fabric is a uniweave system in which the relatively large-diameter SiC monofilaments are held straight and parallel, collimated at 100 to 140 filaments per inch and held together by a cross-weave of a low-density yarn or metallic ribbon. There are now two types of looms that can be specially modified to produce the uniweave fabric. The first is a single-arm Rapier-type loom capable of producing continuous 60 in. wide fabric with the SiC filament oriented in the "fill" (60 in. width) direction. The other is a shuttle-type loom in which the SiC monofilaments are oriented in the continuous direction with the light-weight yarn a metal ribbon in the "fill" axis. The shuttle loom can weave fabric up to 6 in. wide. Various types of cross-weave materials have been used, such as titanium, aluminum, and ceramic yarns.

PROCESSING METHODS

"Investment casting" is a fabrication technique that has been used for many years but is still universally accepted as a very cost effective method for producing complex shapes.

The aerospace business has for some time rejected aluminum castings due to the low strengths that are typically achieved; however, with a material that is now fiber dependent and not predominantly matrix controlled, significant structural improvements have been derived so as to revive the interest in this low-cost procedure. The investment casting technique, sometimes called the "Lost Wax" process, utilizes a wax replicate of the intended shape to form a porous ceramic shell mold where, upon removal of the wax (by steam heat) from the interior, a cavity for the aluminum is provided. The mold includes a funnel for gravity pouring, with risers and gates to control the flow of the aluminum into the gage section. A seal is positioned around the neck of the funnel, allowing the body of the mold to be suspended into a vacuum chamber. By a combination of gravity and vacuum (imposed through the porous walls of the shell mold), the total cavity is filled with aluminum.

The SiC fibers are installed in mold using the fabric described above by either first placing the fabric into the wax replica or simply splitting open the mold and inserting the fabric into the cavity after the wax has been removed. At present, the latter approach is usually used due to contamination and oxidation of the fibers during wax burnout. At some future date, the necessary techniques for including the fiber in the wax (thereby reducing the processing costs) will probably be developed.

"Hot molding" is a term coined by Textron to describe a low-pressure hot pressing process that is designed to fabricate shaped SiC aluminum parts at significantly lower cost than the typically diffusion bonding, solid-state process. As stated previously, the SCS-2 fibers can withstand molten aluminum for long periods; therefore, the molding temperature can now be raised into the liquid-plus-solid region of the alloy to ensure aluminum flow and consolidation at low pressure, thereby negating the requirement for high-pressure die molding equipment.

The best way of describing the hot molding process is to draw an analogy to the autoclave molding of graphite epoxy where components are molded in an open-faced tool. The mold in this case is a self-heated, slip-cast ceramic tool embodying the profile of the finished part. A plasma sprayed aluminum preform is laid into the mold, heated to a near molten aluminum temperature, and pressure consolidated in an autoclave by a "metallic" vacuum bag. The mold can be profiled as required to produce near net shape parts including tapered thicknesses and section geometry variations.

"Diffusion bonding of SiC/titanium" is accomplished by hot pressing (diffusion bonding) technology, using fiber preforms (fabric) that are stacked together between titanium foils for consoldiation. Two methods are being developed by aircraft and engine manufacturers to manufacturer complex shapes. One method is based on the HIP technology, and uses a steel pressure membrane to consolidate components directly from the fiber/metal preform layer. The other method requires the use of previously hot pressed SCS/titanium laminates that are then diffusion bonded to a titanium substrate during subsequent super-plastic forming operations.

This is typical of the first fabrication procedure noted above. The fiber preform is placed onto a titanium foil. This is then spirally wrapped, inserted, and diffusion bonded onto the inner surface of a steel tube using a steel pressure membrane. The steel is subsequently thinned down and machined to form the "spline attachment" at each end. Shafts are also being fabricated for other engine fabricators without the steel sheath.

The concept developed for superplastic forming of hollow engine compressor blades. Here the SCS/titanium laminates are first diffusion bonded in a press. These are then diffusion bonded to form monolithic titanium sheets, with "stop-off" compounds selectively positioned to preclude bonding in desired areas. Subsequently, the "stack-up" is sealed into a female die. By pressurizing the interior of the "stack-up", the material is "blown" into the female die to form the desired shape, stretching the monolithic titanium to form the interal corrugations.

These processes typcially require long times at high temperature. In the past, all of the materials used have developed serious matrix-to-fiber interactions that seriously degrade composite strength. SCS-6, however, due to its unique surface characteristics, delays intermetallic diffusion and retains its strength up to 7 hours in contact with titanium at 925°C (1700°F).

COMPOSITE PROPERTIES

Since continuous SiC reinforced metals have been in existence for a relatively short period of time, the property data base has been developed sporadically over this period depending on funded applications.

SiC/ALUMINUM

The most mature of the SiC reinforced aluminum (SiC/Al) consolidation approaches is hot molding, and therefore the greatest mechanical property data base has been developed using this material. The design data base for hot molded SCS-2/6061 aluminum includes static tension and compression properties, in-plane and interlaminar shear strengths, tension-tension fatigue strengths (SN curves), flexure strength, notched tension data, and fracture toughness data. Most of the data have been developed over a temperature range of -55°C to 75°C (-65°F to 165°F) with static tension test results up to 480°C (900°F). As can be seen from these data, the inclusion of a high performance, continuous SiC fiber in 6061 aluminum yields a very high strength 1378 MPa (+200 ksi) high-modulus 207 GPa (30 msi) anisotropic composite material having a density just slightly greater, 2.85 g/cm^3 (0.103 lb/in.^{3}) than baseline aluminum. As in organic matrix composites, cross or angle plying produces a range of properties useful to the designer.

The property data developed to date for investment cast SCS/aluminum have been limited to static tension and compression. Fiber volume fractions are lower (40% maximum) than the hot molded laminates (47% typical) due to volumetric constraints in dry loading the shell molds; however, good rule-of-mixture (R.O.M.) tensile strengths and excellent compression strengths (twice the tensile strength) are being achieved.

The use of 6061 aluminum as the matrix material and the capability of the SiC fiber to withstand molten aluminum has made conventional fusion melding a viable joining technique. Although welded joints would not have continuous fiber across the joint to maintain the very high strengths of the composite, baseline aluminum weld strengths can be obtained. In addition to fusion welding traditional molten salt bath dip brazing has been demonstrated as an alternative joining method.

An important consideration for emerging materials is corrosion resistance. Testing has been performed on SCS-2/6061 hot molded material at the David W. Taylor Naval Ship R&D Center[9] under marine atmosphere, ocean splash/spray, alternate tidal immersion, and filtered seawater immersion conditions for periods of 60 to 365 days. The SCS/aluminum material performed well in all tests, exhibiting no more than pitting damage comparable to the baseline 6061 aluminum alloy.

SiC/TITANIUM

SCS-6/Ti 6-4 composites were originally developed at high temperature. There has been a successful program to reinforce the beta titanium alloy 15-3-3-3 with SCS fiber and superior composite properties have been achieved at 1585 to 1930 MPa (230 to 280 ksi) tensile strengths[10].

Fabrication of titanium parts has been accomplished by diffusion bonding and HIP. The HIP technique has been particularly successful in the
forming of shaped reinforced parts (e.g., tubes) by the use of woven
SiC fabric as a preform. The high-strength, high-modulus properties of
SCS-6/Ti represent a major improvement over B_4C-B/Ti composites in which
the modulus of composite is increased relative to the matrix, but the
tensile strength is not as high as would be predicted by the rule of
mixture.

SiC/MAGNESIUM AND SiC/COPPER

SCS-2 has been successfully cast in magnesium.[11] Under a recent
Naval Surface Weapons Center (NSWC) program,[12] development of SiC-re-
inforced copper has been initiated. At present, about 85% of R.O.M.
strengths have been achieved at a volume fraction of 20 to 33%.

APPLICATIONS

The very high specific mechanical properties of SiC reinforced metal
matrix composites have generated significant interest within the aerospace industry, and as a result many research and development programs
are now in progress. The principal area of interest is for high-performance structures such as aircraft, missiles, and engines. However, as
more and more systems are developing sensitivities to "performance"
and "transportation weight", other and less sophisticated applications
for these newer materials are being considered. The following paragraphs
describe a few of these applications.

"SiC/aluminum wind structural elements" are currently being developed.
Ten foot long "Zee" shaped stiffeners are to be hot molded and then subsequently riveted to wing planks for full-scale static and fatigue testing. Experimental results obtained to date have verified material performance and the design procedures utilized.

"SiC/aluminum bridging elements" are being developed for the Army
to be used for the lower chord and the king post of a 52m assault bridge.
Future plans call for development of the top compression tubes of the
new Tri-Arch bridge being developed by Fort Belvoir.

"SiC/aluminum internally stiffened cylinders" are being developed
using the previously discussed investment casting process. A wax replica
is first fabricated that incorporates the total shape of the shell including internal ring stiffeners and the end fittings. The fabric containing
the SiC fibers is then wound onto the inner shell mold, the two halves
of the shell are remated and sealed, and infiltration of the aluminum
is then accomplished.

"SiC/aluminum fins" for high-velocity projectiles are in the process
of evaluation.

"SiC/aluminum missile body casings" have been fabricated utilizing
a unique variation of filament winding. An aluminum motor case is first
produced in the conventional manner; this time, however, with significantly less wall thickness than normally required. The casing is then over-
wrapped with layers of SiC fibers, where each layer is sprayed with a
plasma of aluminum to build up the matrix thickness. No final consolidation of the 90% dense system is required, for the hydrostatic internal
pressure on the circular body imposes no (or very minimal) shear
stresses on the matrix. It is hoped that further development of this

technique will permit full consolidation of the matrix by vacuum bagging the total section and hot isostatic pressing.

"Sic-titanium drive shafts" are being developed and fabricated by the hot isostatic pressing process described previously. These are generally for the core of an engine, requiring increased specified stiffness to reduce unsupported length between bearings and also to increase critical vibratory speed ranges. SiC-Ti tubes up to 5 ft. in length have been fabricated and have incorporated into their ends a monolithic load transfer section for ease of welding to the splined or flanged connections.

'SiC discs for turbine engines" are currently under development. Initially discs were made by winding SiC-Ti monolayer over a mandrel followed by hydrostatic consolidation (hot isostatic pressing). The concept now being developed utilizes a "doily" approach where single fibers are hoop wound between titanium metal ribbons to be subsequently pressed together in the axial direction, reducing the breakage of fibers and simplyfing the production of tapered cross-sections.

"Selectively reinforced SiC-titanium hollow fan blades" are being developed.

"SiC/copper materials" have been fabricated and tested for high-temperature missile applications. Also, SiC/bronze propellers have been case for potential Navy applications where more efficient/quiet propellers are required.

FUTURE TRENDS

The SiC fiber is qualified in aluminum, magnesium, and titanium. Copper matrix systems are under development and reasonably good results have been obtained using the higher temperature titanium aluminides as matrix materials. The SCS-6 fiber demonstrates high mechanical properties to above 1400°C (2550°F). It is natural, then, to project systems such as SiC-nikel aluminides/iron aluminide/superalloys, etc., all of which, on an R.O.M. basis at least, project very useful properties for "engine" and hypersonic vehicle" applications. Work required in this area includes diffusion barrier coatings and matrix alloy modifications to facilitate high-temperature fabrication processes. Also required is the detailed investigation of any detrimental thermal/mechanical cycling effects that may occur as a result of the mismatch in thermal expansion coefficients between matrix and fiber.

REFERENCES

1. DeBolt, H., "Boron and Other Reinforcing Agents," in Lubin, G., ed., Handbook of Composites, Van Nostrand Reinhold Co., New York, 1982, Chapter 10.

2. Krukonis, V. J., "Boron Filaments," in Milweski, J.V., and Katz, H.S., ed., Handbook of Fillers and Reinforcements for Plastics, Van Nostrand Reinhold Co., New York, 1977, Chapter 28.

3. McDaniels, D.L., and Ravenhall, R., "Analysis of High-Velocity Ballistic Impact Response of Boron/Aluminum Fan Blades," NASA TM-83498, 1983.

4. Salamme, C. T., and Yokel, S.A. "Design of Impact-Resistant Boron/ Aluminum Large Fan Blades," NASA CR-135417, 1978.

5. Brantley, J.W., and Stabrylla, R.G., "Fabrication of J79 Boron/ Aluminum Compressor Blades," NASA CR-159566, 1979.

6. Wolff, E., "Boron Filament, Metal Matrix Composite Materials", AF33 (615)3164.

7. Suplinskas, R. J., "High Strength Boron". NAS-3-22187, 1984.

8. Aylor, D. M., "Assessing the Corrosion Resistance of Metal Matrix Composite Materials in Marine Environments," DTNSRDC/SMME-83/45, 1983.

10. Kumnick, A.J., Suplinskas, R.J., Grant, W.F., and Corine, J.A., "Filament Modification to Provide Extended High Temperature Consolidation and Fabrication Capability and to Explore Alternative Consolidation Techniques," N00019-82-C-0282, 1983.

11. Cornie, J.A., and Murty, Y., "Evaluation of Silicon Carbide/Magnesium Reinforced Castings," DAAG46-80-C-0076, 1983.

12. Marzik, J.V., and Kumnick, A.J., "The Development of SCS/Copper Composite Material," N60921-83-C-0183, 1984.

Surface Modification of Ultra-High Modulus Polyethylene Fibers and its Characterization by FT-IR

AHMED TABOUDOUCHT AND **HATSUO ISHIDA**

Department of Macromolecular Science
Case Western Reserve University
Cleveland, OH 44106

Abstract

A surface modification of polyethylene fibers using fuming nitric acid (FNA) has been explored for short reaction times, in the range of 0 to 30 minutes, at 85°C. Using a new technique, Fourier transform infrared diffuse transmittance spectroscopy, the fiber surface was successfully characterized before and after FNA treatment. The main chemical species newly formed after FNA treatment are the following: COOR (1740 cm^{-1}), C=O and COOH (1713 cm^{-1}), O-NO_2 (1630 cm^{-1}) and NO_2 (1554 cm^{-1}) compounds. Carboxylic species have been identified by reaction with butanol (esterification) and epoxy resin. Using model compounds, the identity of the newly formed species has been confirmed.

Introduction

The combination of high specific modulus and strength makes extended-chain polyethylene fibers a very attractive candidate as a new reinforcement for high performance polymer composites. However, this material presents some problems related to its chemical nature, specifically, a chemical inertness and the absence of polar groups which severely limits its adhesion to organic matrices. This limitation leads to a poor composite interface and, hence, to a very low composite strength. It is well-known that the performance of a composite depends critically on the properties at the interface between the fibers and the matrix and its ability to transfer load from the matrix to the fibers. It is evident that, the success of this organic fiber is related directly to the success in improving its adhesion to various matrices.

It is evident that, to understand and control any surface modification, we need an analytical tool capable of following the chemical changes occurring at the surface of the fibers as well as at the interface of its own composites. Fourier transform infrared spectroscopy has been proven to be a very powerful surface technique. But, to our knowledge, no infrared spectrum of ultra-high modulus polyethylene fibers has been reported. The first objective of our present work is to explore the oxidative properties of fuming nitric acid as a method for generating polar functional groups on the ultra-high modulus polyethylene fibers capable of reacting with the organic matrices or for derivatizing into useful reactive sites. The second objective is to apply a new FT-IR spectroscopic technique to investigate the chemical changes occurring on the surface of the polyethylene fibers after the treatment.

EXPERIMENTAL

Fiber Cleaning

Ultra-high modulus polyethylene fabric, Spectra 900, was kindly supplied by Dr. H. Nguyen from Allied-Signal Co. In order to eliminate the surface contaminations, small strips of the fabric (1.5 cm x 3.0 cm) were extracted in methylene chloride (CH_2Cl_2) for a minimum of 48 hours using a Soxhlet apparatus. The washed samples were then dried for 24 hours in an air oven at $60^\circ C$. The dried samples were stored in stoppered vials until needed.

Surface Treatment

The acid treatment was performed by dipping the polyethylene samples in an open glass flask containing fuming nitric acid (90%) heated to $85^\circ C$ by means of a temperature controlled heating mantle. To keep the temperature uniform, a magnetic stirrer was used. The immersion time was varied from 0 to 30 minutes. After the acid treatment, the samples were immediately immersed in deionized water and then washed repeatedly with deionized water. The clean samples were then dried in an air oven at $65^\circ C$ for 48 hours.

FT-IR

Infrared spectra were recorded on a Fourier transform infrared spectrophotometer (Digilab FTS-20E) equipped with a liquid-nitrogen cooled mercury-cadmium-telluride (MCT) detector in the spectral range of $4000-650$ cm^{-1}. The spectrometer was continuously purged with nitrogen gas to minimize the atmospheric water vapor. For all spectra, 400 scans were coadded at a resolution of 4 cm^{-1}, with the optical velocity of 0.3 cm/sec, and the spectra are shown in absorbance mode. The difference between the maximum and the minimum absorbances of the spectrum is designated as ΔA.

Reaction with butanol and epoxy resin.

1. An FNA treated sample was dipped in a glass flask containing an mixture of butanol(30ml) with 2 drops of concentrated H_2SO_4. This system was gently refluxed for 30hours. The sample was then washed for 5 min in acetone using an ultrasonic bath and dried at $65^\circ C$ in an oven.

2. An FNA treated sample was coated with an epoxy(epon 828) by dipping into a 0.2% (by weight) acetone solution of the epoxy. The coated sample was transfered in a crystallization dish and the solvent evaporated at room temperature at atmospheric pressure over night then for 2 h under vacuum. The dried sample was then subjected to a heat treatment at $100^\circ C$ for 3 h in an oven.

Model compounds

In order to identify the new chemical species formed after the acid treatment namely: (C=O), (-COOH), (-NO$_2$), (-O-N=O), (-O-NO$_2$), the following model compounds were used : 4-decanone, 12-nitrododecanoic acid, isoamylnitrite and isoamylnitrate. In this case, all spectra of the model compounds were recorded in the spectral range 4000-550 cm^{-1}, in absorbance mode using a Fourier transform infrared spectrometer (Bomem, Michelson-110) equipped with a nitrogen cooled MCT detector. The spectral resolusion was 4 cm^{-1} with 100 coadded scans.

RESULTS   AND   DISCUSSION

The Diffuse-Transmittance spectroscopy.

To understand and to optimize any surface chemical modification of the fibers, it is essential to have an analytical tool sensitive enough to the chemical changes occurring directly on the surface. No such surface studies have been published on ultra-high modulus polyethylene fibers. In the present work we introduce a new FT-IR technique: Diffuse-Transmittance[1]. The reason of this appelation is related to the fact that this method uses the same set-up as the transmission technique but its principle is based , as we believe, not on a direct light transmission through the sample , but on a light scattering phenomenon occurring through or from the fiber surfaces. The light hitting the sample is isotropically multi-scattered on the surface of the fibers before it reaches the detector.

Surface modification of the polyethylene fibers.

In this study, our main concern is the surface modification of the polyethylene fibers by fuming nitric acid without adverse effect on the bulk fiber properties and placing the emphasis on short treatment time. This is the reason of selecting a maximum treatment time of 30 minutes (85°C). The great reactivity of fuming nitric acid towards PE fibers is evident from fig.1 where the spectra of acid treated and non-treated samples are compared. The spectra (A) and (B) represent, respectively, untreated and FNA treated at 85°C for 30 minutes. The major new bands, indicated by arrows, are located respectively around: 1713cm^{-1}, 1650cm^{-1}, 1630cm^{-1}, 1554cm^{-1}, 1277cm^{-1} and 858cm^{-1}. The spectrum of the untreated fibers does show two unexpected bands for polyethylene around 1634 cm^{-1} and 1740cm^{-1}, probably due to the presence of processing aids in the bulk of the fiber which have resisted our washing process. Except for the band at 1630cm^{-1}, our results are in fair agreement with already published data on reaction of FNA with polyethylene[2, 3]. Their studies did not show an absorbance at this frequency.

This FNA reactivity is also clear even at very short treatment times (less than one minute) as shown in fig.2 where the spectra (A), (B) and (C) represent untreated, treated for 0.5 min and the subtracted spectrum of the untreated from the treated polyethylene fibers respectively. Several new bands appear mainly around 1713cm^{-1}, 1630cm^{-1} and 1554cm^{-1} . These new features are better seen in fig.2 (C) where the spectra subtraction technique has been used. This subtraction was based on the fact that, the absorbance ratio of small polyethylene overtone bands at $2018cm^{-1}$ and $1897cm^{-1}$ did remain practically constant after the FNA treatment indicating no change in crystallinity has occurred. Therefore, the same criterion used by Tabb et al.[4] is applied here, that is the elimination of the absorbance bands in the 1800-2400cm-1 region, assigned mainly to combination modes of crystalline vibration[5].

The effect of treatment time is illustrated in fig.3 where difference spectra at various treatment times are shown. The spectra (A), (B), (C) and (D) stand for the fibers treated for 2, 5, 15 and 30 min, respectively. The effect of the reaction time is obvious from the increase in the band intensities as the treatment time is increased.

The reaction mechanism of FNA with polyethylene is not well understood. Trilla et al.[3] have postulated a radical mechanism. The reaction products expected would consist mainly of the following: hydroxyl, aldehyde, carbonyl, carboxylic acid, nitroso, nitrite, nitrate and nitro groups. Our tentative band assignments will be based on the same assumption.

The $1554cm^{-1}$ band: It appears as a shoulder and rapidly grows as the reaction time is increased. Rueda et al. [2] and Trilla et al. [3] assigned it to a $-NO_2$ group. Generally, aliphatic nitro compounds show two characteristic bands falling in the following ranges :1567-1550 cm^{-1} due to the antisymmetric stretching mode and 1379-1368 cm^{-1} due to the symmetric stretching mode. In our case, the symmetric band is missing from all the spectra of the treated samples. The same observation was reported by Rueda[11] who attributed it to the weakness of this band. Bellamy [6] also stated the instability of this symmetric band and suggested that the antisymmetric mode is more reliable for the identification purpose of nitro compounds. As Rueda et al. [2] and Trilla et al. [3], we also believe that the band at 1554 cm^{-1} corresponds to the $-NO_2$ group.

The bands at 1630 cm^{-1}, 1278 cm^{-1} and 858 cm^{-1}: They appear simultaneously and continue to grow at the same rate as the reaction time is increased. This implies that their origin is due to the same chemical species. These bands appear in the same frequency ranges as those characteristic of the $-O-NO_2$ group of organic nitrates reported by Carrington [7]. The following characteristic frequency ranges were found: 1632-1626 cm^{-1} due to the antisymmetric stretching mode, 1282-1272 cm^{-1} due to the symmetric stretching mode, and 870-855cm-1 due the O-N stretching mode.

The 1650 cm^{-1} and the 1640 cm^{-1} bands: They appear at the early stage of the reaction first as shoulders of the $1630cm^{-1}$ band. Their identities were not clear. The 1640 cm^{-1} band is relatively more intense than the 1650cm-1 band at the beginning of the reaction, and decreases in intensity as the reaction progesses. Meanwhile, the intensity of the $1650cm^{-1}$ band tends to increase. The 1650 cm^{-1} band was reported for the first time and attributed to the $-O-NO_2$ group by Rueda et al. [2]. The same band observed later by Trilla et al.[3] was assigned to the $-O-N=O$ functionality, and a new band at 1640 cm^{-1} was assigned to $-O-NO_2$

by the same author. He noticed that the 1650 cm^{-1} band, assumed to be due to nitrite, disappeared after few hours under their reaction conditions. The nitrite forms an intermediate product in the FNA/polyethylene reaction. We believe that the nitrites are present, and it would be realistic to link them to the 1640 cm^{-1} band, which, in our case, showed a decrease in intensity during extended reaction.

The 1650cm^{-1} band has been observed and assigned to organic nitrate by Rueda et al.[11] Based on our kinetic studies, the growth rate of this band is the same as the band at 1630 cm^{-1}, within our reaction time interval. This may mean that as the reaction proceeds, the increased concentration of the other species formed modifies the immediate environment of part of the nitrates present. This type of interaction may lead to a shift of the 1630 cm^{-1} band to higher frequencies, in this case to 1650cm^{-1}. This view is probable if we compare our reaction times with that used by Rueda et al. [2]. The reaction times he used are very long compared to ours, so that the shift of the band was already completed. This may explain the reason why they did not observe the 1630cm^{-1} band.

The band around 1713cm^{-1}: This region is typical of C=O functionalities, the more probable beeing carbonyl and carboxylic groups. No aldehyde nor hydroxyl absorptions have been observed after FNA treatment. The band observed at 1710 cm^{-1} has been attributed to carboxylic groups[12], based on the fact that the spectrum did show absorption bands characteristic of this functionality ie. bands at 1710 cm^{-1}(νC=0), 1430–1400 cm^{-1}(νC-0), 1280 cm^{-1} and 930 cm^{-1}. The spectra of treated fibers did show a band around 1410 cm^{-1} which is attributed to a carbonyl between two methylene groups [8] but the bands at 1280 cm^{-1} and 930 cm^{-1} were missing. Rueda et al.[2] reported that the 1280 cm^{-1} band appeared only after a long treatment time (more than one hour at 70°C. This may suggest that, in our case, the presence of carboxylic functionalities is at very low concentration compared to that of carbonyl. In order to verify this view, two reactions characteristic of the –COOH group i.e. an esterification reaction with butanol and a reaction with epoxy resin were attempted. The results are shown in fig. 4 and 5. In fig. 4 the spectra A and B represent, respectively, the FNA treated sample before and after the esterification reaction with butanol, and spectrum C their difference. The appearance of a new band around 1740 cm^{-1} characteristic of esters, and the decrease of the band around 1713 cm^{-1} prove the occurence of an esterification reaction between butanol and –COOH groups located on the FNA treated polyethylene fibers. In fig. 5, the spectra of FNA treated polyethylene fibers which are coated with a thin layer of epoxy resin are illustrated. They are before heat treatment (A), after heat treatment (B) and their difference. Again, there is a newly formed band around 1740 cm^{-1} and a decrease of the 1713 cm^{-1} band. This confirms the reaction between oxirane group and the –COOH groups of the FNA treated fibers. These two reactions show, indeed, the 1713 cm^{-1} band is due mainly to the presence of –COOH groups produced after the treatment of the polyethylene fibers with fuming nitric acid.

The 1740cm^{-1} band: This band, although unexpected for unoxidized polyethylene, is present in all our washed untreated polyethylene samples as shown in fig.1 (A). The effect of FNA treatment on this band is obvious from the difference spectra after different FNA treatment times shown in fig. 2. At the beginning of the reaction, this band shows a negative absorbance meaning that the concentration of the chemical species responsible for this band has decreased due to the oxidative effect of the fuming nitric acid. As the reaction time increases, the absorbance of this band changes from a negative to a positive value. This change is related to the two following facts: a) the formation of a new ester product resulting from oxidation of the polyethylene fibers; b) the diseapperance of low molecular weight esters, probably some processing aids remaining in the fibers, which are washed out by the FNA treatment. The sign of this composite band in the difference spectra depends on the ratio of the concentrations of the contributing chemical species, namely the newly formed esters and the disappearing low molecular weight esters.

Bands assignment using model compounds

In order to confirm the identity of the main oxidation products, namely C=O, COOH, O–N=O, O–NO_2 and NO_2, four model compounds were used. Their chemical structures and their characteristic bands, in the frequency range 1800 cm^{-1} – 1530 cm^{-1} are reported in table 1. To simulate the actual surface chemical structure of the polyethylene fibers and any interaction between present chemical groups which may lead to IR band shifting, an IR spectrum of the model compound mixture cast on KBr plates was recorded. The observed IR bands of the mixed model compounds as well as that of the actual FNA treated polyethylene are reported in table 2. A band broadening is noticeable in the case of the bands around 1713 cm^{-1} and 1630 cm^{-1} in the spectrum of mixed model compounds when compared to spectra of neat model compounds. It is clear that more than one chemical group is contributing to these bands: the 1713 cm^{-1} band is due to C=O and COOH, meanwhile the 1630 cm^{-1} arises from O–N=O and O–NO_2 contribution. The band around 1565 cm^{-1} due to NO_2 in neat nitro dodecanoic acid shifts to 1554 cm^{-1} when model compound mixture is used, indicating an interaction between this group and O–N=O and/or O–NO_2 groups. In fig 6, IR spectrum of the model compound mixture is compared to actual subtracted spectra of oxidized polyethylene fibers. Here, spectra A, B, C represent the mixture of model compounds, the subtracted spectra of polyethylene fibers after 2 and 30 min FNA treatment, respectively. The obvious similarity of these spectra confirms the expected presence of C=O, COOH, O–N=O, NO_2 and O–NO_2 groups on FNA treated polyethylene fibers.

Conclusion

This premilinary study has shown two outcoming results:

1) The effectiveness of fuming nitric acid for the surface modification of ultra–high modulus Polyethylene fibers: introduction of C=O, COOH, NO_2, O–NO_2 groups, following very short treatment times.

2) The newly introduced Fourier transform infrared spectroscopic technique, the diffuse–transmission, has been proven to be a very promising analytical tool because of its simplicity and its high sensitivity for surface studies of this polymeric fiber.

References

1. A. Taboudoucht and H. Ishida, to be published.
2. D.R. Rueda, E. Cagiao and F.J. Balta Calleja, Makromol. Chem., **182**, 2705, (1981)
3. R. Trilla, J.M. Perena and J.G. Fatou, Polymer J., **15**, 803, (1983)

4. D.L.Tabb, J.J. Sevcik and J.L. Koenig, J. Polym. Sci., Polym. Phys. Ed., **13**, 815, (1975)
5. S. Krimm, Fortschr. Hochpolym.-Forsch., **2**, 51, (1960)
6. L.J. Bellamy, "The Infrared Spectra of Complex Molecules", Third ed., Chapman and Hall Ltd, New York, (1985)
7. R.A.G. Carrington, Spectrochimica Acta, **16**, 1279, (1960)
8. J.P.Luongo, J. Polym. Sci., **42**, 139, (1960)

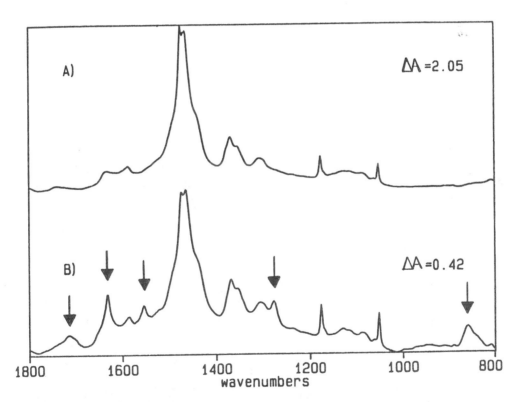

Fig.1. Infrared spectra of polyethylene fibers:
(A) untreated, (B) FNA treated (30 min, 85°C)

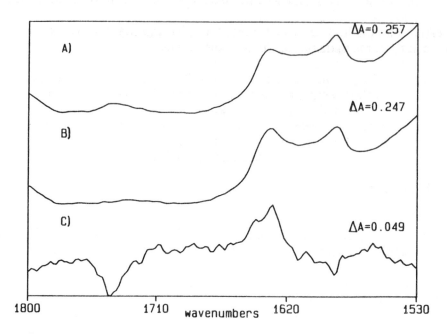

Fig.2. Infrared spectra of polyethylene fibers:
(A) untreated, (B) FNA treated (0.5 min, 85°C)
(C) their difference

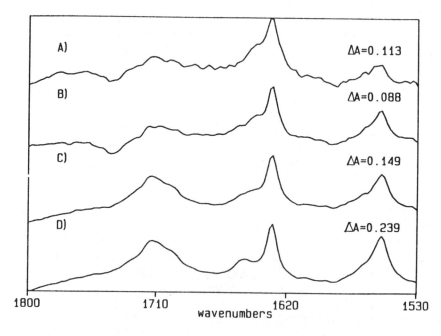

Fig.3. Subtracted infrared spectra of polyethylene fibers
after different FNA treatment times: (A) 2 min,
(B) 5 min, (C) 15 min and (D) 30 min.

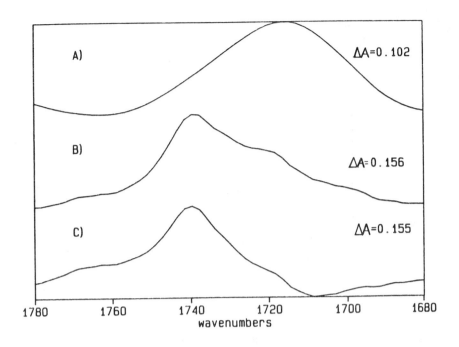

Fig.4.   Infrared spectra of FNA treated polyethylene
         fibers: (A) before reaction with butanol, (B) after
         reaction with butanol and (C) their difference.

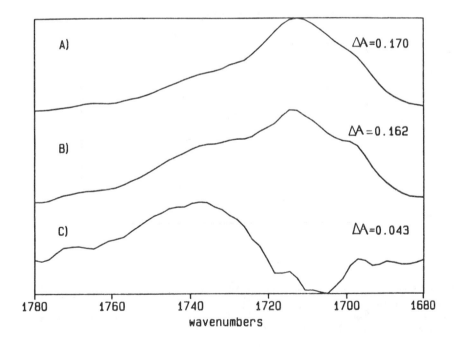

Fig.5. Infrared spectra of FNA treated polyethylene
       fibers: (A) before reaction with epoxy, (B) after
       reaction with epoxy and (C) their difference.

413

Table 1: Observed infrared absorption bands of neat model compounds

		freq.(cm^{-1})	assignment
4-decanone	$CH_3(CH_2)_5CO(CH_2)_2CH_3$	1715	C=O
12-nitrododecanoic acid	$O_2N(CH_2)_{11}COOH$	1698 1565	COOH NO_2
isoamyl nitrite	$(CH_3)_2CHCH_2CH_2ON=O$	1654 1603	ON=O, trans ON=O, cis
isoamyl nitrate	$(CH_3)_2CHCH_2CH_2ONO$	1630	ONO_2

Table 2: Observed infrared absorption bands of model compound mixture

model compound mixture		FNA treated polyethylene fibers	
frequency(cm^{-1})	assignment	frequency(cm^{-1})	assignment
1713	C=O and COOH	1713	C=O and COOH
1630	ON=O , ONO_2	1630 and 1650 1640	ONO_2 ON=O
1554	NO_2	1554	NO_2

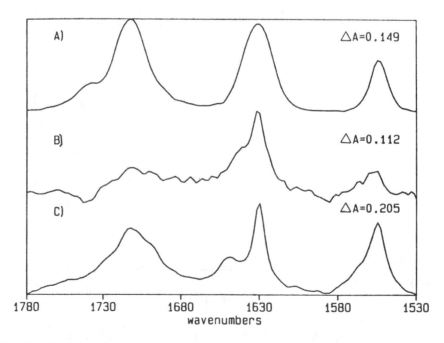

Fig.6. Infrared spectra of (A) model compound mixture, subtracted infrared spectra of polyethylene fibers FNA treated for 2 min (B) and 30 min (C).

Enhancement of Spectra® Brand Polyethylene Fiber Adhesion to Epoxy Matrix

RAM S. RAGHAVA

ABSTRACT

This paper discusses ways of increasing adhesion between ultra-high molecular weight polyethylene fibers (Allied Corporation's Spectra^R brand polyethylene fibers) to an epoxy matrix. The effects on mechanical properties of plasma treatment of polyethylene fiber surface and of hybridization with S-2 glass fibers on mechanical properties have been examined through preparation and testing of polyethylene fiber reinforced epoxy matrix composites. Mechanical properties of polyethylene fiber-epoxy matrix composites before and after exposing to boiling water were measured. The effect of hybridization of Spectra^R fibers with S-2 glass fibers on the magnitude of water absorbed by hybrid fiber epoxy composites in a boiling water test has been evaluated. The effects of various hybridization architectures (inter- and intra-ply; clustered and interspersed) on flexural properties have been measured.

INTRODUCTION

Synergistic mechanical properties in advanced composites have been achieved previously through a combination of dissimilar fibers possessing complementary mechanical properties. High modulus pitch based carbon fibers have been combined [1] with Kevlar^R fibers in interlaminar and intralaminar configurations to enhance impact strength of composites made from rather brittle pitch-based carbon fiber-epoxy composites. Similarly, ultra-high molecular weight polyethylene fibers have been combined with glass [2] and carbon [3,4] fibers in epoxy matrix composites. Hybridization of ultra-high molecular weight polyethylene fibers has been achieved with other advanced fibers in interlaminar [2,3] and in intralaminar configurations. The purpose of the above studies was to enhance impact strength of brittle carbon composites by combining them with tough and ductile untra-high molecular weight polyethylene fibers. Also high

Spectra^R is a trademark of Allied Corporation
Kevlar^R is a trademark of DuPont Corporation

R. S. Raghava, Sr. R&D Associate, BFGoodrich Research & Development Center, 9921 Brecksville Road, Brecksville, Ohio 44141

modulus carbon fibers enhance the stiffness of polyethylene fiber composites. Combining ultra-high molecular weight polyethylene fibers with high modulus anisotropic carbon fibers has deleterious effects since composites made from these hybrid fibers result in high thermal residual stresses.

The purpose of this study was to enhance adhesion of non-polar polyethylene fibers to epoxy matrix by combining them with S-2 glass fibers. Thermal residual stresses developed in hybrid S-2 glass/polyethylene fiber composites are balanced since S-2 glass and polyethylene fibers possess compensating linear coefficients of thermal expansions in the fiber directions. Moreover, S-2 glass fibers possess excellent adhesion to epoxy matrix. Thus, by combining S-2 glass and ultra-high molecular weight polyethylene fibers in intralaminar or interlaminar configurations, good adhesion between polyethylene fibers and epoxy matrix could be achieved through mechanical interlocking with neighboring S-2 glass fibers. In order to achieve the above objective, hybrid fabrics possessing differing volume fractions of S-2 glass and Spectra-1000[R] fibers were woven in a plain weave. Fibers, S-2 glass and Spectra-1000[R], were also combined in a unidirectional epoxy prepreg in intralaminar architecture.

EXPERIMENTAL

Resin Impregnation and Laminate Preparation

An epoxy resin (Epon-826) was combined with a curing agent in the following proportions by weight:

    Epon-826 -- 100 parts
    Carboxyl terminated
    Butadiene nitrile (CTBN)
    Reactive liquid polymer -- 15 parts
    Dicyandiamide (Curing Agent) -- 6 parts
    Fi Kure 62-u (Curing Agent) -- 2 parts

The above ingredients were mixed thoroughly on an ink mill and applied to different fabric forms as described below. Hybrid fabrics were coated with a paint brush after diluting the epoxy resin with methyl ethyl ketone to a pre-determined amount. Prepregs were B-staged for 40 minutes at 117°C in a convection oven. B-staged hybrid prepreg fabrics were cured in a 6"x6" steel mold under 60 psi and 120°C for two hours in a micro-processor controlled hydraulic press. Unidirectional hybrid prepregs were made using a prepregging machine supplied by Research Tool Corp., Owasso, MI. Cross-ply laminates were prepared according to curing cycle discussed above. Surface modification using plasma treatment of Spectra-1000[R] and S-2 glass/Spectra-1000[R] hybrid fabrics was conducted under proprietary plasma treatment procedures. Plasma treated fabrics were impregnated with an epoxy resin as described above.

Mechanical and Physical Property Evaluation

Flexural properties of composites were measured on 1/2" wide and 5" long specimens with a 4" support span and 1-1/3" loading/span in a four point bend test according to ASTMD-790 test method. Relative density of tested samples was measured using the water displacement method according to ASTM D-792. Calculated fiber volume fraction varied between 50% and 55%.

Exposure to Boiling Tap and Salt Water
    Hybrid S-2 glass and Spectra-1000^R and S-2 glass fiber-epoxy matrix
test samples were exposed to boiling tap and salt water baths. Salt
water was prepared using a formulation given by ASTMD 1141-75 (Reapproved
1980). Specimens were withdrawn after a pre-determined exposure duration
and weighed on an electronic balance. Specimens withdrawn from boiling
water baths were tested immediately.

RESULTS AND DISCUSSION

Flexural Properties at Ambient Conditions
    Flexural moduli and flexural strengths of Spectra-1000^R /epoxy, S-2
glass/epoxy and hybrid fabric/epoxy composites were measured as a
function of composite specific gravity. Assuming that voids in these
laminates are negligible since laminates are thin, it is easy to represent composite fiber volume fraction in terms of composite specific
gravity directly. Since the relative ratios of the two fibers are fixed
in the fabrics (nominal volume fractions of Spectra-1000^R and S-2 glass
fibers were 27, 48 and 75), knowledge of individual fiber and resin
volume fractions can be obtained from composite density measurements.
    Flexural moduli and strength values have been plotted in Figure (1).
Also superimposed in Figure (1) are flexural moduli and flexural
strengths of plasma and corona treated Spectra-1000^R fabric-epoxy
composites. Symbols E_{SP}, E_{SC} ,and σ_{SP} and σ_{SC} denote flexural moduli and
flexural strengths of plasma and corona treated Spectra-1000^R fabric-
epoxy composites, respectively. As seen in Figure (1), flexural moduli
of hybrid fabric composites can be approximated by a straight line.
However, flexural strength values are approximated by two straight lines
possessing different slopes. Flexural strength values measured
experimentally are significantly lower than theoretically calculated
values based on rule of mixtures and shown by a broken line. It is
believed that the discrepancy between theoretical and experimental values
may be ascribed to premature failure of Spectra-1000^R fibers from micro
buckling due to low adhesion between the polyethylene fibers and the
epoxy matrix. This failure mode is further promoted due to plain weave
of fabrics. It was observed that strain at maximum load carried by
flexural specimens made from hybrid fabric composites was less than half
the value calculated for hybrid fiber laminate specimens in ASTMD790
test.

Effect of Boiling Water on Flexural Properties
    The effect of boiling tap water on flexural modulus and flexural
strength of hybrid fabric-epoxy composites has been summarized in Figures
(2) and (3), respectively. Flexural properties retained after exposure
to boiling tap water for up to 96 hours are given in Figures (2) and (3).
A drop of less than 20% in flexural modulus and flexural strength is
observed after exposure to boiling tap water.
    Percent water absorbed by Spectra-1000^R/S-2 glass hybrid fabric-
epoxy composites has been plotted as a function of (exposure time)$^{1/2}$ in
Figure (4), where results are compared with S-2 glass/epoxy composites.
It is interesting to note that the data can be approximated by two
curves; the higher curve gives water absorbed by composites made of S-2
glass/epoxy, and hybrid fabric, with 73% Spectra-1000^R and 27% S-2 glass.
The magnitude of water absorbed by composites made from hybrid fabric
with 48% and 25% Spectra-1000^R fibers lie on the lower curve. It is

possible that higher amounts of water absorbed in S-2 glass/epoxy composite could be due to generation of micro-cracks between fiber and matrix interface after a critical amount of water has been absorbed. Similarly, higher magnitude of water absorbed in the composite made of hybrid fabric with 73% Spectra-1000[R] and 27% S-2 glass may be from interlaminar delamination of test samples during a boiling water test. As discussed later, although polyethylene is considered hydrophobic (non-wettable), Spectra-1000[R]/epoxy composite laminates absorb up to 10% tap water upon exposure to 48 hours in a boiling water test. Large amount of water absorbed by Spectra[R] fiber/epoxy composites can only be explained due to interlaminar delamination, since polyethylene exhibits a low bond strength with an epoxy matrix.

The effect of boiling salt water on flexural properties of hybrid Spectra-1000[R]/S-2 glass fabric and epoxy composites is shown in Figure (5) are a function of percent water absorbed. Superimposed in Figure (5) is flexural properties along 45° to fiber direction of S-2 glass fabric-/epoxy composites. The percent drop in flexural modulus and flexural strength of hybrid Spectra-1000[R]/S-2 glass fabric epoxy composites is small. Comparatively, a significant reduction (up to 40%) in flexural properties is observed for S-2 glass fabric/epoxy composites test samples taken at 45° to fiber direction. The percent of flexural modulus and strength retained for the hybrids lie in a 90% to 105% broad band.

Percent sea water absorbed in a boiling salt water test is shown in Figure (6) as a function of (exposure time)$^{1/2}$. Results are plotted for S-2 glass/epoxy and hybrid fabric/epoxy composites. The trend of these results is similar to observed in a boiling tap water test. The magnitude of sea water absorbed in boiling water test is smaller than the amount of tap water absorbed under same test conditions. Higher amount of sea water is absorbed in S-2 glass/epoxy composites and hybrid fabric composites containing 27% S-2 glass and 73% Spectra-1000[R]. A smaller amount of sea water is absorbed in remainder hybrid fabric/epoxy composites. The explanation advanced to explain absorption of tap water in various classes of composites is also applicable here.

Flexural Properties With Other Reinforcing Arrangements

Table (1) lists flexural properties of corona and plasma treated Spectra-1000[R] fabric reinforced epoxy composite. Flexural properties of hybrid laminates containing clustered and interspersed plys of Spectra-1000[R] fabric/epoxy and S-2 glass fabric/epoxy are also shown in Table (1). Also listed for comparison are flexural properties of hybrid fabric/epoxy composites.

As shown in Table (1), flexural modulus and flexural strength of corona treated Spectra[R] fiber-epoxy are very low. The reason for low flexural strength is poor compressive properties of Spectra-1000[R]/epoxy composites. The compression strength measured according to ASTMD 695 for Spectra-1000[R]/epoxy composite was 4500 psi. This compares favorably with 5200 psi reported in reference [3]. Low compression strength, in part, results from low adhesion between polyethylene fibers and epoxy matrix. Plasma surface modification of polyethylene fibers enhances flexural strength and flexural modulus. However, increase in both values is inadequate for these materials to be considered suitable for structural applications.

S-2 glass/epoxy and Spectra-1000[R]/epoxy plys can be combined in a hybrid laminate in several ways. Plys of a material can be clustered or interspersed through a laminate's thickness. The position of these plys in a laminate's thickness, outside or inside, can be varied. Table (1) lists flexural properties of 24 ply hybrid laminates at ambient conditions and after boiling in tap water for 48 hours. As shown in Table (1), use of S-2 glass/epoxy plys outside in a laminate give optimum mechanical properties. Clustered S-2 glass/epoxy plys present outside in a laminate give highest flexural modulus. However, flexural strength is not high since laminate failure occurs prematurely in Spectra-1000^R/epoxy plys near neutral axis due to interlaminar delamination since Spectra-1000^R/epoxy composites exhibits low shear strength. Interspersed ply configuration with S-2 glass/epoxy plys being present outside in a laminate give optimum flexural modulus and flexural strength.

The last item in Table (1) give the flexural properties of hybrid fabric/epoxy composites. As shown in Table (1), an overall balanced flexural properties can be achieved by varying relative volume fractions of Spectra-1000^R and S-2 glass fibers. For equal volume functions of two fibers, flexural moduli of 3.0 msi each and flexural strengths of 15.3 ksi and 18.0 ksi can be achieved from clustered hybrid ply and hybrid fabric laminates, respectively. Use of higher S-2 glass volume fraction gives still higher flexural moduli and flexural strengths.

Flexural properties of hybrid ply and hybrid fabric composites after exposure to boiling tap water for 48 hours are also given in Table (1). Hybrid fabric composites retain superior flexural moduli and flexural strengths after exposure of flexural specimens to 48 hours in boiling tap water.

The amount of water absorbed in Spectra^R/epoxy hybrid ply, clustered and interspersed, and hybrid fabric and epoxy matrix composites is shown in Table (2). The highest amount of water is absorbed by corona treated Spectra^R fiber/epoxy composite laminates after exposure to boiling tap water for 48 hours. The smallest amount of water is absorbed by laminates possessing lower Spectra-1000^R (48% to 25%) volume fraction. The position of Spectra^R fiber ply in a clustered or in an interspersed ply laminate also affects the amount of water absorbed after exposure of 48 hours. S-2 glass plys present externally in a clustered or in an interspersed ply laminate absorb smaller amount of tap water in a boiling water test than the water absorbed when Spectra^R fiber plys are present outside of a laminate. The explanation for this results is not known at this time.

SCANNING ELECTRON MICROSCOPY STUDIES

Figure (7) shows optical and scanning electron micrographs of a Spectra-1000^R/epoxy test sample failed in a four-point bend test. In Figure (7a) interlaminar delamination and lamina buckling are shown. In Figure (7b) fiber debonding and micro-buckling is clear. It can be seen in Figure (7b) that fiber has yielded prior to buckling as shown by shear bands.

Scanning electron micrographs of S-2 glass and CTBN toughened epoxy composites after boiling for 96 hours in tap water are shown in Figure (8). As shown in this figure, extensive micro-cracking between glass fibers and epoxy matrix is present. Figure (9) shows micro-cracking in hybrid fabric/epoxy composites after exposure to boiling water for 96 hours. By comparing Figures (8) and (9), it is clear that as the fiber volume fraction of Spectra-1000^R increases in S-2 glass/ Spectra-1000^R fabric, the propensity for micro-cracking in S-2 glass/epoxy interphase decreases. The reasons for this failure behavior are not known at the present time.

CONCLUSIONS

1. Spectra-1000^R/epoxy composites do not possess adequate flexural properties for use in structural applications.

2. Hybridization of Spectra^R fibers with S-2 glass fibers either in a fabric or in a laminate through interlaminar stacking results in improved flexural properties.

3. The effect of boiling tap and sea waters on flexural properties of hybrid fabric-epoxy composites after extended exposure is minimal.

4. Hybrid fabric/epoxy composites absorb significantly smaller amount of water than water absorbed by clustered or interspersed hybrid ply and Spectra-1000^R/epoxy composites.

5. The relative position of S-2 glass plys in clustered or in interspersed ply laminates, outside or inside, affects the amount of water absorbed. S-2 glass plys present outside in a laminate provide protection against water permeation.

6. Low mechanical properties in Spectra^R fiber/epoxy composites results from interlaminar delamination and buckling and fiber debonding, yielding and micro-buckling.

7. Presence of Spectra-1000^R fibers in a hybrid fabric reduces the propensity of micro-cracking in S-2 glass/epoxy interphase after exposure to boiling water.

8. Plasma treatment of Spectra-1000^R fibers increases adhesion between the fibers and epoxy. However, the increase in flexural properties is inadequate to render polyethylene fiber composites suitable for structural applications.

9. Hybridization with S-2 glass fibers enhances adhesion between Spectra-1000^R fibers and epoxy. Hybrid fiber composites possess adequate mechanical properties for their consideration in structural applications.

REFERENCES

1. R. S. Raghava and S. T. Peters, SAMPE Quarterly, 7 (Jan. 1987)

2. N. H. Ladizesky and I. M. Ward, Composites Science and Technology, 26, 199 (1986).

3. D. F. Adams and R. S. Zimmerman, Proceedings of 31st International SAMPE Symposium and Exhibition held in Anaheim, CA, April 7-10, 1986.

4. A. Poursartip, G. Riahi, E. Teghtsoonian and N. Chinatambi, 19th International SAMPE Technical Conference held in Washington, D.C., Oct. 13-15, 1987.

TABLE 1.- COMPARISON OF FLEXURAL PROPERTIES OF SPECTRA-EPOXY AND S-2 GLASS-SPECTRA-EPOXY AT AMBIENT CONDITIONS AND UNDER HOSTILE ENVIRONMENTS

MATERIAL SPECTRA-EPOXY	FLEXURAL PROPERTIES, AMBIENT		EXPOSURE TO BOILING WATER FOR 48 HOURS	
	MODULUS MSI	STRENGTH KSI	MODULUS MSI	STRENGTH KSI
1. SPECTRA-EPOXY	0.63	5.1		
2. PLASMA TREATED SPECTRA-EPOXY	1.3	13.0	--	--
3. SPECTRA/S-2 GLASS-EPOXY HYBRID LAMINATES				
CLUSTERED				
A) 6s/12G/6s	0.87	21.0	0.65	19.8
B) 6G/12s/6G	3.00	25.3	2.56	22.5
INTERSPERSED				
A) (s/G/G/s)6	1.7	20.3	1.45	14.45
B) (G/s/s/G)6	2.2	23.0	1.6	14.7
4. S-2 GLASS/SPECTRA-HYBRID FABRIC-EPOXY COMPOSITES				
A) SPECTRA/GLASS VOLUME FRACTIONS 75:25	2.5	12.0	2.3	11.0
B) SPECTRA/GLASS VOLUMES 48:52	3.0	18.0	2.5	18.2
C) SPECTRA/GLASS VOLUMES 27:73	3.4	31.0	2.7	28.0

TABLE 2. PERCENT WATER PICK-UP OF SPECTRA EPOXY SPECTRA-GLASS-EPOXY COMPOSITES AFTER EXPOSURE TO 48 HOURS TO BOILING WATER

MATERIAL	EXPOSURE TIME HOURS	WATER PICK-UP %
SPECTRA-EPOXY	48.0	10.0
INTERPLY HYBRID OF SPECTRA-S-2 GLASS EPOXY		
1) CLUSTERED		
A) 6s/12G/6s	48.0	5.7
B) 6G/12s/6G	"	3.4
2) INTERSPERSED		
A) (s/G/G/s)6	48.0	6.5
B) (G/s/s/G)6	"	3.9
S-2 GLASS/SPECTRA HYBRID FABRIC		
A) SPECTRA/GLASS VOLUME FRACTION 73:27	48.0	1.85
B) SPECTRA/GLASS VOLUME FRACTIONS 48:52	48.0	1.4
C) SPECTRA/GLASS VOLUME FRACTIONS 25:75	48.0	1.4

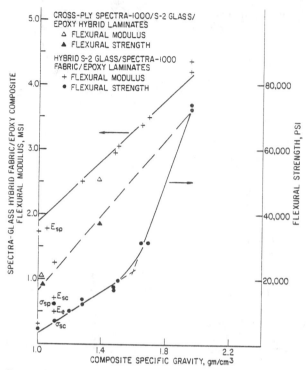

Figure I. VARIATION of FLEXURAL MODULI and STRENGTHS of SPECTRA/
EPOXY and HYBRID FABRIC (Spectra and S-2 Glass) EPOXY
COMPOSITES

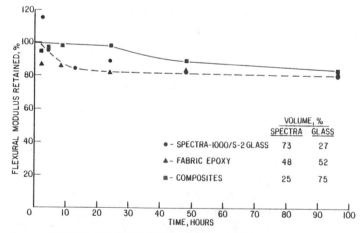

Figure 2. VARIATION of FLEXURAL MODULI of SPECTRA-1000/GLASS FABRIC/EPOXY
COMPOSITE as a FUNCTION of EXPOSURE to BOILING WATER

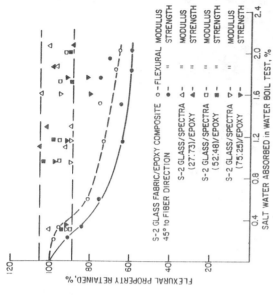

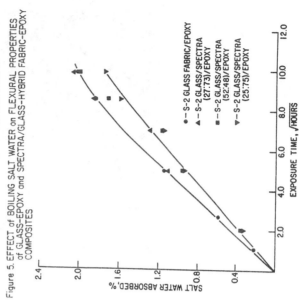

Figure 5. EFFECT of BOILING SALT WATER on FLEXURAL PROPERTIES of GLASS-EPOXY and SPECTRA/GLASS-HYBRID FABRIC-EPOXY COMPOSITES

Figure 6. ABSORPTION of SEA WATER as a FUNCTION of EXPOSURE TIME in BOILING WATER TEST in EPOXY MATRIX COMPOSITES

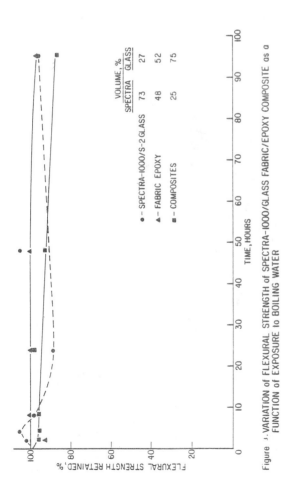

Figure 3. VARIATION of FLEXURAL STRENGTH of SPECTRA-1000/GLASS FABRIC/EPOXY COMPOSITE as a FUNCTION of EXPOSURE to BOILING WATER

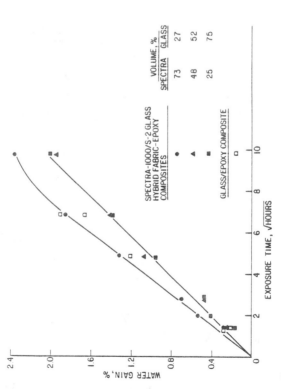

Figure 4. WATER ABSORPTION in SPECTRA-1000/GLASS/EPOXY COMPOSITES in BOILING WATER vs √TIME

423

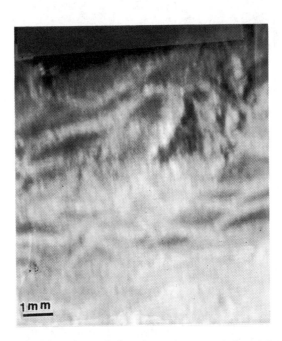

a. Optical micrograph showing delamination and buckling failure modes in Spectra-1000/epoxy composite in four point bend test.

b. Fiber yielding and microbuckling in four point bend test of Spectra-1000/epoxy composite.

Fig. 7. Failure modes in Spectra-1000/epoxy composite in a four point bend test.

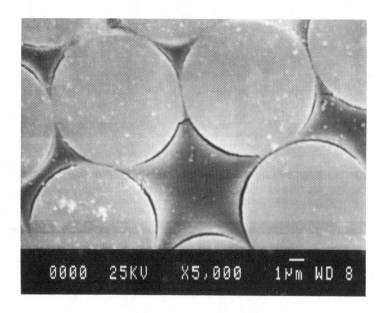

a. SEM showing micro-cracking around glass fibers.

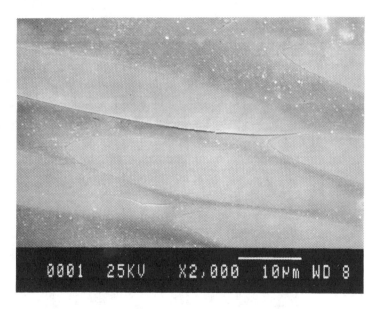

b. SEM showing micro-cracking along glass fibers.

Fig. 8. Scanning electron micro-graphs (SEM) showing micro-cracking in glass-epoxy (Hexcel-F155) composites exposed to boiling water for 96 hours.

b. SEM showing micro-cracking around glass fibers in Spectra-glass (25:75) hybrid fabric-epoxy composite exposed to boiling water for 96 hours.

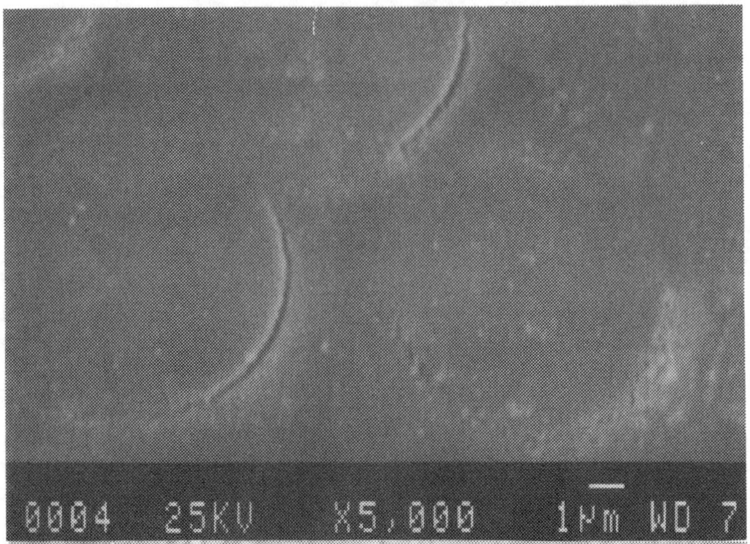

c. SEM showing micro-cracking around glass fibers in Spectra-glass (50:50) hybrid fabric-epoxy composite exposed to boiling water for 96 hours.

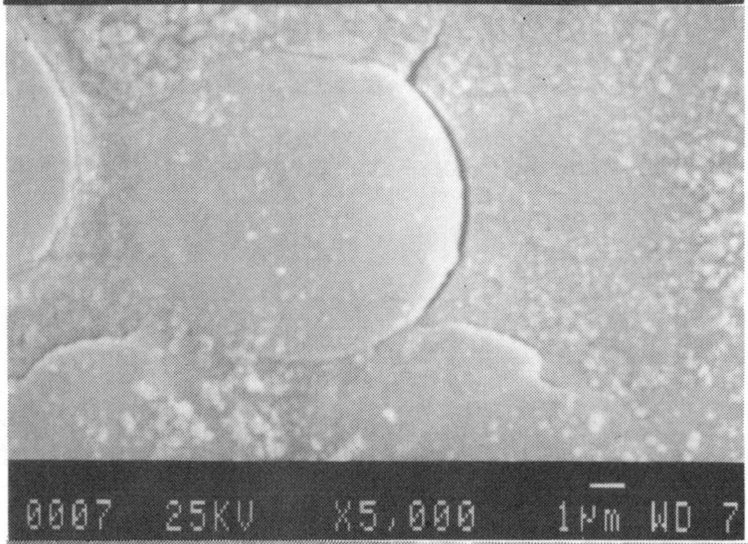

d. SEM showing glass & Spectra fibers in Spectra-glass (75:25) hybrid fabric-epoxy composite exposed to boiling water for 96 hours.

Fig. 9. Effect of boiling water exposure for 96 hours Spectra-glass hybrid fabric-epoxy composites a volume fractions of glass & Spectra.

Modelling of Elastic Properties of 3-D Textile Structural Composites

THOMAS J. WHITNEY AND TSU-WEI CHOU

ABSTRACT

A model originally proposed for 2-D textile composites is extended to a class of 3-D materials to predict their inplane properties. Numerical results agree in trend with preliminary experimental results, but an assumption of incompressible yarns apparently causes underestimation of axial modulus. Studies carried out to show the effect of varying geometric parameters demonstrate a wide range of variability in the properties of these materials.

INTRODUCTION

Recent advances in textile manufacturing technology have introduced various preforms as possible 3-D reinforcements for composites. These materials have the potential to overcome inherent weaknesses in composite laminates, such as poor interlaminar shear strength, weak impact resistance, and low thickness direction stiffness. However, analytical models and data bases for some of these preforms are lacking or are not well developed, yet are necessary in order to finally utilize the materials' unique mechanical properties [1-3]. In addition, material property dependence on geometric and manufacturing parameters (tow size, number of yarns per unit length and width, number of layers penetrated by each yarn, etc.) needs to be investigated.

The first step in the analysis of these materials is the determination of in-plane and bending properties and the development of analytical data bases. Several models have already been advanced to predict these properties in 2-D woven composites [4-6]. Procedures to analyze 3-D braided structures, based on energy principles [7] and lamination theory [8-10] have also been proposed. Those models extending lamination theory to two and three dimensionally reinforced composites show much promise for accurately predicting thermo-elastic properties, as they are based on well-known theory and methodology.

The scope of this paper is the extension of a model for thermo-elastic properties, based on lamination theory, to the geometries of 3-D woven textile composites and to add to the analytical data base for these materials. An example of such materials will be introduced, followed by the modeling and mathematical development. Numerical results will be compared to preliminary test results from the literature. The effects of varying several geometric parameters will also be discussed. This work is an extension to 3-D geometries of the model proposed by Ishikawa and Chou.

ANGLE-INTERLOCK GEOMETRIES

Figure 1 shows schematics of one type of 3-D woven preform, known as "angle-interlock" materials. They are differentiated from conventional 2-D weaves by the presence of interlacing

Thomas J. Whitney, Graduate Student, and Tsu-Wei Chou, Professor, Center for Composite Materials and Department of Mechanical Engineering, University of Delaware, Newark, DE 19716

weft yarns which may penetrate more than one layer of warp yarns. *Plain weave* refers to the interlacing of consecutive layers of warp yarns, while *satin weaves* encompass geometries that interlace multiple warp layers.

Many variations in this basic geometry are possible, depending on the number of layers interlaced, the pattern of repeat, and the presence of in-laid weft stuffers (normally done to reduce Poisson's effects). To strictly define the architecture of these materials, three parameters have been identified: n_{ft}, the number of warp layers interlaced by a weft yarn; n_{rt}, the number of warp rows between weft yarns within a weft plane; and n_{st}, the number of warp rows shifted by adjacent weft planes. Figure 2 demonstrates these parameters for the satin and plain weaves. These parameters are also used as a convenient nomenclature system, as Figure 3 demonstrates.

STRUCTURE AND UNIT CELL

The actual structure of an angle-interlock preform may vary greatly from the simplified, exploded views shown in Figures 1-3. In general, the tension in weft yarns during the weaving process will induce some undulation in the warp yarns, while the weft yarns themselves have much less undulation than idealized previously. Referring to Figure 4, two different types of warp undulations are present: the black yarns, representing warp yarns actually interlaced by adjacent weft yarns; and the grey yarns, representing non-interlaced warp yarns, which undulate due to the displaced black yarns. The number of each of these types of yarns per unit cell interlaced between adjacent weft yarns (white in Figure 4) is a function of the previously described parameters (n_{ft}, n_{st}, and n_{rt}).

Figure 5 shows a schematic of a unit cell for these structures. The undulating shape of the yarns is described by assuming the yarns to be perfectly flexible and to possess circular cross sections which remain circular. By assuming no out-of-plane crimping, the undulating shape is taken to be a circular arc while passing over the transverse yarns, and straight between these regions. The warp undulation is described by the parameter z_1, which is a measure of the warp undulation caused by the weft yarn straightening during the weaving process, and z_2, which is a measure of the vertical distance between adjacent weft yarns in a given weft plane and is determined by weaving machine parameters. Weft undulation is also determined by the parameters z_1 and z_2. In this model, several geometric constraints, given as functions of n_w and n_{ws} (number of warp and warp stuffer yarns, respectively), and d_w, d_{ws}, and d_f (diameter of warp, warp stuffer, and weft yarns, respectively) are necessary to ensure smooth undulation and prevent overlapping of yarns. In the warp direction, they include relationships between $(z_2-z_1)^2$ and l^2 (unit cell length); $(z_1)^2$ and l^2; and z_1 and z_2. In the weft direction, a relationship between $(w/2)^2$ (unit cell width) and $f(z_1)$ is imposed, where $f(z_1)$ is also a function of n_{ft} and is derived from an assumption that undulation in the interlaced warp and weft yarns are inversely related.

LAMINATION PROCEDURE AND ANALYSIS

The analysis is carried out by assuming that each yarn and the adjacent resin constitute a "lamina". Figures 6a and 6b portray these laminae for the warp and weft yarns, respectively. In this model, regions of pure matrix exist between the fibers resulting in a higher fiber volume fraction for the laminae than for the composite as a whole. It is further assumed that the stiffness properties of the unit cell are a "volume averaged" superposition of the stiffnesses of the warp and weft lamina, and that lamination theory is applicable to each incremental length along the x axis.

From lamination theory, the A, B, and D matrices are given by

$$[A, B, D]_{ij} = \int_{\text{lower surface}}^{\text{upper surface}} Q_{ij}[1, z, z^2]dz \qquad (1)$$

Referring to Figure 5, the A matrix of a strip, for example, is given by

$$[A]_{ij} = \int_{h_{mw1}}^{h_{uw1}} Q^w_{ij} dz + \int_{h_{uw}}^{h_{lws}} Q^r_{ij} dz + \int_{h_{lws}}^{h_{uws}} Q^{ws}_{ij} dz + \int_{h_{uws}}^{h_{lw2}} Q^r_{ij} dz + \int_{h_{lw2}}^{h_{mw2}} Q^w_{ij}$$

$$= Q^r_{ij}(h_{lws}-h_{uw}+h_l-h_{ls})+Q^w_{ij}(h_{uw1}-h_{mw1}+h_{mw2}-h_{lw2})$$

$$+Q^{ws}_{ij}(h_{uws}-h_{lws}) \tag{2}$$

where

h_{lws} = height of lower trace of warp stuffer surface
h_{uws} = height of upper trace of warp stuffer surface
h_{lw} = height of lower trace of warp surface (1- lower warp, 2 - upper warp)
h_{uw} = height of upper trace of warp surface (1- lower warp, 2 - upper warp)
h_{mw} = height of middle trace of warp surface (1- lower warp, 2 - upper warp)

and the superscripts on Q refer to warp lamina (w), warp stuffer lamina (ws) and resin (r) properties. The stiffnesses of the yarn laminae are reduced because of the inclination of the yarns in the thickness direction. The angle through which the local properties must be transformed is given by

$$\theta = \tan^{-1}\left(\frac{dh}{dx}\right) \tag{3}$$

in which h is taken to be the height of the trace of the center of the given yarn. This angle is therefore a function of x. Recall again that the properties transformed are local lamina properties and not those of the yarns themselves. The elastic properties become [11]:

$$E_x = \frac{1}{\dfrac{\cos^4\theta}{E_1} + \left(\dfrac{1}{G_{13}} - \dfrac{2v_{31}}{E_1}\right)\cos^2\theta\sin^2\theta + \dfrac{\cos^4\theta}{E_3}} \tag{4}$$

$$E_y = E_2 \tag{5}$$

$$G_{xy} = G_{12}\cos^2\theta + G_{23}\sin^2\theta \tag{6}$$

$$v_{yx} = v_{13}\cos^2\theta + v_{21}\sin^2\theta \tag{7}$$

The local Q matrix is then given by

$$[Q]_{ij} = \begin{bmatrix} \dfrac{E_x}{D} & \dfrac{E_y v_{yx}}{D} & 0 \\[2mm] \dfrac{E_y v_{yx}}{D} & \dfrac{E_y}{D} & 0 \\[2mm] 0 & 0 & G_{xy} \end{bmatrix} \tag{8}$$

in which i and j are considered to be 1, 2 and 6 and

$$D = 1 - \frac{v_{yx} E_y}{E_x} \tag{9}$$

To incorporate weft yarns, a similar procedure is used to calculate the A, B, and D matrices for the weft laminae. These matrices are superimposed by direct volume averaging with warp lamina. For example:

$$A_{ij} \text{increment} = \frac{(A_{ij}{}^w)(z_2 w - d_f l_f) + A^f{}_{ij} d_f l_f}{z_2 w} \tag{10}$$

in which l_f is the length of the weft yarn in the unit cell, superscripts w and f refer to warp and weft directions, respectively, and the superscript "increment" indicates that this averaging takes place at each incremental slice in the x direction. A similar procedure can be utilized to take into account lamina associated with weft stuffers.

Under a uniformly applied in-plane load in the x-direction, the compliances may be averaged along the length:

$$[a^T{}_{ij}, b^T{}_{ij}, d^T{}_{ij}] = \frac{1}{l} \int_0^l [a_{ij}, b_{ij}, d_{ij}] dx \tag{11}$$

in which a_{ij} is the inverse of A_{ij} and superscript T refers to a "total" compliance. The properties in the "x direction" (those obtained by an x-direction analysis) may be obtained from these compliances:

$$E^{uc}{}_x = \frac{1}{z_2 \, a^T{}_{11}} \tag{12}$$

$$v^{uc}{}_{xy} = \frac{-a^T{}_{12}}{a^T{}_{11}} \tag{13}$$

in which superscript uc refers to properties for the unit cell. In the y direction, a uniformly applied displacement along the length leads to averaging the stiffnesses through the width:

$$[A^T{}_{ij}, B^T{}_{ij}, D^T{}_{ij}] = \frac{1}{w} \int_0^w [A_{ij}, B_{ij}, D_{ij}] dx \tag{14}$$

Inverting these quantities to arrive at compliances permits evaluation of "y direction" properties (those obtained by an analysis in the y direction):

$$E^{uc}{}_y = \frac{1}{z_2 \, a^T{}_{22}} \tag{15}$$

$$G^{uc}{}_{xy} = \frac{1}{z_2 \, a^T{}_{66}} \tag{16}$$

EXPERIMENTAL COMPARISON

Numerical results based on the model were compared to preliminary test data on satin and plain weave angle-interlock commingled graphite/PEEK composites [12,13]. Unit cell dimensions and the resulting parameters are listed in Table 1, and were determined by measuring cell lengths and widths on the consolidated panel surfaces. The value of (z_2-z_1), which represents undulation in the stuffer yarn, was set at 0.18mm, or 90% of the maximum. This value represents a stuffer yarn with relatively small amounts of undulation as has been seen in photomicrographs [14]. The value of $f(z_1)$ was set at 3.81mm, since this was the smallest value that satisfied the constraints and yielded positive values for z_1 and z_2 for both geometries. Although positive values are not strictly necessary from a constraint viewpoint, photomicrographs have shown positive values to be the case for these geometries. However, this value results in higher weft undulation than is seen in photomicrographs [14]. A weft stuffer yarn was also present in each unit cell in these materials.

The results in Table 2 show some agreement in Poisson's ratio, while the model underpredicts axial modulus. It has been shown that during consolidation, pressure causes the yarn cross sections to deform into elliptical shapes and some out-of-plane crimping to occur, thereby reducing the undulation and increasing the stiffness [12]`. The assumption of circular cross-sections thus forces higher inclination angles than actually exist in the material. However, the comparison shows an encouraging trend that supports the geometric modelling constraints. In both experiment and model, the plain weave modulus is higher, while the Posson's ratio is lower. This demonstrates the constraints' ability to acount for lower weft inclination in the plain weave than in the satin weave to achieve the same value of (z_2-z_1).

PARAMETRIC STUDY

Parametric studies were conducted to evaluate the effect of the parameters n_w and n_{ws} on the elastic properties. The values for these parameters were allowed to vary independently from 0 to 5. As a reference point, the value of z_2 was set so that the gap between weft and warp yarns at the unit cell edges would be constantly equal to d_f. In addition, a warp direction constraint was satisfied by setting

$$z_2 - z_1 = C \, (z_2\text{-}z_1)_{max} \tag{17}$$

in which $(z_2\text{-}z_1)_{max}$ is given by the constraint relationship and $0 \leq C \leq 1$. A similar constant, B, was developed for the unit cell length, in which $B \geq 1$. Combining (17) and the B constraint yields a minimum allowable value of C which was met by setting C to a value mid-way between 1 and the minimum value. A similar constraint on B was also developed, but was automatically satisfied by setting B = 1.1. The value for w (unit cell width) was obtained by solving for n_{ft} based on n_w and n_{ws}. This value was used to obtain $f(z_1)$, and finally the value of w was set to 1.5 times the minimum value obtained from the weft constraint. With this formulation and the assumption for the value of z_2, structures with the same relative degree of jamming were achieved.

Figures 7-10 show the results of the study for a graphite/epoxy system with 1 weft stuffer and with the following parameters and properties: $d_f = d_w = d_{ws} = 1.27$mm; $E_f = 234.4$ GPa; $v_f = 0.22$; $E_m = 4$ GPa and $v_m = 0.37$. Figure 7 demonstrates a large increase in axial modulus with increases in both n_w and n_{ws}, most likely due to the fact that larger numbers of yarns produce similar degrees of jamming with lower inclination angles. Figure 8 shows large decreases in transverse modulus with increases in n_w and n_{ws}. In this case, larger numbers of warp yarns require the weft yarns to incline more in order to interlace the structure and cause the undulation necessary to satisfy the prescribed values of z_2 and C. Similar reasoning may explain the shear behavior. The Poisson's ratio calculation increases due to reasons similar to axial modulus, but flattens due to the increasing angle in the transverse (weft) direction.

CONCLUSIONS

A model to predict elastic properties in 2-D woven textile composites has been extended to a class of 3-D composites known as "angle interlock" geometries. The model accurately predicts trends in properties in comparison with preliminary experiments. However, assumptions concerning the incompressibility of the bundles apparently hampers the effort to predict properties for materials processed by thermoforming, as the constraint conditions may need to be modified to include deformed elliptical cross-sections. Parametric studies were also conducted to determine the effects of n_w and n_{ws} on the elastic properties. E_y and G_{xy} both appear to reduce with increasing n_w and n_{ws}, approaching a minimum value at $n_w = n_{ws} = 5$.

E_x and v_{xy} increase with increasing n_w and n_{ws}, with v_{xy} approaching a maximum at $n_w = n_{ws} = 5$. Significant variability in the properties of these architectures exist with the variance of the geometric parameters.

ACKNOWLEDGEMENTS

The authors wish to acknowledge the support of the National Science Foundation through the Center for Composites Manufacturing Science and Engineering at the University of Delaware. They also wish to express their gratitude to Mr. Leo Taske and Dr. Azar P. Majidi for their assistance in this work and their discussions on weaving and processing of angle-interlock geometries.

REFERENCES

1. Chou, T-W., McCullough, R.L., and Pipes, R.B. "Composites", *Scientific American*, Oct. 1986.

2. Macander, A.B., Crane, R.M., and Camponeschi, E.T. "Fabrication and Mechanical Properties of Multidirectionally Braided Composite Materials", *Composite Materials: Testing and Design (Seventh Conference), ASTM STP 893*, J.M.Whitney, Ed., American Society for Testing and Materials, Philadelphia, Pa., 1986.

3. Crane, R.M., and Camponeschi, E.T. "Experimental and Analytical Characterization of Multidirectionally Braided Graphite/Epoxy Composites", *Experimental Mechanics*, Vol 26, No 3, September 1986.

4. Ishikawa,T., and Chou, T-W. "Stiffness and Strength Behavior of Woven Fabric Composites", *Journal of Materials Science*, Vol 17, 1982.

5. Ishikawa, T., and Chou, T-W. "Elastic Behavior of Woven Hybrid Composites", *Journal of Composite Materials*, Vol 16, January 1982.

6. Ishikawa, T., and Chou, T-W. "Nonlinear Behavior of Woven Fabric Composites", *Journal of Composite Materials*, Vol 17, September 1983.

7. Ma, C-L., Yang, J-M., and Chou, T-W. "Elastic Stiffness of Three-Dimensional Braided Textile Structural Composites", *Composite Materials: Testing and Design (Seventh Conference), ASTM STP 893*, J.M.Whitney, Ed., American Society for Testing and Materials, Philadelphia, Pa., 1986.

8. Camponeschi, E.T., and Crane, R.M. "A Model for Fiber Geometry and Stiffness of Multidirectionally Braided Composites", *3-D Composite Materials*, NASA Conference Publication #2420, 1986.

9. Yang, J-M., Ma, C-L., and Chou, T-W. "Fiber Inclination Model of 3-D Textile Structural Composites", *Journal of Composite Materials*, Vol 20, September 1986.

10. Whitney, T.J., Chou, T-W., Taske, L.E., and Majidi, A.P. "Performance Maps of 3-D Textile Structural Composites", to be published in *Fiber-Tex 87*, NASA Conference Publication.

11. Lekhnitskii, S.G. *Theory of Elasticity of an Anisotropic Body*. Mir Publishers, Moscow, 1981.

12. Taske,L.E., and Majidi, A.P. "Performance Characteristics of Woven Carbon/PEEK Composites", *Proceedings of The American Society for Composites*, Second Technical Conference, Technomic Publishing Co., 1987.

13. Majidi, A.P., Rotermund, M.J., and Taske, L.E. "Thermoplastic Preform Fabrication and Processing", *SAMPE Journal*, Vol 24, January/February, 1988.

14. Taske, L.E. Master's Thesis under preparation, University of Delaware, 1988.

Table 1. Unit Cell Parameters*

Geometry	l(mm)	w(mm)	z1(mm)	z2(mm)
Satin Weave	5.00	3.00	1.44	6.02
Plain Weave	2.79	1.78	2.56	7.13

* All yarn diameters = 1.27mm

Table 2. Comparison to Preliminary Experiments*

Geometry	E_x(GPa)	E_y(GPa)	ν_{xy}	G_{xy}(GPa)
Satin Weave (Exper.)	39	N/A	0.16	N/A
Satin Weave (Model)	23.0	11.2	0.13	1.20
Plain Weave (Exper.)	40	N/A	0.08	N/A
Plain Weave (Model)	25.0	19.8	0.10	2.02

* All experimental values are averages of 3 tests [14].

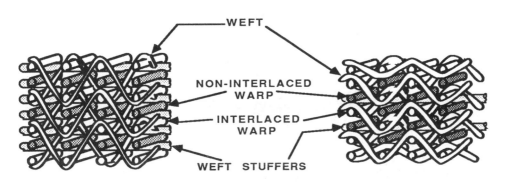

FIGURE 1. SATIN AND PLAIN WEAVE PREFORMS

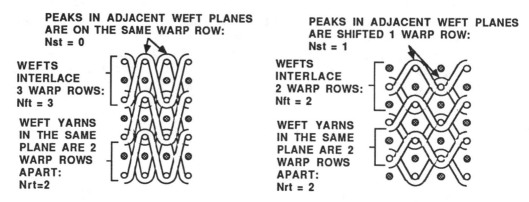

FIGURE 2a and 2b. GEOMETRIC PARAMETERS: SATIN AND PLAIN WEAVES

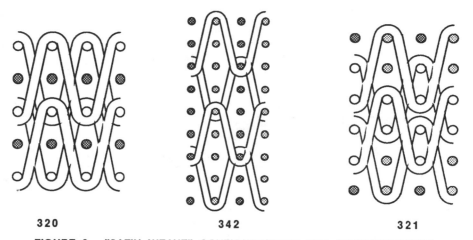

320   342   321

FIGURE 3. "SATIN WEAVE" CONFIGURATIONS AND CORRESPONDING
PARAMETERS (nft,nrt,nst)

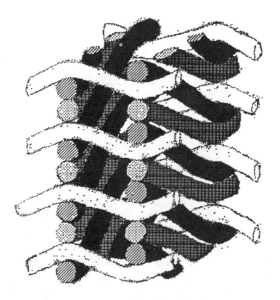

FIGURE 4. IDEALIZED UNDULATION IN ANGLE-INTERLOCK ARCHITECTURE

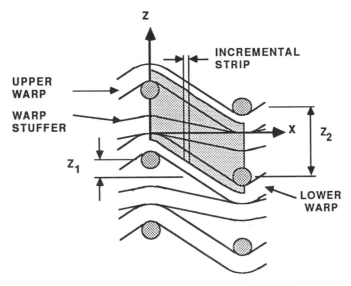

FIGURE 5. UNIT CELL SCHEMATIC

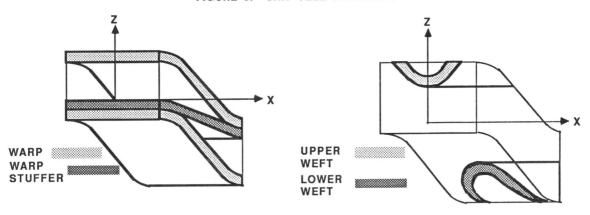

FIGURES 6A. LAMINAE ASSOCIATED
WITH WARP YARNS

FIGURE 6B. LAMINAE ASSOCIATED
WITH WEFT YARNS

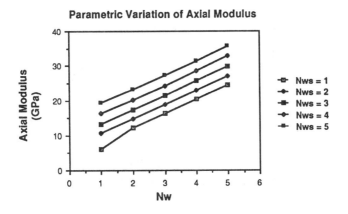

FIGURE 7. PARAMETRIC BEHAVIOR OF E_x

Parametric Variation in Transverse Modulus

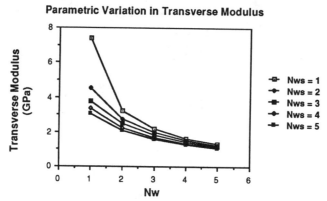

FIGURE 8.   PARAMETRIC BEHAVIOR OF E_y

Parametric Variation in Poisson's Ratio

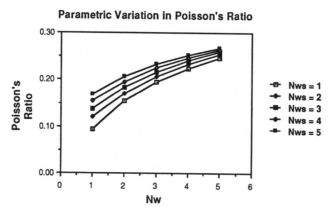

FIGURE 9.   PARAMETRIC BEHAVIOR OF ν_{xy}

Parametric Variation in Shear Modulus

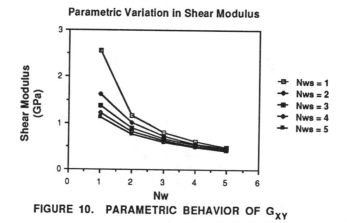

FIGURE 10.   PARAMETRIC BEHAVIOR OF G_{XY}

SYMPOSIUM X

Internal Stresses on Composite Materials

Solidification Processes in Thermoplastics and Their Role in the Development of Internal Stresses in Composites

PAUL ZOLLER

ABSTRACT

The solidification behavior of thermoplastics is discussed and illustrated with new dilatometric data on poly(ether ether ketone) (PEEK). The buildup of residual stresses caused by a mismatch between expansion/shrinkage properties of the matrix and the reinforcing fibers is discussed, using experimental data on the curvature of unbalanced cross-ply strips.

INTRODUCTION

Currently the majority of advanced polymeric composites, i.e composites containing continuous fibers, are made with a thermoset as the matrix. Most common are the typical 350°C cure epoxy matrix resins. Composites made with these resins suffer from a number of undesirable properties. For one thing, they are not "tough", nor "damage tolerant", by whatever tests and methods one might want to define these terms. In addition to modifying epoxies to overcome these shortcomings, the idea of using thermoplastics as matrix resins has been popular for a number of years. To be sure, most thermoplastics are in fact "tough", although this is perhaps most evident only in their neat form, and does not translate simply and fully into the "toughness" of a composite [1]. In addition, thermoplastic matrix composites are often thought to offer processing advantages (no autoclave, thermoformability, etc.), but this author has argued for a number of years that such a processing advantage does not exist, and this view has now found considerable support.

Whatever the relative competitive position of thermosets and thermoplastics in advanced composites is, it is a fact

Paul Zoller, Professor, University of Colorado, Dept. of Mechanical Engineering, Campus Box 427, Boulder, CO 80309

that the two classes of resins differ in many fundamental
aspects. This paper will concentrate on one of these - the
changes in volume occurring during solidification - and
explore the consequences of this behavior for the processing
of composites, in particular with respect to the development
of internal stresses.

SOLIDIFICATION BEHAVIOR OF THERMOPLASTICS - GENERAL

At sufficiently high temperatures, thermoplastics exist
in a thermodynamic equilibrium state: the melt. This state can
be attained repeatedly and reproducibly, if no chemical change
(degradation) occurs with time. This is not always easy to
achieve because of the relatively high temperatures involved.
The melt state is involved at one time or another during all
processing operations for thermoplastics (except of course for
those processes which involve solutions of the thermoplastic
resins). On cooling, thermoplastics solidify either by forming
a glass, or by (partial) crystallization, over fairly sharply
definable temperature intervals. The solid state of polymers
which results from these processes is not an equilibrium
state: the structure and properties of solid polymers depend
not only on the values of the current external variables
(temperature, pressure,..) but also on the formation history
of the material, that is the history that it experienced while
going from the melt state to the solid state. Clearly
important are the speed and pressure with which the material
is cooled. It is known, e.g., that a polymer which (partially)
crystallizes on solidification at a certain slow cooling rate,
may remain entirely amorphous if cooled at the highest
possible rate, while, at intermediate rates, it might assume a
"degree of crystallinity" which depends on the cooling rate,
and also on the pressure under which it was cooled. Naturally,
properties such as stiffness, strength, toughness, etc. depend
strongly on the degree of crystallinity. In most cases the
simple concept of a degree of crystallinity is not even
sufficient to describe the state of a material. A finer
description of the structure ("morphology") is required to
assess the properties. Similarly, heat treatments in the solid
state can affect the morphology and properties of
crystallizable resins. In addition to cooling rate and
pressure, the presence of additives ("crystallization
promoters"), also affects properties. With respect to the
formation of glasses, the situation is perhaps even more
complex, since no convenient method (comparable to scattering
techniques for crystalline materials) is available to assess
the "structure" of glassy materials, but it is known that
properties of glasses, too, depend markedly on their formation
history, and on heat treatment in the glassy state
("annealing"). One of the simplest aspects of polymer
solidification is the question of the volume changes involved
in the process.

VOLUME CHANGES ASSOCIATED WITH SOLIDIFICATION

This topic is of concern, because of the vastly different volume changes with temperature exhibited by thermoplastics and the fibers used in advanced composites. Thermoplastics experience a volume shrinkage of 5 to 25% between the melt and the solid state. The lower values are generally found for amorphous resins, the higher ones for crystallizable resins, with a part of the volume change caused by the volume change associated with the crystallization process (Fig. 1). By comparison, epoxies generally experience a net volume shrinkage of 1-6% between the uncured and cured states at ambient temperature, but larger volume changes may exist at intermediate times in the cure cycle [2].

Fibers typically used in advanced composites have a negative coefficient of (linear) thermal expansion in the fiber direction (approximately -0.3×10^{-6} °C^{-1} for graphite, and 2×10^{-6} °C^{-1} for Kevlar aramid fibers), but much higher coefficients in the transverse direction (18×10^{-6} °C^{-1} for graphite, 60×10^{-6} °C^{-1} for Kevlar). Thus, over a temperature range of, say, 300°C, the volume change of a graphite fiber will be around 1%, that of a Kevlar fiber about 3-4%. These volume changes are smaller than those of the matrices (Fig. 1), but not by all that much. The real problem is in the anisotropy of the dimensional changes of the fibers. There is a tremendous mismatch between the expansion in the fiber direction and the liner dimensional changes of all polymeric matrix resins. The buildup of residual stresses as a consequence of this mismatch will be discussed below, after a discussion of the volume change properties of a specific material: poly(ether ether ketone).

VOLUME CHANGES WITH TEMPERATURE IN POLY(ETHER ETHER KETONE)

These data on the thermophysical behavior of ICI´s poly(ether ether ketone) (PEEK) are reported here because PEEK is one of the leading candidates for a thermoplastic matrix resin. To study volume changes, we use a commercial high pressure dilatometer. [3] This apparatus uses a sample of 1-2 g, and can measure volume changes with an absolute accuracy of better than 0.003 cm^3/g over a temperature range to 400°C and at pressures to 200 MPa (= 2000 bar $\approx$ 29,000 psi). Of course, the high pressures are of no direct interest for the processing of thermoplastics into composites, since these operation rarely involve pressure above 200 psi (or 1.4 MPa), but high pressures play a considerable role in more conventional processing operations, such as injection molding. Also, the use of pressure in dilatometric experiments provides distinct experimental advantages, as it keeps "false" volume changes caused by outgassing of a sample to a minimum. Fig. 2 shows the PVT relationship of PEEK as determined by our techniques. Concentrating only on the zero-pressure isobar, a glass transition is clearly observed near 150 °C as a break in the isobar. A second break occurs at 335-340 °C, marking the end of the melting process. The appearance of the isobars is very typical of semicrystalline polymers, except that the strength of the break at the glass transition indicates that PEEK is a material of relatively low crystallinity [4].

The following fits describe the zero pressure isobars of the solid state and the melt to about 0.0002 cm³/g:

$$V_o = 0.7641 \exp(1.61 \times 10^{-4} T) \quad \text{for } T < T_g$$

$$V_o = 0.7518 \exp(6.69 \times 10^{-4} T) \quad \text{for } T > T_m$$

(V_o in cm³/g, T in °C)

These fits represent temperature-independent volume expansivities of 1.61×10^{-4} °C^{-1} for the solid state below T_g and 6.69×10^{-4} °C^{-1} for the melt. If we assume that PEEK is processed at 380°C, the volume difference between this processing temperature and the sample at 30°C is 0.202 cm³/g, or about 26% of the volume at 30°C, very much in line with the other semicrystalline thermoplastics in Fig. 1.

A value for the melting point for PEEK was obtained by performing isobaric melting and crystallization experiments at different pressures. An example of such data is given in Fig. 3. As a sample is heated at a constant pressure (10 MPa = 145 psi and 100 MPa = 14,500 psi) the end of the melting interval is clearly visible as a sharp break in the curves. When the sample is cooled from the melt (at 2.5°C/min in these examples) the heating and cooling isobars superimpose perfectly in the melt (as they must because of the equilibrium nature of the melt state), but on cooling the melt state is maintained to temperatures below the melting point for the particular pressure. Supercooling is, of course, observed for all polymers, as well as for low-molecular weight substances. By plotting the melting point T_m and the crystallization temperature T_c as a function of pressure, and extrapolating to zero pressure, a melting point of 338 °C and a crystallization temperature of 301 °C has been determined.

BUILDUP OF RESIDUAL STRESSES IN COMPOSITES WITH THERMOPLASTIC MATRICES

Naturally a melt of a thermoplastic is not able to support non-hydrostatic stresses for any length of time: instead it flows to relieve shear stresses. In order to judge a material for its potential to build up residual stresses due to a mismatch between matrix and fiber thermal expansion, the volume shrinkage below the temperature at which the matrix assumes enough of a solid-like character to build up stresses should be considered. This temperature is certain to be near the glass-transition temperature for a glass-forming material, and probably somewhere in the crystallization region for crystallizable resins. Experimentally this has been confirmed in a series of experiments on both amorphous and crystallizable resins, using photoelastic measurements of the induced stresses (for amorphous, i.e. transparent resins) [5], and studying the curvature of cross-ply (0/90°) laminates for both amorphous and semicrystalline resins [6,7].

With respect to amorphous thermoplastic resins, the situation is particularly simple. In all experiments [5,7] stresses began building up at the glass transition temperature. In semicrystalline matrices, stresses began to build in or slightly below he crystallization regions (as determined by differential thermal analysis), but individual materials seemed to differ slightly in the details. The comparison of the onset of stress buildup and the crystallization region in a neat resin by DSC must be taken with a grain of salt (although the cooling rates were comparable), because the fibers, or debris from fibers, might act as crystallization agents.

The exact calculation of residual stresses as a function of temperature would be a formidable task. In principle it would require a non-isothermal viscoelastic analysis of a material, the viscoelastic descriptors of which (typically the shear and bulk relaxation modulus) change continually as solidification progresses, because of the ongoing development and refinement of internal structures. The data for such an analysis are simply not available. Instead a much simpler approach has been developed.

The curvature in cross-ply composite strips can be modeled very simply and accurately using an elastic analysis based on the bimetallic strip equation, modified to include temperature dependent properties of the composite plies. The thermal expansivities of the plies required in the bimetallic strip equation were calculated from the Schapery equations [8], and the elastic properties of the plies by using a mechanics of materials approach found in most textbooks (e.g. [9]). Needed for this calculation are the volume fraction of fibers, the longitudinal and transverse moduli of the fibers, the longitudinal and transverse expansion coefficient of the fibers, the Poisson ratio of the fiber and matrix, the Young´s modulus of the matrix (which drops with temperature, and becomes essentially zero above Tg or Tm), and the temperature-dependent thermal expansivity of the matrix, which can be taken from PVT measurements.

The result is a zero-parameter fit of the curvature of cross-ply strips as a function of temperature. For an amorphous matrix (polysulfone/graphite) cross-ply strip [7] the agreement between theory and experiment is nearly perfect (Fig. 3). In semi-crystalline matrix composites, the behavior is more complex. The range of phenomena is illustrated by Fig. 4 and 5. In Fig. 4 data on crossplies made from poly(ethylene terephthalate) and graphite are presented. First it is quite apparent that the results from the different experiments show a lot of scatter in the region below 200°C. During the experiments acoustic emissions were detected at the lower temperatures. These are a result of cracking in the matrix of the transverse ply. These cracks also are responsible for the relatively poor agreement with the bimetallic strip theory at low temperatures. One of the characteristics of all curves, however, is the steep increase in curvature right below the onset. This must be associated with the steep decrease in volume as the material crystallizes. Apparently, for PET, the morphology that forms very early in the crystallization process is able to support stresses. Similar behavior was detected in composite strips with a Nylon 66 matrix [6]. In other materials the morphology may be such that it is unable to support residual stresses until crystallization is almost complete, and the steep increase in curvature just below the onset is basically absent. This was found, e.g. for J-polymer composites [6]. Yet another type of behavior is represented by PEEK [6] (Fig. 5). On cooling from the melt the dimensionless curvature began at about 319°C. Apparently this sample crystallized at a somewhat higher temperature than the PEEK used in the PVT studies discussed above. This could be explained, for example, by the conjecture that the grade used in composites might have a lower molecular weight to give it better flow properties, or by a nucleating effect of the fibers. It seems likely that curvature in the cross-ply strips begins somewhere in the crystallization region. However, the steep increase in curvature just below the onset is absent. We have speculated that this could be associated with the formation of a type of crystal orientation relative to the fiber that would decrease crystallization-induced stresses. The only real feature in this cooling curve is a slight break near 150°C, associated with the glass transition of PEEK, which, as we mentioned above, is unusually pronounced for a semicrystalline material.

CONCLUSION

In the experiments described, the presence of residual stresses was obvious, due to the appearance of curvature in the cross-ply strips. The curvature develops because of a mismatch in the properties of the 0 and 90° plies. Even in balanced composites there will be residual stresses due to the mismatch in expansion properties of the fibers and matrix. The onset of this type of residual stress can be expected to be at the same temperatures as those observed here [5,7], and models can be used to calculate these stresses [10]. They can also be observed directly by photoelastic techniques in transparent matrices [5]. It is also apparent that in semicrystalline materials, which have the ability to develop a wide range of internal structures, the buildup of residual stresses is influenced by the details of the solidification process and the resulting morphology. In amorphous resins, which are relatively free of structure, the effect of processing variables is likely to be much less pronounced. What is not at all clear is how residual stresses will affect the properties of a composite. Naturally, the cases in which matrices crack simply because of internal stresses are to be avoided. In addition there may well be more subtle effects of internal stresses, such as diminished chemical resistance (environmental stress cracking), effects on fatigue properties etc. Finally, the large volume changes (5-25%) that thermoplastic resins experience between melt state and solid state, are a problem all by themselves, since they will make the manufacture of large, precise parts rather difficult.

There is thus ample incentive for understanding the interaction between processing variables, structures, residual stresses, and resulting properties.

REFERENCES

1. Hunston, D.L., Comp. Tech. Rev., Vol. 2, 1984, pp. 176-180.

2. Yates, B., McCalla, B.A., Phillips, L.N., Kingston-Lee, D.M., and Rogers, K.F., J. Mater. Sci., Vol. 14, 1979, pp. 1207 ff.

3. GNOMIX RESEARCH, 3809 Birchwood Drive, Boulder, CO 80302.

4. Zoller, P., Starkweather, H.W., Jones, G.A., "The Equation of State and Heat of Fusion of Poly(ether ether ketone)", J. Polym. Sci. Polym. Phys. Ed., to be published.

5. Nairn, J.A. and Zoller, P., J. Mater. Sci., Vol. 20, 1985, pp. 355-376.

6. Nairn, J.A., Zoller, P., "Residual Thermal Stresses in Semicrystalline Thermoplastic Matrix Composites", Proc. of the Fifth International Conference on Composite Materials, San Diego, 1985, pp. 931-946.

7. Nairn, J.A. and Zoller, P., "The Development of Residual Thermal Stresses in Amorphous and Semicrystalline Thermoplastic Matrix Composites", Toughened Composites, ASTM STP 937, Norman J. Johnston, Ed., American Society for Testing and Materials, Philadelphia, 1987, pp. 328-341.

8. Schapery, R.A., J. Comp. Matl., Vol 2, 1968, pp. 380-404.

9. Jones, J.J., Mechanics of Composite Materials, McGraw-Hill Book Co., New York 1975, pp. 90-98.

10. Nairn, J.A., Polymer Compos., Vol. 6, 1985, pp. 123-130.

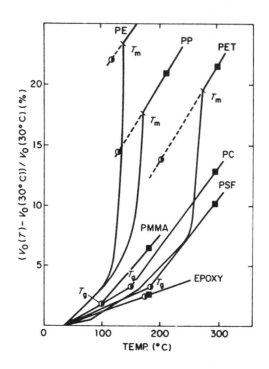

Figure 1: Zero-pressure volume change (relative to volume at 30°C) of several thermoplastics and a typical epoxy. Squares designate typical processing temperatures, circles the onset of solidification (crystallization or glass formation) at slow cooling rates (1-2°C/min) appropriate to some forms of composites processing.(PE=Polyethylene, PP=Polypropylene, PET=Poly(ethylene terephthalate), PC=Polycarbonate, PMMA=Poly(methylmethacrylate, PSF=Polysulfone).

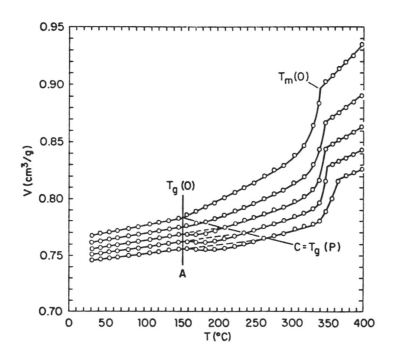

Figure 2: Isobars of the PVT relationship of PEEK at pressure increments of 40 MPa (5800 psi). These isobars were cross-plotted from isothermal data. Line A is an isotherm through the zero-pressure glass transition, line C marks the pressure dependence of the glass transition temperature. This aspect of the PVT relationship of PEEK is of no particular interest to composites, and is discussed in a forthcoming publication [4].

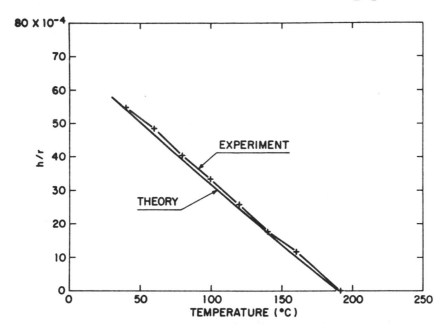

Figure 3: Dimensionless curvature of a polysulfone/graphite cross-ply composite strip as a function of temperature during cooling from 200°C at 1.5°C/min. "Theory" curve is prediction of the bimetallic strip equation.

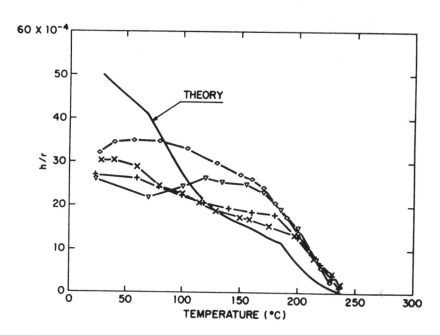

Figure 4: Dimensionless curvature as a function of temperature while cooling from the melt for four poly(ethylene terephthalate)/graphite cross-ply-composites, and the prediction of the bimetallic strip equation. Cooling rate 1.5°C/min.

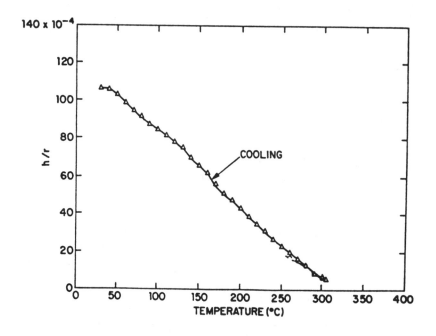

Figure 5: Dimensionless curvature as a function of temperature while cooling a PEEK/graphite cross-ply composite strip. Cooling rate 1-3°C/min.

Thermal Skin/Core Residual Stresses Induced During Cooling of Thermoplastic Matrix Composites

T. J. CHAPMAN, J. W. GILLESPIE, JR., J.-A. E. MANSON, R. B. PIPES AND J. C. SEFERIS

ABSTRACT

With increased usage of high performance thermoplastic composite materials as a result of growing cost-effective manufacturing processes, such as hot stamping of pre-consolidated laminates and laser consolidation, a need for a fundamental understanding of these methods has emerged. Specifically, knowledge of the influence of the rapid cooling rates inherent in these techniques and subsequent internal stress generation is essential to enable widespread use of thermoplastic composites through these processing methods. The primary effect investigated is the formation of residual stresses as a laminate is cooled rapidly. An analytical methodology, based on lamination and plate theory, and incorporating the anisotropic thermal properties and viscoelastic material behavior, was formulated. Experimentally, the creep and stress relaxation properties of PEEK-based carbon fiber-reinforced composites (APC-2) were measured. Residual stress measurements were performed with the Process Simulated Laminate and an alternate suface machining technique. Results of testing of APC-2 supported by the computer model revealed a skin/core profile of residual stresses with compression at the surface and tension in the center of the laminate. For an initial cooling rate of 35°C/s at the surface, 25% of the ultimate tensile strength in the 90 degree direction was measured at the center of the APC-2 composite laminates. Subsequent annealing and quenching of preconsolidated laminates indicate that stress relaxation at temperatures far above the glass transition causes most residual stress build-up to occur near the transition temperature.

INTRODUCTION

With the increasing interest focused on high performance thermoplastic composite materials by many industries including those related to aerospace applications, many unique manufacturing and processing methods have been developed. Although the primary goal of these techniques has been cost effectiveness through simplified, rapid production,

T. J. Chapman, Graduate Research Assistant, Center for Composite Materials, University of Delaware, Newark, DE, 19716, and Visiting Graduate Student, Polymeric Composites Laboratory, University of Washington, Seattle, WA, 98195.
J. W. Gillespie, Jr., Assistant Director for Research, Center for Composite Materials, University of Delaware, Newark, DE, 19716.
J.-A. E. Manson, Professor, Department of Chemical Engineering, University of Washington, Seattle, WA, 98195.
R. B. Pipes, Dean, College of Engineering, University of Delaware, Newark, DE, 19716.
J. C. Seferis, Director, Polymeric Composites Laboratory, University of Washington, Seattle, WA, 98195.

consideration of the resulting properties of the material is necessary since structural performance is the ultimate requirement.

Of the numerous important material properties, the formation of internal stresses, caused primarily by non-isothermal conditions such as those encountered during some processing techniques, is the concentration of this investigation. These processes include hot stamping of pre-consolidated laminates, diaphragm-forming, and laser consolidation, each a potential alternative in the production of thermoplastic composite components. An extensive classification of sources and driving forces for internal stress development has been previously proposed [1]. Included in these are intrinsic material properties which account for the anisotropy and heterogeniety inherent in continuous-fiber composites, processing conditions including post-processing heat treatments, and environmental conditions present in structural applications. Furthermore, all driving forces are generated through the combination of ply orientations, microscopic fiber/matrix effects, and skin/core profiles. The latter term refers to the existence of non-uniform conditions between skin or surface and core or center. The internal stresses generated by all these sources appear in various forms such as fiber buckling and void formation during which stresses are relieved and residual stresses which are those which remain at ambient conditions following processing or heat treatment.

This project focuses on residual stress development caused by thermal skin/core effects, the time-dependent volume change induced in the material during cooling. With a concentration on macroscopic stress formation, each ply undergoes a unique volume history during any non-isothermal cooling. The presence of temperature gradients and associated behavior are, therefore, the factors of interest.

Another factor of potential importance addressed in this investigation is the viscoelastic behavior of the thermoplastic matrix during cooling from melt conditions. Since solidification occurs well above the glass transition temperature in the case of most materials, the potential for recovery in this temperature range may result in relieving of stress through creep or stress relaxation. Because viscoelastic recovery is highly dependent upon the time available for such behavior, the specific cooling rate achieved during processing or following heat treatment becomes important, particularly at temperatures near or above the glass transition.

In order to characterize the effect of thermal skin/core differences upon the development of residual stresses in continuous-fiber thermoplastic matrix composite materials, a combination of modelling and experimenting is necessary. The modelling consists of a combined analytic and numerical approach through a FORTRAN computer code to account for various cooling conditions, ply orientations, and material systems. Experimental measurements of residual stresses can be accomplished through monitoring of strain changes as layers of material are successively removed from a sample surface. This surface removal may be performed through the Process Simulated Laminate (PSL) technique proposed previously [1,2] or by machining the surface with water cooling provided. A variety of heat treatments including annealing and quenching can be employed to further characterize the formation of residual stresses during rapid cooling cycles. Although the particular composite considered in this investigation is carbon/polyetheretherketone (APC-2), the methods presented should be applicable to any continuous-fiber thermoplastic composite.

COMPUTER MODEL

The initial stage of the modelling of laminate cooling is calculation of temperature as a function of position and time. A convenient and commonly-employed technique of heat transfer analysis is the finite difference method in which differential terms are approximated using finite increments and length dimensions are divided through placement of nodes.

The cooling of a thermoplastic matrix composite such as carbon/PEEK is complicated by the fact that crystallization occurs following solidification of the polymer and continues through the non-isothermal process. Combined with the fact that the nucleation and growth of crystals causes heat transfer, this temperature-dependent crystallization requires the simultaneous modelling of temperature and crystallization histories. Therefore, the energy equation [3] which relates these two variables must be transformed to finite difference form.

$$\frac{\delta}{\delta t}(\rho C T) = \frac{\delta}{\delta x}\left(K\frac{\delta T}{\delta x}\right) + H \tag{1}$$

where T = temperature, C = specific heat, K = thermal conductivity, x = coordinate normal to the plate, H = heat generated per unit volume due to crystallization, and ρ = density.

For the purposes of the model developed for this investigation, a one-dimensional analysis in the thickness direction was chosen with assumption that all properties, including mechanical properties and residual stresses, are constant across the thickness of each single ply. Furthermore, the temperature and crystallinity are specified as uniform in all in-plane directions. In this way, the finite difference formulation is completed by specifying an initial crystallinity (0% for processing cycles) in each ply, the time-dependent temperature at the laminate surface, and an assumption of symmetric cooling about the laminate midplane. The latter is applicable for most processing methods including those in a hot press such as that employed in this study. The prediction of crystallinity requires the inclusion of a sub-model to correlate this variable with the laminate cooling rate. Such a model for APC-2 proposed by Velisaris and Seferis [4] has shown excellent agreement with experimental results. Finally, the temperature distributions enable prediction of solid/melt interface location by comparison of ply temperatures with an assumed solidification at 340°C.

With the complete temperature and crystallinity profiles known as a function of time, the mechanical properties required for residual stress modelling, elastic moduli and Poisson's ratios in both the longitudinal and transverse directions, can be calculated. The initial step of this process is an analysis of thermoplastic matrix properties based upon temperature and crystallinity. This is accomplished by treating the matrix as a composite, with the amorphous phase acting as a matrix and crystalline phase as reinforcement. Assuming a temperature dependence of amorphous properties but none in those of the crystals, a randomly-oriented short fiber composite code can be utilized with degree of crystallinity serving as fiber volume fraction. Combining the predicted matrix properties with assumed temperature-independent carbon fiber properties, those of the composite material are easily found through use of micromechanics. In the case of PEEK, the degree of crystallinity probably has a negligible influence upon mechanical properties, but this may not be true for other thermoplastic matrix systems.

The previous analysis of composite properties assumes completely elastic behavior of the thermoplastic matrix, an unrealistic approximation at temperatures near or above the glass transition. In this range, time becomes an important variable since polymer chains are permitted to reorient and cause partial or complete relaxation of any applied or induced stresses. Obviously, the time available for relaxation is directly dependent upon the specific cooling rate. That is, gradual cooling, such as that in an autoclave processing, allows sufficient time for full relaxation at most temperatures while rapid cooling or quenching may prohibit this. For this reason, characterization of the viscoelastic response of polyetheretherketone is provided to the model as a result of stress relaxation and creep testing of fully-crystallized single ply samples processed in a hot press.

Following the prediction of temperature and crystallinity histories, interface location, and mechanical properties as a function of time, the thermal stress state at any stage of the cooling cycle can be determined. A final requirement to facilitate this analysis is the thermal expansion properties in both material directions. Included in this expansion behavior is that caused by changes in temperature as well as that related to volume changes due to crystallization of the thermoplastic matrix. This volume contraction is dependent upon the laminate cooling rate since this variable directly influences the degree of crystallinity. A volume change of approximately 25% is assumed to occur in each ply between solidification and equilibrium temperatures based on previous experiments of numerous thermoplastic matrices [5].

Since the orientation of each ply in the laminate should be capable of assuming any value, the mechanical properties must be transformed to Q and $\overline{Q}$ matrix forms. These transformations are performed for each ply individually through sine and cosine equations as specified by Vinson and Sierakowski [6]. Following the subsequent technique outlined by the latter, the laminate ABD stiffness matrix, based upon the individual ply $\overline{Q}$ matrices, and the plate thermal loads, functions of the temperature change, thermal expansion coefficients, and $\overline{Q}$ matrix of each ply, are computed. These plate lamination theory equations relate the contributions of the properties of each ply to those of the solidified plate. For this reason, the strains calculated from the ABD matrix and thermal loads represent those which are induced in all plys simultaneously. This fact concerning uniform straining across the thickness of the laminate is due to the assumed constraint provided to each ply by all other solidified plys.

However, the fact that each ply possesses a unique temperature at each particular time step indicates that the unconstrained contraction of each ply would differ from that actually induced in the plate. These individual contractions are simply calculated from the temperature change, thermal contraction coefficients, and $\overline{Q}$ matrix of each ply. The differences between the unconstrained theoretical strains and those actually caused by thermal loads on the plate are the foundation of the thermal stresses in each ply. During cooling, these stresses can lead to void formation or fiber buckling at elevated temperatures, thus relieving the stresses, or remain at room temperature in the form of residual stresses.

The final assumptions employed in the FORTRAN computer model included in this study concern the conditions under which stress build-up occurs. That is, stress formation in the outermost ply, the first to solidify, is not assumed to commence until the second ply solidifies since no constraint is applied until this time. Furthermore, any viscoelastic relaxation of stresses is assumed to occur in the form of stress relaxation rather than creep such that strains remain uniform across the laminate thickness and no inter-ply sliding takes place.

To summarize the capabilities of the computer model discussed above, the available outputs at each time step include temperature, degree of crystallinity, elastic mechanical properties, and thermal stress state of all plys as well as the location of the solid/melt interface. Although this investigation is limited to carbon/PEEK composite laminates, the model is applicable to any continuous-fiber thermoplastic composite material. The required input parameters include fiber, crystalline phase, and temperature-dependent amorphous phase elastic properties, composite viscoelastic response, thermal contraction behavior including crystallization shrinkage, thermal conductivity and heat capacity as functions of temperature, surface temperature history, and individual ply orientations.

EXPERIMENTAL THEORY AND TECHNIQUES

The experimental measurement of residual stresses in a composite can be accomplished through various methods. Two separate but related techniques were employed to gather the data included in this report. Both of these alternatives involve the application of a strain gage to the surface of the composite test sample. Through the successive removal of layers of material from the specimen surface opposite to the strain gage, the residual stress profile can be calculated based upon the composite stiffness and observed change in strain. An equation developed by Lee et. al. [7] relates these changes in strain to the residual stress present at each level of removal.

$$\sigma_r(z) = -\frac{E}{2}\left[(d/2 - z)\frac{\Delta\varepsilon_s(z)}{\Delta z} - 4\varepsilon_s(z) + 6(d/2 - z)\sum_{-d/2}^{z}\frac{\varepsilon_s(\Phi)}{(d/2 - \Phi)^2}\Delta\Phi \right] \quad (2)$$

where σ_r = residual stress, z = distance from the center of mass, Δz = thickness of removed layer, E = elastic modulus, d = laminate thickness, ε_s = absolute strain at the top surface, and $\Delta \varepsilon_s$ = strain difference upon removal of layer.

This formula is based upon the induced beam strains, including a visual curvature toward the surface from which layers are removed, as a result of the onset of a non-uniform stress state. That is, with the removal of material, the stresses across the plate thickness no longer add to zero. The test specimens, all unidirectional to allow accurate strain monitoring, possessed in-plane dimensions of 2-3 cm in width and approximately 7 cm in length. The strain gages were oriented in the direction transverse to the fibers since the individual strains are expected to be more significant than in the longitudinal direction and, therefore, are more accurately measured. This statement does not imply that the stresses along the fibers are similarly less severe since the stiffness in the longitudinal direction is an order of magnitude higher than the transverse stiffness.

The first technique of layer removal available for residual stress measurement is the Process Simulated Laminate (PSL) in which single layers of Kapton film are inserted to facilitate easy separation of adjacent plys. The PSL specimens utilized for previous stress measurements by Manson et. al. [1] consisted of 40 plys with ten 4-ply Constitutive Laminates (CL). This type of sample was employed to measure the residual stress state in test samples processed in a quench-mold with no subsequent heat treatment.

The second layer removal method, used for all tests to provide data for this project, consists of successive grinding of the surface using an end mill and water cooling to prevent heating and any resulting stress relief. By clamping to the milling table to maintain a constant condition of flatness, a uniform specimen thickness could be produced at each grinding increment. This method of testing was used to assess the validity of the PSL technique in the case of quench-mold processing and to determine the effects of various annealing and quenching heat treatments.

The first set of samples were obtained from a 40-ply laminate processed in the quench-mold used in previous PSL studies with an initial surface cooling rate of 35°C/s. With one sample exposed to no treatment, three others were annealed at 300, 177, and 125°C respectively for 30 minutes and slowly cooled to room temperature at a rate of approximately 5°C/min using the press platen internal water-circulation cooling system. Annealing temperatures were selected to include ranges between melting, glass transition, and room temperatures. Since the original quench-mold process causes significant residual stresses, these annealing procedures are expected to partially or fully relieve stresses with duration and temperature of annealing as the important variables.

The second set of samples exposed to heat treatment consisted of 50 plys and were originally processed in an autoclave with nearly isothermal cooling, thereby inducing no residual stresses since non-isothermal temperature distributions are assumed to account for any stress build-up. These stress-free samples were then heated to 300°C for 30 minutes to create uniform crystallinity profiles in each. From this state, four different treatments were employed: quenching to room temperature, slow-cooling to room temperature, slow-cooling to 177°C for 30 minutes followed by quenching to room temperature, and slow-cooling to 125°C for 30 minutes and quenching. Quenching was accomplished through direct immersion in water.

RESULTS AND DISCUSSION

The results of the residual stress tests of quench-mold-processed specimens, using the Process Simulated Laminate configuration, yielded the initial indication of this property in carbon/PEEK composites. Cooled at an initial rate of 35°C/s, the degree of crystallinity measured through differential scanning calorimetry was found to be approximately 28% by volume. As displayed in Fig. 1, the stress curve is parabolic in form with maximum values of 45 MPa in compression at the surface and 20 MPa in tension at the center [1]. Although these particular results were obtained using strain gages and successive peeling of constitutive laminates, similar responses were recorded through grinding of PSL samples with strain gages to assess the effect of the debonding, and through measurement of curvatures and in-plane length changes with a corresponding conversion to stress [2]. Of these three variations of the fundamental technique, the first is optimal in terms of time and ease of performance as well as the possibility of further testing through the semi-nondestructive nature of the test.

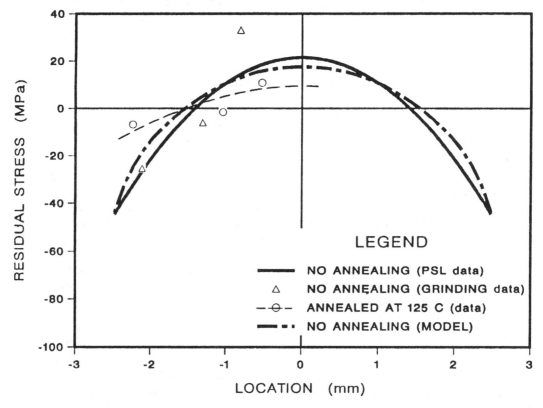

Fig. 1. 90 degree direction residual stress in quench-mold test
samples and model prediction.

A study of these initial results reveals that the tensile stress present at the center of the laminate constitutes 25% of the 90 degree direction ultimate tensile strength of APC-2. Obviously, the quench-mold cooling rate is sufficient to induce very significant stresses, indicating that faster cooling common in other manufacturing methods would likely produce more severe stress states. Since a constant crystallinity across the laminate thickness was observed in DSC tests, it is apparent that morphological skin/core effects are negligible while a thermal skin/core profile, caused by non-isothermal cooling and resulting time-dependent volume changes, is the primary driving force of the residual stress development. As in the case of these PSL tests, all subsequent residual stress measurements using the surface layer grinding method concentrated on the macro scale. However, a complete understanding of internal stresses requires consideration of the micro scale to include stress conditions in the constituent materials as well as the fiber/matrix interface region.

In order to further quantify the PSL stress measurement technique and assess the influence of the Kapton material, the data obtained by milling of quench-mold samples along with direct strain measurement is included in Fig. 1. Comparison to the parabolic response of the PSL indicates that comparable values have been obtained. Although some small variation is apparent, the inherent difficulties encountered in the grinding method (maintaining consistent clamping conditions in each layer removal, possible heat generation, and surface degradation and abrasion) seem to be responsible and limit this technique to qualification of stress conditions. For these reasons as well as the ease and time of experimentation, the Process Simulated Laminate with a strain gage mounted on one surface is presently best for measurement of residual stresses.

Since the quench-mold allows monitoring and recording of surface temperature during cooling, the computer model described previously can be executed with input of material parameters and elastic and viscoelastic behavior of APC-2 to assess the validity of assumptions made concerning residual stress build-up. The output of the model for the 90 degree direction is plotted in Fig. 1 along with data from experiments. With a similar parabolic form and skin/core effects indicated by compression at the surface and tension at the center, the slight deviation of the model predictions from the PSL stress values are attributable to imperfect material parameters provided as input. That is, much of the time- and temperature-dependent properties of the composite, and the matrix in particular, were measured for fully-crystallized material. In reality, the continuous crystallization caused during process cooling results in changing degree of crystallinity and properties in each ply, dependent on the specific cooling rate. Nonetheless, the basic assumptions appear to be applicable and the somewhat limited material property data produce predictions which agree well with experiments. For completeness, the output for the 0 degree direction is presented in Fig. 2. Although no experimental data is available for comparison, the results indicate that the high stiffness provided by the fibers and the absence of stress relaxation causes significant residual stress levels.

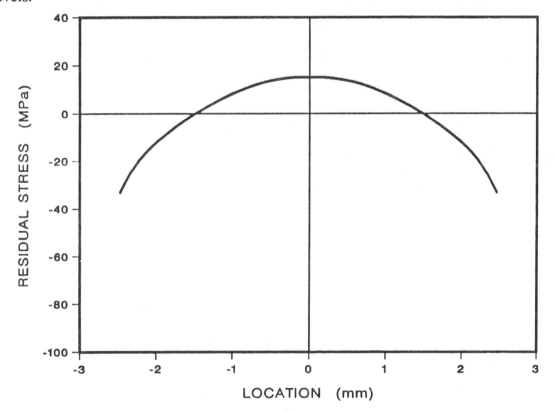

Fig. 2. 0 degree direction residual stress model prediction for
quench-mold test samples.

The tests of quench-mold samples annealed at 177 and 300°C and slowly cooled to room temperature produced strain levels below the accuracy of the strain gages, indicating that no significant stresses were present. Therefore, the conclusion can be drawn that annealing for 30 minutes at these elevated temperatures has allowed sufficient polymer chain reorientation to relieve the residual stresses induced by rapid cooling during processing. Conversely, the sample annealed for 30 minutes at 125°C displayed a stress profile with maximum absolute values below those measured prior to annealing (Fig.1). This fact appears to suggest that this elevated temperature allowed some stress relieving to take place, but longer annealing times would be necessary to produce the effects observed for higher annealing temperatures.

Finally, the autoclave-processed samples which were exposed to various heat treatments provided additional insight into the development of residual stresses. The specimen annealed at 300°C for 30 minutes, cooled to 125°C for 30 minutes, and quenched to room temperature as well as that slowly cooled from 300°C to room temperature showed no significant stress state as indicated by negligible strain measurements. The latter sample served as proof that holding and slowly cooling from 300°C after the previous processing cycle produces no stresses. The other specimen shows that rapid cooling from 125°C, only a portion of the full cooling in a quench-mold processing, causes no significant stresses. In comparison, those samples quenched from 177 and 300°C resulted in very pronounced stress profiles as shown in Fig. 3. The results of these two tests are approximately equal with magnitudes slightly larger in the case of the higher temperature. These results suggest that cooling during a quench-mold processing produces only small stress build-up as the temperature drops from the melting region to just above the glass transition temperature (143°C). Because of previously-listed reservations concerning the surface grinding methods and the small difference in cooling rate between the two quenches, only qualitative conclusions can be made based upon the available data.

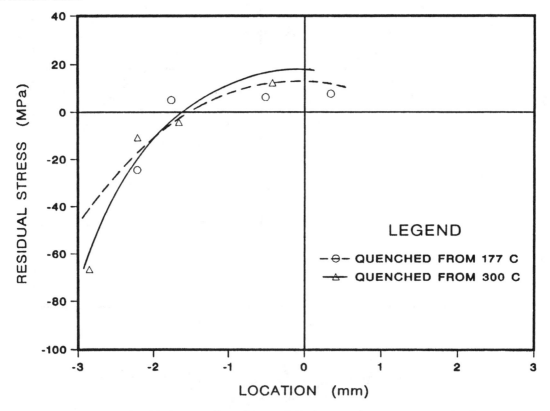

Fig. 3. 90 degree direction residual stress in water-quenched test samples.

The results for these four test samples annealed at 300°C suggest a theory concerning the development of residual stresses during quench-type cooling. Presented in Fig. 4, the data seems to indicate that a majority of stresses are induced in the region of the cooling stage near the glass transition temperature. More specifically, stresses at elevated temperatures are permitted to be relieved with time by means of viscoelastic behavior of the matrix while the approach to room temperature is accompanied by isothermal conditions, thereby producing little stress effects. Obviously, the specific cooling rate determines the range of significant stress build-up since this variable affects viscoelastic response. Based upon known relaxation behavior, faster cooling rates will result in stress development at higher temperatures since less time is provided for relaxation.

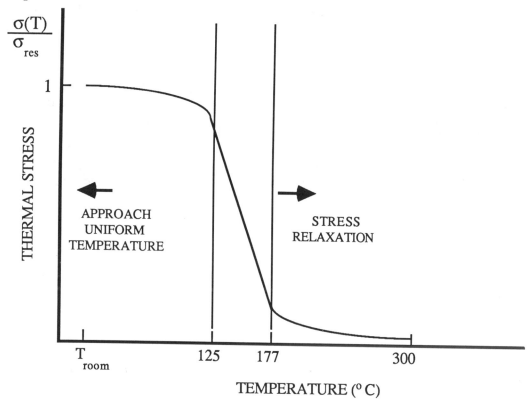

Fig. 4. Residual stress development during quench cooling.

CONCLUSION

Of the numerous classifications of sources and driving forces which produce internal stresses in polymer-based composite materials, an attempt has been made to isolate two in particular: thermal skin/core effects caused by time-dependent volume changes and morphological skin/core stresses resulting from total volume changes. Although both are often encountered in processing, the specific cooling rates studied produced no crystallinity variations and thus negligible morphological skin/core effects in carbon/PEEK.

A parabolic residual stress profile with compressive stresses at the laminate surface and core tension stresses approximately 25% of the ultimate material strength in the 90 degree direction were measured. Annealing of samples above the glass transition temperature provided adequate conditions for complete stress relief while annealing at a lower temperature allowed partial stress relaxation. Quenching of stress-free specimens from various temperature values indicated that most residual stress development during rapid cooling occurs in a temperature range near the glass transition. Dependent upon the cooling rate, stress relaxation is permitted at elevated temperatures while isothermal conditions are approached as the ply temperatures decrease to near ambient. From a combination of this type of testing, a procedure can be developed to eliminate the residual stresses induced by fast-cooling processing techniques such as hot stamping, diaphragm-forming, and laser consolidation.

Two experimental methods of quantifying residual stresses in the 90 degree direction have been presented with the common feature being placement of a strain gage on the laminate surface to observe strain changes as ply layers are removed. These are the Process Simulated Laminate with inserted Kapton film to facilitate debonding and water-cooled grinding of surface layers. While both produce comparable results, the PSL technique appears preferable in terms of ease and time of experimentation as well as its semi-nondestructive nature. Grinding has a few inherent problems, but nonetheless provides an adequate qualitative indication of the material response.

Finally, a finite difference-based FORTRAN computer model has been developed to predict residual stresses in both in-plane directions with required inputs including surface cooling history, ply orientations, constituent elastic properties, heat transfer parameters, and viscoelastic and thermal contraction behavior of the composite. In the case of carbon/PEEK, the degree of crystallinity is an insignificant factor in the mechanical properties. Other materials may exhibit different behavior and should be included in the model. Available output include temperature, crystallinity, mechanical properties, and thermal stresses of each ply as well as solid/melt interface location at any time increment during the cooling of a laminate. Applicable to any continuous-fiber thermoplastic composite material, the model produces good correlation with experimental results for APC-2 in the 90 degree direction and additionally provides predictions for the 0 degree direction.

REFERENCES

1.   Manson, J.-A. E., Copeland, S. D., and Seferis, J. C., "Intrinsic Process Characterization and Scale Up of Advanced Thermoplastic Structures," Proceedings of SPE ANTEC '88, Atlanta, GA, April, 1988.

2.   Manson, J. A., and Seferis, J. C., "Internal Stress Determination by Process Simulated Laminates," Proceedings of SPE ANTEC '87, Los Angeles, CA, May, 1987.

3.   Lee, W. I., Talbott, M. F., Springer, G. S., and Berglund, L. A., "Effects of Cooling Rate on the Crystallinity and Mechanical Properties of Thermoplastic Composites," Advances in Modeling of Composite Processes, 1986.

4.   Velisaris, C. N., and Seferis, J. C., "Crystallization Kinetics of Polyetheretherketone (PEEK) Matrices," Polymer Engineering and Science, Volume 26, Number 22, December, 1986.

5.   Nairn, J. A., and Zoller, P., "Matrix Solidification and the Resulting Residual Thermal Stresses in Composites," Journal of Material Science, 20, 1985.

6.   Vinson, J. R., and Sierakowski, R. L., The Behavior of Structures Composed of Composite Materials, Martinus Nijhoff Publishers, Dordrecht, the Netherlands, 1986.

7.   Lee, E. H., Rogers, T. G., and Woo, T. C., "Residual Stresses in a Glass Plate Cooled Symmetrically from Both Surfaces," Journal of American Ceramic Society, 48(9), 1965.

Modelling of Heat Transfer During the Processing of Thermoplastic Composites

TONY E. SALIBA AND **RONALD A. SERVAIS**
School of Engineering
1350 North Fairfield Rd.
Dayton, OH 45432-2698

DAVID P. ANDERSON
University of Dayton
Research Institute
Dayton, OH 45469

ABSTRACT

Heat transfer and crystallization kinetics during the processing of thermoplastic composite materials are modeled mathematically in order to predict the temperature and degree of crystallinity as a function of position and time. The model can accommodate complex geometries, variable properties of the anisotropic medium as well as a choice of boundary conditions. The model was verified by comparison with experimental results.

INTRODUCTION

The use of thermoplastic composite materials in various industries is rapidly expanding. The attractive properties include a higher service temperature, increased impact resistance (damage tolerance), and improved solvent resistance. In addition, thermoplastic composites adaptability to metal forming techniques and faster processing present a potential for low cost manufacturing. However, several issues need to be resolved including maintaining consistent preprocessing material quality, determining processing conditions, understanding fatigue and creep behavior, and managing the morphology [1]. Process modeling offers a means of evaluating processing conditions and the determination of the effect of process variables on the composite material properties.

Several physical and chemical phenomena take place during an autoclave or press processing of thermoplastic composites. These include heat transfer, resin flow (to a small extent), chemical reactions (for some amorphous polymers), and other physical changes such as crystallization. Several one-dimensional mathematical models describing thermoplastic processing have been developed [2,3]. However, models that can predict heat transfer and crystallization in two-dimensional complex shaped composites were not available. In this paper, a computer model describing the transient two-dimensional heat transfer coupled with the crystallization phenomena is described.

HEAT TRANSFER MODEL

Two-dimensional transient heat transfer in anisotropic composite materials is described by the energy equation:

$$\frac{\partial}{\partial x}\left(k_x \frac{\partial T}{\partial x}\right) + \frac{\partial}{\partial y}\left(k_y \frac{\partial T}{\partial y}\right) + S = \rho\, C\, \frac{\partial T}{\partial t} \qquad (1)$$

where:

k_x is the thermal conductivity in the x-direction (W/m°C)

k_y is the thermal conductivity in the y-direction (W/m°C)

ρ is the density of the composite (kg/m³)

C   is the heat capacity of the composite (J/kg°C)

S   is the heat generation due to exothermic reactions or to crystallization (W/m³)

T   is the temperature (°C)

t   is time (sec)

Three boundary condition options have been explored with the above equation: a temperature specified condition characteristic of a press processing; a convective condition characteristic of an autoclave process; and, an insulated boundary.

CRYSTALLIZATION MODELS

Three crystallization submodels [2,3,4] have been incorporated in the heat transfer code. Springer et al. [2] uses a phenomenological model capable of predicting crystallinity rates only. Velisaris and Seferis's [3] model does provide other morphological data but does not account for changes in the superstructure nucleation density. The time and temperature in the melt can make a dramatic change in bulk nucleation [5] making it an important processing parameter. The primary model [4] derived from Hillier [6] includes this bulk nucleation parameter. In addition, Anderson's model accounts for the temperature dependence of the crystallization rate constant, differentiates between volume fraction and mass fraction crystallinity resulting from the density difference between the various phases, accounts for various aspects of the crystallization mechanism, and has the ability to predict morphology.

SOLUTION TECHNIQUE

The solution technique of the energy equation consists of grid generation, equation discretization, equation assembly, and evaluation. The grid generation technique is based on the method of Thompson et al. [7]. The method consists of specifying the boundary points of the system under study and generating the interior points of the grid by solving the following equation:

$$
\left\{
\begin{array}{l}
\xi_{xx} + \xi_{yy} = P(\xi, \eta) \\[2ex]
\eta_{xx} + \eta_{yy} = Q(\xi, \eta)
\end{array}
\right.
\tag{2}
$$

where $P(\xi,\eta)$ and $Q(\xi,\eta)$ are coordinate control functions used for managing grid points location and density. Solution of the previous equation establishes the correspondence between the coordinates of every points in the physical domain (x,y) and those of its image in the rectangular computational domain (ξ,η).

The energy equation is then transformed from the physical domain to the computational field where it is discretized using an implicit second-order accurate in space and first-order accurate in time finite difference technique.

The resulting difference equation is then applied to every computational node to obtain a system of algebraic equations. The boundary conditions are then implemented and the system of equations is solved using the successive-over-relaxation method to obtain the temperature at every node. The temperature dependent crystallization parameters are evaluated and the degree of crystallinity at every node is calculated. The procedure outlined above is repeated at every time step to obtain the temperature and degree of crystallinity distributions as a function of time. The details of the code development and the solution technique can be found in Reference 8.

RESULTS

Experimentally measured temperatures were used to validate the heat transfer model. Two processing cycles of a 32-ply, 6 inch by 4 inch, ICI's APC-2 (Poly-Ether-Ether-Ketone (PEEK) / carbon fibre composite) panels were run and the temperatures measured at seven locations within the composite panel at 30 second intervals. The measured temperature profiles at various positions in the composite are shown in Figure 1. Thermocouples placed on the panel boundaries provided the temperature profiles used as boundary conditions. The temperature profiles at various interior positions were then predicted and compared with measured temperatures at the same position. Agreement between predicted and measured temperatures was observed at all locations under study for both runs as can be seen from Figure 2 and 3.

Using the described code, heat transfer through complex shaped systems can be modeled. Figure 4 shows the grid generated by the code to have an appropriate grid point distribution. Figure 5 shows the predicted temperatures at various positions in the rib indicating the two-dimensional nature of heat transfer in such a shape.

The degree of crystallinity was also predicted as a function of position and time. Figure 6 shows the effect of cooling rate on the crystallization rate. These predicted degree of crystallinity show agreement with other published predictions [3]. In addition, the final degree of crystallinity for PEEK composite materials is not greatly affected by the processing except for very fast and very slow cooling.

Several computer runs indicated that for the APC-2 thermoplastic system used in this study, the heat generation due to crystallization was negligible compared to the external heat transfer rate. Thus the heat transfer is independent of the crystallization. The opposite is obviously not true since heat transfer and the temperature distribution strongly affect the crystallization mechanism. This result was also reported by Velisaris and Seferis [3].

The morphology of the crystalline superstructure is dependent on the nucleation density which is strongly dependent on the thermal history of the sample during and after melting. Table I shows the crystallization maximum temperature observed in the Differential Scanning Calorimeter (DSC) for APC-2 and PEEK neat resin 380 film samples. This effective crystallization temperature [9] is reflective of the nucleation density; the higher the melt temperature or longer time at any one temperature the lower the nucleation density.

CONCLUSIONS

A computer model describing the transient two-dimensional heat transfer in complex-shaped thermoplastic composite materials was developed. Temperature and degree of crystallinity can be predicted as a function of position and time during a press or autoclave processing of a poly-ether-ether-ketone based composite. The heat transfer model was verified using experimental data generated in the press. The computer model is general, runs on a personal computer, and can be easily adapted to handle other thermoplastic composites.

AKNOWLEDGEMENT

This work was partially supported by the Air Force Wright Aeronautical Laboratories, Materials Laboratory (AFWAL/MLBM) under contract No. F33615-84-C-5070, with Mr. Marvin Knight as Contract Monitor.

REFERENCES

1. Johnston, N. M., and Hergenrother, P. M., "High Performance Thermoplastics: A Review of Neat Resin and Composite Properties", 32nd International SAMPE Symposium, Anaheim, CA, 1987, pp. 1400-1412.

2. Springer, G. S., Burwassser, J., Lee, W., Miller, A., and Talbott, M., "Processing Thermoplastic Composites", Mechanics of Composites Review, Dayton, OH, October 22-24, 1985, pp. 22-25.

3. Velisaris, C. N., and Seferis, J. C., "Crystallization Kinetics of Polyetheretherketone (PEEK) Matrices", Poly. Engr. and Sci., Vol. 26, No. 22, Dec 1986, pp. 1574-1581.

4. Anderson, D. P., "Process Modeling of Thermoplastic Matrix Composites I: Survey of Crystallization Models", AFWAL-TR-86-4099, Air Force Wright Aeronautical Laboratory, WPAFB, OH, 1986.

5. Lee, Y. C., and Porter, R. S., "Crystallization of Poly(ether ether ketone) (PEEK) in Carbon Fiber Composites", Polym. Eng. Sci, Vol. 26, 1986, p. 633.

6. Hillier, I. H., "Modified Avrami Equation for the Bulk Crystallization Kinetics of Spherulitic Polymers", J. Polymer Sci.: Part A, Vol. 3, 1965, pp. 3067-3078.

7. Thompson, J. F., Warsi, Z. U., and Mastin, C. W., Numerical Grid Generation: Foundation and Applications, North Holland, NY, 1985.

8. Saliba, T. E., "Process Modeling of Heat Transfer During the Curing of Advanced Composites", Ph.D. Dissertation, University of Dayton, August, 1986.

9. Fraiser, G. V., Keller, A., and O'Dell, J. A., "The Influence of Nucleation Density and Cooling Rate on Crystallization of Polyethylene from the Melt," J. Appl. Polymer Sci., Vol. 22, 1978, p. 2979.

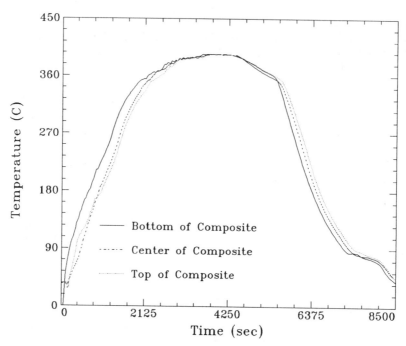

Figure 1: Experimentally Measured Temperatures For Run 2

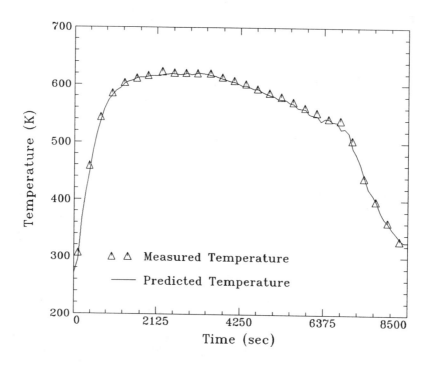

Figure 2: Heat Transfer Code Verification Using Run 1

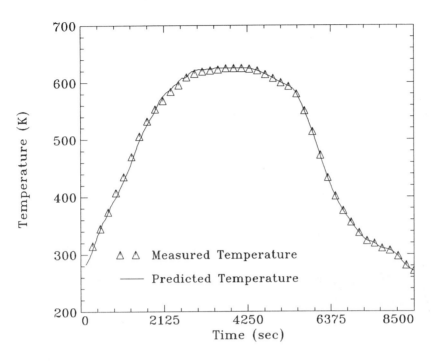

Figure 3: Heat Transfer Code Verification Using Run 2

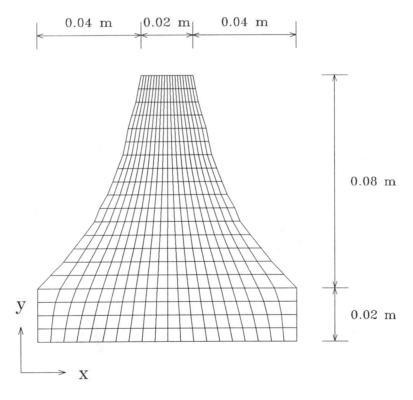

Figure 4: Physical Coordinate System

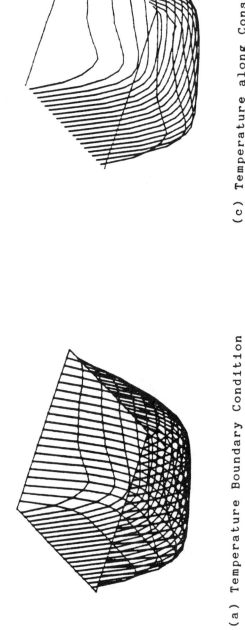

(a) Temperature Boundary Condition

(b) Convection Boundary Condition

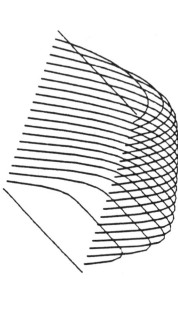

(c) Temperature along Constant x-lines

(d) Temperature along Constant y-lines

Figure 5: Temperature Distribution in a Complex
Shaped Homogeneous Material

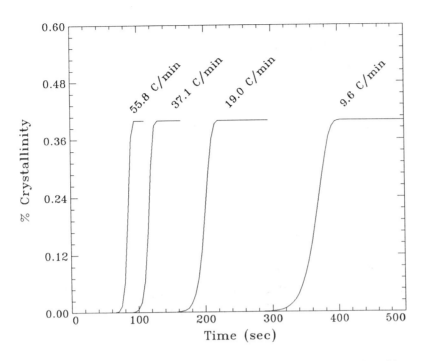

Figure 6: Percent Crystallinity at Various Cooling Rates

Table I: Crystallization Temperature (°C) as a Function of Melt Hold
Time and Temperature

Time at 400°C (min)	APC-2	380 film
5	326	314
15	319	308
45	311	306
120	310	302

Temp for 5 min (°C)	APC-2	380 film
350	311	306
370	310	306
400	308	305
420	307	301

A Model for the Prediction of Ply Stresses in a Composite Laminate During Forming

T. G. GUTOWSKI* AND A. S. TAM*

ABSTRACT

This paper presents an analytical method for determining ply-slip during the deformation forming of composites. A comparison of the model with data shows qualitative and quantitative agreement.

INTRODUCTION

A serious problem which occurs during the deformation forming of composite prepregs is unwanted fiber motion. For example, during the bending of thermoplastic laminates, compressive stresses can build up on the inside radius of the part causing fibers to buckle and wrinkle. Bent parts with this wrinkling have substantially poorer mechanical properties and reduced dimensional accuracy as compared to parts without the wrinkling [1,2]. These stresses can be relieved however, if the composite is allowed to shear and slip between plies and fibers. In some cases this slip is the only mode of deformation available to the composite to accommodate the deformed shape. Therefore, ply slip is an important deformation mode for composites.

ANALYSIS

The slip process can be studied by modeling the composite as a structure with alternating linear elastic, and Newtonian viscous layers. Elastic layers may represent plies or individual fiber layers, whereas the viscous layers represent the resin acting between them.

This problem can be analyzed by applying equilibrium to a small

---

*Timothy G. Gutowski, Associate Professor of Mechanical Engineering and Director of the MIT-Industry Composites and Polymer Processing Program, and Albert S. Tam, Research Assistant, Laboratory for Manufacturing and Productivity, Massachusetts Institute of Technology, Cambridge, MA 02139

element of the elastic layer and then applying the constitutive equations for the two component materials. The problem is best cast using an Eulerian coordinate system which is commonly used for the description of viscous flows. In our case, the application of this coordinate system to the elastic deformation of the fibers is not a problem due to the small strains involved. The result of this procedure is a set of N coupled parabolic partial differential equations in terms of the fiber displacements. Here N is the number of elastic layers. The equations can be decoupled using the singular value decomposition of the system matrix. This leads to a new set of equations using "principal coordinates". Now, if the boundary conditions and initial conditions are also converted to the principal coordinates, the resulting set of equations can be solved in a straightforward manner. This solution procedure does require an assumption however, as to how the imposed deformation produces the initial ply displacements; in our solution we assume the initial displacements are elastic. The obtained solutions in principal coordinates can now be converted back to real coordinates from which the real strains can be calculated. Multiplication of these strains by the individual ply moduli, E_i, will yield the axial stresses in each ply everywhere in the composite.

This yields the solution to a type of stress relaxation problem. For real problems where the imposed deformation is time varying, one may apply the super-position integral. In [3] we have done this for the three-point bending of an arbitrary composite laminate which is deformed at a constant rate. The final solution, which is given in Eq. 1, involves a double summation over the number of plies N, and over an infinite number of Eigenvalues.

$$\sigma_i(x,t) = \frac{3HE_i}{a^3L} \frac{\Delta_f}{t_f} \sum_{j=1}^{N} \sum_{m=0}^{\infty} \frac{S_{ij}Q_j^*\lambda_j[1-\cos(k_m a)]\cos(k_m x)}{k_m^4} \left[1-\exp\left(-\frac{k_m^2 t}{\lambda_j} \right) \right]$$

(1)

The stress $\sigma_i(x,t)$, is the stress in the ith ply which may vary both spatially and as a function of time. Here, "H" is the total composite thickness, "a" is one-half the span length, whereas "L" is one-half the length of the composite. Δ_f/t_f equals the deflection rate. S_{ij} is the transformation matrix made up of the normalized Eigenvectors for the initial system matrix. Q_j^* is a normalizing vector which maps the real strains into the principal coordinate system. λ_j are the Eigenvalues for the original system matrix, and

$$k_m = \frac{2m + 1}{2} \frac{\pi}{L}.$$

When compared with experiments conducted by Soll [1-2], this equation agrees both qualitatively and quantitatively. Fig. 1 illustrates both the short rise time for the stresses during the forming of a right angle part, and the order of magnitude of the induced maximum compressive stresses. For example, when forming parts with a twenty second forming time, Soll found that an applied tension of 95 psi was sufficient to suppress buckling and produce a good part, whereas an applied tension of 10 psi was insufficient to suppress buckling. This agrees very well with the data for the twenty second forming time shown in the Figure.

REFERENCES

1.   Soll, W., "Behavior of Advanced Thermoplastic Composites in Forming", S.M. Thesis, Mech. Eng. Dept., M.I.T., Cambridge, MA, 1986.

2.   Soll, W., and Gutowski, T.G., "Forming Thermoplastic Composite Parts", Proceedings of the 33rd International SAMPE Symposium, Anaheim, CA, March 1988.

3.   Tam, A. S., and Gutowski, T. G., "Ply-Slip During the Forming of Thermoplastic Composite Parts", Submitted to the Journal of Composite Materials, April, 1988.

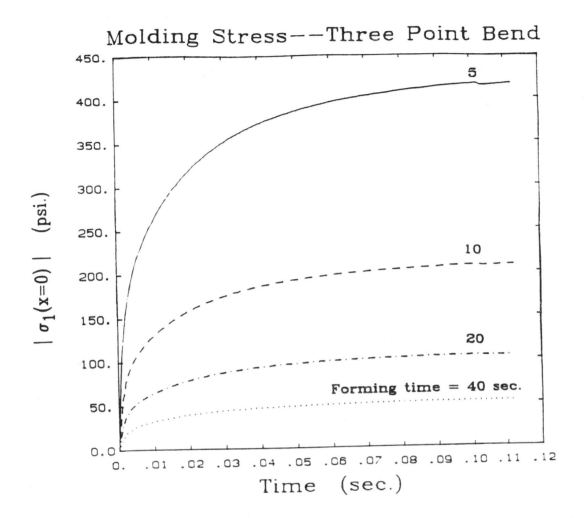

Fig. 1:   Maximum Compressive Stresses During Forming
Thermoplastic Composite Bends, See [3].

The Initiation of Microcracking in Cross-Ply Laminates: A Variational Mechanics Analysis

JOHN A. NAIRN

ABSTRACT

A variational mechanics analysis is used to get an equation for the energy release rate due to the initiation of microcracking in cross-ply laminates. The new theory is compared to previous theories based on the shear-lag model. By comparison to experimental data and by a consideration of the approximations in the various theories, the variational mechanics theory is shown to provide the best analysis of microcrack initiation. An analysis of two sets of graphite/epoxy data yields a microcracking fracture toughness of 89 J/m^2.

INTRODUCTION

Many observations have confirmed that the initiation of damage in cross-ply laminates is by microcracks in the 90° plies running parallel to the fibers [1-12]. Although the formation of microcracks does not precipitate into catastrophic failure, their presence can be very undesirable. For example, the microcracks cause a loss of stiffness [4, 12-15]; stiffness degradation occurs under both tensile and shear loading [14,15]. In the walls of pressure vessels or liquid containers, leaks might develop. In wet or corrosive environments, microcracks provide pathways which can accelerate corrosion [10]. Finally, the microcracks change the thermal expansion coefficients [16,17]; this effect is important in applications relying on dimensional stability.

Several analytical models have been proposed as attempts to predict the initiation of microcracking. Some of these models have been based on lamina strength [2,6,16,17] (e.g. first-ply failure type analysis) while others have been based on fracture mechanics or strain energy release rate [5,7,8,18-20]. The strength approach suffers from a lack of fundamental applicability. As experimentally determined by Flaggs and Kural [11], the stress in the 90° ply at initiation of microcracking in the 90° ply is not a lamina property but depends on the overall laminate construction. The formation of microcracks can therefore not be interpreted in terms of a unique 90° lamina tensile strength. For this reason, the fracture mechanics approach is preferred. In the fracture mechanics approach it is postulated that a new microcrack will form when it is both mechanistically possible and energetically favorable [7]. It has been argued that the mechanistic possibility of forming microcracks is also guaranteed by the presence of microflaws caused by fiber matrix debonds [6,7,9]; these debonds occur at very low strains (0.1%) and provide a stress singularity which can serve as a microcrack nucleation site [7]. Accepting this hypothesis, the fracture problem is reduced to satisfying energetic favorability. In other words, the microcrack will form when the strain energy

John A. Nairn, Assistant Professor of Materials Science and Engineering, University of Utah, Salt Lake City, Utah 84112

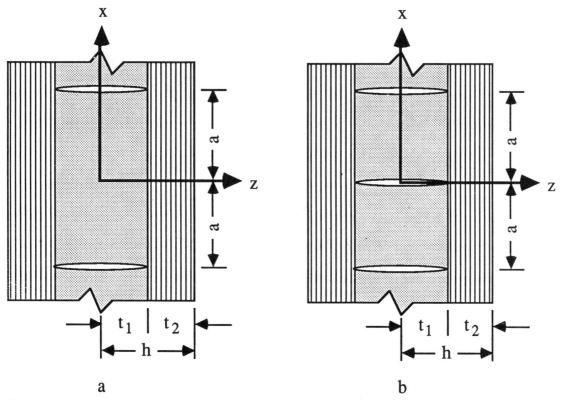

Figure 1: Side view of cross-ply laminates with microcracks. a. Two microcracks in the 90° plies. b. A new microcracked formed midway between two existing microcracks.

released is greater than some critical strain energy release rate called the microcracking fracture toughness. Note, the microcracking fracture toughness is distinct from the more commonly measured delamination fracture toughness.

The energy release rate analyses cited above [5,7,8,18-20] are too qualitative to be accurate. Their have two major problems. First, they all incorporate some form of shear-lag approximation. Second, the incorrectly include the effect of residual thermal stresses. In a recent paper [21], we described a new analysis for the energy release rate due to formation of microcracks. The new analysis uses variational mechanics and can be shown with mathematical rigor to be more accurate than previous attempts at the microcracking problem. The previous paper derived general expressions for the energy release rate as a function of density of microcracks. In this paper, by taking the limit as the microcrack density approaches zero, we consider specifically the energy release rate for the initiation of microcracking. The new microcrack initiation theory is compared to previous theories as well as to experiments.

THERMOELASTIC VARIATIONAL ANALYSIS

In this section, we summarize the thermoelastic variational analysis used in Ref. [21] to get the strain energy release rate for the formation of a microcrack midway between two existing microcracks (See. Fig 1). Consider the cross-ply laminate in Fig. 1a and consider it to be residual stress free at temperature T_0. We begin with an uncracked composite at T_0 having an applied load of σ_0. The load is an axial load applied parallel to the fibers in the 0° plies. By a constant strain assumption, the applied stress is partitioned into the 90° plies (ply 1) and the 0° plies (ply 2) according to:

$$\sigma_{x0}^{(1)} = \frac{E_T}{E_0} \sigma_0 \qquad \text{and} \qquad \sigma_{x0}^{(2)} = \frac{E_A}{E_0} \sigma_0 \qquad (1)$$

Here E_0 is the composite modulus, and E_T and E_A are the transverse and axial moduli of the lamina. Now change the temperature to T_S and form the microcracks illustrated in Fig. 1a These changes will cause the stresses in the lamina to change. Following Hashin [14], we make one and only one assumption: we assume that the x axis tensile stresses in each ply are independent of the z coordinate and only depend on the x coordinate. After applying thermal loads and opening cracks, the stresses in the 90° and 0° plies will change to:

$$\sigma_x^{(1)} = \sigma_{x0}^{(1)} - \psi_1(x) \qquad \text{and} \qquad \sigma_x^{(2)} = \sigma_{x0}^{(2)} - \psi_2(x) \qquad (2)$$

Inserting the stresses in Eq. (2) into the stress equilibrium equations, using the appropriate boundary conditions, and requiring force balance in the axial directions, it is possible to express $\sigma_x^{(1)}$, $\sigma_{xz}^{(1)}$, $\sigma_z^{(1)}$, $\sigma_x^{(2)}$, $\sigma_{xz}^{(2)}$, and $\sigma_z^{(2)}$ in terms of $\psi(x) = \psi_1(x)$ [14].

$$\sigma_x^{(1)} = \sigma_{x0}^{(1)} - \psi(x) \qquad \sigma_{xz}^{(1)} = \psi'(x)z \qquad \sigma_z^{(1)} = \frac{1}{2}\psi''(x)(ht_1 - z^2) \qquad (3)$$

$$\sigma_x^{(2)} = \sigma_{x0}^{(2)} + \frac{t_1}{t_2}\psi(x) \qquad \sigma_{xz}^{(2)} = \frac{t_1}{t_2}\psi'(x)(h-z) \qquad \sigma_z^{(2)} = \frac{t_1}{2t_2}\psi''(x)(h-z)^2 \qquad (4)$$

These stresses constitute an admissible stress state; they obey all stress equilibrium conditions, traction boundary conditions, and interface continuity. By the principal of minimum complementary energy, the function $\psi(x)$ which produces the minimum value for the complementary energy will give the best approximation to the composite strain energy. The function $\psi(x)$ is found by plugging the admissible stress state into the complementary energy function and finding the minimum by using the calculus of variations. For thermoelastic analyses, a variational solution based on the principle of minimum complementary energy is found by minimizing the function Γ given by

$$\Gamma = \frac{1}{2}\int_V \sigma K \sigma \, dV \quad + \quad \int_V \sigma \alpha T \, dS \quad - \quad \int_{S_1} \sigma \hat{u} \, dS \qquad (5)$$

where σ is the stress tensor, $\mathbf{K}$ is the compliance tensor, α is the thermal expansion coefficient tensor, T is $T_S - T_0$, V is the volume of the composite, and S_1 is that part of the composite surface subjected to fixed displacement of $\hat{u}$ [22]. In this problem S_1 is null; i.e. there are no displacement boundary conditions.

The thermoelastic variational problem was solved in Ref. [21]. The final result for $\psi(x)$ is:

$$\psi(x) = (\sigma_{x0}^{(1)} - \frac{\Delta \alpha T}{C_1})\phi(x) + \frac{\Delta \alpha T}{C_1} \qquad (6)$$

where $\Delta\alpha = \alpha_T - \alpha_L$ (difference between transverse and longitudinal thermal expansion coefficients) and ϕ is

$$\phi(x) = \frac{2(\beta \sinh \alpha\rho \cos \beta\rho + \alpha \cosh \alpha\rho \sin \beta\rho)}{\beta \sinh 2\alpha\rho + \alpha \sin 2\beta\rho} \cosh \alpha\xi \cos \beta\xi$$

$$+ \frac{2(\beta \cosh \alpha\rho \sin \beta\rho - \alpha \sinh \alpha\rho \cos \beta\rho)}{\beta \sinh 2\alpha\rho + \alpha \sin 2\beta\rho} \sinh \alpha\xi \sin \beta\xi \qquad (7)$$

where $\rho = a/t_1$ and $\xi = x/t_1$. The α and β terms in Eq. (7) are:

$$\alpha = q^{1/4}\cos \frac{\theta}{2} \quad , \qquad \beta = q^{1/4}\sin \frac{\theta}{2} \quad , \quad \text{and} \quad \tan \theta = \sqrt{\frac{4q}{p^2} - 1} \qquad (8)$$

where $p = (C_2 - C_4)/C_3$ and $q = C_1/C_3$. The constants C_1 to C_4 are:

$$C_1 = \frac{hE_0}{t_2 E_A E_T} \qquad\qquad C_2 = \frac{\nu_T}{E_T}(\lambda + \frac{2}{3}) - \frac{\nu_A}{E_A}\frac{\lambda}{3} \qquad (9)$$

$$C_3 = \frac{\lambda+1}{60E_T}(3\lambda^2 + 12\lambda + 8) \qquad\qquad C_4 = \frac{1}{3}(\frac{1}{G_T} + \frac{\lambda}{G_A}) \qquad (10)$$

In Eqs. (9) and (10), G_A, G_T, ν_A, and ν_T are the axial and transverse shear moduli and Poisson's ratios respectively, and $\lambda = t_2/t_1$. Note that this solution is specific for $4q/p^2$ greater than 1. This inequality holds for most composites. The solution for $4q/p^2$ less than 1 is given in Ref. [21].

Substitution of $\psi(x)$ defined in Eq. (6) into Eqs. (3) and (4) completely defines the stress state including the residual thermal stresses. The residual thermal stresses alone can be found be setting the applied stress σ_0 to zero. Substituting the stress state into the strain energy equation:

$$U = \frac{1}{2}\int_V \sigma \, \mathbf{K} \, \sigma \, d \, V \qquad (11)$$

allows us to calculate the strain energy between any pair of microcracks in a cracked cross-ply laminate. It then is a straightforward problem to find the total amount of strain energy released on forming a new microcrack, or in going form the state in Fig 1a to the state in Fig. 1b. The result from Ref. [21] is:

$$G_m = (\sigma_0^2 \frac{E_T^2}{E_0^2} + \frac{\Delta\alpha^2 T^2}{C_1^2}) t_1 C_3 (2\chi(\rho/2) - \chi(\rho)) \qquad (12)$$

where

$$\chi(\rho) = 2\alpha\beta(\alpha^2 + \beta^2) \frac{\cosh 2\alpha\rho - \cos 2\beta\rho}{\beta \sinh 2\alpha\rho + \alpha \sin 2\beta\rho} \qquad (13)$$

In Eq. (12), the term $\sigma_0 E_T/E_0$ is just the stress in the 90° plies - $E_T \epsilon_0$. From Eqs. (3) and (6) and the fact that $\phi(x)$ approaches zero far from the microcracks, the term $\Delta\alpha T/C_1$ is the initial thermal stress in 90° plies of the uncracked laminate - $E_T \epsilon_{th,0}$. Eq. (12) can then be rewritten as:

$$G_m = (\epsilon_0^2 + \epsilon_{th,0}^2) E_T^2 t_1 C_3 (2\chi(\rho/2) - \chi(\rho)) \qquad (14)$$

Eq. (14) gives the energy release rate for the formation of a new microcrack midway between two existing microcracks. To get the energy release rate for the formation of the first microcrack, we take the limit as the microcrack spacing, ρ, approaches infinity. Simple evaluation reveals

$$\lim_{\rho \to \infty} (2\chi(\rho/2) - \chi(\rho)) = 2\alpha(\alpha^2 + \beta^2) = 2\alpha\sqrt{\frac{C_1}{C_3}} \tag{15}$$

and the energy release rate reduces to

$$G_m = (\varepsilon_0^2 + \varepsilon_{th,0}^2) E_T^2 \, 2t_1 \alpha \sqrt{C_1 C_3} \tag{16}$$

This expression can be simplified further under the assumption that p<0. Although p will not be less than zero for the general laminate, it does hold for graphite/epoxy and glass/epoxy laminates. In general p should be less than zero for any cross-ply laminate made with high modulus fibers. Substituting the expression for α and simplifying results in the final variational mechanics energy release rate equation for the initiation of microcracking:

$$G_m = (\varepsilon_0^2 + \varepsilon_{th,0}^2) E_T^2 t_1 \sqrt{\frac{1+\lambda}{\lambda} \frac{E_0}{E_A E_T C_V}} \tag{17}$$

where

$$C_V = \frac{1}{C_4 - C_2 + 2\sqrt{C_1 C_3}} \tag{18}$$

COMPARISON WITH SHEAR-LAG THEORIES

Bailey et. al. [8], Han et. al. [19], and Laws [20] have used the shear-lag approximation to analyze microcracking in cross-ply laminates. Putting their resulting expression for the energy release rate due to the formation of the first microcrack into the nomenclature of this paper, their equations are similar and are as follows: Bailey's result is (Eq. (21) of Ref. [8]):

$$G_m = (\varepsilon + \varepsilon_{th,0})^2 E_T^2 t_1 \sqrt{\frac{1+\lambda}{\lambda} \frac{E_0}{E_A E_T G_T}} \tag{19}$$

Han's result is (Eq. (19) of Ref. [19]):

$$G_m = (\varepsilon + \varepsilon_{th,0})^2 E_T^2 t_1 \sqrt{\frac{1+\lambda}{\lambda} \frac{E_0}{3E_A E_T G_T}} \tag{20}$$

Laws result is (Eq. (20) of preprint for Ref. [20]):

$$G_m = (\varepsilon + \varepsilon_{th,0})^2 E_T^2 t_1 \sqrt{\frac{1+\lambda}{\lambda} \frac{E_0}{Kt_1 E_A E_T}} \tag{21}$$

Comparing Eqs. (19)-(21) to the variational mechanics result in Eq. (17), we see two major differences. First is the incorporation or residual thermal strains. In the variational mechanics solution, the energy release rate is proportional to the sum of the squares of the the applied strain and the thermal strain, In the shear-lag theories, it is proportional to the square of the total strain. The difference is that the variational analysis is a true thermoelastic analysis and handles the residual thermal stresses correctly, while the shear-lag model does not. In the variational analysis, the residual thermal stresses are explicitly calculated. As described above, an initially uncracked composite is subjected to the formation of cracks and a change in temperature. The stress analysis gives both the appliedstresses and the residual

Table I: Lamina properties used in the calculations in this paper. Fiberite T300/934 data is from Ref. [7] and the graphite/epoxy data is from Ref. [11]. The properties marked with * are not provided in those references and have been estimated by comparison to typical lamina properties.

Property	Fiberite T300/934	Graphite/Epoxy
Axial Modulus (GPa)	137.9	127.0
Transverse Modulus (GPa)	11.7	8.3
Axial Shear Modulus (GPa)	4.55	4.0*
Transverse Shear Modulus (GPa)	6.34*	6.0*
Axial Poisson's Ratio	0.29	0.29
Transverse Poisson's Ratio	0.40*	0.40*
Axial Thermal Expansion Coef (ppm/°C)	0.09	-0.3*
Transverse Thermal Expansion Coef (ppm/°C)	28.2	28.0
Effective $T_S - T_0$ (°C)	-65.0*	-75.0*

thermal stresses. By setting σ_0 to zero in the stress state in Eqs. (3) and (4), the resulting stress state is the residual thermal stresses in the cracked composite. In the presence of cracks there are axial, transverse, and shear residual thermal stresses. The distribution of these thermal stresses changes as new microcracks are formed and the thermal strain energy release must be accounted for. The variational approach explicitly accounts for the release of thermal strain energy.

In contrast, the existing-lag analyses are not thermoelastic analyses and therefore do not explicitly calculate the residual thermal stresses. Instead a residual thermal stress term is added to the shear-lag solution. This approach, however, is not accurate. One reason is that the shear-lag model is a one-dimensional analysis, and as discussed above, there are transverse and shear thermal stresses in the microcracked laminate. By ignoring some of the residual thermal stresses, the shear-lag model apparently does not correctly account for the release of thermal strain energy upon creating a microcrack.

The second difference between Eq. (17) and Eqs. (19) to (21) is the term under the square root. Each equation is identical except for one term. Where the variational analysis has the term $1/C_V$, Bailey's analysis has $1/G_T$, Han's analysis has $1/3G_T$, and Laws analysis has $1/Kt_1$. G_T is the shear modulus of the 90° plies and K is a shear-lag parameter. K is a material property and gives the shear-stiffness of a hypothetical shear-stress transfer layer of undetermined thickness between the 0° and 90° plies. Unfortunately, K cannot be determined from lamina properties and must be treated as an adjustable parameter and determined from laminate data such as microcracking data [20].

To see the effects of the different terms under the square root, we plot $1/C_V$, $1/G_T$, $1/3G_T$, and $1/Kt_1$ (the "shear" terms) versus number of 90° plies in $(0_2/90_n)_s$ as a function of n in Fig. 2. The plot is for Fiberite T300/934 graphite epoxy laminates and the lamina properties are listed in Table I. The "shear" terms for the four theories are different and in particular have a different dependence on the thickness of the 90° plies. The "shear" term in the Bailey and Han theories are independent the thickness of the 90° plies. In the Laws and the variational theories the "shear" terms decrease as a function of the 90° plies thickness. To plot the Laws "shear" term, we had to know K. Because K is an unknown adjustable parameter, we picked K to make the variational theory and the Laws theory agree when n=1. For small n (less than 10) a K could be picked which would make the Laws theory similar to the variational theory. We will therefore expect similar predictions for the two theories at small n unless the discrepancies in handling the residual thermal stresses become important.

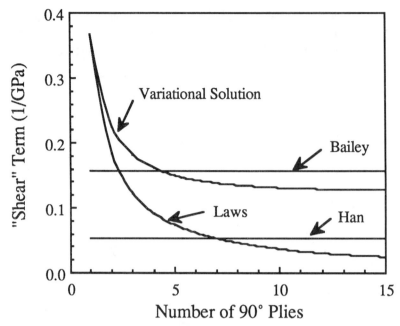

Figure 2: Plot of the "shear" terms as a function of n in $(0_2/90_n)_s$ laminates. The "shear" term in the variational solution is $1/C_V$, in the Bailey solution is $1/G_T$, in the Laws solution is $1/Kt_1$, and in the Han solution is $1/3G_T$.

COMPARISON WITH EXPERIMENT

A useful microcracking theory should be able to predict the strain to formation of the first microcrack as a function a laminate structure. We compare the above theories to two sets of literature data on cross-ply graphite/epoxy laminates. The first set of data is for graphite/epoxy laminates of the type $(0_4/90_n/0_4)$ made from commercially available (but unspecified) pre-preg [7]. The second set of data is on Fiberite T300/934 graphite/epoxy laminates of the form $(0_2/90_n)_s$ [11]. The lamina properties for these two materials are given in Table I. Some of the lamina properties were supplied in Refs [7] and [11]. Unknown lamina properties were estimated by comparing to typical graphite/epoxy lamina properties.

Solving Eq. (17) and Eqs. (19)-(21) for applied strain, gives an expression for failure strain as a function of laminate properties, residual thermal strains, and a critical energy release rate or a microcracking fracture toughness - G_{mc}. These equations can be used to analyze experimental data by finding the constant G_{mc} that best fits the data. The experimental data of Ref. [7] along with least squares fits to each of the four theories is given in Fig. 3. The Bailey and Han theories give identical fits because of their similar form. The Bailey/Han fit is significantly worse than the other two fits. The Bailey fit yields a $G_{mc} = 59$ J/m^2, and the Han fit yields $G_{mc} = 34$ J/m^2. The Laws theory and the variational theory are much better and in fact predict the experimental data well. the variational theory yields $G_{mc} = 88$ J/m^2. The Laws theory cannot give a fracture toughness because of the unknown parameter K. The fit using Laws equation can only give $G_{mc}\sqrt{K}$.

The fits in Fig. 3 required a knowledge of the initial residual thermal strains in the 90° plies. By the variational analysis, the thermal strain is

$$\varepsilon_{th,0} = \frac{\Delta\alpha T}{E_T C_1} \tag{22}$$

If we know T, we can get the thermal strain for use in the data analysis. The analysis for Eq. (22) assumed temperature independent properties of the lamina. When moduli decrease at

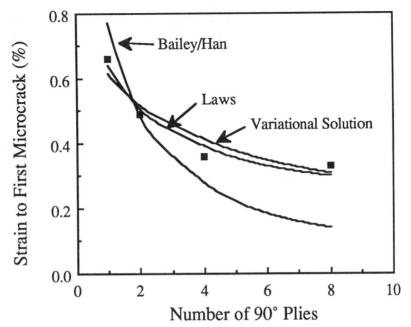

Figure 3: Plot of the strain to initiation of microcracking in $(0_4/90_n/0_4)$ graphite/epoxy laminates as a function of the number of 90° plies (n). The data is from Ref. [7]. The fits are from the indicated energy release rate analyses.

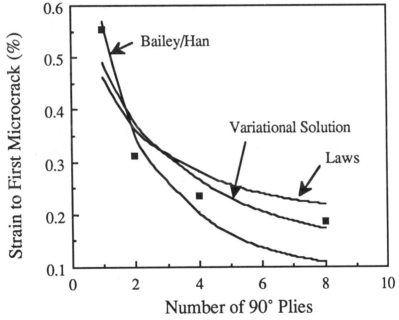

Figure 4: Plot of the strain to initiation of microcracking in $(0_2/90_n)_s$ fiberite T300/934 laminates as a function of the number of 90° plies (n). The data is from Ref. [11]. The fits are from the indicated energy release rate analyses.

elevated temperature, Eq. (22) can be expected to overestimate the residual thermal strains. Instead of picking T from a stress free-temperature, we pick an effective T. The effective T is picked such that the energy release rate at failure by Eq. (17) is closest to constant. The effective T's are given in Table I.

The experimental data from Ref. [11] along with least squares fits to the four theories is in Fig. 4. Again the Bailey and Han theories give relatively poor fits. The Bailey fit yields G_{mc} = 120 J/m^2 and the Han fit yields G_{mc} = 69 J/m^2. The Laws and the variational analysis again give good fits. To the extent than that the variational analysis predicts a steeper dependence on the number of 90° plies, it provides a better fit than the Laws fit. The variational analysis yields G_{mc} = 89 J/m^2. Again the Laws fit only gives $G_{mc}\sqrt{K}$ and therefore does not specify G_{mc}.

We might also compare these four theories to initiation of failure as predicted by laminated plate theory using a first ply failure analysis. First ply failure analysis, however, is not a useful fracture criterion and its predictions in cross-ply laminates are totally at odds with experimental observations. By a set a simple laminated plate calculations, it is easy to show that all the stresses in the 90° plies for a given level of applied strain increase as the number of 90° plies decreases. A first ply failure analysis would therefore predict that the strain to first microcrack would increase as the number of 90° plies increases. This prediction is the opposite of the experimental observation. We note that the fundamental problems with first-ply failure analysis cannot be remedied by selecting elaborate lamina failure criteria such as quadratic failure criterion, rather than simple maximum stress or maximum strain failure criteria. By any rational failure criterion, first-ply failure analyses are useless in cross-ply laminates.

DISCUSSION

From the results in Figs. 3 and 4, the Bailey and Han analyses give relatively the poorest fits. Although these theories provide fracture toughness of reasonable magnitude (34-120 J/m^2), they probably do not give an accurate critical energy release rate. The Laws analysis and the variational analysis both give good fits to the experimental data. Although the Laws fits and the variational analysis fits are different, it is not possible to pick the better analysis by a comparison with experimental data. We must consider other factors.

First consider the approximations made in the two theories. In the variational analysis there is only one assumption - that the axial stresses in the plies are independent of z. The Laws shear-lag analysis make this same assumption along with several other assumptions. The major additional assumption in the Laws analysis is the shear-lag approximation. By the shear-lag approximation, the shear stresses in the plies are ignored and all shear strain is assumed to be in a hypothetical shear stress transfer region between the 0° and 90° plies. This approximation is good only when the stiffness of the plies is much greater than the stiffness of the shear stress-transfer region. Because the shear-stress transfer region must be matrix material and because the stiffness of the 90° plies is not much greater than the stiffness of the matrix, the shear-lag approximation cannot be expected to be quantitatively accurate in cross-ply laminates.

Because stress-states in the Laws analysis has axial stresses that are independent of z, they can be said to be a subset of the admissible stress states used in the variational analysis (Eqs. (3) and (4)). By the theorem of minimum complementary energy [22], the strain energy found by the variational analysis is a minimum and is the closest to the true strain energy of all the admissible stress states under consideration. In other words, it is a mathematically rigorous statement to say that the strain energy found by the variational analysis is closer to the true strain energy than the strain energy found by the Laws shear-lag analysis. Logically, the strain energy release rate found by the variational analysis will also be closer to the true strain energy release rate than the shear-lag analysis.

Finally, there is a significant difference in the way the shear-lag theories and the variational analysis handle residual thermal stresses. By the comments in the preceding paragraphs, we expect that the variational analysis provides the correct incorporation of residual thermal stresses. For the temperature independent data analyzed in this paper, the effect of mishandling residual thermal stresses was not apparent. For interpreting temperature dependent data, such as microcracking caused by thermal cycling [23], the variational analysis will probably be preferred.

In conclusion, we expect that Eq. (17) gives the most accurate energy release rate for the formation of the first microcrack in cross-ply laminates. If the thermal strain is known or can be estimated, than a measure of the strain to the first microcrack can be used to measure the critical microcracking energy release rate in any composite. The measured critical energy release rate, G_{mc}, has the physical interpretation as the microcracking fracture toughness. Because the microcracks are contained within the 90° plies, the microcracking fracture toughness can be called the intralaminar fracture toughness. The microcracking fracture toughness is distinct from the interlaminar fracture toughness and provides new and probably important composite fracture characterization information. From the data analyzed in this paper, the microcracking fracture toughness in graphite/epoxy composites is 89 J/m^2.

ACKNOWLEDGEMENTS

This work was supported in part by a gift from the Materials Academic Board of E. I. duPont deNemours & Co., monitored by Dr. Alan R. Wedgewood

REFERENCES

1. H. T. Hahn and S. W. Tsai, J. Comp. Mat., 8, 288 (1974).
2. K. W. Garrett and J. E. Bailey, 12, 157 (1977).
3. K. W. Garrett and J. E. Bailey, 12, 2189 (1977).
4. A. L. Highsmith and K. L. Reifsnider, ASTM STP 775, 103 (1977).
5. A. Parvizi, K. W. Garrett, and J.E. Bailey, J. Mat Sci, 13, 195 (1978).
6. A. Parvizi and J. E. Bailey, J. Mat. Sci., 13, 2131 (1978).
7. M. G. Bader, J. E. Bailey, P. T. Curtis, and A. Parvizi, Proc. 3rd Int'l Conf. on Mech. Behavior of Materials, 3, 227 (1979).
8. J. E. Bailey, P. T. Curtis, and A. Parvizi, Proc. Roy. Soc. London, A366, 599 (1979).
9. J. E. Bailey, and A. Parvizi, J. Mat. Sci., 16, 649 (1981).
10. F. R. Jones, A. R. Wheatley, and J. E. Bailey, in Composite Structures, p415 (Ed. I. H. Marshall, Appl. Sci. Publ., Barking, England, 1981).
11. D. L Flaggs and M. H. Kural, J. Comp. Mat., 16, 103 (1982).
12. S. E. Groves, C. E. Harris, A. L. Highsmith, D. H. Allen, and R. G. Norvell, Exp. Mech, March, 1987, p73.
13. R. Talreja, J. Comp. Mat., 19, 355 (1985).
14. Z. Hashin, Mechanics of Materials, 4, 121 (1985).
15. Z. Hashin, Eng. Fract. Mech., 25, 771 (1986).
16. P. W. Manders, T. W. Chou, F. R. Jones, and J. W. Rock, J. Mat. Sci., 18, 2876 (1983).
17. H. Fukunaga, T. W. Chou, P. W. M. Peters, and K. Schulte, J. Comp. Mat., 18, 339 (1984).
18. D. L. Flaggs, J. Comp. Mat., 19, 29 (1985).
19. Y. M. Han, H. T. Hahn, and R. B. Croman, Proc. of the Amer. Soc. Composites, Second Technical Conf., Newark, DE, September 23-25, 1987.
20. N. Laws, J. Comp. Mat., in press (1988).
21. J. A. Nairn, J. Comp. Mat, in press (1988).
22. D. E. Carlson, in Mechanics of Solids: Volume II, p325 (Ed. C. Truesdell, Springer-Verlag, Berlin, 1984).
23. C. T. Herakovich and M. W. Hyer, Eng. Fract. Mech., **25**, 779 (1986).

SYMPOSIUM XI

Composite Material Design and Characterization

Design of Laminated Composite with Controlled-Damage Concept

C. T. SUN* AND T. L. NORMAN**

ABSTRACT

Graphite/epoxy laminates with adhesive strips at the 0/90 interface are investigated. Specimens were impacted under approximately fixed-fixed boundary conditions. Comparisons were made between the specimens with and without the adhesive from X-radiographs. Delamination plotted against velocity shows that specimens with adhesive have a substantial reduction in delamination area as compared to specimens without adhesive. It was observed that below a certain velocity the adhesive acts as a softening strip which confines the delamination to the area of the mesh formed by the adhesive. Bending residual strength tests indicate that the specimens without adhesive experience continuous reduction in strength whereas the strength for the specimens with adhesive remains nearly constant, until reaching a certain threshold velocity where there is a sudden decrease. Damage mechanisms were examined through the use of microphotographs.

INTRODUCTION

Delamination is a failure mode commonly found in graphite/epoxy laminated composite materials subjected to impact loading [1-2]. This failure mode occurs at the lamina interface and can propagate, affecting the structural integrity. Although delamination can lead to widespread damage within the structure, it contributes to absorbing impact energy. If delamination is suppressed, impact energy could result in fiber breakage, which would have a greater effect on laminate strength.

One method being considered for interface toughening is the use of adhesive layers. By placing tough adhesive films along the interfaces of the laminas, Chan et al [3] demonstrated that edge delamination in coupon laminated specimens subjected to in-plane tension could be suppressed. Sun and Rechak [4] employed a similar method to delay the initiation of and reduce the extent of delamination induced by impact loading. In [4], the mechanisms provided by the presence of interlaminar adhesive layers in reducing impact damage was studied. It was found that adhesive layers included in a composite laminate could increase the contact area and, as a result, reduce the transverse shear stress concentration effect. This, together with the toughened interfacial property, resulted in less delamination. In a follow-on paper, Rechak and Sun [5] developed a guide for optimal use of adhesive layers to optimize the use of adhesives.

*Professor and **Graduate Student, School of Aeronautics and Astronautics, Purdue University, West Lafayette, Indiana 47907.

Although adhesive layers increase the interfacial toughness, they also add weight and reduce stiffnesses and compressive strength. Moreover, the suppression of delamination may result in massive fiber breakage when impact velocity exceeds a certain threshold velocity.

The goal of this research is to develop a new concept of laminate design which will allow delamination to occur in the region of impact, but will keep it from propagating excessively. This is done by the use of adhesive strips in the graphite/epoxy laminate. This paper consists of delamination investigations and bending residual strength tests of graphite/epoxy laminated composites with adhesive strips subjected to transverse impact loading. Results are compared with graphite/epoxy laminated composites without adhesive.

LAMINATE DESIGN

Figure 1 shows the controlled-damage design concept for the graphite/epoxy laminate with adhesive strips for a [0/90/0] laminate. As seen in the figure, the outer ply of each lamina contains the FM-1000 adhesive and AS4 graphite/epoxy strips. The ply containing the adhesive is formed by alternating graphite/epoxy with adhesive strips, the width of the strips depending on the particular design chosen. When 0 and 90 degree lamina are placed together, the adhesive strips form a square mesh arrangement at the interface.

Two different controlled-damage laminate designs were used, one containing 2.54 cm (1.0 in) width graphite/epoxy alternating with 1.27 cm (0.5 in) width adhesive strips, and the other containing 1.27 cm width graphite/epoxy alternating with 0.635 cm (0.25 in) width adhesive strips on the outermost layer. Laminates with 1.27 cm width adhesive strips have a layup geometry of $[90/A/0_a/0/0_a/90_a/90_3 /90_a/0_a/0/0_a/A/90]$ where subscript "a" denotes adhesive strips present in the ply and A indicates the use of a full adhesive ply. Laminates with 0.635 cm wide adhesive strips have a layup geometry of $[0_a/0/0_a/90_a/90/90_a/0_a/0/0_a/90_a/90/90_a/0_a/0/0_a]$. Plain graphite/epoxy panels with layup geometries of $[90_2/0_3/90_5/0_3/90_2]$ and $[0_3/90_3/0_3/90_3/0_3]$ were fabricated for comparison with the controlled-damage laminates with similar layup geometries listed previously.

EXPERIMENTAL PROCEDURE

Two controlled-damage composite laminates, one with 0.635 cm and the other with 1.27 cm width adhesive strips, were cut into four 7.62 cm (3.0 in) width specimens. The lengths of the specimens correspond to a span of 10.16 or 15.24 cm (4 or 6 in). In addition, a controlled-damage laminate with 0.635 cm width adhesive strips was cut into 3.81 cm (1.5 in) width specimens.

Each specimen was tested under approximately fixed-fixed boundary conditions with 10.16 or 15.24 cm span. A 2.22 cm (0.875 in) diameter steel ball was used as the impactor. Damage was assessed through X-radiographs for delamination and microphotographs of the polished cross-section for damage mechanisms. To aid in the dispersion of penetrant throughout the multi-layered laminate, a small hole was drilled at the impact site after impact, where penetrant was injected with a syringe. The 3.81 cm width specimens where subjected to three point bending test after impact for residual strength.

RESULTS

In this study, delamination control and residual strength of laminated composites with adhesive strips due to impact loading have been investigated.

For all impact cases, the ball was directed to the center of the specimen, so that the impact site would be in the center of the square mesh formed by the adhesive strips at the lamina interfaces. For the specimens with 0.635 cm adhesive strips, multiple adhesive squares encompass the impact site. Figures 2 and 3 show delamination area plotted against velocity. Residual strength is plotted against velocity in Fig. 4. These figures represent impact cases discussed below.

Delamination

Figures 5 and 6 show the progression of damage for the 3.81 cm width plain and controlled-damage specimens with layup geometry of $[0_3/90_3/0_3/90_3/0_3]$ and $[0_a/0/0_a/90_a/90/90_a/0_a/0/0_a/\ 90_a/90/90_a/0_a/0/0_a]$, respectively. Adhesive strips used in the controlled-damage specimens were of 0.635 cm width. Impact velocities for these specimens ranged from approximately 14.8-20.0 m/s (581-789 in/s) and specimens were subjected to a 15.24 cm span. As can be seen from the X-radiographs, delamination is much more severe in the specimens without the adhesive. For the impact cases shown in Fig. 5, we find that the delamination area for the controlled-damage specimens is contained within the 1.27 cm square mesh formed by the adhesive strips while delamination is much more extensive in the plain specimens. For the impact cases of Fig. 6, delamination in the plain specimens extends to the edges of the specimens. At these velocities, the delamination in the controlled-damage specimens was contained to a region at the center, with the exception of the impact velocity of 19.8 m/s (780 in/s). At this velocity, a bending crack formed on the side opposite impact has cracked into the laminate and initiated delamination at the 0/90 degree interface across width of the specimen. Also seen in Figs. 5 and 6 are matrix cracks along the fiber direction on surface ply for both plain and controlled-damage specimens. These will be discussed in detail later. It should also be observed that transverse shear cracking is much more severe in the plain specimens than in the controlled-damage specimens as seen by the transverse lines in the X-radiographs.

Figure 2 shows the delamination area plotted against velocity for the above configuration. In the figure, the dashed line indicates the area encompassed by the adhesive mesh around the impact site. This area is 1.61 cm^2 (0.25 in^2) corresponding to a mesh size of 1.27×1.27 cm (0.5×0.5 in). The delamination area for most of the cases of the controlled-damage specimens is below the dashed line indicating that the delamination is contained to within the adhesive mesh around the impact site. For the case of impact at 19.8 m/s, as observed in Fig. 6, delamination is quite extensive. The delamination area for all cases of the plain specimens is quite extensive, resulting in areas much higher than for the controlled-damaged specimens at approximately the same velocity.

Figures 7 and 8 show the progression of damage in the plain and controlled-damage specimens with layup geometry of $[90_2/0_3/90_5/0_3/90_2]$ and $[90/A/0_a/0/0_a/90_a/90_3/90_a/0_a/0/0_a/A/90]$, respectively, with the controlled-damage specimens consisting of adhesive strips 1.27 cm in width. This particular laminate design was chosen to suppress the longitudinal matrix crack observed in Figs. 5 and 6. Impact velocities for these cases range from approximately 14.2-26.5 m/s (560-1042 in/s) and specimens were subjected to a 10.16 and 15.24 cm span in Figs. 7 and 8, respectively. X-radiographs show the a similar trend as seen for previous layup geometries. The controlled-damage specimens suppress delamination to the 2.54 cm square mesh encompassed by the adhesive strips, with the exception of the specimen at 26.5 m/s of Fig. 8. As observed previously, the delamination across the width of the specimen was initiated from a bending crack in the transverse direction on the side opposite impact. Cross-sectional photographs indicate that the bending crack cuts through the matrix over a distance into the laminate and branches into delamination at the 0/90 interface. For this type of failure, the adhesive strips are not effective. However, delamination is suppressed. Figure 3 shows delamination area plotted against velocity for the impact cases of Figs 7 and 8.

Figures 9 and 10 show X-radiographs and transverse and longitudinal cross sections of the plain and controlled-damage specimens with layup geometries of $[0_3/90_3/0_3/90_3/0_3]$ and $[0_a/0/0_a/90_a/90/90_a/0_a/0/0_a/90_a/90/90_a/0_a/0/0_a]$, respectively, with 0.635 cm adhesive strips used in the controlled-damage specimens. Note that the site of impact is at the top of the cross-section. Comparing the X-radiographs of the plain and controlled-damage specimens for the same span and approximate velocity, the same trend observed in previous cases is repeated here. The X-radiographs of of the controlled-damage specimens show that delamination is confined to the region of the boundary formed by the adhesive strips in the transverse direction, but propagates along the matrix cracks in the fiber direction of the surface ply. This was observed in the specimens of identical layup in Figs. 5 and 6. As in the case of the bending cracks, matrix cracks originate at the surface can initiate and propagate delamination along the fiber direction, rendering the adhesive strips less effective there.

Residual Strength

Figure 4 shows the results from the 3-point bending residual strength test for the specimens of Figs. 5 and 6. Additional plain and controlled-damage specimens have been included in the residual strength tests of Fig. 4 consisting of cases before impact and after impact velocities above 20 m/s (787 in/s). As seen in the figure, the plain graphite/epoxy specimens experience increasing damage with increasing impact velocities, whereas the strength of the controlled-damage specimens remains fairly constant until approximately 20 m/s, where there is a sudden decrease. This corresponds to Fig. 2 where it was seen that at approximately 20 m/s the delamination area of the controlled-damage specimen increases suddenly. The strength of the plain specimens before impact is in the range of approximately 2100-2400 MN/m^2 (305-348 ksi), which is higher than that of the controlled-damage specimens with a range of approximately 1700-1900 MN/m^2 (247-276 ksi), due to the percentage of adhesive in the controlled-damage specimens. However, residual strength test after impact at the higher velocities considered show that the strength of the controlled-damage specimens is higher than the plain specimens. Results also indicate that the threshold velocity, the velocity after which the specimen has zero residual strength, is approximately 23 m/s (906 in/s)for the controlled specimens specimens and approximately 21 m/s (827 in/s) for the plain specimens.

Damage Mechanisms

The transverse and longitudinal cross-sections of Fig. 9 show the extend of delamination between the 0/90 degree interfaces of a plain specimen. Specifically, delamination in the longitudinal cross-sections is between the second and the fourth 0/90 interface of four interfaces counting from the side of impact. As expected, delamination is quite severe for these cases. Also seen in the longitudinal cross-sections is the staircase appearance of the matrix cracks commonly found in impact loading problems. Delamination in the transverse cross-sections is between the first and third 0/90 interface and is quite extensive for some cases at the third interface. Matrix cracks are also observed in the transverse cross-sections.

The transverse and longitudinal cross-sections of Fig. 10 for the controlled-damage specimen show the extent of delamination and the effect of the adhesive strips. Delamination is between the second and fourth 0/90 interface in the longitudinal cross-sections and between the first and third 0/90 interface in the transverse cross-sections. These locations are identical to those observed in Fig. 9 of the plain specimen. However, delamination is much less severe. Matrix cracks are also observed in Figs. 10.

The cross-sectional microphotographs of Figs. 10 show the effect of the adhesive strips. Delamination cracks are seen propagating along 0/90 degree interfaces up to the adhesive strip, at which point the adhesive strip act as a softening strip to arrest the delamination.

SUMMARY AND CONCLUSION

In this study, delamination control and residual strength resulting from impact loading of laminated composites with adhesive strips have been investigated. A comparison of delamination area from X-radiographs for specimens of two different layup geometries and spans has clearly indicated that delamination is substantially controlled by the use of adhesive strips in laminated composites. Cross-sectional microphotographs show that the adhesive acts as a softening strip to arrest the delamination. It has been shown that although the strength of the composite is less with the inclusion of the adhesive strips before impact, residual strength after impact at the higher velocities in the controlled damage specimens is higher than in the plain specimens. Although the use of adhesive strips does suppress delamination to a region specified by the spacing of the strips, there are limitation to its effectiveness at this time. At higher velocities, matrix and bending cracks can form on the surface ply, and propagate into the laminate initiating delamination at the 0/90 degree interface. Delamination is suppressed in the direction transverse to the crack, however, in some cases propagates for a distance along the length of the crack. Further design modifications must be made to remove these surface cracks or control them.

ACKNOWLEDGEMENT

This work was supported by Office of Naval Research under Contract No. N00014-84-K-0554. The technical monitors are Dr. Y. Rajapakse (ONR) and Mr. Lee Gause (NADC).

REFERENCES

1. Takeda, N., Sierakowski, R. L., Malvern, L. E., "Microscopic Observations of Cross Sections of Impacted Composite Laminates," *Composite Technology Review,* Vol. 4, pp. 40-44 (1982).

2. Joshi, S. P., and Sun, C. T., "Impact Induced Fracture in a Laminated Composite," *J. Composite Materials,* Vol. 19, pp. 51-66 (1985).

3. Chan, W. S., Rogers, C., and Aker, S., "Improvement of Edge Delamination Strength Using Adhesive Layers," in *Composite Materials Testing and Design, 7th Conference,* (Editor, J. M. Whitney). ASTM STP893, American Society for Testing and Materials, pp. 266-285 (1976).

4. Sun, C. T., and Rechak, S., "Effect of Adhesive Layers on Impact Damage in Composite Laminates," in *Composite Materials Testing and Design, 8th Conference,* (Editor J. D. Whitcomb), American Society for Testing and Materials, ASTM STP972 (1988).

5. Rechak, S. and Sun, C. T., "Optimal Use of Adhesive Layers in Reducing Impact Damage in Composite Laminates," *Composite Structures 4,* Proceedings of the 4th International Conference on Composite Structures, Paisley, Scotland, July 27-29, 1987.

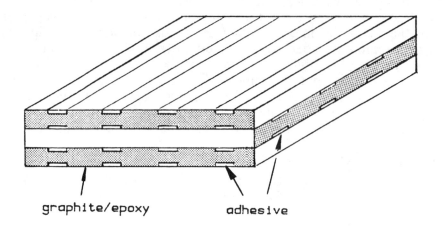

graphite/epoxy   adhesive

Fig. 1   Controlled damage concept for [0/90/0] laminate.

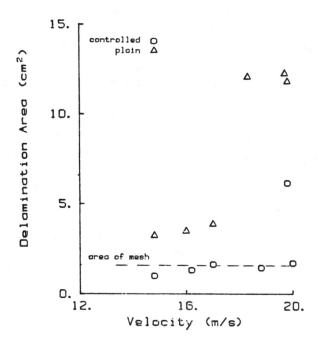

Fig. 2   Delamination area plotted against velocity for $[0_3/90_3/0_3/90_3/0_3]$ controlled and plain specimens. The width of the specimens is 3.81 cm, the span is 15.24 cm and the width of the adhesive is 0.635 cm.

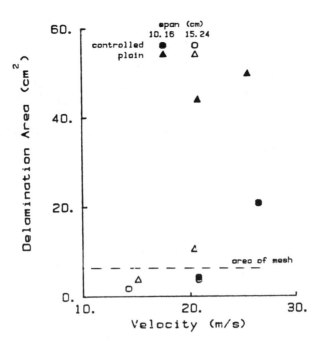

Fig. 3   Delamination area plotted against velocity for $[90_2/0_3/90_5/0_3/90_2]$ controlled and plain specimens. The width of the specimens is 7.62 cm and the width of the adhesive is 1.27 cm.

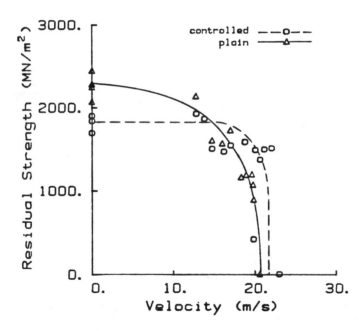

Fig. 4   Residual strength plotted against velocity from three point bending tests results of $[0_3/90_3/0_3/90_3/0_3]$ controlled and plain specimens. The width of the specimens is 3.81 cm, the span is 15.24 cm and the width of the adhesive is 0.635 cm.

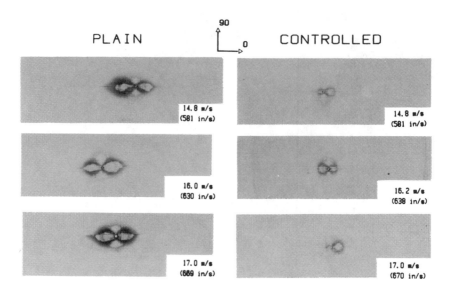

Fig. 5   X-rays of [0₃/90₃/0₃/90₃/0₃] plain and controlled specimens after impact velocities 14.8-17.0 m/s. The width of the specimens is 3.81 cm, the span is 15.24 cm and the width of the adhesive is 0.635 cm.

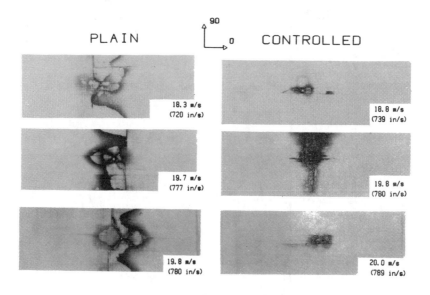

Fig. 6   X-rays of [0₃/90₃/0₃/90₃/0₃] plain and controlled specimens after impact velocities 18.3-20.0 m/s. The width of the specimens is 3.81 cm, the span is 15.24 cm and the width of the adhesive is 0.635 cm.

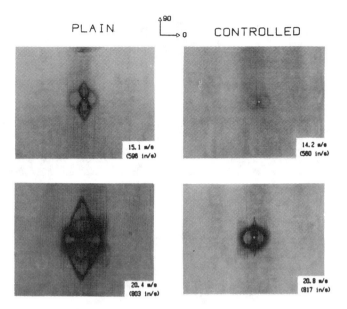

Fig. 7   X-rays of $[90_2/0_3/90_5/0_3/90_2]$ plain and controlled specimens after impact velocities 15.1-20.8 m/s. The width of the specimens is 7.62 cm, the span is 15.24 cm and the width of the adhesive is 1.27 cm.

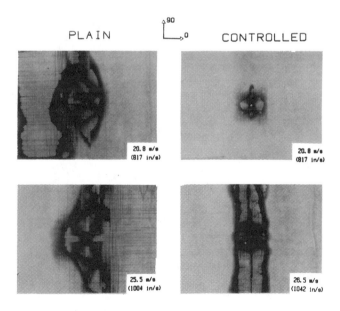

Fig. 8   X-rays of $[90_2/0_3/90_5/0_3/90_2]$ plain and controlled specimens after impact velocities 20.8-26.5 m/s. The width of the specimens is 7.62 cm, the span is 10.16 cm and the width of the adhesive is 1.27 cm.

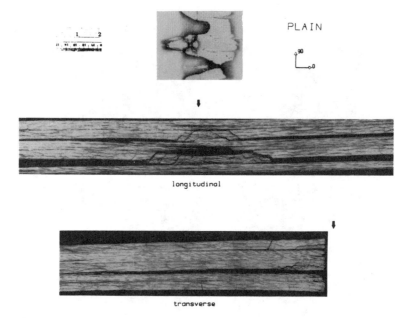

Fig. 9   X-ray and cross-sectional photographs of a $[0_3/90_3/0_3/90_3/0_3]$ plain specimen after impact. Impact velocity is 25.3 m/s and magnification for the cross-section is 25x. The specimen width is 7.62 cm and the span is 10.16 cm.

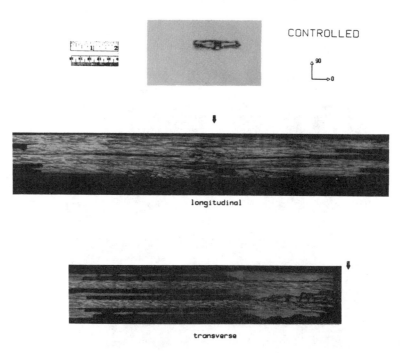

Fig. 10   X-ray and cross-sectional photographs of a $[0_3/90_3/0_3/90_3/0_3]$ controlled specimen after impact. Impact velocity is 25.0 m/s and magnification for the cross-section is 25x. The specimen width is 7.62 cm, the span is 10.16 cm and the width of the adhesive is 0.635 cm.

The Toughness of Damaged or Imperfect Laminates

WILLIAM M. JORDAN

ABSTRACT

The effect of voids and prior loading history upon mode I and mode II delamination fracture toughness of two quasi-isotropic fiber reinforced composite systems was examined. The two systems studied were AS4/3501-6, a relatively brittle grapite/epoxy system, and AS4/APC2 (PEEK), a very ductile graphite/thermoplastic system. Voids were introduced into the AS4/3501-6 system during curing and they occurred at the ply interfaces. Up to 6% voids were found to increase both the mode I and mode II delamination fracture toughness.

The effect of prior loading history on delamination fracture toughness was examined for both the AS4/3501-6 and AS4/APC2 systems. Previous mode II crack extension was found to increase the subsequent mode I delamination fracture toughness. Previous mode I crack extension increased the subsequent mode II delamination fracture toughness.

INTRODUCTION

In design of fiber reinforced polymeric composites for aerospace applications, an important parameter that must be evaluated is the delamination fracture toughness of the composite. One of the problems in obtaining accurate fracture toughness properties is the difficulty in obtaining test laminates that accurately reflect what would be used in real world applications. Two aspects of this problem were examined in the present study. The first aspect was to examine whether specimens with voids can be used to accurately predict the toughness of a defect free laminate (and the reverse, can a defect free specimen predict the toughness of a structural laminate that contains voids). This was done by examining the effect of voids upon delamination fracture toughness. The second aspect was to examine whether (or how) prior loading affects delamination fracture toughess. This dealt with the issue of whether a virgin specimen can be used to predict the toughness of a laminate that has been damaged by permanent deformation (and the reverse, can a previously tested laminate be used to predict the toughness of an unloaded structural laminate).

The first aspect examined was the effect of voids upon delamination fracture toughness. A series of four AS4/3501-6 laminates were cured with three different void

Dr. William M. Jordan is an Assistant Professor in the Mechanical and Industrial Engineering Department at Louisiana Tech University, Ruston, LA, 71272

contents. Some voiding might be expected to improve toughness. For example, one of the suggested explanations for how rubber particles toughen epoxy resins is that they allow local voiding, which changes the nature of the state of stress near the crack tip [1,2].

The second aspect examined was the effect of loading history on delamination fracture toughness. Specimens that had been tested under mode II conditions were later tested in mode I conditions. Other specimens that had been tested under mode I conditions were subsequently tested under mode II conditions. Some preliminary work by Cohen [3] appears to indicate that the history of previous loading can have a significant affect upon mode I delamination fracture toughness.

EXPERIMENTAL PROCEDURE

Quasi-isotropic panels of AS4/3501-6 were layed up according to the specified orientations in 25 square centimeter panels (See Table I for fiber orientations). A strip of teflon was placed at the midplane about 4 cm into one edge of the laminate to provide a starter crack. This was would provide a starter crack that would grow in the 90 degree direction (parallel to the fibers at the center plane). Quasi-isotropic panels of AS4/APC2 were layed up in 10 by 15 cm laminates with an aluminum insert to provide a starter crack at the midplane. The starter cracks went into the laminate a distance of approximately 4 cm.

One difficulty in designing the choice of stacking sequence relates to the problem of twisting occurring during testing. During mode I testing, only one half of each laminate is actually being bent. To eliminate coupling between twisting and bending each half laminate has to be symmetric. However, during mode II testing, the entire laminate is bending and to eliminate coupling between bending and twisting an overall symmetric layup is required. It is possible to fabricate a symmetric layup that has each half laminate also symmetric. This would produce a layup that is not commonly used in industry and would defeat part of the goal of this study which was to examine toughness on the types of quasi-isotropic layups actually in common use. The choice made during this project was to use a symmetric stacking sequence, knowing there will be some slight twisting during mode I testing. (A calculation of the laminate stiffness showed that for the systems used in this study, the values of the [B] matrix are small with respect to that of the [A] matrix, indicating twisting should not be a large problem).

The laminates were cured in an autoclave according to standard practices for such materials. The cured laminates were C-scanned and X-rayed to establish that there were no internal defects other than the cracks caused by the edge inserts or the voids deliberately induced on selected laminates.

The mode I double cantilevered beam (DCB) test specimens were cut to be approximately 2 to 2.5 cm wide and up to 20 cm long. The mode II end notch flexure specimens were cut to be approximately 2 to 2.5 cm wide and about 18 cm long.

The mode I tests were performed on a screw driven Instron tensile test machine operating in displacement control. The crosshead rate used was .127 cm/min. The load and displacement were continually recorded on a strip chart. The side of the specimen was coated with a brittle white coating to aid in determining the growth of the crack. The crack length was monitored at discrete intervals (every 1.27 cm of crack growth) and this length was marked on the load-displacement chart. The crack was grown from its original length to a length of 11.4 cm on each specimen. At two points during the crack growth period, the specimen was unloaded so that the shape of the unload/load curve could be determined. At least two specimens from each layup were tested in mode I conditions. There were at least six separate locations on each specimen where values of G_{Ic} could be calculated.

The mode II tests were performed on a screw driven Instron tensile test machine operating in displacement control and with a crosshead displacement rate of .127 cm/minute. To provide a sharp starter crack (for the crack tip at the end of the insert is rather dull), the original crack was grown (in mode I conditions) about 1 cm before mode II testing began. The test fixture used for mode II testing had a three point bend geometry. The span of the two bottom supports was 10.16 cm, with the load being applied at the center of the specimen. For most of the tests, the crack tip was placed one-half way between one of the lower supports and the applied center load (this provided a starter crack of 2.54 cm). For tests involving specimens that had been previously loaded under mode I conditions, the initial crack length was 3.81 cm. The load was slowly applied until the crack began to grow. Crack growth was usually unstable, with the crack growing in one step to the center loaded region. One of the advantages of this test geometry is that once the crack has grown, the specimen can be slid over so that a crack length of 2.54 cm is again obtained, and another test can be performed on the same specimen. This is why the mode II specimens were cut considerably longer than the span of the three point test fixture. There were at least 2 specimens tested for each layup under mode II conditons, with at least 3 crack growth episodes per specimen tested.

Metallographic specimens were cut from the untested portion of the panels and examined within a metallograph. This was used to help verify the nature and amount of the voids in the panels where voids were deliberately obtained.

EXPERIMENTAL ANALYSIS AND RESULTS

Analysis of Mode I and Mode II Tests

Mode I tests using Double Cantilevered Beam test specimens have been done by a number of investigators, including this one [1,2]. The most common analysis has been one using linear elastic beam theory [1,2,3,4,5]. Devitt, et al, [6] allowed for non-linear elastic behavior but not for permanent damage in the composite caused by crack growth. The results in this study were all in the linear elastic region as defined by Devitt, and the test results were calculated using linear elastic beam theory. For an example of the calculation method used, see references [1] and [2].

Two common methods for performing mode II delamination fracture toughness tests are the end loaded split laminate test and the end notch flexure test (E.N.F.). The end-loaded split laminate method has been used by the author [1,2]. In a recent conference presentation Corleto and Bradley [7] reported a comparison of these two test methods for use with a brittle and a ductile resin system. Their conclusion was that these two methods give similar results. Since the E.N.F. test method uses simpler fixtures (a three point bend test geometry) that is the method that was used in this study. There were non-linearities in the load/unload curve for some of the systems. To deal with this issue, this study followed the example of Corleto and Bradley [7] and calculated G_{IIc} using the area within the load/unload curve as the energy required to grow the crack. This was divided by the specimen width and the length of crack extension to get a value for G_{IIc}. This assumes no far field damage and all of the input energy is used to grow the crack.

Results of Mode I and Mode II Tests for Systems with Voids

Voids were deliberately put into two AS4/3501-6 laminates with different stacking sequences in an attempt to determine their impact upon delamination fracture toughness. Following the example of Harper, et al [8] this was accomplished by varying the time and amount of pressure applied to the laminate during curing. The standard temperature and vacuum cycle was applied to these laminates.

Mode I test results are reported in Table I. The fracture toughness values reported in this study are for steady state crack growth. In each system, there was a smaller value of an apparent initiation fracture toughness. There was then a region where the value of G_{Ic} did not change with respect to crack length. This is the value of G_{Ic} reported in this study. For long crack lengths (on the order of 9-10 cm) there was then an increase in fracture toughness. This may be related to permanent damage occurring within the laminate as the crack grows.

The layup marked in Table I (with two asterisks) had significant permanent deformation as a result of crack growth. (This permanent deformation is in the form of a permanent opening displacement.) The result for this layup was calculated using linear elastic beam theory and labelled as an upper bound limit to G_{Ic}. This layup with the permanent deformation was the one where there were four plies of each orientation grouped together. The remaining layups did not have significant permanent deformation and the fracture toughness results calculated using linear elastic beam theory can be considered valid. The system that was more heavily spliced (the one where each ply had a differring orientation from the adjacent plies) had a higher mode I toughness and no significant permanent deformation.

The presence of voids increased the mode I toughness. As the percentage of the voids increased, so did the mode I toughness. An increase in voids from 0 to 3.6% increased the mode I toughness by 84% while 6% voids increased the mode I toughness by at least 137% .

Mode II results are also shown in Table I. The presence of voids increased the mode II toughness significantly. As the percentage of voids increased, so did the mode II toughness. Increasing from 0 to 3.6% voids increased the mode II toughness by 39%, while increasing to 6% voids increased the mode II toughness by at least 55%.

The addition of voids increased both the mode I and mode II delamination fracture toughness for both layups. This result was expected, for Harper et al [8] and Tang et al [9] both reported that the presence of voids decreased the strength and transverse modulus of graphite/epoxy systems. Frequently the micro-mechanisms which lead to a decrease in strength and modulus will also increase toughness. This apparently occurred in this case as well. As shown in Table I, the laminates with voids had a lower bending modulus and a higher toughness.

The voids typically occurred at the interfaces of the original plies. An example of this is shown in the photomicrographs in Figure I. Volume percentage of voids were calculated from the difference in specific gravity and are reported in Table I.

Results of Mode I and Mode II Tests for Systems with Prior Loading
Preliminary work by Cohen [3] indicated that the loading history may have an affect upon mode I delamination fracture toughness. In this project some specimens were delaminated under mode I conditions (crack grown to about 11 cm length) and then delaminated under mode II conditions. Other specimens were delaminated under mode II conditions and then delaminated under mode I conditions.

Test results for this trial are shown in Table II. Previous delamination increased both the mode I and mode II fracture toughness. The effect of prior mixed mode delamination on mode I toughness had been reported by Cohen [3]. In both cases, the subsequent mode I delamination fracture toughness was increased. This may be related to internal damage caused by the mode II loading. The mode I crack is now growing through damaged material instead of virgin material. For the AS4/3501-6 system the effect of prior loading was greater on the layup where there was more grouping of like oriented plies together. For the more highly spliced layup, there was less of a prior loading effect. This is consistent with the results reported above that the more highly spliced laminate had less permanent deformation when tested in mode I conditions than did the less spliced laminate.

More surprising was the result that prior mode I delamination crack growth increased subsequent mode II toughness. Apparently mode I crack growth also damages the laminate, resulting in the mode II crack growing through a damaged region.

DISCUSSION OF RESULTS

Effect of Voids on Delamination Fracture Toughness
The presence of voids (in the amounts indicated in Table I) increased both the mode I and mode II delamination fracture toughness. This is not surprising when the effect of voids on other mechanical properties is examined. The presence of voids decreases transverse strength and modulus as was previously noted [8,9]. Typically a material that has lower modulus and strength is also tougher, and that was the situation in this case. Harper, et al [8] found for a unidirectional AS4/3502 system that about 3.4% voids decreased E_{22} about 25% and had almost no effect upon E_{11}. This is not surprising for the transverse E_{22} modulus is much more sensitive to the resin properties than the E_{11} modulus which is more dependent on the fiber properties. These results are quite consistent with the results of this study which indicated that 3.6% voids would decrease the E_{90} modulus about 10%. The system used in this study was a quasi-isotropic layup so that the resin properties would dominate for one fourth of the plies (the ones at 90), would play a role for 50% of the plies (the plus and minus 45 plies), and be less significant for the zero direction plies.

There are several possible mechanisms by which voids could produce an increase in toughness. The first one might simply be that the presence of the voids decreases the modulus and therefore increases the inherent toughness of the resin system. A more ductile system is typically a more tough system, and this may be an illustration of that.

A second possibility is that the presence of the voids may change the state of stress near the crack tip, thus postponing the time when crack propagation will begin. A variation of this has been suggested for the reason why rubber particles toughen epoxy resins. It has been suggested that there are small voids near the vicinity of the rubber particles that relax the stresses and make the system more ductile (and therefore tougher) [1,2]. However, it must be noted that any voids occurring near rubber particle additions are several orders of magnitiude smaller than these voids.

A variation of the second possibility is that the presence of the voids changes the stress concentration near the crack tip. A system without voids would tend to have a very sharp crack tip (with a high stress concentration at that location). With large numbers of voids, the crack may grow from void to void, always keeping a relatively blunt crack tip (the edge of the void is now the crack tip). An examination of Figure 1 shows how very blunt a crack tip would be if the crack grew from void to void. This blunt crack tip now has a much lower stress concentration. Therefore a higher stress must be applied before a critical value of stress is reached at the crack tip that will allow continued crack extension.

The designer must be cautioned, in that it is not recommended that voids be deliberately added to a resin system in order to improve toughness. This study only dealt with a very specific type of void (at the ply interface) and only with a void content up to 6%

by volume. A different type of void or an increase in void content above 6% may have very different results.

The results of this study on voids indicate two positive benefits for the designer. The first positive benefit is that the mere presence of interfacial voids may not require that a composite part must be rejected. A few voids may indeed increase the composite delamination toughness. The presence of the voids may, however, lower the stiffness of the composite laminate below acceptable levels. The second positive benefit is that if the designer uses a void free laminate toughness for design, his design will always be conservative, for a part with voids will be at least as tough as the void free laminate he used for design purposes. The designer must be cautioned in that the use of a system with voids to determine the design toughness will give a misleadingly high value for the toughness.

Effect of Loading History on Delamination Fracture Toughness

Prior loading of the laminate increased the fracture toughness. This was true irrespective of whether the mode I delamination crack extension was done before or after the mode II delamination crack extension. This is in accord with the results of Cohen [3]. He delaminated some specimens at about 35% mode II conditions and then found that the subsequent mode I delamination fracture toughness was larger than before. In this portion of the study laminate layups were chosen so that there would be no significant permanent deformation resulting in the specimens after crack extension had occurred. This was verified by the fact that all of the mode II unload curves reached a zero load at the same time they reached a zero extension. This does not necessarily mean that there was no permanent damage within the laminate.

Prior mode I loading increased the resulting the mode II toughness. This appears to indicate that there was significant permanent damage in the laminates before the second set of tests were begun. This damage must be more than just microcracking near the crack tip (previously reported by the author [1,2]), for that mode I microcracking was on a relatively small scale very near the crack tip.

These results raise again the question of whether the reported fracture toughness results are the intrinsic fracture toughness or merely reflect the far field damage that was caused by the initial test. It was not possible to give a definitive answer from these current test results. However, if a system is loaded more than one time, it may be damaged before the second loading has begun. Then these fracture toughnesses may reflect what is actually occurring in the laminate, even though it might not be the toughness in a virgin, undamaged laminate.

These results indicate a potential problem with some mode II delamination fracture toughness results reported previously by other investigators. When using E.N.F. test specimens a starter crack is typically made through the use of some type of insert into the edge of the laminate at the midplane. This results in a resin pocket at the end of the insert. Many investigators typically grow a mode I crack through this resin pocket before the mode II test is begun. This study indicates that investigators should not grow the mode I crack too long or the laminate may become damaged and the calculated mode II toughness may be larger than the correct value. A standard short mode I starter crack grown through the resin pocket should be employed.

One positive aspect of the history effect is that prior loading will provide a laminate that will be tougher than one that had not been loaded or tested before. This means that if a designer uses results from a test on an virgin laminate he will be using a conservative value of the fracture toughness. However, the designer should be cautioned in that using a previously loaded laminate to produce the fracture toughness will give a misleadingly high value for the toughness.

CONCLUSIONS AND SIGNIFICANCE

Several significant results have been obtained in this research program:

(1)   The presence of voids at the ply interface may actually increase the mode I and mode II fracture toughness. This is useful in that the manufacturer does not necessarily have to remove all voids from the laminate for it to be successfully used in a real world application. Designing from void free laminate test results will provide a conservative estimate of fracture toughness.

(2)   Prior loading (particularly with prior crack growth) tends to increase the fracture toughness of the composite. This appears to be related to permanent damage that has been created within the laminate.   This is useful to the designer in that a previously damaged laminate may still be usable in many applications where toughness is the controlling parameter. Designing from previously unloaded laminate test results will provide a conservative estimate of fracture toughness.

Conclusions (1) and (2) above must be tempered with the fact that this increased toughness may be accompanied by a decreased stiffness that may make a laminate unusable for that reason (even though the toughness is still acceptable).

ACKNOWLEDGEMENTS

I wish to thank the Air Force Office of Scientific Research, the Air Force Systems Command, and the Materials Laboratory for their sponsorship of this research. Personnel at Universal Energy Systems, Inc., are thanked for their work in the administrative details of this project. Particular thanks go to the technicians under the direction of Tim Hartness who helped with fabrication, curing, and testing of the composite laminates.

I wish to thank the Halliburton Foundation and Louisiana Tech's College of Engineering for providing financial support to allow for the writing and presenting of this paper. This work could not have been done without the love and encouragement of my wife, Gail, and sons Robbie and Steven. Without their willingness to move for the summer of 1987 this never would have been possible.

REFERENCES

1.   Jordan, W.M., "The Effect of Resin Toughness on the Delamination Fracture Behavior of Graphite/Epoxy Composites", Ph.D. Dissertation, Texas A & M University, December 1985.

2.   Jordan, W.M., and Bradley, W.L., "Micromechanisms of Fracture in Toughened Graphite-Epoxy Laminates", in <u>ASTM STP 937 Toughened Composites</u>, American Society for Testing and Materials, 1987, pp. 95-114.

3.   Cohen, R.N., "Effect of Resin Toughness on Fracture Behavior of Graphite/Epoxy Composites", M.S. Thesis, Texas A & M University, December 1982.

4.   Wilkins, D.J., Eisenmann, J.R., Camin, R.A., Margolis, W.S., and Benson,
      R.A., "Characterizing Delamination Growth in Graphite-Epoxy", in
      <u>ASTM STP 775 Damage in Composite Materials,</u> American Society for
      Testing and Materials, 1982, pp. 168-183.

5.   Whitney, J.M., Browning, C.E. and Hoogsteden, W., "A Double Cantilever Beam
      Test for Characterizing Mode I Delamination of Composite Materials",
      <u>Journal of Reinforced Plastics and Composites</u>, Vol. 1, (October 1982),
      pp. 297-313.

6.   Devitt, D.F., Schapery, R. A., and Bradley, W.L., "A Method for Determining the
      Mode I Delamination Fracture Toughness of Elastic and Viscoelastic
      Composite Materials", <u>Journal of Composite Materials</u>, Vol. 14,
      October 1980, pp. 270-285.

7.   Corleto, C.R., and Bradley, W.L., "Mode II Delamination Fracture Toughness of
      Unidirectional Graphite/Epoxy Composites", presented at an A.S.T.M.
      Symposium entitled "Second Symposium on Composite Materials: Fatigue
      and Fracture", Cincinnati, Ohio, April 1987.

8.   Harper, B.D., Staab, G.H., and Chen, R.S., "A Note on the Effects of Voids
      Upon the Hygral and Mechanical Properties of AS4/3502
      Graphite/Epoxy", <u>Journal of Composite Materials</u>, Vol. 21, March 1987,
      pp. 280-289.

9.   Tang, J., Lee, W.I., and Springer, G.S., "Effects of Cure Pressure on Resin
      Flow, Voids, and Mechanical Properties", <u>Journal of Composite Materials,</u>
      Vol. 21, May 1987, pp. 421-440.

Table I

Effect of Voids on the Delamination Fracture Toughness of AS4/3501-6

Layup *	Percent Voids	G_{Ic} (J/m²)	Modulus E_{90} From Mode I Test (GPa)	G_{IIc} (J/m²)
[-45(4)/0(4)/45(4)/90(4)] S				
	0	133 **	40.1***	853 **
	6.0	316	38.4	1325
[-45/0/45/90] 4S				
	0	152	55.4	564
	3.6	281	50.7	785

* Systems delaminated at midplane in 90 degree direction.
** Upper bound estimate because of large degree of permanent deformation.
*** Lower bound estimate because of large degree of permanent deformation

Table II

Effect of Loading History on Delamination Fracture Toughness

Material	Layup *	G_{Ic} (J/m²)	G_{Ic} (J/m²) After Mode II Crack Growth	G_{IIc} (J/m²)	G_{IIc} (J/m²) After Mode I Crack Growth
AS4/3501-6					
	[-45(2)/0(2)/45(2)/90(2)] 2S	158	190	434	904
	[-45/0/45/90] 4 S	152	140	564	742
AS4/APC2					
	[-45(2)/0(2)/45(2)/90(2)] 3 S	1523	1864		
	[-45/0/45/90] 6 S	1509	2084		

* Systems delaminated at midplane in 90 degree direction.

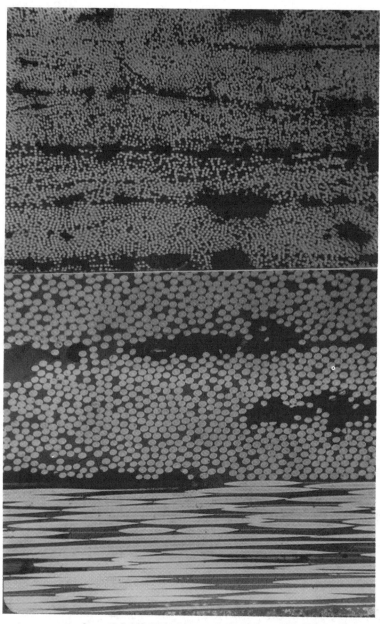

Figure I. Microstructure of an AS4/3501-6 panel with intentional voids. Both photomicrographs show a panel with [-45(4)/0(4)/45(4)/90(4)] S layup. Top photomicrograph is at 100 x and bottom photomicrograph is at 250 x. Voids at the ply interfaces are clearly visible.

Cosserat Micromechanics of Structured Media Experimental Methods

RODERIC LAKES

Department of Biomedical Engineering
Department of Mechanical Engineering
The University of Iowa
Iowa City, IA 52242

ABSTRACT

We describe experimental techniques suitable for study of structured media as Cosserat elastic materials. Results of experiments disclose Cosserat elastic behavior in a variety of materials: foams of high and low density, and human compact bone, which has fibrous, particulate, and porous structural features. As for the isotropic foams, all six of the Cosserat elastic constants were found. The measured characteristic lengths were comparable to the size of the structural elements.

INTRODUCTION

Cosserat elasticity and related theories are thought to offer advantages over classical elasticity in the prediction of stresses in materials with microstructure. Cosserat elasticy is a continuum theory with rotational degrees of freedom associated with points, as well as the usual translational degrees of freedom. A torque per unit area [couple stress] can be transmitted across a differential area, as well as the usual force per unit area [stress] considered in classical elasticity.

THEORETICAL ASPECTS

The question of how much freedom is to be incorporated in an elasticity theory must ultimately be decided by experiment. For example,the early rariconstant theory of Navier, in which the forces act along the lines joining pairs of atoms and are proportional to changes in distance between them, given by the following[1], entails a Poisson's ratio of 1/4. Symbols are defined below.

$$t_{kl} = G\, \varepsilon_{rr}\, \delta_{kl} + 2G\, \varepsilon_{kl} \qquad (1)$$

This was rejected based on experimental measurements of Poisson's ratio. We now use the following constitutive equation for *classical* isotropic elasticity[2,3].

$$t_{kl} = \lambda \varepsilon_{rr}\, \delta_{kl} + 2G\, \varepsilon_{kl} \qquad (2)$$

More freedom is incorporated in the Cosserat theory of elasticity, also known also as micropolar elasticity[4,5,6]. This incorporates a local rotation of *points* as well as the translation assumed in classical elasticity. In the isotropic Cosserat solid there are six elastic constants, in contrast to the classical solid in which there are two. Certain combinations of Cosserat elastic constants have dimensions of length and are referred to as characteristic lengths. The constitutive equations for a linear isotropic Cosserat elastic solid are[5] :

$$t_{kl} = \lambda \varepsilon_{rr}\, \delta_{kl} + (2\mu + \kappa)\varepsilon_{kl} + \kappa \varepsilon_{klm}(r_m - \phi_m) \qquad (3)$$

$$m_{kl} = \alpha\, \phi_{r,r}\, \delta_{kl} + \beta \phi_{k,l} + \gamma \phi_{l,k} \qquad (4)$$

in which t_{kl} is the asymmetric force stress, m_{kl} is the couple stress, $\varepsilon_{kl} = (u_{k,l} + u_{l,k})/2$ is the small strain, u is the displacement, and e_{klm} is the permutation symbol. The microrotation ϕ in Cosserat elasticity is kinematically distinct from the macrorotation $r_k = (e_{klm}u_{m,l})/2$. In three dimensions, the isotropic Cosserat elastic solid requires six elastic constants λ, μ, α, β, γ, and κ for its description. Technical elastic constants are of use conceptually. These are [6,7,8]: Young's modulus $E = (2\mu+\kappa)(3\lambda+2\mu+\kappa)/(2\lambda+2\mu+\kappa)$, shear modulus $G = (2\mu+\kappa)/2$, Poisson's ratio $\nu = \lambda/(2\lambda+2\mu+\kappa)$, characteristic length for torsion $l_t = [(\beta+\gamma)/(2\mu+\kappa)]^{1/2}$, characteristic length for bending $l_b = [\gamma/2(2\mu+\kappa)]^{1/2}$, coupling number $N = [\kappa/2(\mu+\kappa)]^{1/2}$, and polar ratio $\Psi = (\beta+\gamma)/(\alpha+\beta+\gamma)$. When α, β, γ, κ vanish the solid becomes classically elastic.

The constitutive equations may also be written in terms of the technical constants:

$$t_{kl} = 2G[(\nu/(1-2\nu))\varepsilon_{rr}\delta_{kl} + \varepsilon_{kl} + [N^2/1-N^2]\varepsilon_{klm}(r_m - \phi_m)] \tag{5}$$

$$m_{kl} = 2Gl_t^2[(1/\Psi - 1)\phi_{r,r}\delta_{kl} + [1-2(l_b/l_t)^2]\phi_{k,l} + 2(l_b/l_t)^2\phi_{l,k}] \tag{6}$$

When the characteristic lengths vanish, these reduce to Eq. 2, the constitutive equation for classical elasticity. It is expected that the characteristic lengths are comparable to the size of the principal structural elements in the material.

Generalized continuum theories: a comparison

In the area of generalized continuum mechanics, literally hundreds of papers have been published, many of them dealing with the solution of specific boundary-value problems in Cosserat-type theories. In addition, more general theories have been evolved to include anisotropy [9], nonlinear effects [10], and viscoelastic behavior [11]. Nonlocal theories have been developed in which the stress at a point is expressed as a functional of the deformation histories of all material points in the solid [12]. Another type of generalization involves a hierarchy of higher order kinematical and dynamical variables [3-15].

The structure of available microcontinuum theories is summarized in table 1. These data provide a portion of the rationale for the use of Cosserat (micropolar) elasticity in the present proposal. The theories listed are shown in order of increasing sophistication; each theory is a special case of the one following it. Cosserat elasticity is sufficiently sophisticated to account for many of the phenomena observed in composites, but sufficiently simple that all the elastic constants can be extracted from feasible experiments.

TABLE 1: GENERALIZED CONTINUUM THEORIES

Theory	Kinematical variables	Dynamical variables	Available boundary value-problems	Available finite element methods	Number of elastic constants for isotropic solid
Classical elasticity	displacement	stress	yes, many	yes	2
Couple stress elasticity [16]	displacement	stress, couple-stress	yes, many	yes	4

Cosserat (micropolar) elasticity [5]	displacement, local rotation	stress, couple-stress	yes, many	yes	6
Micro-structure (micro-morphic) elasticity [13,14]	displacement, local rotation local micro strain	stress, couple-stress, double stress	one	yes, 2d	18
Multi-polar theories [15]	displacement higher micro-deformations	stress, higher multipoles of stress	no	no	Depends on on grade of theory.

Theoretical predictions of elastic constants

Explicit calculations of Cosserat elastic constants of various model materials with microstructure have been carried out by a number of investigators. For example in a two dimensional model composed of orientable points, joined by extensible and flexible rods, the equations obtained in a continuum approximation are identical with those of Cosserat elasticity [17]. Both the classical elastic constants and the couple-stress coefficients can be expressed in terms of the properties of the classically elastic beams composing the latticework [17]. In another model, a three dimensional honeycomb structure consisting of thin-walled cubical cells, the model material behaves like a Cosserat solid and the elastic coefficients are calculated from structural considerations [18]. In composites with a laminated structure, microelastic effects also are predicted to occur and their magnitude is a function of the classical elastic constants of the composite's constituents [19]. Similarly, in fibrous composites, the new microelastic constants have been predicted on the basis of the classical elastic constants of the fiber and matrix [20].

EXPERIMENTAL METHODS FOR COSSERAT SOLIDS

A variety of experimental methods may be considered for examination of microelastic solids. We choose Cosserat elasticity from among the above theories since it has sufficient freedom to account for many of the phenomena in composites, and it is simple enough that many analytical solutions are available. The methods are based on the availability of an analytical solution for the geometry in question, in which one or more of the Cosserat elastic constants depends on a measurable quantity. To be successful, a method must be sensitive to Cosserat elastic effects and must be insensitive to the effects of confounding variables.

Method of size effects

The method of size effects is based on the size dependence of torsional rigidity[7] and bending rigidity [8] which is predicted to occur in the bending and torsion of circular cross section rods of a Cosserat elastic solid. In these predictions, the rigidity deviates from the classical proportionality to the fourth power of the diameter. Slender rods appear more rigid than would be expected classically. The method of size effects is capable of determining all six of the Cosserat elastic constants of an isotropic solid. The method has been used both in earlier attempts [7,21-23] to explore Cosserat elastic behavior and in the writer's laboratory [24-29].

The rationale tor choosing the method of size effects is as follows. The method has been used successfully by the writer both in demonstrating the existence of nonclassical effects in certain materials and in determining the Cosserat elastic constants.

Instrumentation to be used in the method of size effects must be capable of measurements upon specimens with a wide range of rigidity and must not contribute errors due to friction or other causes, which might obtrude in the data. In addition it is desirable that the same configuration be used for both torsion and bending of the same specimen so that apparatus calibration is identical for both. The apparatus shown in figure 1 meets these requirements [29]. The apparatus is capable of creep, dynamic, constant load rate, and resonance experiments, depending on the electrical signal input to the Helmholtz coil. Calibration of the apparatus and error analysis are discussed in [31].The apparatus makes use of a disc-shaped permanent magnet attached to the end of a specimen to apply a torque to the specimen. The magnet is placed at the centre of a Helmholtz coil, which produces a very uniform magnetic field B in response to an electric current. The magnet is made of samarium cobalt alloy which sustains an exceptionally high level of magnetization, or magnetic dipole moment per unit volume. The torque on the specimen is proportional to the current in the Helmholtz coil. Bending and torsion are achieved by properly orienting the Helmholtz coil with respect to the magnet, so that the torque vector is in the correct direction. The angular displacement measurement is performed via an interferometric method. The laser beam passes through a Ronchi ruling, 300 lines per inch, designated by A in figure 1; interference fringes thus produced are reflected from the specimen mirror and pass through an identical ruling located at B. The second ruling is located so that its line spacing exactly matches the spacing of the fringes in the laser beam at that point. As the specimen end rotates, the fringe pattern moves, generating an oscillating light intensity behind ruling B. A light detector consisting of a photodiode mounted behind a narrowband interference filter transforms the light signal to an electrical signal. The filter passes only the laser wavelength and excludes room light. One fringe corresponds to 140×10^{-6} rad. Resolution is 0.05 fringe or better. In the configuration shown in figure 1 the magnet torque is such that torsional loading occurs. To perform a bending experiment, the Helmholtz coil is rotated 90 deg. and the rulings are correspondingly rotated. The analysis scheme [7,8] for the method of size effects is illustrated in Figs. 2 and 3. Fig. 2 shows how the Cosserat elastic constants are extracted from torsional size effect data and Fig. 3 shows how they are extracted from pure bending size effect data.

Wave and vibration mode methods

Several authors have suggested the use of wave propagation methods in this context. These methods are based on the fact that both propagating acoustic waves and standing waves exhibit dispersion in the presence of Cosserat elastic and other generalized continuum effects [5,10,12]. We are cautious in applying such methods since both viscoelastic (time-dependent) and Cosserat (spatial or gradient sensitive) effects will contribute to the dispersion of waves. Viscoelastic behavior, therefore constitutes a confounding variable in wave methods. By contrast, in the method of size effects, viscoelasticity can be effectively decoupled by using isochronal (constant time) or constant frequency data.

Strain distribution methods

One such method is based on the predicted warp and strain distributions on the surface of a square cross section prismatic bar in torsion. In a Cosserat solid, nonzero stress and strain are predicted at the corners of the cross section, in contrast to the results of classical elasticity[30]. The peak strain at the center of the lateral surfaces is reduced below classical expectations. The warp of cross sections is predicted to be reduced in Cosserat solids[30]. Evidence of stress at the corner can be sought by examining the displacement of a small corner notch directly in the case of compliant materials[31] and by strain gages[32] or by holographic interferometry[33] in the case of rigid materials. The warp itself can also be measured. The physical origin of couple stresses in the square bar is depicted in Fig. 4.

EXPERIMENTAL RESULTS

Several of the above experimental methods have been applied to a variety of materials, such as polyurethane foams of high and low density, human bone, and new foam materials with negative Poisson's ratios. Results for several materials are as follows.

Material	Elastic constants	Phenomena
Dense polyurethane foam, 0.34g/cc	$E=300MPa$, $G=104MPa$, $\nu=0.4$ $l_t=0.62mm$, $l_b=0.33mm$, $N^2=0.04$, $\Psi=1.5$ [29]	Size effects[29], corner crack motion in torsion of square bar[33]
Syntactic foam 0.59g/cc	$E=2758MPa$, $G=1033MPa$, $\nu=0.34$ $l_t=0.065mm$, $l_b=0.032mm$, $N^2=0.1$, $\Psi=1.5$ [29]	Weak size effects: material is nearly classical [29]
Human compact bone, 2g/cc	$E=12GPa$, $G=4GPa$, $l_t=0.22mm$, $l_b=0.45mm$, $N^2 \geq 0.5$	Anisotropy, size effects 25-27] Nonclassical strain distribution in torsion of square bar[32]
Low density foams	$E=1.1MPa$, $G=0.6$ MPa, $N=0.07$ $l_t=3.8mm$, $l_b=5.0$ mm, $N^2=0.09$[28]	Corner crack motion in torsion of square bar[31], warp reduction, size effects[28]

Internal consistency test

Internal consistency tests for Cosserat elastic constants determined via the method of size effects in rods of circular cross section are as follows. For example, E and G may be used to calculate Poisson's ratio as in the classical case, and the shape of the bending theoretical curve depends upon Poisson's ratio. In addition, the torsional and bending characteristic lengths determine β/γ uniquely. The ratio β/γ however, governs the shape of the bending curve in the vicinity of the origin. Only for the values $\beta/\gamma = \pm1$ does the theoretical bending curve pass through the origin. The coupling number N influences the shape of both torsion [7] and bending [8] theoretical curves. Specifically, for $\Psi = 1.5$, its maximum thermodynamically permissible value, N governs the maximum apparent stiffening of thin specimens in torsion. Physically, arbitrarily large stiffening effects are not to be expected since any continuum theory will break down if the specimen size becomes equal to or less than the structure size. These consistency tests have been applied by the author to experimental results for several isotropic cellular solids. In a low density polystyrene foam, the Cosserat continuum was found to be only an approximate description of the material, based on failure of these consistency tests over a certain region [28]. In a high-density polyurethane foam, satisfaction of the consistency tests led to the conclusion that this material behaves as a Cosserat solid [29].

DISCUSSION
Composites with stress raisers: significance of Cosserat mechanics

One of the major predictions [34,6] of Cosserat elasticity is a reduction of stress/strain concentration factors in many geometries. Such predictions, which have been confirmed in bone[35] may be useful in dealing with phenomena in composites which are currently dealt with empitically. There currently exists a major body of empirical data dealing with fracture of composite materials in the presence of stress raisers [36,37]. A common theme is found in results in this area. The strength of a composite specimen with a hole of notch cannot be correctly predicted by means of stress concentration factors or stress intensity factors derived from the classical theory of elasticity. Moreover, the observed strength of specimens with holes is found to depend on the size of the hole [36,37,38]. For example in graphite-epoxy, a large hole, e.g., 1 inch diameter, results in an observed stress concentration which agrees with the elasticity theory prediction [39]. Smaller holes exhibit less stress concentration. For a sufficiently small hole, there is essentially no reduction in strength [39], so that the effective stress concentration for failure is unity. Such behavior cannot be accounted for under the theory of elasticty, which predicts stress concentrations to be independent of hole size. The materials examined in these studies tend to be brittle and to exhibit negligible plastic deformation. An elasto-plastic continuum model is therefore not likely to be of use. Some aspects of the hole size effect phenomenon have been treated using statistical analysis of flaw populations in the vicinity of the hole [36]. Analyses of this type, e.g., Weibull analysis, can account for some but not all of this size dependence.

Empirical fracture models

A variety of fracture models have been proposed to describe the observed phenomena. For example, Whitney and Nuismer proposed the following two criteria. In the first, the criterion for fracture is that fracture occurs if the average stress over some distance a_0 becomes equal to the ultimate strength of the material [36,37,40]. In the second criterion, fracture occurs when the stress some distance d_0 from the stress raiser becomes equal to the ultimate stress. This is the so-called point stress criterion. It is tacitly assumed that these 'characteristic distances' are material parameters and are thus independent of the size of the discontinuity. The various empirical models are generally found to agree with the experimental results, for particular configurations. This is not surprising since the procedure is basically one of curve fitting. Karlak [41] performed fracture tests on quasi-isotropic graphite-epoxy laminates of different stacking sequences, with holes of different size. He concluded that the strength depends on the stacking sequence. Moreover, he conclude that the Whitney-Nuismer point-stress 'characteristic distance' d_0 is not a material constant but is related to the square root of the hole radius. Karlak then proposed a two-parameter model to fit the data.

While the empirical models discussed above are superior to the incorrect predictions of classical elasticity, they are essentially ad hoc in that reference is not made to constitutive behavior, composite microstructure, or to any physical arguments.

Cosserat elasticity and related theories are thought to offer advantages over classical elasticity in the prediction of stresses in materials with microstructure. In particular, analytical solutions for stress concentration around circular holes and elliptic holes in plates disclose smaller stress concentration factors in a Cosserat solid[6,34] as opposed to a classical solid. Similar results have been obtained for predicted stress-intensity factors for cracks in plates as well as planar circular cracks in three-dimensional solids. The deviations are most significant when the hole size or crack length is no more than ten times the Cosserat characteristic length. In composite materials with stress concentrations due to holes or cracks, the observed fracture behavior is not correctly predicted by the classical theory of anisotropic elasticity. The experimental stress concentrations are consistently less than the theoretical ones, even under small strain conditions.

Strain distribution around holes and cracks

The experimental determination of the distribution of strain around an inhomogeneity is considerably more difficult than a simple strength measurement. Consequently, relatively few studies of this type have been reported. For example, the strain field around comparatively large holes (2.54 cm diameter) in a strip of boron-epoxy composite and glass epoxy composite was explored [42,43]. Strain distributions obtained experimentally again were in substantially good agreement with theory [43]. The situation is different in the case of small holes. Observed maximum strain is less than the strain anticipated on the basis of classical elasticity. For example in a quasi-isotropic graphite-epoxy plate with an elliptic hole, strain concentrations were about a factor of six less than the classical prediction [44].

In a study of human bone conducted in the author's laboratory, strain concentrations around holes in a strip under tension were found to be less than classically predicted values [35]. Strains in the direction perpendicular to the applied tension, by contrast, were greater than the classically predicted values.

Summary

We have seen that there is a major body of experimental evidence that the strength of composite materials with stress raisers cannot be correctly predicted by the theory of elasticity or by linear elastic fracture mechanics. Empirical fracture models can describe some of the observed behavior by curve fitting procedures. The empirical models are ad hoc. When additional experimental variables are introduced, e.g. specimen width, further adjustable parameters must be introduced. It would therefore by unwise to rely on the empirical fracture models in crucial design decisions, particularly when the geometry of the designed structural component differs from that of the test specimens. Moreover, the fracture models do not address the issue of strain distributions around inhomogeneities and stress raisers.

Classical elasticity is unable to correctly predict stresses and strains in the vicinity of small stress raisers and interfaces in composites and other structured materials. This shortcoming persists when anisotropy is properly taken into account. Design of composite structures in the presence of this lack of predictive power can be problematical. Design is further complicated by the fact that many problems which currently arise in composites are associated with interfaces and stress concentrators. Cosserat elasticity and other generalized continuum theory is capable of explaining some of the phenomena observed in composites.

CONCLUSION

We perceive a need for a method for the rational prediction of strain and stress fields in composites, as well as failure processes. In view of the experimental evidence, we consider generalized continuum theories, supported by experiment, to be of use in such a predictive method.

REFERENCES

[1] Timoshenko, S.P., *History of Strength of Materials*, Dover, (1983).

[2] Sokolnikoff, I.S., *Mathematical Theory of Elasticity*, Krieger, (1983).

[3] Fung, Y.C., *Foundations of Solid Mechanics*, Prentice Hall, (1968).

[4] Cosserat, E. and Cosserat, F., *Theorie des Corps Deformables*, Hermann et Fils, Paris, (1909).

[5] Eringen, A.C. Theory of micropolar elasticity. In Fracture Vol. 1, pp. 621-729 (edited by H. Liebowitz), Academic Press (1968).

[6] Cowin, S. C., An incorrect inequality in micropolar elasticity theory, *J. Appl. Math. Phys. (ZAMP)* Vol. 21,1970, pp. 494-497.

[7] Gauthier, R.D. and Jahsman, W.E., A quest for micropolar elastic constants, *J. Applied Mechanics*, Vol. 42,1975, pp. 369-374

[8]Krishna Reddy, G.V. and Venkatasubramanian, N.R., "On the flexural rigidity of a micropolar elastic circular cylinder", *J. Applied Mech.*, Vol. 45, 1978, pp. 429-431

[9] Tiersten, H.F. and Bleustein, J.L., "Generalized Elastic Continua", *R.D. Mindlin and Applied Mechanics*, G. Hermann, Ed., Pergamon, N.Y., 1974.

[10] Toupin, R.A., "Elastic Materials with Couple-Stresses", *Arch Rational Mech. Anal.*, Vol. 11, 1962, pp. 385

[11] Eringen, A.C., "Linear Theory of Micropolar Viscoelasticity", *Int. Jnl. Engrg. Sci.,* Vol. 5, 1967, pp. 191-204

[12] Eringen, A.C., "Linear Theory of Nonlocal Elasticity and Dispersion of Plane Waves", *Int. Jnl. Engrg. Sci.*, Vol. 10, 1972, pp. 425-535

[13] Mindlin, R.D., "Microstructure in Linear Elasticity", *Arch. Rational Mech. Anal.*, Vol. 16, 1964, pp. 51-78

[14] Eringen, A.C., "Balance laws of micromorphic mechanics", *Int. J. Engng. Sci.*, Vol. 8, 1970, pp. 879-828

[15] Suhubi, E.S. and Eringen, A.C., "Nonlinear Theory of Micro-elastic Solids-II", *Int. J. Engng. Sci.*, Vol. 2, 1964, pp. 389-404

[16] Mindlin, R.D. and Tiersten, H.F., "Effects of Couple-Stresses in Linear Elasticity", *Arch. Rational Mech. Anal.*, Vol. 11, 1962, pp. 415-448

[17] Askar, A. and Cakmak, A.S., "A Structural Model of a Micropolar Continuum", Int. Jnl. Engrg. Sci., Vol. 6, 1968, pp. 583-589

[18] Adomeit, G., "Determination of Elastic Constants of a Structured Material", *Mechanics of Generalized Continua*, E. Kröner, ed. Iutam Symposium, Freudenstadt, Stuttgart, Springer-Verlag, 1967.

[19] Herrmann, G. and Achenbach, J.D., "Applications of Theories of Generalized Continua to the Dynamics of Composite Materials", *Mechanics of Generalized Continua*, E. Kröner, ed., Iutam Symposium, Freudenstadt, Stuttgart, Springer-Verlay, 1967.

[20] Hlavacek, M, "A continuum theory for fibre-reinforced composites", *Int. J. Solids Structures*, Vol. 11, 1975, pp. 199-211

[21] Ellis, R.W. and Smith, C.W., "A Thin-Plate Analysis and Experimental Evaluation of Couple-stress Effects", *Experim. Mech.*, Vol. 7, 1967, pp. 372-380

[22] Schijve, J., "Note on Couple Stresses", *J. Mech. Phys. Solids*, Vol. 14, 1966, pp. 113-120

[23] Perkins and Thomson, "Experimental evidence of a couple stress effect", *AIAA J.* Vol. 11, 1973, pp. 1053-1059

[24] Lakes, R.S., "Dynamical Study of Couple Stress Effects in Human Compact Bone", *J. Biomechanical Engineering*, Vol. 104, 1982, pp. 6-11

[25] Yang, J.F.C. and Lakes, R.S., "Transient Study of Couple Stress in Compact Bone: Torsion", *J. Biomechanical Engineering*, Vol. 103, 1981, pp. 275-279

[26] Yang, J.F.C. and Lakes, R.S., "Experimental Study of Micropolar and Couple Stress Elasticity in Bone in Bending", *J. Biomech.*, Vol. 15, 1982, pp. 91-98

[27] Lakes, R.S. and Yang, J.F.C., "Micropolar elasticity in bone: rotation modulus κ. Proceedings, 18th Midwest Mechanics Conference, Developments in Mechanics, Vol. 12, 1983, 239-242

[28] Lakes, R.S., "Size effects and micromechanics of a porous solid", *J. Materials Science*, Vol. 18, 1983, pp. 2572-2581

[29] Lakes, R.S., "Experimental microelasticity of two porous solids", *International Journal of Solids and Structures*, Vol. 22, 1986, pp. 55-63

[30] Park, H.C. and Lakes, R.S., "Torsion of a Cosserat elastic prism of square cross section", *Int. J. Solids, Structures*, Vol. 23, 1987, pp. 485-503

[31] Lakes, R.S., "Demonstration of consequences of the continuum hypothesis", *Mechanics Monograph*, Vol. M5, 1985, pp. 1-5

[32] Park, H.C. and Lakes, R.S., "Cosserat micromechanics of bone: strain redistribution by a hydration-sensitive constituent", *J. Biomechanics*, Vol. 19, 1986, pp. 385-397

[33] Lakes, R.S., Gorman, D., and Bonfield, W. "Holographic screening method for microelastic solids", *J. Materials Science*, Vol. 20, 1985, pp. 2882-2888

[34]Kaloni, P. N. and Ariman, T. Stress Concentration in Micropolar Elasticity, *J. Applied Mathematics, Physics (ZAMP)* , Vol. 18, 1967, pp. 136-141.

[35]Lakes, R.S. and Yang, J.F.C., "Concentration of strain in bone", Proceedings 18th Midwest Mechanics Conference, Iowa City, Developments in Mechanics Vol. 12, 1983, pp. 239-242

[36]Whitney, J.M., Daniel, I.M., and Pipes, R.B., *Experimental Mechanics of Fiber Reinforced Composite Materials*, SESA/Prentice Hall (1982)

[37]Awerbuch, J. and Madhukar, S., "Notched strength of composite laminates: predictions and experiments: a review", *J. Reinforced Plastics and Composites* , Vol. 4, 1985, pp. 3-159

[38]Daniel, I.M., Biaxial testing of graphite/epoxy composites containing stress concentrations", *AFML-TR-76-244* Part 1 (Dec.1976), Part 2 (June 1977)

[39]Waddoups, M.E., Eisenmann, J.R. and Kaminski, B.E., "Macroscopic fracture mechanics of advanced composite materials", *J. Composite Materials* , Vol. 5, 1971, pp. 446-454

[40]Whitney, J.M. and Nuismer, R.J., "Stress fracture criteria for laminated composites containing stress concentrations", *J. Composite Materials* , Vol. 8, 1974, pp. 253-275

[41]Karlak, R.F., "Hole effects in a related series of symmetrical laminates", in *Proceedings of failure modes in composites, IV*, The metallurgical society of AIME, Chicago,1977, pp. 106-117

[42]Daniel, I.M., Rowlands, R.E. and Whiteside, J.B., "Deformation and failure of a boron-epoxy plate with a circular hole", in *Analysis of the test methods for high-modulus fibers and composites*, ASTM Publishing (1973)

[43]Rowlands, R.E., Daniel, I.M. and Whiteside, J.B., "Stress and failure analysis of a glass-epoxy composite plate with a circular hole", *Experimental Mechanics* , Vol. 13, 1973, pp. 31-37

[44]Daniel I.M., "Strain and failure analysis in graphite/epoxy plates with cracks", *Experimental Mechanics, Vol.* 18, 1978, pp. 246-252

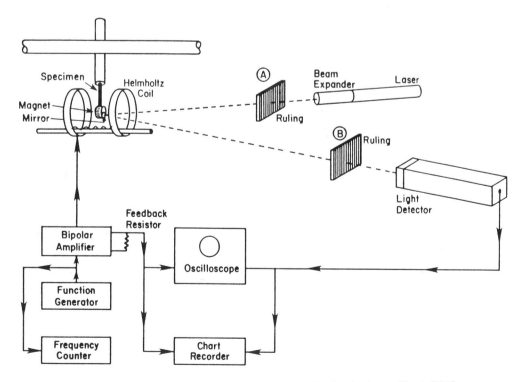

Fig. 1   Experimental configuration for the method of size effects[29].

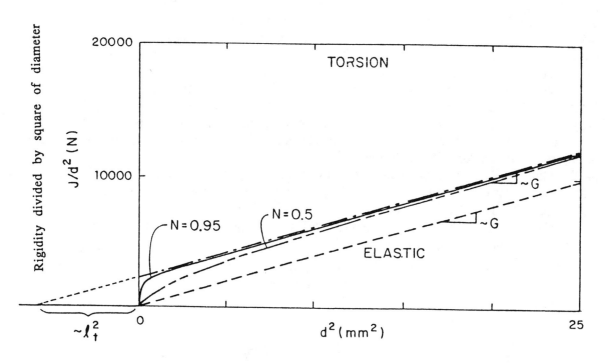

Fig. 2 Determination of Cosserat elastic constants from torsional size effect data, based on an analytical solution[7].

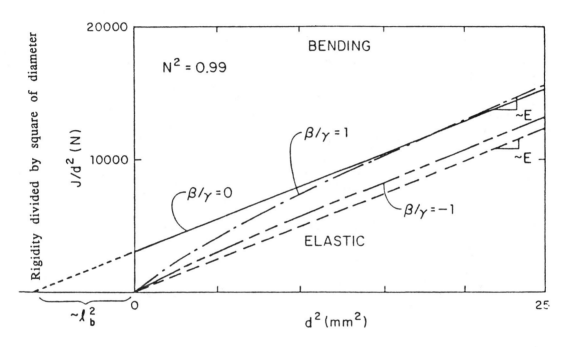

Fig. 3 Determination of Cosserat elastic constants from bending size effect data, based on an analytical solution[8].

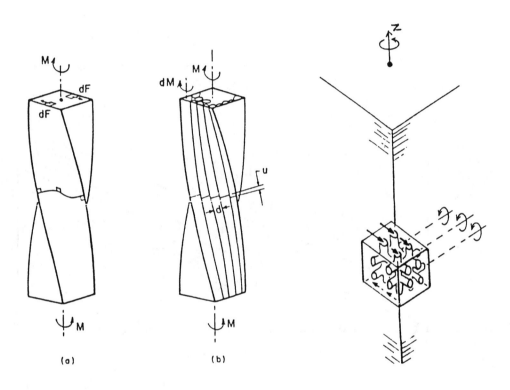

Fig. 4   Physical origin of couple stresses in the torsion of a square bar.
(a) Classically elastic bar, (b) Fibrous bar[32], (c) Foam bar [31].

History Dependent Free Edge Stresses in Composites

KUEN Y. LIN, I. H. HWANG AND LARRY B. ILCEWICZ

ABSTRACT

Generalized plane strain finite elements are formulated using linear viscoelastic theory. Analyses are performed to study changes in interlaminar stresses at the free edge of multidirectional laminates subjected to cyclic load conditions. Results indicate that several variables can affect peak amplitude in interlaminar stresses and cycle to cycle variations in stress states. These parameters include frequency, stress ratio, relative dwell times and environment.

INTRODUCTION

Dimensional stability and internal stress states of polymeric matrix composites have been found to depend on load and environmental history, e.g. in [1-5]. This history dependence has been linked to the viscoelastic behavior of the matrix. The dimensional stability of multidirectional laminates can be controlled within close tolerances by orienting fibers in the principal directions of load. However, such design practices do not eliminate history dependent changes in the internal stress. These changes can affect strength and durability of a composite by altering failure mechanisms and local stress concentrations near existing damages such as matrix cracks and fiber breaks. Therefore, it is important to consider such viscoelastic effects on the internal stress histories in performing accelerated laboratory tests to evaluate the durability of composite materials.

In previous papers, special finite element procedures were developed for hygrothermal viscoelastic analysis of composite laminates with complicated geometry and boundary conditions [6-8]. Both two-dimensional laminated plates and quasi three-dimensional (generalized plane-strain) cases were treated. The finite element methods developed have been used to solve for both inplane and interlaminar stresses and histories in laminated composites.

The current work analytically evaluates changes in free edge stress distribution as a function of load and environment history. The analysis involves the use of generalized plane-strain finite elements. Several examples are given to demonstrate history dependence of internal stresses. Variables studied include cycling frequency, stress ratios, relative dwell times, environment, and number of cycles. Parameters contributing to changes in peak internal stress conditions will be identified. Results from this study are intended to provide insight on using short-term creep and fatigue tests to predict long-term durability behavior.

K. Y. Lin, Associate Professor; I. H. Hwang, Research Assistant, Department of Aeronautics and Astronautics, FS-10, University of Washington, Seattle, WA 98195

L. B. Ilcewicz, Senior Specialist Engineer, Advanced Programs-Composite Group, Boeing Commercial Airplanes, Seattle, WA 98124

FORMULATION

The geometry considered in the present analysis is a long composite laminate subjected to a uniform extensional strain ε_0 (t) as shown in Figure 1a. It is assumed that the laminate is in the state of generalized plane strain, that is, the stresses and strains are independent of the coordinate x. With these assumptions, the viscoelastic solution of interlaminar stresses near the free edge was obtained by Lin and Yi [7,8] using a special finite element formulation. The solution involves the use of the integral form of viscoelastic constitutive equations [9] in which the relaxation moduli are expressed in terms of exponential series. The finite element procedures [10] and the variational theorem are employed to derive an element stiffness matrix and Euler's equation for the variation. Assembly of the element stiffness matrices over the entire domain leads to the following global equations for nodal displacements u_n:

$$\int_{\tau=0}^{\tau=t} K_{mn}(\zeta - \zeta)\ \frac{\partial u_n}{\partial \tau}\ d\tau = F_m(t) - F_m^0(t) \tag{1}$$

where m,n =1,2,3,......,NDT. NDT is the total degrees of freedom.

In the above, K_{mn} is the assembled stiffness matrix and ζ is the reduced time. $F_m(t)$ represents the "residual" nodal force vector due to the applied strain ε_0 and $F_m(t)$ are the nodal forces which are zero everywhere except the boundary where the tractions are prescribed.

A direct integration of Eq.(1) would require enormous computing time and memory storage since the stiffness matrix K_{mn} depends on past histories. To overcome such difficulties, a recursive formula was developed for the numerical solution of Eq. (1). With the recursive algorithm, the displacements u_n at time t_p can be integrated by using only the solution at time t_{p-1} rather than all previous solutions for $t<t_p$. The complete finite element procedure for thermo-viscoelatic problems can be found in [7,8].

RESULTS AND DISCUSSION

The current work addresses a three-dimensional stress state at the free edge of a laminate using linear viscoelasticity. The authors recognize the fact that most composite materials currently used in the aerospace industry have some nonlinear viscoelastic behavior [2-4], particularly when the application involves extreme environments, long times or high strain levels. Attempts to generalize the present finite element procedure to model this behavior as applied to the problem of free edge stresses will be made in the future. The complex nature of this generalization and the projected computing costs for such an analysis did not appear warranted until first gaining experience with linear viscoelasticity.

Example analyses were performed to demonstrate the effects of cyclic load conditions on interlaminar stresses in composite laminates. The problem analyzed was a $(45/90/-45/0)_s$ quasi-isotropic laminate subjected to an applied cyclic strain $\varepsilon_0(t)$ in the x direction (see Fig. 1a). Square wave forms as shown in Fig. 1b were used for strain cycles. Cyclic parameters studied include frequency, R-ratio (ratio of minimum to maximum strain values), relative dwell times (hold times at minimum and maximum strain conditions), equilibrium environments and number of cycles. The viscoelastic properties obtained from the literature [11] for a graphite/epoxy composite, GY70/339, were used. This material exhibits strong viscoelastic behavior in high temperature and humidity environments.

The finite element model used in the analysis consists of 30 mesh divisions in the y-direction and 12 divisions along the z-axis which makes a total of 360 elements on the yz plane. An individual element has four corner nodes each containing three displacement freedoms (u,v,w) to represent 3-D deformations (12 dof's per element). A total of 1209 degrees of freedom is used for the entire model. Typical computing time on the Cray X-MP is 16 seconds per iteration. Depending on parametric variables, up to eight iterations were used for each cycle. The strong viscoelastic effects of GY70/339 materials subjected to extreme environmental conditions allow trends in behavior to be demonstrated in a small number of cycles, holding computing costs down.

Before reviewing the effects of various cyclic parameters on free edge stresses three important points should be made. First, history dependent changes in interlaminar stresses, as judged by normalizing results with elastic predictions, were generally found to be different than those predicted from the one-dimensional analysis. Second, history dependent changes in different components of interlaminar stresses such as σ_z, τ_{xz}, and τ_{yz} were each found to be different. Finally, a maximum of 8 cycles was used in each example to indicate trends in viscoelastic response. Some of the examples ceased to have changes with additional cycles, while others exhibited a decreased rate of change with additional cycles. The latter cases would best be illustrated by converting the time scale from time t to log(t) in the x-axis of plots.

Free Edge Stress Distributions: The results of interlaminar stress distributions along the 90/-45 interface in a $(45/90/-45/0)_s$ laminate are shown in Figs. 2 and 3 at three specific times: t=0, 100, and 1400 seconds. Note that time t=100 seconds corresponds to minimum strain condition ($\varepsilon_0 = -0.006$ in/in) and t=1400 is associated with the maximum strain ($\varepsilon_0 = 0.006$ in/in) in the applied load cycle. The environmental conditions used in these analyses are: temperature T=120°F and moisture content (*mc*) M=1.4%. The loading conditions are: frequency f=0.005 cycles/sec., maximum applied strain $\varepsilon_0 = 0.006$ in/in with R= -1, dwell times t_1/t_2=1. Similar to the elastic case, large interlaminar shear stress τ_{xz} (see Fig. 2) and normal stress σ_z (Fig. 3) exist near the free edge y/b=1 due to mismatch in Poisson's ratios and other material properties between layers. The differences in stress distributions at t=100 and 1400 seconds are due to the different strain levels applied as well as the history effects.

Effect of cyclic frequency f : Viscoelastic changes in free edge stresses are expected to depend on the cycle frequency *f*. In order to evaluate this effect, analyses were made with two different frequencies holding other variables constant. The data base used for viscoelastic properties of GY70/339 limited studies to values of *f* less than 0.1 cycles/sec. Values for *f* of 0.005 cycles/sec and 0.00005 cycles/sec were chosen. The R-ratio was held at 0.1 (i.e., tension/tension fatigue), relative dwell times were equivalent (t_1/t_2=1), and environment was 120°F/1.4% *mc*. Figs. 4 and 5 compare the results of interlaminar stress σ_z at a point close to the free edge and the 90/-45 interface for f=0.005 and 0.00005 cycles/sec. These stress values have been normalized with respect to the corresponding elastic solutions at time t=0. Figs. 6 and 7 show normalized τ_{xz} values at the same location as a function of time.

For cyclic variables used in Figs. 4-7, peak interlaminar stresses are less than elastic predictions due to viscoelastic effects. This is typical of cyclic conditions when both maximum and minimum applied strain conditions are of the same sign (i.e., R > 0). As expected, viscoelastic effects are strongest for the lowest frequency (*f* = 0.00005 cycles/sec. in Figs. 5 and 7), causing the greatest drop in peak interlaminar stresses. Note that interlaminar stresses relax during the constant strain portions of each cycle. As a result, the peak absolute values of interlaminar stress during a cycle always occured immediately after elastic steps in the square wave (i.e., steps from maximum or minimum conditions). This trait is characteristic of all cycles evaluated in this study.

It is noted that viscoelastic effects are very strong in Figs. 4-7 due to the severe environment assumed. Strong environmental effects are typically indicated by large fluctuations within a cycle. Following the first cycle, viscoelastic changes continue to occur, but cycle to cycle differences diminish. By the eighth cycle the interlaminar stress envelope appears to be the same as that of the previous cycle. Also note that although maximum and minimum cycling strains are of the same sign in Figs. 4-7 (i.e., R = 0.1), viscoelastic effects cause sign reversal in interlaminar stresses. The sign reversal is important to the condition by which interlaminar stress envelopes become constant (i.e., repeat with increasing numbers of cycles). Interlaminar stress envelopes in these figures tend toward a repeating sequence because equivalent dwell times for opposite signs of interlaminar stress eventually result in a balanced cycle of relaxation. The balanced cycle is reached quickly in these examples due primarily to the extreme environmental conditions which force a sign reversal in interlaminar stress during the first cycle. Note that unequal dwell times will not allow a balanced cycle of relaxation. This will be discussed later.

Effect of R-ratio: Peak interlaminar stress depends directly on R-ratios. Figs. 4-7 show that the interlaminar stresses are generally less than the elastic predictions for R=0.1. For cases with R-ratio approaching 1, longer time is needed before interlaminar stress envelops become constant. The amplitude of peak interlaminar stresses would continue to decrease with time until the equilibrium condition is reached. For example, consider two cycles with equivalent maximum applied strains, but having different R-ratios of 0.1 and 0.8. Significantly more relaxation and lower interlaminar stresses would occur in the latter case before reaching a repeated stress envelope.

In order to demonstrate the effects of applied strain reversal (R < 0), the case of R=-1 was studied. The relative dwell times were held equivalent and environment was 120°F/1.4% *mc*. The results of normalized τ_{xz} values at a point close to the free edge and the 90/-45 interface are plotted in Fig. 8 for f=0.005 cycles/sec. and in Fig. 9 for f=0.00005 cycles/sec. This provides a direct comparison with the results of R =0.1 shown in Figs. 6 and 7. It is seen that, for the case of negative R-ratios, viscoelastic effects cause higher peak interlaminar stresses than elastic predictions. This is opposite to the trend observed for positive R-ratios and may contribute to higher growth rates for edge delamination. This may partially explain why tension/compression fatigue tends to be the worst case in experimental fatigue studies with multidirectional laminates. The difference between viscoelastic and elastic peak stresses depends directly on the magnitude of relaxation during periods of constant applied strain as affected by cyclic frequency, dwell times and environment. For example, Fig. 9 has higher peak interlaminar stresses than Fig. 8.

Strong environmental effects are again indicated in Figs. 8 and 9 by large fluctuations within a cycle. Cycle to cycle differences diminish even more rapidly than was the case in Figs. 6 and 7. This is due to the R = -1 condition whereby a sign reversal in interlaminar stress is insured during the first cycle, forcing balanced stress relaxation. This occurs independently of the strength of viscoelastic effects. Consequently, severe environmental effects only have an influence on the magnitude of changes within a cycle. In general, cycle to cycle differences cease very rapidly when R < 0, provided that the relative dwell times at maximum and minimum applied strains are equivalent.

Effect of dwell times: When relative dwell times at maximum and minimum strain conditions are not equivalent, balanced relaxation of opposite signs of interlaminar stress within the same cycle is not possible. This is true regardless of the number of cycles. As a result, both cycle to cycle variations and fluctuations within a cycle depend on viscoelastic effects. Fig. 10 shows interlaminar shear stress τ_{xz} calculated from cycles with unbalanced dwell times (t_1/t_2=0.01). Other parameters were held constant, i.e. R= -1, f=0.00005 cycles/sec, and the environment was 120°F/1.4% *mc* Note the cyclic variation of maximum and minimum interlaminar stresses. This trend of increasing maximum interlaminar stress will continue until the minimum interlaminar stress approaches zero.

Effect of environmental conditions: Two different environmental conditions were studied: one in the 70°F/0.8% *mc* condition and the other under 120°F/1.4% *mc* environment. Other cyclic parameters remain the same, that is, R = -1, t_1/t_2=0.01, and f = 0.00005 cycles/sec. The normalized interlaminar shear stresses τ_{xz} in the 70°F/0.8% *mc* environment are shown in Fig. 11. These results follow the same trend as Fig. 10, but at a much slower rate due to relatively weak viscoelastic behavior in this environment. The concept of time/temperature/moisture superposition suggests that if this cycling is allowed to continue for a long period of time, the same peak values as those shown in Fig. 10 for more severe environments can be attained. Note that this situation occurs only with unequal dwell times.

Effect of Multiaxial Stresses: The stress histories in Fig.9 for the generalized plane strain state are replotted in Fig. 12 (solid line) and compared with the results obtained from the 1-D viscoelastic analysis (dotted line). It is evident that the time dependency of a multiaxial stress problem can not be well predicted through the use of a simplified 1-D model. In general, each component of relaxation moduli has a different time function. As a result, the viscoelastic response in 1-D problems is significantly different from that of 3-D problems.

The mechanisms that degrade life and residual strength of composite materials are quite different from those of metals; therefore, the approach to durability as applied to composites should reflect these differences. The current study has identified some of the cyclic variables that change the magnitude of free edge stresses. The question of time-dependent changes in interlaminar stresses, following the initiation of delamination, must be addressed in the future. Experimental studies on other types of damage accumulation and analysis of viscoelastic changes in stress distribution near existing damage are also needed. Such studies are expected to yield other clues on factors that affect the long-term durability of composite materials.

CONCLUSIONS

A linear viscoelastic constitutive law was incorporated with generalized plane strain finite elements to study the effects of cyclic variables on free edge stresses in multidirectional laminates. Variables studied include cycling frequency, R-ratio, relative dwell times, environment, and number of cycles.

Viscoelastic changes in interlaminar stresses were shown to be significant. In general, positive R-ratios were found to decrease peak interlaminar stresses, while negative R-ratios result in an increase. The absolute value of change is dependent on other factors affecting viscoelastic response such as frequency and environment.

Cycles with equivalent dwell times for maximum and minimum applied strains tend to shift the interlaminar stress state toward a condition of balanced relaxation. This allows viscoelastic behavior to occur within a cycle but eliminates cycle to cycle variations. Since a condition of balanced relaxation requires that dwell periods within a cycle have opposing signs of interlaminar stress, the number of cycles necessary to reach such a state depend on frequency, R-ratio, and environment.

Cycles not having equivalent dwell times for maximum and minimum applied strains tend to shift the interlaminar stress associated with the long dwell towards zero. When R-ratio is negative, the magnitude of interlaminar stress associated with the short dwell is highest of all cases studied.

ACKNOWLEDGEMENTS

The work reported in this paper is sponsored by the Boeing Company. The authors would like to thank Dr. John T. Quinlivan, Robert D. Wilson, and Jay M. Hopper of Boeing Commercial Airplanes for their support and helpful disussions.

REFERENCES

1. Flaggs, D.L., and Crossman, F.W., "Analysis of the Viscoelastic Response of Composite Laminates During Hygrothermal Exposure," J. Composite Materials, Vol. 15, 1981, p. 21.

2. Tuttle, M.E., and Brinson, H.F., Accelerated Viscoelastic Characterization of T300/5208 Graphite/Epoxy Laminates, NASA CR-3871, 1985.

3. Harper, B.D., and Weitsman, Y., "On the Effects of Environmental Conditioning on Residual Stresses in Composite Laminates," Int. J. Solids Structures, Vol. 21, 1985, p. 907.

4. Cardon, A.H., Hiel, C.C., and Brouwer, H.R., "Nonlinear Viscoelastic Behavior of Epoxy-Matrix Composites Under Combined Mechanical and Environmental Loadings," Proceedings of International Symposium on Composite Materials and Structures, Beijing, China, Technomic Publ. Co., 1986, p. 619.

5. Rothschilds, R.J., Ilcewicz, L.B., Nordin, P., and Applegate, S.H., "The Effect of Hygrothermal Histories on Matrix Cracking in Fiber Reinforced Laminates," J. of Eng. Materials and Tech., Trans. of ASME, 110(2), 1988, pp 158—168.

6. Lin, K.Y., and Hwang, I.H., "Thermo-Viscoelastic Response of Graphite/Epoxy Composites," J. of Eng. Materials and Tech., Trans. of ASME, 110(2), 1988, pp 113-116.

7. Lin, K.Y., and Yi, S., Time-Dependent Behavior of Laminated Composites in Hygrothermal Environments, Final Report on Contract # Y-422563-OD56N to the Boeing Commercial Airplane Co., 1987.

8. Lin, K. Y., and Yi, S., "Analysis of Time-Dependent Free Edge Stresses in Composites," paper in preparation.

9. Schapery, R.A. "Stress Analysis of Viscoelastic Composite Materials," J. Composite Materials, Vol. 1, 1967, p. 228.

10. Wang, A.S.D., and Crossman, F.W., "Some New Results on Edge Effect in Symmetric Composite Laminates", J. Composite Materials, Vol. 11, 1977, pp. 92-106

11. Crossman, F.W., Mauri, R.E., and Warren, W.J., "Moisture—Altered Viscoelastic Response of Graphite/Epoxy Composites," Advanced Composite Materials — Environmental Effects, ASTM STP 658, J.R. Vinson, ed., American Society for Testing and Materials, 1978, pp 205-220.

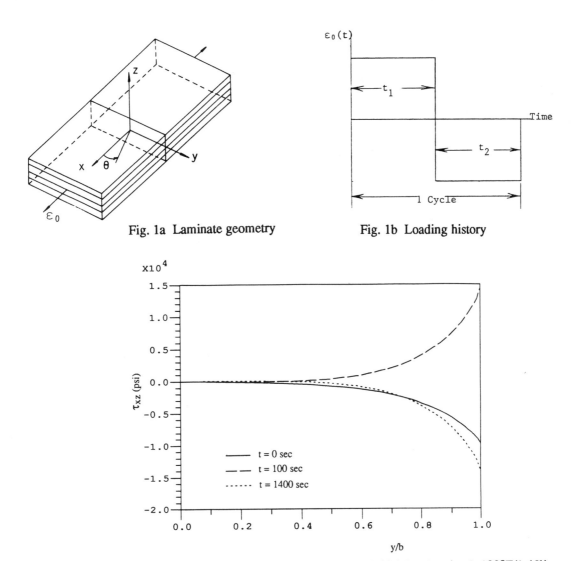

Fig. 1a Laminate geometry

Fig. 1b Loading history

Fig. 2 Distributions of interlaminar shear stress τ_{xz} (f=0.005, R=-1, t_1/t_2=1, 120°F/1.4%)

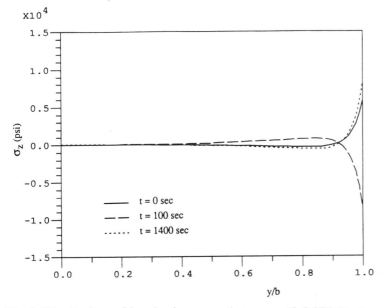

Fig. 3 Distributions of interlaminar normal stress σ_z (f=0.005, R=-1, t_1/t_2=1, 120°F/1.4%)

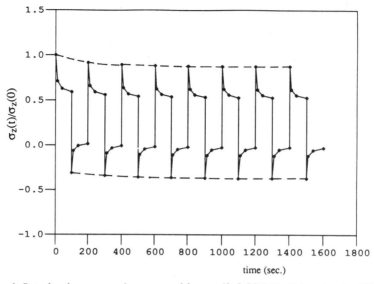

Fig. 4 Interlaminar normal stress σ_z history (f=0.005, R=0.1, t_1/t_2=1, 120°F/1.4%)

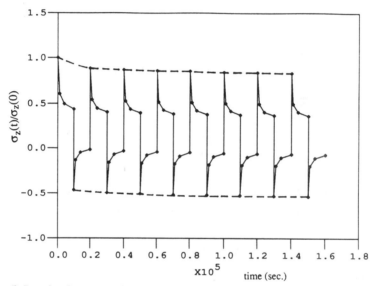

Fig. 5 Interlaminar normal stress σ_z history (f=0.00005, R=0.1, t_1/t_2=1, 120°F/1.4%)

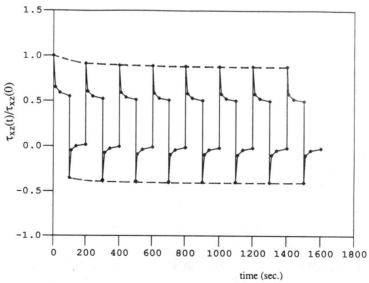

Fig. 6 Interlaminar shear stress τ_{xz} history (f=0.005, R=0.1, t_1/t_2=1, 120°F/1.4%)

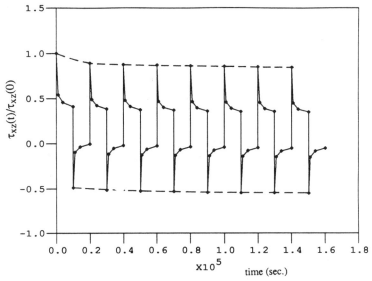

Fig. 7 Interlaminar shear stress τ_{xz} history (f=0.00005, R=0.1, t_1/t_2=1, 120°F/1.4%)

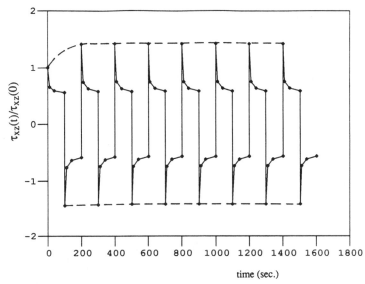

Fig. 8 Interlaminar shear stresses τ_{xz} (f=0.005, R=-1, t_1/t_2=1, 120°F/1.4%)

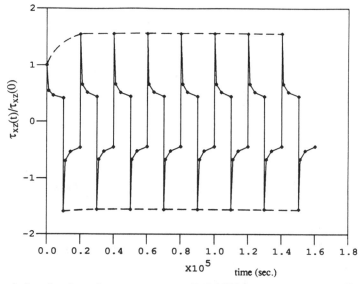

Fig. 9 Interlaminar shear stresses τ_{xz} (f=0.00005, R=-1, t_1/t_2=1, 120°F/1.4%).

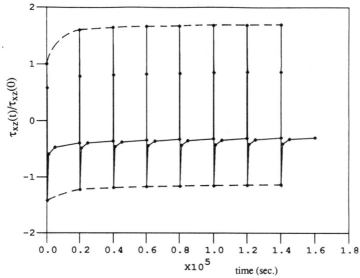

Fig. 10 Interlaminar shear stresses τ_{xz} (f=0.00005, R=-1, t_1/t_2=0.01, 120°F/1.4%)

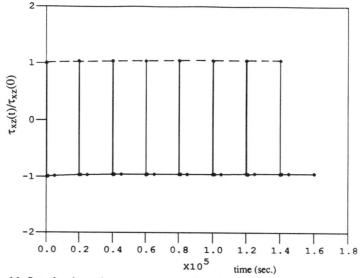

Fig. 11 Interlaminar shear stresses τ_{xz} (f=0.00005, R=-1, t_1/t_2=0.01, 70°F/0.8%)

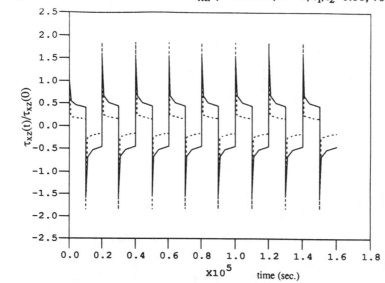

Fig. 12 Comparison of 1-D and 3-D viscoelastic solutions (f=0.00005, R=-1, t_1/t_2=1, 120°F/1.4%)

Materials with Negative Poisson's Ratios: Dependence of Properties on Structure

R. S. LAKES AND J. B. PARK
Department of Biomedical Engineering
Department of Mechanical Engineering
The University of Iowa
Iowa City, IA 52242

E. A. FRIIS
Orthopaedic Research Institute
929 N. Francis St.
Wichita, KS 67214

ABSTRACT

A new class of materials exhibiting negative Poisson's ratios has been developed. The negative Poisson's ratio arises from a 're-entrant' cell structure. Various features of the re-entrant cell shape are controllable by the processing techniques.

INTRODUCTION

The theoretically allowable range of Poisson's ratio is from -1.0 to +0.5, in a linearly elastic, isotropic material. In common materials, only values in the range 0 to +0.5 are, however, observed [1]. Examples of Poisson's ratios are: approximately 0.5 for rubbers and for soft biological tissues, 0.45 for lead, 0.33 for aluminum, 0.27 for common steels, 0.3 ± 0.1 for conventional polymer foams, and nearly zero for cork.

Several cases of negative values of Poisson's ratio are known: (i) in certain highly anisotropic materials in certain directions, and (ii) in certain macroscopic structures. As for (i), 'single crystal' pyrite[2] or single crystal cadmium[3] may exhibit negative Poisson's ratio in some directions. Macroscopic, anisotropic, two-dimensional flexible models of certain honeycomb structures (not materials) have exhibited negative Poisson's ratios in some directions[4]. A structure of rods, springs, and hinges has been proposed[5], which exhibits a negative Poisson's ratio of -1. The materials described below, by contrast, exhibit the negative Poisson's ratio in all directions, they need not be anisotropic, and they do not depend on large size assembled structural elements.

METHODS

The principle underlying the materials with negative Poisson's ratios is that of a 're-entrant' unit cell[6]. Idealized structures of a normal foam cell and re-entrant cell are shown in Figures 1 and 2 respectively. In the case of a normal foam, as tension is applied in one direction, the cells elongate in that direction and constrict in the orthogonal directions, so that the cells deform into ovoids. Poisson's ratio is therefore positive. By contrast, in the re-entrant foam cell, tension applied to one pair of protruding ribs will cause the inwardly directed ribs to unfold, resulting in an expansion of the cell in all directions. Poisson's ratio is consequently negative. In real foams, the cell structure is more complex, as shown in figures 3 and 4. Nevertheless, the same processes occur.

Foams with negative Poisson's ratios can be produced from a variety of conventional foams, including thermoplastic polymer foams such as polyester polyurethane, polyether polyurethane[6], and viscoelastic elastomers, from metallic foams based on copper and aluminum[6,7] and from thermosetting polymer foams such as silicone rubber[7].

Re-entrant foams based on thermoplastic polymers were produced by a thermomechanical technique, from conventional low density open-cell polymer foams[6]. Prismatic specimens of square cross section were cut and placed in a mold 22 mm by 22 mm by 127 mm in size, to achieve a triaxial compression. The mold was then heated to above the softening temperature of the foam, held at constant temperature for a period of time, and cooled. The foam cell structure was transformed by this procedure into a re-entrant structure[6]. The required times and temperatures were determined empirically for each type of foam examined[6,8]. For example, a commercially available low density [0.043 g/cm^3] polyester polyurethane foam was subjected to 170°C for about eighteen minutes.

Specimens were then tested in tension in a MTS servohydraulic testing machine. Longitudinal and lateral engineering strain was determined from magnified video recordings of the deformation.

RESULTS

As shown in Fig. 5, polyurethane foams subjected to different permanent volumetric compressions were found to exhibit negative Poisson's ratios and reduced Young's moduli. Poisson's ratio decreased with increasing permanent volumetric compression. It was observed that the cell structure became more convoluted and more compact as the permanent volumetric compression was increased. For sufficiently high permanent volumetric compression [>3], large compressive strain resulted in cell ribs contacting each other. The negative Poisson's ratio was then reduced in magnitude and the apparent stiffness increased.

In tension and compression, the transverse expansion and contraction due to the negative Poisson's ratio was sufficient to be obvious to the unaided eye. Moreover, effects of the negative Poisson's ratio were observable in the bending of bars of re-entrant foam. The lateral surfaces [intersecting the plane of bending] of a bent prismatic beam of a conventional material with a positive Poisson's ratio assume a saddle shape: the 'anticlastic curvature' of bending, in which the transverse curvature is opposite the principal curvature of bending. The converse effect, synclastic curvature, was observed in bent bars of re-entrant foam: the corresponding lateral surfaces assumed an ellipsoidal shape.

Results of uniaxial tensile load-deformation tests revealed significant differences between re-entrant foam and conventional foam, as shown in Fig. 6. The re-entrant foams exhibited lower Young's moduli than the conventional foam from which they were derived. Moreover, the tensile stress-strain curves exhibited a much wider range of linear behavior. Such behavior is similar to what was earlier noted in compression[6].

These foam materials can be subjected to large strains. We have observed that the negative Poisson's ratios are maintained over a large strain range in tension if the original foam was given a relatively large permanent volumetric compression. In comparison, foams given smaller permanent volumetric compression will exhibit a negative Poisson's ratio over a range of strain in compression.

DISCUSSION: STRUCTURAL ASPECTS

The foams considered here have a relatively large cell size. Large cells are not, however, required to produce negative Poisson's ratios. There is no characteristic length scale in the [classical] theory of elasticity by contrast to Cosserat elasticity and related theories. Poisson's ratio is an elastic constant of the classical theory of elasticity. Consequently, the structure responsible for the negative Poisson's ratio could be on the atomic scale.

The role of structural assumptions in the prediction of Poisson's ratio can be traced to the foundations of the theory of elasticity. In 1821, Navier proposed a theory of interatomic interaction in which the forces are central, and act along the lines joining pairs of atoms and the deformation is affine. Poisson himself soon after concluded from this theory that Poisson's ratio must be 0.25 for all materials. This view was accepted for many years until experiments disclosed different Poisson's ratios for various materials. In common materials for which Poisson's ratio differs from 0.25, the interatomic forces must be non-central[9,10] and/or non-affine[11] deformation occurs. The existence of couple stresses [and Cosserat elastic effects] in the former case is indicated, to satisfy the condition for equilibrium. Consequently, structural considerations indicate the existence of couple stresses and/or non-affine deformation of the unit cells for most materials. In those materials for which the dominant structural features are on the atomic scale, the Cosserat characteristic length would be of atomic size, and consequently would be unobservable in macroscopic experiments. Such materials would appear classically elastic, with a Poisson's ratio differing from 0.25.

The negative Poisson's ratio increases in magnitude with permanent volumetric compression. The structural interpretation of this phenomenon is that the cell ribs become progressively more convoluted and have more tendency to unfold under tension. A sufficient tensile strain can be expected to abolish the effect since the ribs will become straightened. The strain required for this will depend on the permanent volumetric compression used in making the foam.

CONSEQUENCES OF NEGATIVE POISSON'S RATIOS

Many phenomena in the deformation of elastic materials depend on the Poisson's ratio. The simplest is that a material with a negative Poisson's ratio will get fatter in cross section when stretched and thinner when compressed. The relation between the shear modulus G, the bulk modulus B (the inverse of the compressibility) and Poisson's ratio ν is: B = $2G(1+\nu)/(1-2\nu)$. When the Poisson's ratio approaches 1/2, as in rubbery solids, the bulk modulus greatly exceeds the shear modulus and the material is referred to as incompressible When the Poisson's ratio approaches -1, the material becomes highly compressible; its bulk modulus is much less than its shear modulus.

Consider cantilever bending of a beam of circular cross section of radius a. If bending is by a concentrated load P, axial and shear stress components are [12]

$$\sigma_{zz} = -P(L-z)x/[\pi a^4/4] \qquad \sigma_{xz,maximum} = -[P/\pi a^4][(1+2\nu)/2(1+\nu)].$$

As Poisson's ratio approaches -1, the shear stress becomes large, but the deflection curve is unchanged. In the case of a uniformly distributed load [12], the curvature K depends very much on Poisson's ratio: $K = [M/EI][1 - \{\{7 + 12\nu + 4\nu^2\}/6(1+\nu)\}\{a^2/L^2\}]$.

Consider the bending of plates, for which the bending rigidity $D = Eh^3/12(1-\nu^2)$, in which E is Young's modulus and h is the plate thickness. The rigidity becomes large for a given value of E as the Poisson's ratio approaches -1. For pure bending of plates, the rigidity is proportional to $D(1+\nu)$, so that even moderately negative values of the Poisson's ratio will significantly affect the rigidity.

Stress concentrations may depend upon Poisson's ratio. For example, the maximum stress around a spherical cavity in a field of tension T is $\sigma = T[(27-15\nu)/2(7-5\nu)]$, so that the stress concentration factor σ/T is 2.17 for ν =0.5, 1.93 for ν =0, 1.82 for ν =-0.5, and 1.75 for ν =-1. In this case the stress concentration factor does not vary much with Poisson's ratio, even for negative values. In other cases, eg. a hole in a bent bar, or a rigid inclusion in a field of tension, a larger effect of Poisson's ratio may be seen.

In the indentation of a block of material by a localized circular pressure distribution of radius a, the maximum indentation w for a given pressure P is $w/P = 2a(1-\nu^2)/E$, in which E is Young's modulus [12]. Consequently, a material with a negative Poisson's ratio approaching the thermodynamic limit -1 will be difficult to indent even if the material is compliant.

POSSIBLE APPLICATIONS

We envisage three classes of applications of novel foams with negative Poisson's ratios : (i) those based on the isotropic resilience and toughness of the foams and (ii) those based on the negative Poisson's ratio, and (iii) those based on the dynamic properties of the foams.

As for (ii), particular values of Poisson's ratio have been found to be advantageous in existing aplications, eg. the cork of a wine bottle. Cork, with a Poisson's ratio of nearly zero, is ideal in this application. Rubber, with a Poisson's ratio of 0.5, could not be used for this purpose because it would jam as a result of lateral expansion when pressed into the neck of the bottle. Other cases may be suggested by the foregoing.

For a specific example, consider the choice of materials for a wrestling mat. Let us examine the penetration rigidity F/w in which F in the indentation force and w is the maximum displacement, for elastic materials in different geometries. For an elastic half space under a circular pressure distribution [8,12] of radius a,

$$[F/w]_{narrow} = Ga/(1-\nu).$$

Consider now the uniform compression of a layer of thickness H and radius a, in which the lateral Poisson effect is restrained. Let the force F be distributed uniformly over the layer. Then,

$$[F/w]_{wide} = Ga^2/H[2(1+\nu)/(1-2\nu)].$$

In this application, it is desired that the layer reduce impact forces which may be distributed over a wide area or a narrow area. The layer must be sufficiently compliant for distributed forces, yet must be sufficiently rigid that it does not bottom out under a concentrated force. Consequently, the following ratio is to be minimized:

$$[F/w]_{wide}/ [F/w]_{narrow} = [2a/H] [(1-\nu^2)/(1-2\nu)].$$

Materials with negative Poisson's ratios offer the best performance, and rubbery materials are the worst in this application.

The toughness of a material can also depend on its Poisson's ratio. In the context of elasticity theory [13], the critical tensile stress σ for fracture of a solid of surface tension T, Young's modulus E, with a plane circular crack of radius r is $\sigma = \sqrt{[\pi ET/2r(1-\nu^2)]}$. When the Poisson's ratio approaches -1, a material of given Young's modulus is predicted to become very tough. In the context of material microstructure, the presence of many bent ribs with low stiffness but high extensibility may be expected to enhance the toughness.

Potential applications for re-entrant foams include wrestling mats, knee pads, tear resistant sponges, air filters, robust shock absorbing material, helmet liners, packing material, gaskets, vascular implants, fillings for highway joints, pads under carpeting, ankle wrap, lecture demonstrations in elasticity, and sound absorbing layering.

CONCLUSION

Foam materials with negative Poisson's ratios represent a new class of cellular solids which have unusual properties and which may be useful in a variety of application.

REFERENCES

1. Y. C. Fung, *Foundations of solid mechanics*, (Prentice Hall, 1968), p.353.
2. A. E. H. Love, *A treatise on the mathematical theory of elasticity*, (Dover,1944) 4th ed.
3. Y. Li, "The anisotropic behavior of Poisson's ratio, Young's modulus, and shear modulus in hexagonal materials", *Phys. Status Solidi* Vol. 38, 1976, pp. 171-175
4. L. J. Gibson, M. F. Ashby, G. S. Schajer, and C. I. Robertson, "The mechanics of two dimensional cellular solids", *Proc. Royal Society London*, Vol. A382, 1982, pp. 25-42
5. R. F. Almgren, "An isotropic three dimensional structure with Poisson's ratio = -1", *J. Elasticity*, Vol. 15, 1985, 427-430

6. R. S. Lakes, "Foam materials with a negative Poisson's ratio",
    Science , Vol. 235, 1987, pp. 1038-1040

7. Friis, E.A., Lakes, R.S., and Park, J.B., "Negative Poisson's ratio polymeric
    and metallic materials", *Journal of Materials Science,* accepted (1988).

8. Lakes, R.S. "Foam materials with a negative Poisson's ratio,"
    Developments in Mechanics, Vol. 14, p. 758-763, Proceedings, 20th Midwest Mechanics
    Conference, Purdue University, Aug. 31-Sept. 2, 1987.

9. S. Burns, "Negative Poisson's ratio materials, letter", *Science* , Vol. 238, 1987, p. 551

10. R.S. Lakes, "Negative Poisson's ratio materials, reply", *Science,* Vol. 238, 1987, p.
    551

11. A. Ruina, Cornell University, private communication.

12. S. P. Timoshenko and J. N. Goodier, *Theory of Elasticity,* (McGraw Hill, 1982) 3rd Ed.

13. I. N. Sneddon, *Fourier Transforms*, (McGraw Hill, 1951).

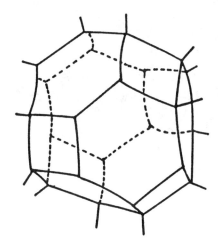

Fig. 1   Schematic unit cell of conventional foam.

Fig. 2   Schematic unit cell of re entrant foam

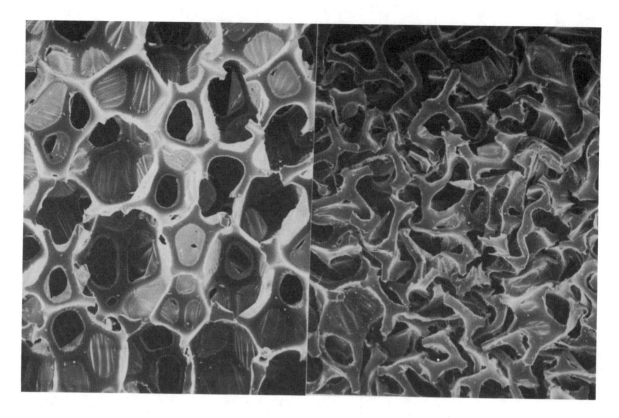

Fig. 3   Cell structure of conventional foam   Fig. 4   Cell structure of re-entrant foam

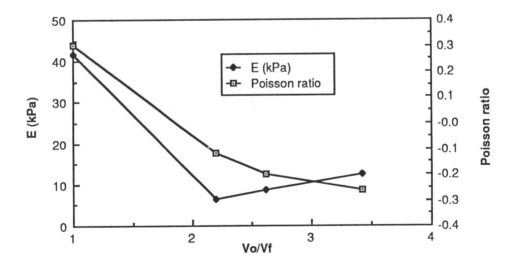

Fig. 5   Young's modulus and Poisson's ratio <u>vs</u> permanent volumetric compression

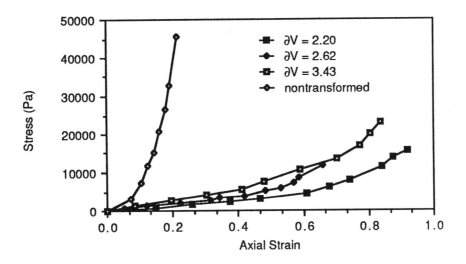

Fig. 6   Stress strain curves for normal and re entrant foams

SYMPOSIUM XII

Durability and Non-Destructive Testing

Engineering Tomography: A Quantitative NDE Technique for Composite Materials

H. M. STOLLER, J. G. CROSE AND W. H. PFEIFER

ABSTRACT

The term Engineering Tomography (ET) is applied to the integration of tomographic inspective techniques, image reconstruction algorithms, data correlation methods and numerical analysis methods to predict the behavior of engineering structures. An example is to measure the spatial variation in density of a material specimen with x-ray computed tomography (CT), translate this data into a finite element analysis (FEA) model and predict the stress and strain states due to a specific loading condition given a set of density-dependent material properties. This method leads to "quantitative" non-destructive evaluation (NDE), a methodology for understanding the effects of defects and enhancement of material characterization data.

CT has been demonstrated to have the resolution and sensitivity capabilities to measure small density variations in composite material systems. Software has been developed to automate the calibration of CT measurements to provide accurate density data. Composite materials have matrix-dependent properties which are sensitive to density variations. The basis exists to quantitatively, on a non-intrusive basis, assess the effect of density inhomogeneity defects on the performance of composite material structures.

Recent efforts, have supported initial quantitative NDE studies. Software is being developed which linked CT scans and CT data to finite element models. Finite element analysis software is being modified to utilize density-dependent material properties. Predicting the effect of density variations on the performance of a composite material structure has been successfully demonstrated. Refinements in the ET methodology appear attainable and are being pursued.

INTRODUCTION

Engineering Tomography (ET) integrates the multi-faceted capabilities of advanced NDI techniques with finite element analysis (FEA) methods. Current reliance is on NDI techniques which provide spatial information on material characteristics. Use of FEA methods provide the means to assess the influence of spatially-resolved material properties on component response. Integrated, the use of tomographic NDI techniques and FEA methods has the capability to provide a quantitative NDE assessment of the performance of composite material components.

One major difficulty in performing quantitative evaluations of composite materials is the limited knowledge that exists on what defects need to be detected and the magnitude of their effect. For example, severe wrinkles (i.e., ply foldbacks) severely degrade longitudinal tensile

H.M. Stoller, J.G. Crose, W.H. Pfeifer, PDA Engineering, 2975 Redhill Avenue, Costa Mesa, CA 92626.

properties whereas low density regions have little effect on fiber-dominated properties [3]. Conversely, density variations have a significant effect upon matrix-dominated properties. Delaminations influence both in-plane and out-of-plane properties.

ET, through the use of NDI techniques which can spatially define material characteristics, has the potential to achieve the desired quantitative, non-intrusive assessment capability. PDA Engineering has been conducting a series of investigations to advance this technology. The effort has focused on density inhomogeneities because of the ability of CT to resolve density variations at a spatial dimension significant to the response of composite materials.

The initial ET studies are reviewed which examined the ability of CT to measure density and density variations in composite materials and developed software to automatically process and display this information. The results of recent investigations are also presented. These developed physical density-material property relationships for representative composite materials; developed software to link the CT information with FEA models; and assessed the ability to predict the performance of composite material structures based on such data.

ENGINEERING TOMOGRAPHY DEFINED

ET is defined as the integration of tomographic (i.e., spatial resolution) NDI techniques, 2D and 3D image reconstruction algorithms, material defect-material property relationships and FEA methods to achieve a quantitative NDE capability. The emphasis is on tomographic NDI techniques because of the advantages of spatially-defined material characteristics. It is recognized that a role exists for non-tomographic techniques in ET such as ultrasonics, thermography, etc., which will provide complementary information on properties.

ET is dependent upon performance-critical relationships between tomographic NDI data and mechanical and thermal properties. To date, this has been accomplished by single energy source CT to deduce physical density and, through characterization programs, to establish density-property relationships. Advances in NDI capabilities, such as increased resolution CT to define voids and delaminations, dual energy CT to determine variations in chemical composition, tomographic nuclear magnetic resonance (NMR) to measure residual volatiles and water absorption, etc., will extend ET to other anomaly regimes.

ET also incorporates advances in computer-aided engineering (CAE) software. This includes the construction of 3D images from a series of 2D NDI scans or slices; the development of FEA models from 2D and 3D NDI images; and linkage software to translate from the CT geometry frame of reference to an FEA geometry frame of reference. CAE software also displays and interrogates the CT information as well as the results of FEA.

The quantitative NDE methodology is summarized in Figure 1.

INITIAL ET STUDIES

Parallel efforts were conducted to develop the engineering and software capabilities required to implement the ET methodology.

In an IRAD study [4,5], the dimensional resolution capability of current medical CT systems and the accuracy of CT density measurements were addressed. A resolution phantom shown in Figure 2 for geometry studies and in Figure 3 for the material studies provided the calibration standards. GE9800 a medical scanner was employed. The dimensional resolution capability was on the order of one-half pixel.* Flaws on the order of one pixel in extent were

---

* CT measurements can be interrogated at the x-ray detector or on a graphics display device. The former will result in greater accuracy, but requires access to the CT data processing algorithms. The latter is limited by the resolutions of the display device (e.g., 512 x 512).

detected. Industrial CT scanners with increased resolution capability will extend the capability of CT to detect very small flaws.

X-ray attenuation is primarily affected by the density and the atomic number of the interrogated material. When single energy source CT scanners are employed, these effects cannot be differentiated. One issue then is to what extent can single source machines be effectively applied to multi-atomic number materials.

Shown in Figure 4 is the correlation achieved between CT measurements and the density of a mono-atomic number material, carbon. The correlation is excellent, with an accuracy to within 1.5%. The physical density measurements were obtained on rods with dimensions on the order of 1 inch in length by 3/8 inch in diameter. The accuracy of CT is probably better since the resolution capability of CT is finer than the dimensions of the density specimens. For more complex material systems, the CT measurement-density correlation results are presented in Figure 5. Several resins and graphite/resin composites produce a correlation response similar to that of carbon. Therefore, based on the carbon CT-density correlation, the results of Figure 4 can be employed for graphite/resin systems.

Concurrently, PDA developed software to perform color graphics imaging and interrogation of reconstructed CT data. Reconstructed CT pixel for each 2D scan slice is interfaced with PDA's PATRAN-II™ software and displayed** . The display allows selection of density calibrations, a density display window, a 15 step color scale and various data interrogation options. These options include image manipulation, image magnification, access to density values and pixel location via cursor movement, and histogram computations of density profiles. This software is undergoing quality assurance testing prior to release as a PATRAN module, designated as P/ET. The P/ET software package constitutes the first generation in engineering tomography capability.

CURRENT ET STUDIES

Current studies in advancing ET technology include: development of software linkages between CT and FEA; development of 3-D image reconstruction algorithms; developing density property relationships for composite materials of interest; and conducting validation studies comparing predicted responses of composite material components, with experimental results from tested components. Status of these studies are summarized below.

Software for CT-FEA Linkage

Software to link CT with FEA has been accomplished for the 2D case. The 2D linkage methodology developed involved generating a 2D CT density data map from a 2D CT image and then integrating the CT density data map with a developed FEA model. It was determined that 2D FEA codes could efficiently accept a sequential list of density values corresponding to each finite element as a means for integrating CT data in FEA codes. This sequential list of density values was selected as the CT data transfer format and is the output of the linkage software. Since the CT grid tends to be finer than the elements of the FEM, a density value averaging routine for each element was incorporated into the linkage software.

The flow chart for the developed software linkage program is illustrated in Figure 6. The results of a typical density value transfer from a CT scan into an FEM is presented in Figure 7.

3D Image Reconstruction from 2D CT Scans

Significant advances have been made to develop 3D images and FEA solid models from 2D CT scans. The medical use of CT has pioneered the development of 3D reconstruction

---

** PATRAN-II™ is a CAE software system which includes pre-processing for finite element model preparation, interfaces to a variety of FEA codes, and post-processing of FEA data.

algorithms, an approach based on the volumetric extension (i.e., into the third dimension) of information contained in the 2D scan. Linkage of a series of 2D scans creates a 3D image. This approach has limitations with respect to accuracy of boundary representations.

The PDA approach uses image processing algorithms to extract edge information from each 2D image. This edge information is then transformed into a graphic entity (a curve); one curve per CT scan. Since each scan image is at a specified location on the subject, the compilation of the edge information defines the outside contour of the subject. The edge information is then ordered by location and a smooth surface is created between adjacent edges. Finally, a global smoothing, or "skinning" process, is conducted over the entire surface. The result is a surface representation approach to 3D reconstruction, compatible with the creation of a 3D FEA model.

Figure 8 shows the edge information for a human pelvis and femur created from the edge information of a series of 2D CT scans. Figure 9 shows the FEA model developed from the edge data shown in Figure 8.

The CT-FEA linkage model has been developed for 2D CT scans into 2D FEA models. The extension of this linkage capability to three-dimensions, while complex in its data handling and book-keeping requirements, is a relatively straightforward exercise.

Physical Density-Material Property Relationships

ET investigations have focused on the effect of density inhomogeneities. In numerous applications, these density variations are critical to the performance of the composite material hardware. An example of typical density variations in a graphite-reinforced resin system (graphite/polyimide) is illustrated by CT data presented in Figure 10. An enlargement of one anomalous region along with a confirming metallograph is presented in Figure 11.

Several composite material systems are undergoing density-dependent characterization studies to generate suitable material property information for FEA. One being studied is graphite/bismaleimide (G/BMI). In a recent effort, four panels of G/BMI-5245 (Hysol) were fabricated with varying porosity levels. The panels, employed unidirectional plies and were fabricated in a multi-layer quasi-isotropic configuration. Bulk density measurements were made on each specimen prior to testing. Testing was conducted in accordance with ASTM specifications. The testing program consisted of short beam shear, interlaminar shear, cross-ply tensile, tensile and compressive strengths as functions of density. Some results are presented in Figures 12-14.

Given this property dependency on density, it is evident the ability to determine non-intrusively and on a spatial basis, the density of a composite material component would aid predicting the performance of the component.

THE APPLICATION AND ASSESSMENT OF ENGINEERING TOMOGRAPHY

Progress is being made in developing an ET capability for composite materials. Two examples are presented and discussed in a generic fashion to demonstrate the feasibility of the methodology and to illustrate the need for further refinements. These examples address only the effect of density variations, but they incorporate all of the methodology elements of ET previously discussed.

CT scans of the two composite material rings are shown in Figure 15. Ring A is characterized as a "medium" quality material. The ring was essentially uniform in density with the exception of two low density areas located at approximately 325°. Ring B is designated as "low" quality. It contained two low density areas, from 0° to 40° and from 170° to 200°, with the most severe low density region being grouped near 200°.

Each ring was tested in a hoop compression test fixture. Strain gages were placed near most of the anomalies identified by CT. The failure pressure was predicted assuming nominal (i.e., typical design parameters) material properties.

A FEA model was constructed of each ring (Figure 16). The CT-FEA linkage software was used to establish the average density for each element. Density-property correlations were developed using limited data available in the literature. These correlations were used to assign material properties to each element. PDA's ROSAAS software code, modified to incorporate density-dependent material properties, was used for the finite element analysis. The Tsai-Wu failure theory was used as the failure criterion.

Figure 17 presents the distribution of the "Tsai-Wu factor of safety" at 100% of ultimate. For the medium quality ring, the minimum Tsai-Wu factor of safety has a value of approximately 0.80. This occurs in the center of the specimen wall at the most severe density variation. Most of the rest of the ring has values greater than 1.0. The expected gradient from low to high exists from the inside to the outside of the specimen. Failure would be predicted to initiate at the severe flaw.

For the "low" quality ring, the Tsai-Wu factor of safety indicates failure could occur in either of the two low density regions. The predicted factor of safety is slightly lower in the 160° - 200° region. The minimum factor of safety is 0.71, indicating that this ring should have a lower failure load relative to the medium quality ring.

X-rays of the two failed rings are presented in Figure 18. The medium quality ring failed at a loading pressure approximately 40% higher than predicted. The ring broke in two places, a clean translaminar failure at 190° and a jagged combination interlaminar/translaminar failure at 150°. These failures are almost 180° away from the predicted site. Two possible explanations exist for this result. The first is that hoop strain at the low density site caused a "clam shell" effect which created a bending stress concentration 180° away. The second is that initial, but partial failure, did occur at the predicted site, but that this part of the ring, like an arch, continued to bear load with catastrophic failure occurring subsequently 180° away.

For the "low" quality ring, failure occurred at two places at a loading pressure approximately 36% higher than predicted. One failure was a clean translaminar break in the first low density region. The second was a massive bending failure at 190° to 205°, in the second low density region. Again, it is not known which failure occurred first; however, both locations correspond exactly with areas of a low Tsai-Wu factor of safety.

In comparing the behavior, both rings had a failure pressure significantly higher than predicted. Subsequently, properties testing confirmed a higher ultimate strength for the material system, bringing the revised predicted failure strengths significantly closer to the experimental values. Relatively speaking, the correlation between the performance prediction for the two rings was good. The low quality ring failed at a load 6.4% less than that of the medium quality ring. The minimum Tsai-Wu factor of safety at failure for the low quality ring was 11.6% less than the medium quality ring. Reiteration of these predictions utilizing better material properties would bring these two parameters even closer into agreement.

CONCLUSIONS

Excellent progress is being made in developing a quantitative NDE capability to predict the performance of composite material systems.

NDI techniques with tomographic capabilities provide the ability to deduce material properties on a spatially-defined basis. CAE software is being developed to link NDI with FEA and provide a method by which an FEA model can be automatically created.

Information is being generated on the effect of defects on material properties. Correlations are being developed for density-property relationships. As the sensitivity of NDI techniques are improved, this capability should be extended to more dimensionally-dependent defects such as delaminations or macropores.

Finally, the first steps have been taken to integrate NDI information with FEA to predict the effects of defects. As illustrated by the presented results, this first effort can be termed a moderate success. While illustrating the need for better material models and more refined analysis procedures, these results indicate that the development of a quantitative NDE capability, which we have labeled Engineering Tomography, is feasible.

ACKNOWLEDGEMENTS

The investigations documented in References 1 and 2 provided much of the basis for this paper. Both were funded by the Air Force Wright Aeronautical Laboratories (AFWAL). Dr. Thomas Moran was the technical monitor for both programs. The investigation conducted under subcontract to the Aerojet Strategic Propulsion Company (ASPC) had Harvey Peck as the technical monitor. Dennis Sigel, of PDA's Software Products Division, is responsible for the development of the P/ET software module.

REFERENCES

1.   J.G. Crose, et al, "Exploitations of Advanced X-Ray Computed Tomography Technology", 87-PDA-FR-1220-30, September 1987, Submitted to Aerojet Strategic Propulsion Company under ASPC P.O. Nos. 410034 and 409451.

2.   W.H. Pfeifer, R.L. Hack and H.M. Stoller, "Assessment of Computed Tomography as an Improved Quantitative NDE Technique for Composite Materials", 87-PDA-FR-5335-00, May 1987, AFWAL Contract No. F33615-86-C-5084.

3.   J. Berry, et al, "IUS Carbon-Carbon Nozzle Development and Qualification", 1982 JANNAF Rocket Nozzle Technology Subcommittee Meeting, Monterey, CA., CPIA No.367.

4.   W.H. Pfeifer, "Computed Tomography of Advanced Composite Materials", Presented at the ASM/ESD Advanced Composites Conference, Dearborn, MI, December, 1985.

5.   R.L. Hack, D.K. Archipley-Smith, and W.H. Pfeifer, "Engineering Tomography: A Quantitative NDE Tool", Review of Progress in Quantitative nondestructive Evaluation, ed. D.L. Thompson, and D.E. Chimenti, Plenum Press, NY., 1987, pp 401-410, Vol. 6A.

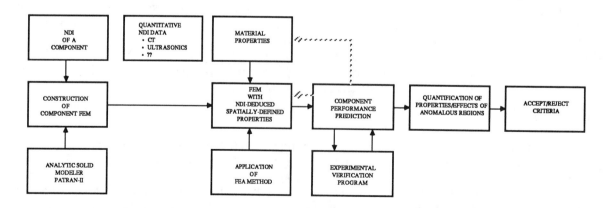

Fig. 1. Engineering Tomography Methodology: A Quantitative Approach to Non-Destructive Evaluation

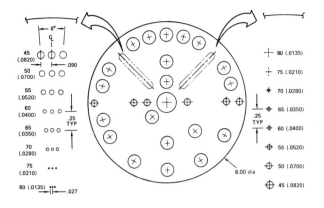

Fig. 2. Resolution Phantom: Geometric Design for Dimensional Resolution Studies

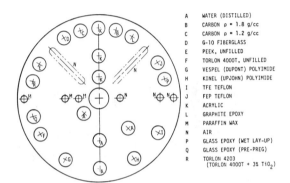

A	WATER (DISTILLED)
B	CARBON ρ = 1.8 g/cc
C	CARBON ρ = 1.2 g/cc
D	G-10 FIBERGLASS
E	PEEK, UNFILLED
F	TORLON 4000T, UNFILLED
G	VESPEL (DUPONT) POLYIMIDE
H	KINEL (UPJOHN) POLYIMIDE
I	TFE TEFLON
J	FEP TEFLON
K	ACRYLIC
L	GRAPHITE EPOXY
M	PARAFFIN WAX
N	AIR
P	GLASS EPOXY (WET LAY-UP)
Q	GLASS EPOXY (PRE-PREG)
R	TORLON 4203 (TORLON 4000T + 3% TiO$_2$)

Fig. 3. Resolution Phantom: Material Description for Density Correlation Studies

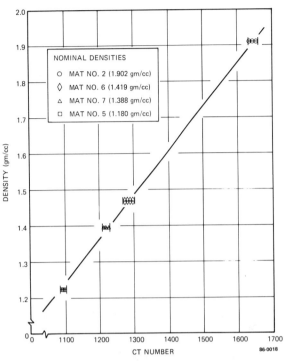

NOMINAL DENSITIES
○ MAT NO. 2 (1.902 gm/cc)
◇ MAT NO. 6 (1.419 gm/cc)
△ MAT NO. 7 (1.388 gm/cc)
□ MAT NO. 5 (1.180 gm/cc)

Fig. 4. CT Number Versus Absolute Density Correlation for Mono-Atomic Number Material - Carbon

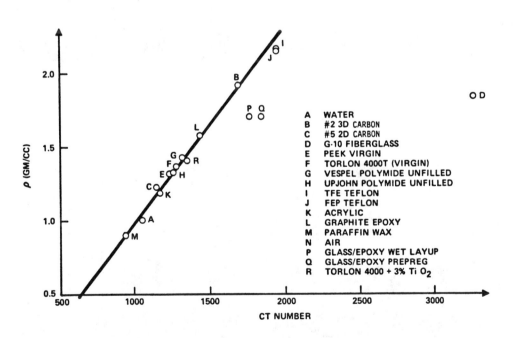

A	WATER
B	#2 3D CARBON
C	#5 2D CARBON
D	G-10 FIBERGLASS
E	PEEK VIRGIN
F	TORLON 4000T (VIRGIN)
G	VESPEL POLYMIDE UNFILLED
H	UPJOHN POLYMIDE UNFILLED
I	TFE TEFLON
J	FEP TEFLON
K	ACRYLIC
L	GRAPHITE EPOXY
M	PARAFFIN WAX
N	AIR
P	GLASS/EPOXY WET LAYUP
Q	GLASS/EPOXY PREPREG
R	TORLON 4000 + 3% Ti O$_2$

Fig. 5. CT Number Versus Absolute Density Correlation for a Range of Material Systems

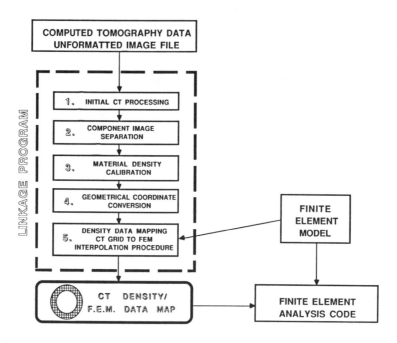

Fig. 6. Computed Tomography to Finite Element Analysis Linkage Program Design Flow Chart

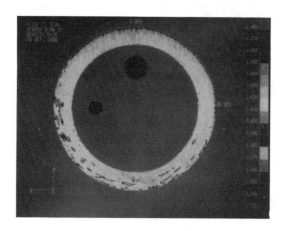

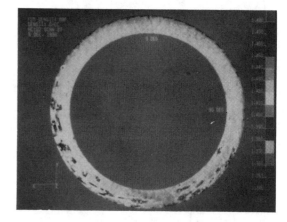

CT Scan of Ring Specimen Showing Density Variations

Finite Element Model Density Map of Ring Specimen Employing CT-FEA Software Linkage

Fig. 7. CT Density Scan and Corresponding Finite Element Analysis Model Density Distribution for Composite Material Ring Specimen

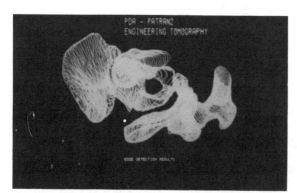

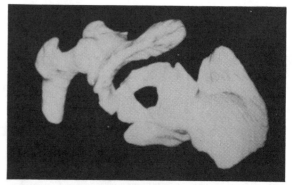

Fig.8. 3-D Surface Model of a Hip Bone Created from 2-D Scan Images

Fig. 9. Solid Model of a Hip Bone Created from 2-D Scan Images and 3-D Surface Model

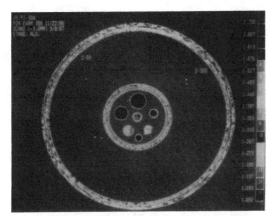

Fig. 10. Density Variations in a G/PI Ring Specimen

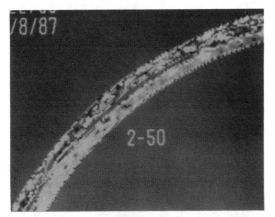

Fig. 11-a. Enlargement of a Porous Region
in a G/PI Ring Specimen

Fig. 11-b. Metallograph of Porous Region
in a G/PI Ring Specimen

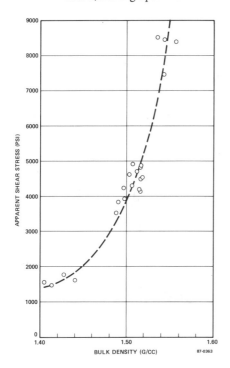

Fig. 12. Short Beam Shear Strength
as a Function of Density

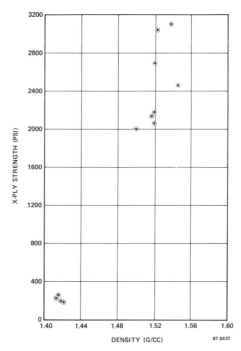

Fig. 13. Cross-Ply Tensile Strength
as a Function of Density

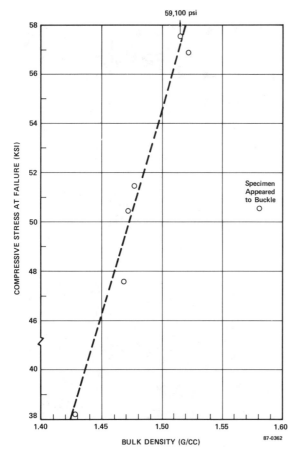

Fig. 14. Compressive Strength
as a Function of Density

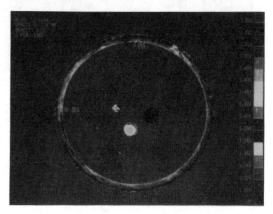

Fig. 15-a. CT Scan of Medium Quality Ring (B)

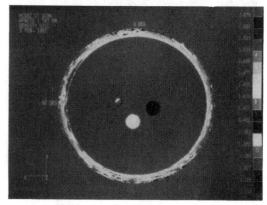

Fig. 15-b. CT Scan of Low Quality Ring (A)

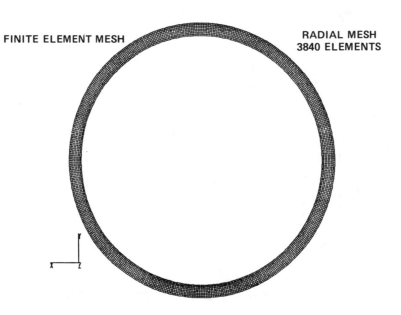

Fig. 16. Finite Element Model of Ring

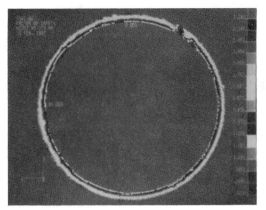
Tsai-Wu Factor of Safety - Medium Quality Ring

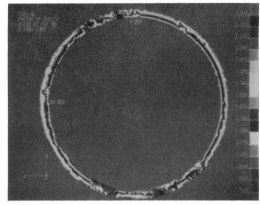
Tsai-Wu Factor of Safety - Low Quality Ring

Fig. 17. Tsai-Wu Factors of Safety for Low Quality and Medium Quality
Ring Specimens at Predicted Ultimate Strength

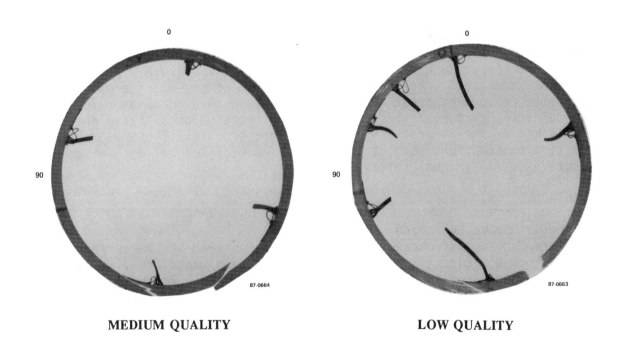

MEDIUM QUALITY **LOW QUALITY**

Fig. 18. X-Ray of Failed Ring

Effect of Thermal Spike on Composite Moisture Capacity

M. J. ADAMSON

ABSTRACT

As part of a continuing study of the effects of thermal spike on composite physical properties, severals sets of 16-ply TGDDM-DDS epoxy/graphite fiber composites were first saturated with moisture, and then individual specimens were subjected to thermal spikes of up to 240C for periods of up to 25 minutes. At the cure temperature, 177C, all sets show a change in rate of mass lost during spike. Post-spike moisture reabsorption capacity data provides evidence of temperature dependent molecular structural dilation and consolidation of the epoxide matrix. This evidence suggests that either a thermoreversible phenomenon such as physical aging or a nonreversible additional cure becomes active near the upper limit of the aircraft supersonic flight profile. These and other important observations suggest that one or more of several mechanisms may be active during thermal spike.

1. INTRODUCTION

1.1 Thermal Spike Phenomena

Thermal spikes are thermal excursions encountered by the surfaces of high performance aircraft during a supersonic dash maneuver. Typically, surface temperatures increase at about 50C/min. from a low of -60C at subsonic cruise conditions to as high as 149C during supersonic flight. The surfaces continue to experience the high temperatures until subsonic flight cools surfaces at about 34C/min. once again to -60C [1].

TGDDM-DDS epoxides are the polymeric matrix material of many composites used in the construction of high performance aircraft. These epoxides are cured at 177C in an attempt to ensure that the softening point (glass transition) is well above the expected highest service temperature. Because cured epoxies are nonhomogeneous materials, the less dense regions may soften at lower temperatures than the more dense material. Moisture swells these (and other) polymers by dilating the molecular structure, thus causing substantial changes in physical properties [2]. One of these changes is an apparent reduction of the glass transition temperature (Tg). Furthermore, thermal spike of the swollen epoxide can magnify the effects

M. J. Adamson, Research Engineer, Ames Research Center, National Aeronautics and Space Administration, Moffett Field, CA 94035.

of swelling dilation. The extent to which these effects represent serious degradation of the molecular structure is unknown, but the possibility is great that the highest service temperature for swollen epoxides may drop below 149C.

1.2 Reversible and Irreversible Dilation

The dilation exists in two forms: reversible and irreversible. Both forms increase the moisture absorption rate and ultimate absorption capacity by increasing local intermolecular spacing. Regardless of reversibility, the net effect is a lower density of the polymeric matrix material that binds the fibers together. It is usually assumed that reversible dilation heals viscoelastically, that is, the consolidation process depends on both time and temperature [3]. It is assumed also that irreversible dilation is caused by intermolecular spacing too large to be reduced by the molecular mobility below the softening temperature.

Some studies have implied that irreversible effects are caused exclusively by microcrack formation (a form of irreversible dilation)[4,5], but it is likely that other mechanisms are active below the 149C thermal spike temperature. Dynamic mechanical analysis, DMA, of cured TGDDM-DDS epoxy resins shows, for example, that the glass transition occurs over a temperature range that begins at about 120C and ends at about 210C [6]. Whenever the glass transition temperature (Tg) is exceeded, additional irreversible cure is possible because of the enhanced mobility; therefore, localized cure in the lower density regions of the epoxide could take place during a thermal excursion anywhere within that range. In addition, it has been demonstrated that small molecules (such as water) retard the consolidation of dilated polymers [3,7]; therefore, the quantity of moisture desorbed during a spike should have some discernible effect on the ability of a polymer to densify. In view of these observations, the active mechanisms should be identified and considered before conclusions are drawn about the effects of thermal spike on physical and mechanical properties of composites.

1.3 Origins of Dilation

The four known causes of molecular dilation are: mechanical, thermal, swelling, and hygrothermal. Mechanical dilation is not addressed in this work. Thermal- and swelling dilation are included only to the extent that they relate to the subject of this investigation, hygrothermal dilation.

Thermal dilation is described by the Williams, Landel and Ferry (WLF) free volume concept of the glass transition [8]. The WLF argument for thermal dilation is based principally on the observation that the thermal expansion coefficient is greater above Tg than below. The essence of the dilation concept is that above Tg the unit volume fraction of free volume contained within the glassy material exceeds 0.025 and this value increases with temperature. Above Tg the influences of molecular entanglement, crosslinking and hydrogen bonding are small and thermal dilation increases molecular mobility to the extent that the material loses rigidity.

Swelling dilation does not require a thermal excursion to increase free volume in the polymer network structure of TGDDM-DDS epoxies. The increase of free volume due to swelling does not add to the mobility of polymer molecular segments unless the swelling agent is removed and dilation recovery kinetics are slower than desorption kinetics. Swelling begins immediately as moisture enters the epoxide molecular structure. Initially, the volume increase reflects only about one-half of the volume of moisture that is absorbed, but as the free volume becomes filled the swelling efficiency approaches unity. Swelling efficiency also increases slightly with temperature between +1 and 74C [2]. Below about 100C, swelling is a long-term process that is characterized by increased moisture capacity, by volumetric hysteresis, and by increased diffusivity. The swelling effects are reversed by physical aging (consolidation) of the dilated glassy structure after the swelling agent is removed [3,7,9]. Consolidation requires either long term low temperature aging or short term elevated temperature aging.

Hygrothermal dilation of moisture laden composites and neat resins is caused by thermal excursions into the range from slightly above the temperature at which the material was saturated to as high as the glass transition temperature of the unswollen resin. Sub-Tg thermal spike dilation is comparatively instantaneous and exceeds that caused by the long-term swelling process. Sub-Tg hygrothermal dilation is not reversed by short term drying to 82C [1], nor by comparatively long term drying at 70C [10], but is partially reversed by vacuum desiccation 7 days at 150C [9]. Thermal spike effects to 200C have been reported by Mijovic and Lin [11] for swollen epoxy resins and composites, but their results did not support any particular damage mechanism.

1.4 Objectives of this Work

The goal of this investigation is to obtain preliminary evidence for active mechanisms over a wide temperature range. This will be accomplished through a presentation of the effects of thermal spikes of up to 25 minutes of heating time and up to 240C on the moisture desorption and post-spike reabsorption capacity of TGDDM-DDS epoxy/graphite-fiber composite laminates laden with moisture. The extended spike temperature range will also provide data above the glass transition where molecular structural equilibration theoretically becomes instantaneous [12] and the duration of the spike permits significant moisture desorption. Differential thermal analysis (DTA) is used to identify the active mechanisms over the entire range of test temperatures. Linear thermal expansion is measured to locate the glass transition temperature.

2. EXPERIMENTAL

2.1 Material

The uncured Hercules 3501-5 resin material, and Hercules AS fibers in Hercules 3501-5 resin prepreg were supplied by Hercules, Inc. The Union Carbide Thornel 300 fiber in Narmco 5208 resin prepreg was supplied by Narmco. One set each of cured Hercules 3501-5 resin, AS/3501-5 composite and the T300/5208 composite were processed by Lockheed, Sunnyvale. The T300/934 composites were processed by Texstar Plastics, Grand Prarie, Texas, from uncured Fiberite HYE 1034C prepreg tape that was supplied by the Fiberite Corporation. All processing was according to the manufacturers' specifications for heated vacuum degassing and stepwise curing. Final cure temperature for the resin and laminates was 177C (350F). The cured epoxy sheet, approximately 0.125 inches thick, and the unidirectional 16-ply laminate sheets were cut by diamond-tip saw into 0.5 by 4 inch specimens. Laminate fibers were parallel to the 4-inch dimension.

2.2 Procedure
Specimen mass was measured using a Mettler analytical balance accurate to within 0.05 mg. Before weighing, wet specimens were wiped dry with a lint-free towel and air dried 5 minutes at room temperature, and dry specimens were exposed 5 minutes to room conditions. This procedure eliminated possible error caused by specimen heat upsetting the balance zero and by uncertainty of surface water quantity.

Initially, residual moisture was removed by storing the specimens in a heated vacuum desiccator kept at 100C until their rate of mass loss became less than 0.05 mg/day.

All composite specimens were moisture-conditioned by immersion in distilled water controlled within 0.5C, using a constant temperature bath.

Thermal spike was accomplished by placement of specimens inside a preheated air circulating oven for the required time interval, followed by removal to room temperature air for cooling. Post-spike mass was determined after 5 minutes of cooling. The thermal spike heating times were chosen to produce a systematic variation in the quantity of moisture lost at each temperature. The test temperature range from 25 to 240C was selected to include the portion of the flight thermal profile where damage is most probable [1]; the cure temperature (177C); the glass transition that by DMA is associated with the cure temperature [6]; and the low temperature portion of a second glass transition that, also according to DMA, begins at about 200C and reaches a maximum at about 260C [6].

Differential thermal analysis (DTA) was performed on cured 3501-5 epoxy resin using a DuPont model 900 Differential Thermal Analyzer.

Linear strain of the resin was measured in the 4 inch dimension using a dial caliper accurate to within 0.0005 inches.

3. RESULTS AND DISCUSSION

3.1 Desorption
In Figure 1, the "Mass Fraction Lost" (mass lost during spike normalized by the mass of water absorbed prior to spike) is presented as a function of the thermal spike temperature. In all cases, a change in rate of desorption corresponds to the cure temperature. Considering the WLF thermal dilation model in which mobilities both of small molecules and of the polymeric structure are significantly greater above Tg, this observation is in agreement with the DMA results that associate the glass transition with the cure temperature.

A second observation is that the mass loss curve between room temperature and 177C is linear regardless of spike duration. Apparently the desorption mechanism is not affected over that temperature interval by any change in phase of the moisture contained within the laminate. Instead, this seems to support the concept that the mobility of absorbed moisture depends upon something more than an equilibrium between liquid and vapor.

A third observation is that a mass equivalent to the entire mass of the moisture that had been absorbed at 74C prior to the application of thermal spike was desorbed after about 20 or 25 minutes of exposure to 240C. No attempt was made to identify by direct means whether the desorbed mass was water, decomposition products, unreacted starting materials, mass loss due to additional curing, or a combination of two or more of these possibilities. Indirect identification of the desorbed material is offered from reabsorption and thermal analysis studies in the following sections of this report.

3.2 Reabsorption

Figure 2 represents post-spike reabsorbed moisture content relative to pre-spike moisture content. These data show the reabsorption capacity after about 5 weeks of total immersion in water held at 74C, the same temperature used for pre-spike absorption. Instead of a simple linear increase of absorption capacity as a function of increasing spike temperature [9], more complex behavior was encountered. Three distinct regions which apparently reflect ranges of differing molecular mobility are observed. Even the 1- and 2-minute specimen curves which are shifted toward higher temperatures (probably because the heating interval simulating the thermal spike was too short for the entire specimen to reach thermal equilibrium) have three regions. In the first region, 100C and below, no significant spike effect is seen. In the second region, roughly 100 to 140C, the moisture capacity increased at each temperature by approximately the same amount regardless of spike duration (except for the 1- and 2-minute spikes). Large differences in reabsorption capacity are evident within the third region, between 140 and 240C. The greatest increase was produced by 1-minute spikes. The 2- and 9-minute spikes caused moisture fraction increases less than those of the short-term 1-minute spikes. By contrast, the longer term 20- and 25-minute spikes produced significantly lower reabsorption capacities, decreasing with spike temperature to the extent that above 200C the moisture capacity was actually reduced to a lower value than before thermal spike.

The decreases in reabsorption capacity suggest that either a thermoreversible phenomenon such as physical aging or nonreversible additional cure becomes active above the cure temperature where molecular mobility is very high. This observation of the reversible nature of at least some of the thermal spike effects is particularly important because it provides two alternatives to the concept of irreversible microcrack damage as the sole cause of increased moisture capacity in spiked graphite/epoxy composite laminates. Microcracking, however, cannot be completely ruled out by these decreases because the effects of additional cure may simply reduce the swelling capacity more than enough to compensate for microcrack enhanced absorption. It is important to note that those specimens that lost the least mass are also those that displayed the least reversibility. This observation supports the concept that small molecules contained within the free volume of a swollen structure retard consolidation.

In a prior work [9], it was demonstrated that the dilation of the epoxy matrix of a composite laminate saturated with moisture at 25C and spiked to 132C could be closely predicted if it is assumed that the increase is directly proportional to the increase in temperature. In the present work, the specimens were saturated with moisture at 74C; therefore, any increase of capacity should be directly proportional to temperature above 74C. Figure 2 reveals the temperature range over which such a first order relationship may be used to estimate an increase has an upper limit at about 140C.

3.3 Thermal Analysis

Differential thermal analysis is a technique for studying the thermal behavior of materials as they undergo physical and chemical changes during heating or cooling. In conventional DTA as used to produce the data for Figure 3, the sample and reference are heated at the same rate. If no physical or chemical changes occur the temperatures of sample and reference are the same and delta T is zero. In the case of polymers, a broad endotherm indicates a broad range of temperatures over which a physical or chemical transition is taking place. In Figure 3 the broad endotherm between approximately 90 and 170C very likely reflects the utilization of energy to break intersegmental hydrogen bonds. Also for polymers a broad exotherm may indicate a phase transition such as crystallization or additional cure. Crystallization in TGDDM-DDS epoxies is unlikely; therefore, the broad exotherm between 170 and 260C is probably caused by curing. In an amorphous polymer the glass transition usually appears as a baseline shift [13] such as shown between 240 and 260C. This shift corresponds to the glass transition of fully cured TGDDM-DDS epoxies found by DMA [6] and differential scanning calorimetry (DSC)[14]. If a baseline shift occurred near the cure temperature, the shift was obscured by the reaction change from endo- to exothermic.

Figure 4 is a plot of strain as a function of temperature for 3501-5 resin over the range from 25 to 270C. The abrupt change in the thermal expansion coefficient at 180C is in agreement with a WLF type Tg of thermal dilation at the cure temperature. Dilation would explain the increased rate of moisture lost during spike above Tg. The absence of a significant change of expansion coefficient at 240C indicates that the shift in baseline at approximately 240C in Figure 3 does not represent a glass transition in the WLF sense but may represent the energy required for a "loosening" of crosslinks.

4. CONCLUDING COMMENTS

From these data, it is not possible to discern which type of dilation, reversible or irreversible, occurred at each spike temperature. It is likely that below the cure temperature zone hygrothermal dilation is the dominant mechanism with little or no contribution from additional cure, loss of unreacted low molecular weight resin, or loss of polymer decomposition products. Within and above the cure temperature zone, a complex combination of mechanisms appears to be active: loss of moisture is rapid; physical aging should be rapid after sufficient moisture loss; additional cure may take place once sufficient moisture is lost to significantly increase polymer molecular mobility; a portion of the polymer may decompose and be lost; or unreacted low molecular weight starting material may also be desorbed. Because the cure zone overlaps the aircraft thermal spike temperature zone, a definite need to study the contribution of each possible mechanism is established.

REFERENCES

1. McKague, E. L., Jr., Halkias, J. E. and Reynolds, J. D., "Moisture in Composites: The Effect of Supersonic Service on Diffusion," Journal of

Composite Materials, Vol. 9, January, 1975, pp. 2-9.

2. Adamson, M. J., "Thermal Expansion and Swelling of Cured Epoxy Resin Used in Graphite/Epoxy Composite Materials," Journal of Materials Science, Vol. 15, 1980, pp. 1736-1745.

3. Stewart, M. E., Hopfenberg, H. B., Koros, W. J. and McCoy, N. R., "The Effect of Sorbed Penetrants on the Aging of Previously Dilated Glassy Polymer Powders. II. n-Propane Sorption in Polystyrene," Journal of Applied Polymer Science, Vol. 34, 1987, pp. 721-735.

4. Whitney, J. M. and Browning, C. E., "Some Anomalies Associated With Moisture Diffusion in Epoxy Matrix Composite Materials," Advanced Composite Materials-Environmental Effects, ASTM STP 658, edited by J. R. Vinson, American Society for Testing and Materials, Philadelphia, 1978, pp. 43-60.

5. Collings, T. A., Mead, D. L. and Stone, D. E. W., "The Effects of High Temperature Excursions on Environmentally Exposed CFC," Composite Structures 4, Vol. 2, Damage Assessment and Material Evaluation, edited by I. H. Marshall, Elsevier Applied Science, London, N. Y., 1987, pp. 2.345-2.374.

6. Keenan, J. D., Seferis, J. C. and Quinlivian, J. T., "Effects of Moisture and Stoichiometry on the Dynamic Mechanical Properties of a High-Performance Structural Epoxy," Journal of Applied Polymer Science, Vol. 24, 1979, pp. 2375-2387.

7. Connelly, R. W., McCoy, N. R., Koros, W. J., Hopfenberg, H. B. and Stewart, M. E., "The Effect of Sorbed Penetrants on the Aging of Previously Dilated Glassy Polymer Powders. I. Lower Alcohol and Water Sorption in Poly(methyl methacrylate)," Journal of Applied Polymer Science, Vol. 34, 1987, pp. 703-719.

8. Williams, M. L., Landel, R. F., and Ferry, J. D., "The Temperature Dependence of Relaxation Mechanisms in Amorphous Polymers and Other Glass-Forming Liquids," Journal of the American Chemical Society, Vol. 77, 1955, pp. 3701-3707.

9. Adamson, M. J., "A Conceptual Model of the Thermal-Spike Mechanism in Graphite/Epoxy Laminates," Long-Term Behavior of Composites, ASTM STP 813, edited by T. K. O'Brien, American Society for Testing and Materials, Philadelphia, 1983, pp. 179-191.

10. Adamson, M. J., Noorda, B. S. and Musich, M., "A Conceptual Model of Free Volume in Glassy TGDDM-DDS Epoxies," Unpublished report submitted for NASA publication, 1987.

11. Mijovic, J. and Lin, K-F, "The Effect of Hygrothermal Fatigue on Physical/Mechanical Properties and Morphology of Neat Epoxy Resin and Graphite/Epoxy Composites," Journal of Applied Polymer Science, Vol. 30, 1985, pp. 2527-2549.

12. Struik, L. C. E., "Physical Aging in Plastics and Other Glassy Materials," Polymer Engineering and Science, Vol. 17, No. 3, March 1977, pp. 165-173.

13. ASTM D3418-75 Standard Method of Test for Transition Temperatures of Polymers by Thermal Analysis.

14. Kong, E. S.-W., Adamson, M. J. and Mueller, L., "Moisture Diffusion and Solid-State NMR Investigations on the Physical Aging Processes in Network Epoxy Glasses," Composites Technology Review, Vol. 6, No. 4, 1984, pp. 170-173.

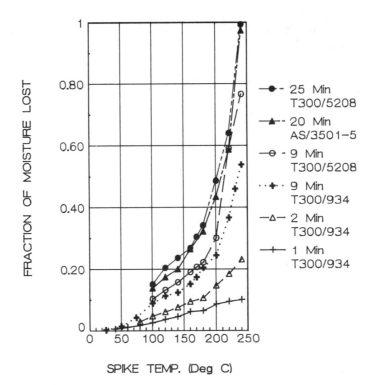

Fig. 1.   Mass lost during thermal
spike of composites.

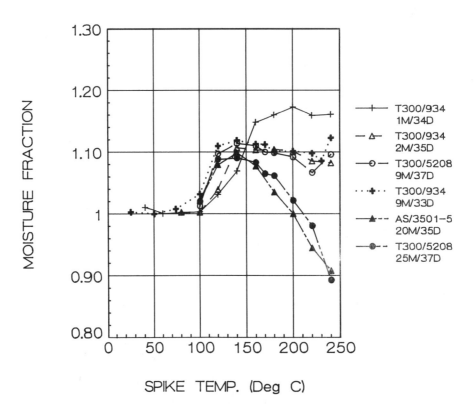

Fig. 2.   Moisture reabsorption at
74C after thermal spike.

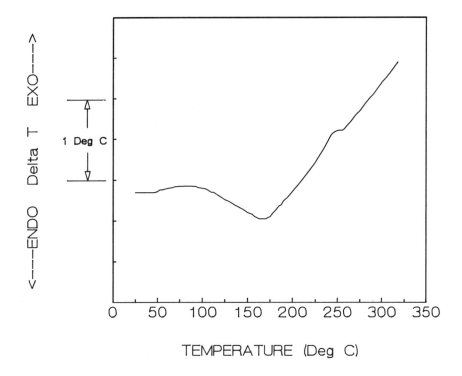

Fig. 3.   Differential thermal
           analysis of 3501-5 resin.

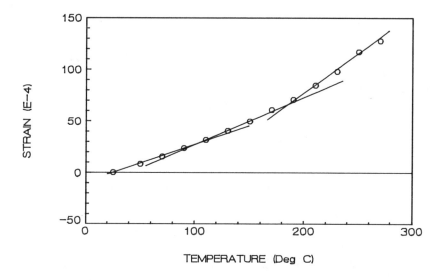

Fig. 4.   Linear strain of 3501-5
           resin.

Creep and Creep-Recovery of a Thermoplastic Resin and Composite

CLEM HIEL

ABSTRACT

The database on advanced thermoplastic composites, which is currently available to industry, contains little data on the creep and viscoelastic behavior. This behavior is nevertheless considered important, particularly for extended-service reliability in structural applications. The creep deformation of a specific thermoplastic resin and composite is reviewed. The problem to relate the data obtained on the resin to the data obtained on the composite is discussed.

INTRODUCTION

Advanced thermoplastic matrix-carbon fiber composites possess a combination of properties that appear to be superior to currently used thermoset composites. Significant attributes include higher heat resistance, greater impact strength, rapid manufacturability and easy repair and recycling.

This potential has led a wide range of industries to consider thermoplastic composites for the design of future loadbearing structures. It is expected that high quality components can be fabricated at rates which are necessary for cost-effective mass-production. This aspect in itself is a reason for industries such as the automotive to consider large scale involvement with advanced thermoplastic composites.

The industrial use of these materials requires a sound and broad data base. A systematic characterization effort has mainly been focussed on the short-term stiffness, strength, impact-resistance and fracture toughness and not on the long-term response such as creep and viscoelasticity. The concern is that an incomplete data base might give a misleading picture of the capabilities of advanced thermoplastic composites, especially when used for highly stressed structural components.

The objective of this paper is to present experimental data on the time-temperature-creep compliance behavior of PEEK- APC-2 composite. This material consists of AS-4 carbon fibers in a polyetheretherketone matrix.

*Currently with NASA-Ames Research Center, Test Engineering and Analysis Branch, Moffett Field, CA 94035

The thermoplastic matrix is an aromatic semi-crystalline polymer which was developed by ICI [1]. The results will complement an existing database and allow us to draw conclusions on the performance of this material at elevated temperatures.

BACKGROUND ON MATERIALS AND LAMINATE FABRICATION

A thermoplastic PEEK resin, designated Victrex 450g was supplied by ICI in the form of dogbone specimens. The degree of crystallinity of this material was estimated at 35%, by using X-ray diffraction as outlined in [2-8].

Sixteen ply thick unidirectional panels, measuring 200 x 200 mm were laid up, using APC-2 prepreg. The layup was subsequently put underneath an aluminum vacuum bag and autoclaved. Applied temperature , pressure and vacuum followed the guidelines summarized in [3]. In situ measurements indicated that the highest temperature in the autoclave was 389 deg. C. and that the cooling rate was 12- 13 deg. C/ min. The crystallinity of the material was estimated at 34 %.

The tensile and flexural properties of the first panel were measured and compared to published data as summarized in table I. Especially the transverse strength and the transverse strain to failure were much lower than expected. The laminate also appeared too thick thus suggesting a lack of consolidation during the autoclave process. A microphotograph of a cross section perpendicular to the fibers does indeed reveal interlaminar voids as seen in Fig.1a.

A second panel was fabricated, this time with the periphery around the vacuum bag tightly clamped to the tool plate. Again specimens were cut and tested, now with special emphasis on the transverse properties. Good results were obtained, as indicated in Table I. Microphotographs now revealed a much better consolidation, as seen in Fig. 1b.

INSTRUMENTATION AND CREEP CHARACTERIZATION

Transverse and 15 degree-off axis tensile specimens were cut from the panels, using a diamond coated saw blade. Specimen dimensions were 14 x 130 mm and 14 x 340 mm respectively.

The transverse composite- as well as the neat resin- specimens were instrumented with uniaxial strain gages (EP- 250BZ-350). The gages were bonded back to back onto each specimen to detect loading anomalies caused by bending. The 15 deg.-off axis specimens were instrumented with a uniaxial gage on one side and a rozette (WA-06-250WR-350) on the other. All gages were bonded using M600 adhesive. The adhesive was cured at 75 deg. C for 8 hours and subsequently postcured at 190 deg. C for one hour.

Creep characterization tests were done on lever arm creep frames, on which a thermal chamber was mounted. Each test was done under isothermal conditions. A constant load was applied for 20 minutes to each specimen. The load was subsequently released and the material was allowed to recover for 200 minutes. The strain gage response during creep and -recovery was

continuously monitored using a Bruel and Kjaer BK1526 strain indicator
with multipoint selector. Temperature compensation was accomplished using
a dummy gage setup.
The creep experiment was repeated at various temperatures between room
temperature and 190 deg.C. A thermocouple was attached to the specimen
surface to monitor the local temperature during each test. These
measurements confirmed that variations in temperature did not exeed + 1.5
deg. C of a preset value.

RESULTS AND DISCUSSION

   The time-temperature-compliance relationship for the neat resin-matrix
material is represented by a 3-dimensional surface, as shown in Fig. 2. As
can be observed, the surface has little curvature below the glass
transition temperature (143 Deg C.) and an increasingly strong curvature
above Tg. This curvature which is representative for the creep deformation
thus identifies two temperature regions. One in which the materials
response is quasi-elastic and thus interesting for structural use. The
other one in which there is increasingly excessive creep deformation thus
pointing to a temperature region were use of the material should be
avoided.

   Creep- compliance measurements transverse to the fibers are summarized
into a similar three-dimensional surface as shown in Fig. 3. Again two
distinct curvatures can be observed respectively for temperatures below
and above 103 deg C. Structural use of the material should thus be avoided
above this temperature. The large discrepancy between the softening
temperatures of the resin and the composite is a rather surprising
result.

   The dependence of the shear creep compliance on time and temperature is
summarized in Fig.4. The figure contains the same features as were
discussed higher. Again the large discrepancy in softening temperature can
be observed as compared to the neat resin.

   The data indicates that the temperature performance of the resin does
not translate into the matrix material of the composite. The reason why
the composite has this poor property retention when loaded transverse to
the fibers and in shear is an unresolved question. Additional insight is
nevertheless provided by questioning the nature of the used resin and the
role of the fiber-matrix interface.

   The grade of PEEK resin which was used for the experiments might have a
different molecular weight as compared to the in-situ matrix material.
Lower molecular weight resins generally have a lower glass transition
temperature and a reduced viscosity. They consequently allow a better
impregnation of the fibers and are thus preferred by the prepreggers.
Whitacker et al. [4] studied a variety of PEEK resins, from different
sources and they found that the glass transition temperature was
practically identical for all of them. The molecular weigth however
turned out to be significantly different.
   Another possibility is that the morphologies of the bulk resin and the
in-situ resin are different. There is indeed evidence that the presence of
the fibers causes an heterogeneous nucleation of the in-situ matrix [5].
This interaction may result in an interphase layer with substantially

different properties. The large interfacial area which is present in the composite might explain the effect on global material properties.

Some recent work by theocaris [6] supports the idea of an interfacial area with different properties. A methodology was developed and used to predict that the glass transition temperature of a metal filled epoxy is 13 deg C lower than the Tg of the matrix. This was also verified experimentally. Righby [7] reported the same trends for a PEEK/glass system.

CONCLUSIONS

The short-term stiffnesses and strength's of fabricated laminates were in good agreement with data from the literature.

Neat PEEK resin creeps extensively above 144 deg C. The composite laminate however already creeps extensively at 103-110 deg C.

The reason for this discrepancy is not clear at this stage. The role of the large interphase area between fibers and matrix needs to be better understood in order to make further progress.

The results suggest that the potential of APC-2 composite for use in primary loadbearing structures should be looked at cautiously for service temperatures in excess of 100 deg C.

REFERENCES
----

1. Belbin,G.R. and Staniland, P.A., "Advanced Thermoplastics and Their Composites," Phil. Trans. R Soc. Lond., A 322, 1987, pp.451-464

2. Blundell, D.J. and Osborn, B.N., "Crystalline Morphology of the Matrix of PEEK-Carbon Fiber Aromatic Polymer Composites," SAMPE Quarterly, Vol 17, No.1, 1985,pp.1

3. Woodward, S., "PEEK/Carbon Composite: ICI-Application Note for Vacuum and Pressure Fabrication," Ref. No.RTBSWHCJ821, May 1986.

4. Whitaker, R.B. et al. "Characterization and Adhesive Bonding of Polyetheretherketone Resin and Composites." 16th National SAMPE Technical Conference, 1984, pp.361-374.

5. Crick, R., Meakin, P., Moore, R. and Leach, D. "Fracture and Fracture Morphology of Aromatic Polymer Composites," Proc. ECCM-1, Bordeaux,1985, pp.253-258.

6. Theocaris, P.S., "The Mesophase and It's Influence on the Mechanical Behavior of Composites," Advances in Polymer Science 66, H.H. Kausch and H.Z. Zachmann eds. Springer Verlag, Berlin-Heidelberg 1985

7. Rigby, R.B., "Polyetheretherketone," Engineering Thermoplastics and Applications, J.M. Margoes ed. New York and Basel, 1985

8. Xiao, X.R., Hiel, C.C. and Cardon, A.H. "Viscoelastic Characterization of Thermoplastic Matrix Composites," Progress Report, Univeristy of Brussels, 1986.

Table I. Properties of Fabricated Laminates, Compared
with literature Data

	reported ref.[8]	laminate panel 1	Laminate panel 2
0-Degree Tensile			
Modulus (Gpa)	134.	131.	–
Strength (Mpa)	2310.	2200.	–
failure strain (%)	1.45	1.50	–
90-Degree Tensile			
Modulus (Gpa)	8.9	9.8	10.1
Strength (Mpa)	80.	58.	79.
failure strain (%)	1.0	.6	.85
0-Degree Flexural			
Strength (Mpa)	1880.	2390.	–
90 Degree Flexural			
Strength (Mpa)	137.	93.	–
Thickness (mm)	2.00	2.36	2.10
Density (g/cm3)	1.60	1.59	1.60

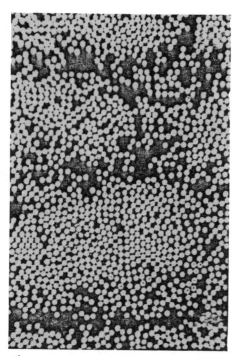

Fig. 1a Cross-Section Laminate I (350X)

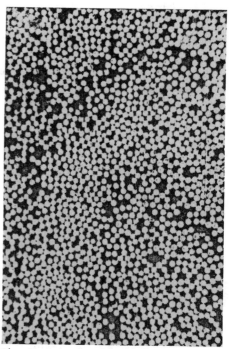

Fig. 1b Cross-Section Laminate II (350X)

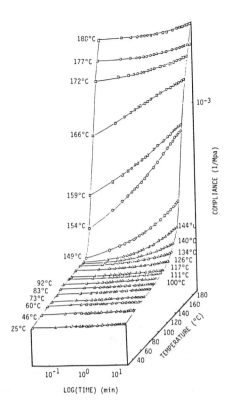

Fig. 2. Temperature Dependence of Creep
Compliance of PEEK.

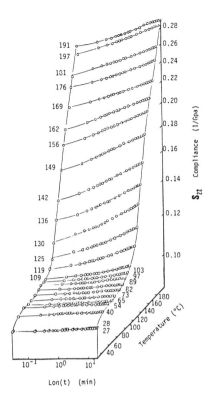

Fig. 3. Temperature Dependence of Transverse
Creep Compliance of APC-2 Laminate.

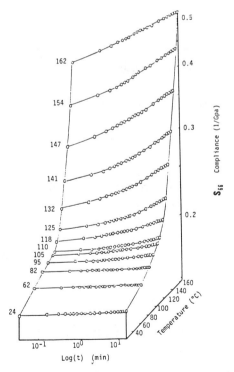

Fig. 4. Temperature Dependence of Shear Creep
Compliance of APC-2 Laminate.

Modeling Creep Behavior of Fiber Composites

J. L. CHEN AND C. T. SUN

Abstract

A micromechanical model for the creep behavior of fiber composites is developed based on a typical cell consisting of a fiber and the surrounding matrix. The fiber is assumed to be linearly elastic and the matrix nonlinearly viscous. The creep strain rate in the matrix is assumed to be a function of stress. The nominal stress-strain relations are derived in the form of differential equations which are solved numerically for off-axis specimens under uniaxial loading. A potential function and the associated effective stress and effective creep strain rates are introduced to simplify the orthotropic relations.

1. Introduction

Creep in fiber composites is an important issue, especially in elevated temperature environments. Since fibers are essentially elastic, creep strains are produced by the matrix. Depending on the loading direction, this time-dependent matrix property may affect macro-composite behavior with different magnitudes. In other words, the creep behavior of composites is anisotropic.

A number of authors have attempted to model the creep behavior of polymer-based fiber composites [1-4]. A good survey on this subject was given in [3]. The main effort in this area has been concentrated on modeling time-dependent behavior using linear or nonlinear viscoelasticity theories.

The objective of this paper is to characterize the nonlinear creep behavior of unidirectional fiber composites consisting of elastic fibers and nonlinear viscous matrices. A micromechanical model is developed based on the typical cell consisting of a fiber and the surrounding matrix. By assuming simplified deformation fields in both fiber and matrix, and using averaged stresses for the composite cell, global stress-strain relations are established. Further simplifications of the constitutive relations are achieved by introducing effective stress and effective creep strain rates.

---

* Research Associate and ** Professor, School of Aeronautics and Astronautics, Purdue University, West Lafayette, Indiana 47907

2. Basic Assumptions

Consider an idealized representative cell of the composite as shown in Fig. 1. As indicated in the figure, the fiber is assumed to have a square cross-section with an area equal to that of the actual cross-section of the cell. Due to symmetry, a quadrant is taken for consideration.

The quadrant of the cross-section is divided in two parts, Part A and Part B, as shown in Fig. 1. Part B is a pure matrix region, while Part A consists of the fiber AF and a matrix region denoted by AM. The composite lamina is assumed to be in a state of plane stress parallel to the $x_1 - x_2$ plane. Thus, $\sigma_{33} = \sigma_{23} = \sigma_{13} = 0$. In addition, the following assumptions are made.

a. In each subregion AF, AM or B, stress and strain fields are uniform.

b. In subregion A, the stress field and strain field in regions AF and AM follow the appropriate constant stress or constant strain assumption, i.e.,

$$\sigma_{12}^{AF} = \sigma_{12}^{AM} = \sigma_{12}^{A} \qquad \text{constant stress}$$

$$\sigma_{22}^{AF} = \sigma_{22}^{AM} = \sigma_{22}^{A} \qquad \text{constant stress} \tag{1}$$

$$\varepsilon_{11}^{AF} = \varepsilon_{11}^{AM} = \varepsilon_{11}^{A} \qquad \text{constant strain}$$

c. For the entire representative volume, the constant strain assumption is adopted, i.e.,

$$\varepsilon_{11}^{A} = \varepsilon_{11}^{B} = \varepsilon_{11}$$

$$\varepsilon_{22}^{A} = \varepsilon_{22}^{B} = \varepsilon_{22} \tag{2}$$

$$\gamma_{12}^{A} = \gamma_{12}^{B} = \gamma_{12}$$

In equations (1-2) superscripts A, B, AF and AM denote subregions A, B, AF and AM, respectively. The stress and strain components without superscripts are the average stresses and strains for the composite. Note that x_1 axis is parallel to the fiber direction.

Since the stress and strain fields are uniform in each subregion, the nominal (average) stress and strain components are given by

$$\sigma_{ij}^{A} = V_1 \sigma_{ij}^{AF} + V_2 \sigma_{ij}^{AM} \tag{3}$$

$$\varepsilon_{ij}^{A} = V_1 \varepsilon_{ij}^{AF} + V_2 \varepsilon_{ij}^{AM} \tag{4}$$

in region A, and

$$\sigma_{ij} = V_1 \sigma_{ij}^{A} + V_2 \sigma_{ij}^{B} \tag{5}$$

$$\varepsilon_{ij} = V_1 \varepsilon_{ij}^{A} + V_2 \varepsilon_{ij}^{B} \tag{6}$$

in the representative cell. In equations (3-6),

$$V_1 = \frac{a_1}{a_1 + a_2} \quad , \quad V_2 = \frac{a_2}{a_1 + a_2}$$

respectively.

The objective here is to derive the relations between the composite nominal stress σ_{ij} and strains ε_{ij} based on the stress-strain relations of the fiber and matrix.

3. Constitutive Equations of Fiber and Matrix

The fiber is assumed to be an elastic material which obeys the generalized Hooke's law, i.e.,

$$\dot{\varepsilon}_{11} = \frac{\dot{\sigma}_{11}}{E^F} - \frac{v^F}{E^F}\dot{\sigma}_{22}$$

$$\dot{\varepsilon}_{22} = \frac{\dot{\sigma}_{22}}{E^F} - \frac{v^F}{E^F}\dot{\sigma}_{11}$$

$$\dot{\gamma}_{12} = \frac{\dot{\sigma}_{12}}{G^F} \tag{7}$$

where E^F, G^F and v^F are Young's modulus, shear modulus and Poisson's ratio of fiber, respectively, and ε_{11}, ε_{22}, and γ_{12} are engineering strains.

For the matrix material, a linearly elastic and non-linearly creep model is adopted here to simplify the analysis. In a state of plane stress the creep strain is derived from a potential function [5]

$$f^M = \frac{1}{3}[\sigma_{11}^2 + \sigma_{22}^2 - \sigma_{11}\sigma_{22} + 3\sigma_{12}^2] \tag{8}$$

with the flow rule given by

$$\dot{\varepsilon}_{ij}^v = \lambda \frac{\partial f^M}{\partial \sigma_{ij}} \tag{9}$$

where $\dot{\varepsilon}_{ij}^v$ is creep strain rate, and λ is a proportionality factor. Combined with the generalized Hooke's law, the constitutive equations of matrix are given by

$$\dot{\varepsilon}_{11} = \frac{\dot{\sigma}_{11}}{E^M} - \frac{v^M}{E^M}\dot{\sigma}_{22} + \frac{\lambda}{3}(2\sigma_{11} - \sigma_{22})$$

$$\dot{\varepsilon}_{22} = \frac{\dot{\sigma}_{22}}{E^M} - \frac{v^M}{E^M}\dot{\sigma}_{11} + \frac{\lambda}{3}(2\sigma_{22} - \sigma_{11})$$

$$\dot{\gamma}_{12} = \frac{\dot{\sigma}_{12}}{G^M} + 2\lambda\sigma_{12} \tag{10}$$

The creep work rate in the matrix is given by

$$\dot{W}^v = \sigma_{ij}\dot{\varepsilon}_{ij}^v = 2\lambda f^M \tag{11}$$

Define effective stress in the matrix as

$$\bar{\sigma} = (3f^M)^{1/2} \tag{12}$$

and the effective creep strain rate $\bar{\dot{\varepsilon}}^v$ according to

$$\bar{\dot{\varepsilon}}^v\bar{\sigma} = \dot{W}^v \tag{13}$$

From equations (11-13) we obtain

$$\bar{\dot{\varepsilon}}^v = \frac{2}{3}\lambda\bar{\sigma} \tag{14}$$

and then

$$\lambda = \frac{3}{2} \frac{\overline{\dot{\varepsilon}^v}}{\overline{\sigma}} \tag{15}$$

For simple loading, the relation between effective stress and effective creep strain rate is approximated by a power law as

$$\overline{\dot{\varepsilon}^v} = \begin{cases} \alpha(\overline{\sigma} - \sigma_o)^m, & \overline{\sigma} \geq \sigma_o \\ 0, & \overline{\sigma} < \sigma_o \end{cases} \tag{16}$$

where σ_o can be considered as the elastic limit. With this relation scalar λ can be rewritten explicitly as

$$\lambda = \frac{3}{2} \alpha \frac{(\overline{\sigma} - \sigma_o)^m}{\overline{\sigma}} \tag{17}$$

It is noted that all the stress components in this section are either in the fiber or the matrix.

The creep equation (16) is by no means a general one. This equation is suitable for materials in which creep is dominated by a constant creep rate behavior.

4. Composite Constitutive Relations

By eliminating all the variables with superscripts AF and AM in the constitutive equations of fiber and matrix by use of the basic assumptions in Section 2, the nominal stress-strain relations of Part A are derived as

$$(1 + \frac{V_1 E^F}{V_2 E^M}) \dot{\varepsilon}_{11}^A = \frac{1}{V_2 E^M} [\dot{\sigma}_{11}^A - (V_1 v^F + V_2 v^M) \dot{\sigma}_{22}^A] +$$

$$\frac{\lambda^A}{3V_2} [2\sigma_{11}^A - (2V_1 v^F + V_2) \sigma_{22}^A - 2V_1 E^F \varepsilon_{11}^A] \tag{18}$$

$$\dot{\varepsilon}_{22}^A = -\frac{v^M}{E^M} \dot{\sigma}_{11}^A + (\frac{V_1}{E^F} - \frac{V_1 v^M v^M}{E^F} + \frac{V_2}{E^M} + \frac{V_1 v^F v^M}{E^M}) \dot{\sigma}_{22}^A - (V_1 v^F - V_1 v^M \frac{E^F}{E^M}) \dot{\varepsilon}_{11}^A$$

$$-\frac{\lambda^A}{3} \sigma_{11}^A + (\frac{2V_2 \lambda^A}{3} + \frac{V_1 v^F \lambda^A}{3}) \sigma_{22}^A + \frac{V_1 \lambda^A E^F}{3} \varepsilon_{11}^A \tag{19}$$

$$\dot{\gamma}_{12}^A = (\frac{V_1}{G^F} + \frac{V_2}{G^M}) \dot{\sigma}_{12}^A + 2V_2 \lambda^A \sigma_{12}^A \tag{20}$$

where

$$\lambda^A = \frac{3}{2} \alpha \frac{(\overline{\sigma}^{AM} - \sigma_o)^m}{\overline{\sigma}^{AM}}$$

Similarly, we have for Part B

$$\dot{\varepsilon}_{11}^{B} = \frac{\dot{\sigma}_{11}^{B}}{E^{M}} - \frac{\nu^{M}}{E^{M}}\dot{\sigma}_{22}^{B} + \frac{\lambda^{B}}{3}(2\sigma_{11}^{B} - \sigma_{22}^{B}) \tag{21}$$

$$\dot{\varepsilon}_{22}^{B} = \frac{\dot{\sigma}_{22}^{B}}{E^{M}} - \frac{\nu^{M}}{E^{M}}\dot{\sigma}_{11}^{B} + \frac{\lambda^{B}}{3}(2\sigma_{22}^{B} - \sigma_{11}^{B}) \tag{22}$$

$$\dot{\gamma}_{12}^{B} = \frac{\dot{\sigma}_{12}^{B}}{G^{M}} + 2\lambda^{B}\sigma_{12}^{B} \tag{23}$$

where

$$\lambda^{B} = \frac{3}{2}\frac{(\overline{\sigma}^{B}-\sigma_{o})^{m}}{\overline{\sigma}^{B}}$$

Using equations (18-20) and (21-33) together with equations (2), the stress and strain components in Part B can be eliminated. For example, from equations (2) and (5), we have

$$\dot{\gamma}_{12}^{A} = \dot{\gamma}_{12}^{B} \tag{24}$$

$$\dot{\sigma}_{12} = V_{1}\dot{\sigma}_{12}^{A} + V_{2}\dot{\sigma}_{12}^{B} \tag{25}$$

Combining equations (20) and (23-25), we obtain

$$\dot{\sigma}_{12}^{A} = q_{31}\dot{\sigma}_{12} + q_{32}\sigma_{12} + q_{33}\sigma_{12}^{A} \tag{26}$$

where

$$q_{31} = \frac{G^{F}}{(V_{1} + V_{2}^{2})G^{F} + V_{1}V_{2}G^{M}} \quad , \quad q_{32} = 2\lambda^{B}G^{M}q_{31}$$

$$q_{33} = -2G^{M}(V_{1}\lambda^{B}+V_{2}^{2}\lambda^{A})q_{31} \tag{27}$$

In a similar manner, the equations for stress rates $\dot{\sigma}_{11}^{A}$ and $\dot{\sigma}_{22}^{A}$ can be derived in terms of the nominal stresses of Part A, and the nominal stress, stress rate and strain of the cell as

$$a_{11}\dot{\sigma}_{11}^{A} + a_{12}\dot{\sigma}_{22}^{A} = q_{11}\sigma_{11}^{A} + q_{12}\sigma_{22}^{A} + q_{13}\dot{\sigma}_{11}$$
$$+ q_{14}\dot{\sigma}_{22} + q_{15}\sigma_{11} + q_{16}\sigma_{22} + q_{17}\varepsilon_{11} \tag{28}$$

$$a_{21}\dot{\sigma}_{11}^{A} + a_{22}\dot{\sigma}_{22}^{A} = q_{21}\sigma_{11}^{A} + q_{22}\sigma_{22}^{A} + q_{23}\dot{\sigma}_{11}$$
$$+ q_{24}\dot{\sigma}_{22} + q_{25}\sigma_{11} + q_{26}\sigma_{22} + q_{27}\varepsilon_{11} \tag{29}$$

where

$$a_{11} = \frac{1}{V_{1}E^{F}+V_{2}E^{M}}+\frac{V_{1}}{V_{2}E^{M}} \quad , \quad a_{12} = -[\frac{V_{1}\nu^{F}+V_{2}\nu^{M}}{V_{1}E^{F}+V_{2}E^{M}} + \frac{V_{1}\nu^{M}}{V_{2}E^{M}}]$$

$$a_{21} = \frac{V_{1}\nu^{M}}{V_{2}E^{M}} + \frac{V_{1}\nu^{F}+V_{2}\nu^{M}}{V_{1}E^{F}+V_{2}E^{M}}$$

$$a_{22} = [\frac{V_1}{V_2 E^M} + \frac{V_1}{E^F}(1 - v^F v^F) + \frac{1}{E^M}(V_2 + V_1 v^F v^M) + \frac{V_1(V_1 v^F + V_2 v^M)(v^F E^M - v^M E^F)}{E^M(V_1 E^F + V_2 E^M)}]$$

$$q_{11} = -\frac{2}{3}[\frac{E^M \lambda^A}{V_1 E^F + V_2 E^M} + V_1 \frac{\lambda^B}{V_2}] \quad , \quad q_{12} = \frac{(2V_1 v^F + V_2)E^M \lambda^A}{3(V_1 E^F + V_2 E^M)} + \frac{V_1 \lambda^B}{3V_2}$$

$$q_{13} = \frac{1}{V_2 E^M} \quad , \quad q_{14} = -\frac{v^M}{V_2 E^M} \quad , \quad q_{15} = \frac{2\lambda^B}{3V_2} \quad , \quad q_{16} = -\frac{\lambda^B}{3V_2}$$

$$q_{17} = \frac{2V_1 E^F E^M \lambda^A}{3(V_1 E^F + V_2 E^M)} \quad , \quad q_{21} = -[\frac{V_1 \lambda^B}{3V_2} + \frac{\lambda^A}{3} + \frac{2V_1 \lambda^A(v^F E^M - v^M E^F)}{3(V_1 E^F + V_2 E^M)}]$$

$$q_{22} = \frac{2V_1 \lambda^B}{3V_2} + \frac{\lambda^A}{3}(2V_2 + V_1 v^F) + \frac{V_1 \lambda^A(2V_1 v^F + V_2)(v^F E^M - v^M E^F)}{3(V_1 E^F + V_2 E^M)}$$

$$q_{23} = \frac{v^M}{V_2 E^M} \quad , \quad q_{24} = -\frac{1}{V_2 E^M} \quad , \quad q_{25} = \frac{\lambda^B}{3V_2} \quad , \quad q_{26} = -\frac{2\lambda^B}{3V_2}$$

$$q_{27} = \frac{1}{3}[V_1 E^F + \frac{2V_1^2 E^F(v^F E^M - v^M E^F)}{V_1 E^F + V_2 E^M}] \tag{30}$$

Equations (26),(28) and (29) are the constitutive equations for the typical composite cell and, thus, for the entire composite. However, these equations still involve nominal stresses for Part A. Theoretically, these stresses σ_{ij}^A should be eliminated further to result in a set of second order ordinary differential equations. However, the present form is more convenient for finite difference solutions and is thus kept for further development of creep analysis.

5. Creep Analysis

For the creep problem, the load is suddenly applied, and the applied stress rates $\dot\sigma_{ij}$ are vanishing. In this case, equations (26),(28) and (29) can be solved to obtain

$$\dot\sigma_{11}^A = (h_{11}\lambda^A + h_{12}\lambda^B)\sigma_{11}^A + (h_{13}\lambda^A + h_{14}\lambda^B)\,\sigma_{22}^A +$$
$$(h_{15}\sigma_{11} + h_{16}\sigma_{22})\,\lambda^B + h_{17}^A \varepsilon_{11}\lambda^A \tag{31}$$

$$\dot\sigma_{22}^A = (h_{21}\lambda^A + h_{22}\lambda^B)\sigma_{11}^A + (h_{23}\lambda^A + h_{24}\lambda^B)\sigma_{22}^A +$$
$$(h_{25}\sigma_{11} + h_{26}\sigma_{22})\lambda^B + h_{27}\varepsilon_{11}\lambda^A \tag{32}$$

$$\dot\sigma_{12}^A = (h_{31}\lambda^A + h_{32}\lambda^B)\sigma_{12}^A + h_{33}\sigma_{12}\lambda^B \tag{33}$$

where

$$h_{11} = \frac{1}{w}(a_{22}r_{11} - a_{12}r_{21}) \quad , \quad h_{12} = \frac{-V_1}{3V_2 w}(V_1 a_{12} + 2a_{22})$$

$$h_{13} = \frac{1}{w}(a_{22}r_{12} - a_{12}r_{22}) \quad , \quad h_{14} = \frac{V_1}{3V_2w}(a_{22} + 2a_{12})$$

$$h_{15} = \frac{1}{3V_2w}(2a_{22} + a_{12}) \quad , \quad h_{16} = \frac{-1}{3V_2w}(2a_{12} + a_{22})$$

$$h_{17} = \frac{1}{w}(a_{22}r_{13} - a_{12}r_{23}) \quad , \quad h_{21} = \frac{1}{w}(a_{11}r_{21} - a_{21}r_{11})$$

$$h_{22} = \frac{V_1}{3V_2w}(2a_{21} + a_{11}) \quad , \quad h_{23} = \frac{1}{w}(a_{11}r_{22} - a_{21}r_{12})$$

$$h_{24} = \frac{-a1}{3V_2w}(2a_{11} + a_{21}) \quad , \quad h_{25} = \frac{-1}{3V_2w}(a_{11} + 2a_{21})$$

$$h_{26} = \frac{1}{3V_2w}(2a_{11} + a_{21}) \quad , \quad h_{27} = \frac{1}{w}(a_{11}r_{23} - a_{21}r_{13})$$

$$h_{31} = \frac{-2V_2}{a_{33}} \quad , \quad h_{32} = -\frac{2V_1}{V_2a_{33}} \quad , \quad h_{33} = \frac{2}{V_2a_{33}}$$

$$a_{33} = \frac{V_1}{G^F} + \frac{V_2}{G^M} + \frac{V_1}{V_2G^M} \quad , \quad r_{11} = -\frac{2E^M}{3(V_1E^F + V_2E^M)}$$

$$r_{12} = \frac{(2V_1v^F + V_2)E^M}{3(V_1E^F + V_2E^M)} \quad , \quad r_{13} = \frac{2V_1E^FE^M}{3(V_1E^F + V_2E^M)}$$

$$r_{21} = \frac{1}{3}[1 + \frac{2V_1(v^FE^M - v^ME^F)}{V_1E^F + V_2E^M}]$$

$$r_{22} = -\frac{1}{3}[2V_2 + V_1v^F + \frac{V_1(2V_1v^F + V_2)(v^FE^M - v^ME^F)}{V_1E^F + V_2E^M}]$$

$$r_{23} = -\frac{V_1E^F}{3}[1 + \frac{2V_1(v^FE^M - v^ME^F)}{V_1E^F + V_2E^M}] \tag{34}$$

in which

$$w = a_{11}a_{22} - a_{12}a_{21} \tag{35}$$

For the creep problem, the constant applied stresses σ_{ij} are given. Equations (31-33) and (18-20) can then be solved successively using the finite difference method. This yields the solutions for σ_{ij}^A and ε_{ij}^A, which combined with equation (5) give σ_{ij}^B. Since $\varepsilon_{ij} = \varepsilon_{ij}^A$, the total creep strain components are thus obtained. Note that in solving equations (31-33), the values of λ^A, λ^B, and ε_{11} at the previous time step are initially used. Iterations within each time step can be performed to improve accuracy.

Figures 2-4 present the creep strains in a metal-matrix composite under off-axis loading. The composite properties are

$$E^F = 55.0 \text{msi} \quad , \quad v^F = 0.1 \quad , \quad G^F = 25.0 \text{msi} \quad , \quad E^M = 9.9 \text{msi} \quad , \quad v^M = 0.3$$

$$G^M = 3.8 \text{msi} \quad , \quad \alpha = 0.25 \text{x} 10^{-10} [\frac{(\text{ksi})^{-m}}{\text{hour}}] \quad , \quad m = 5.0 \quad , \quad \sigma_0 = 0.0$$

$$V_1 = 47\% \qquad \text{(fiber volume fraction)}$$

From Figs. 2-5, it is interesting to note that the composite creep curves consist of two portions; in each portion the creep rate is almost constant. The second portion is more dominant, however.

6. Macromechanics Representation

The micromechanical model developed in the foregoing sections is too complicated to use in structural analysis. In this section, a macromechanical model in which the composite is considered a homogeneous material is developed.

A potential function for the composite is assumed to be of the form

$$f = \frac{1}{2} (\sigma_{22}^2 + 2a_{66} \sigma_{12}^2) \tag{36}$$

from which the creep strain rates are derived, i.e.,

$$\dot{\varepsilon}_{ij}^v = \lambda \frac{\partial f}{\partial \sigma_{ij}} \tag{37}$$

where $\dot{\varepsilon}_{ij}^v$ and σ_{ij} are the nominal (average) creep strain rates and stresses referring to the material coordinate system, respectively. The use of this potential function implies that there is no creep in the fiber-direction. This functional form was also used in [6] to describe orthotropic plasticity in fiber composites with excellent results.

The viscous work rate is

$$\dot{W}^v = \sigma_{ij}\dot{\varepsilon}_{ij}^v = 2\lambda f \tag{38}$$

Defining effective stress as

$$\bar{\sigma} = (3 f)^{\frac{1}{2}} \tag{39}$$

and letting

$$\dot{W}^v = \bar{\sigma} \, \dot{\bar{\varepsilon}}^v \tag{40}$$

the effective creep strain rate can be derived as

$$\dot{\bar{\varepsilon}}^v = \frac{\dot{W}^v}{\bar{\sigma}} = \frac{2}{3}\lambda\bar{\sigma} \tag{41}$$

or

$$\lambda = \frac{3}{2} \frac{\dot{\bar{\varepsilon}}^v}{\bar{\sigma}} \tag{42}$$

The relation between effective stress and effective creep strain rate is approximated by a power law,

$$\dot{\bar{\varepsilon}}^v = \beta \, \bar{\sigma}^n \tag{43}$$

From equation (42) scalar λ can be written explicitly as

$$\lambda = \frac{3}{2} \beta \bar{\sigma}^{n-1} \tag{44}$$

Adding the creep strain rates to the elastic strain rates we obtain the complete constitutive relations

$$\dot{\varepsilon}_{11} = \frac{\dot{\sigma}_{11}}{E_1} - \frac{\nu_{21}}{E_2} \dot{\sigma}_{22}$$

$$\dot{\varepsilon}_{22} = \frac{\dot{\sigma}_{22}}{E_2} - \frac{\nu_{12}}{E_1} \dot{\sigma}_{11} + \frac{3}{2} \beta \bar{\sigma}^{n-1} \sigma_{22}$$

$$\dot{\gamma}_{12} = \frac{\dot{\sigma}_{12}}{G_{12}} + 3\beta a_{66} \bar{\sigma}^{-n} \sigma_{12} \tag{45}$$

For uniaxial off-axis loading, the effective stress and effective creep strain rate can be related to the longitudinal stress σ_x and strain rate $\dot{\varepsilon}_x^v$ as

$$\bar{\sigma} = \sigma_x h(\theta) \tag{46}$$

and

$$\dot{\bar{\varepsilon}}^v = \dot{\varepsilon}_x^v / h(\theta) \tag{47}$$

where

$$h(\theta) = [\frac{3}{2}(\sin^4\theta + 2a_{66}\sin^2\theta\cos^2\theta)]^{\frac{1}{2}} \tag{48}$$

From the creep curves, generated from the micromechanical model, see Figs. 2-4, it is seen that the second portion of the creep curve is much more dominant. Thus, the first portion can be neglected as an approximation.

The value of parameter a_{66} in the potential function f can be determined by requiring the $\dot{\bar{\varepsilon}}^v - \bar{\sigma}$ curves corresponding to all off-axis loadings to collapse into a single curve. Using the results of Figs. 2-4, we obtain

$$a_{66} = 2.1 \tag{49}$$

The $\dot{\bar{\varepsilon}}^v - \bar{\sigma}$ curves for off-axis loadings are shown in Fig. 5. This master curve can be described by the power law of equation (43) with

$$\beta = 0.17 \times 10^{-11} \qquad [\frac{(ksi)^{-n}}{nom}]$$

$$n = 4.93 \tag{50}$$

Note that this power law has a power index almost the same as that for the matrix.

7. Summary

A micromechanical model for creep in fiber composites was developed. The fiber was assumed to be linearly elastic and the matrix nonlinearly viscous with a steady creep rate for a given stress. Creep curves for a composite under off-axis loadings were calculated using this model. For a given stress, the composite was found to creep with a basically constant strain rate.

A macromechanical model was also developed to describe the creep behavior of the composite. A one-parameter potential function was used to establish the relation between strain rates with stresses. Effective stress and effective strain rate were introduced to form a single relation for combined stress states.

Acknowledgement -- This work was partially supported by the National Science Foundation Engineering Research Center for Intelligent Manufacturing Systems at Purdue University and by NASA Langley Research Center under grant No. NAG-1-825.

References

1. Lou, Y.C., and Schapery, R.A., "Viscoelastic Characterization of a Nonlinear Fiber-Reinforced Plastic," *Journal of Composite Materials,* Vol. 5, April 1971, pp. 208-234.

2. Schapery, R.A., "Viscoelastic Behavior and Analysis of Composite Materials," in *Composite Materials,* Vol. 2, G.P. Sendekyj, Ed. Academic Press, New York, 1974, pp. 85-168.

3. Skudra, A.M., and Bulavs, F. Ya., "Strength and Creep Micromechanics of Composites," in *Mechanics of Composites,* I.F. Obraztsov and V.V. Vasil'ev, Ed., MIR Publishers, Moscow, 1982 (English translation), pp. 77-109.

4. Dan Jumbo, E.A., Harbert, B.C., and Schapery, R.A., "Constant Rate, Creep Behavior and the Analysis of Thermoplastic Composite Laminates," presented at the *ASTM 9th Symposium on Composite Materials: Testing & Design,* Reno, Nevada, April 27-29, 1988.

5. Kraus, H., *Creep Analysis,* John Wiley & Sons, New York, 1980.

6. Sun, C.T., and Chen, J.L., "A Simple Flow Rule for Characterizing Nonlinear Behavior of Fiber Composites," *Proceedings of the Sixth International Conference* on Composite Materials, Imperial College, London, England, July 20-25, 1987, pp. 1250-1259.

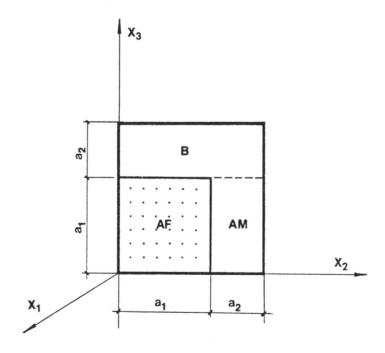

Fig. 1   Representative composite cell

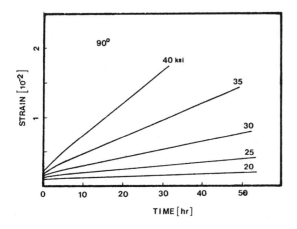

Fig. 2   Creep curves for constant
off-axis loading at θ = 90°

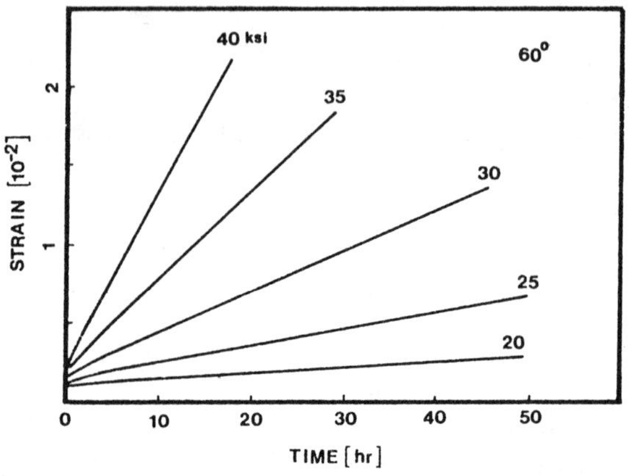

Fig. 3   Creep curves for constant
off-axis loading at for θ = 60°

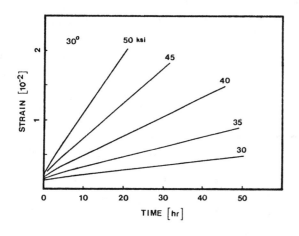

Fig. 4   Creep curves for constant
off-axis loading at for θ = 30°

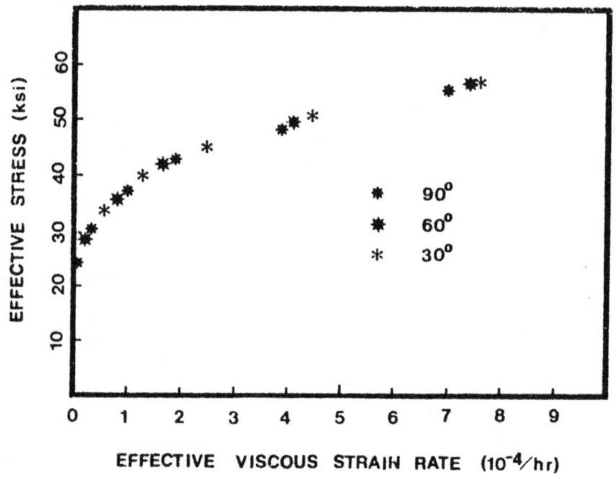

Fig. 5   Effective stress-effective creep
strain rate relation for the
macromechanical model

Thermal Response of Printed Wire Boards

S. CHANDRASHEKARA AND SOM R. SONI

ABSTRACT

The state-of-the-art of thermal and structural response of multilayer printed wire boards is reviewed. The main factors governing the thermal compatibility and material selection for the laminated composites of the board are identified. A simple analytical model with closed form solutions which can effectively predict the thermal response of today's printed board structure is presented. Specifically, both in-plane and out-of-plane thermal expansion coefficients are evaluated for various assemblies of practical interest. The efficiency of the model is validated by comparing the results with those from the more elaborate numerical methods as well as with those obtained by some experiments.

INTRODUCTION

The use of hybrid electronic circuits in communication data processing, industrial testing, medicine, military and aerospace industries has not only been extensive but has led to several challenges in the recent years. The main component of the electronics technology is a multilayer printed wire board (PWB) which is a laminated composite comprising of a number of layers of power planes, dielectric and signal lines. From a technological point of view, the present day laminate materials must serve the needs of such diverse applications as consumer products to computer systems to high-stress environments of military electronics. Further, the need for reducing the package size (consequently reduced space) and the weight is quite obvious. The sophistication in electronics in aircraft, for example, has increased atleast ten times in the past decade whereas the amount of space allotted for electronics has only doubled [1]. This factor has led to high density packaging of integrated circuits and to the advent of leadless chip carriers (LCC). Such a development is successfully meeting the increasing demands of very high speed integrated circuits (VHSIC). A leadless chip carrier, in contrast to a leaded one, has solder pads in place of compliant pins, thus permitting a direct bonding to the contacts on PWB. This results in a reduction of package size in addition to the ease of assembly, shipping and handling of component parts. One of the important features of construction of PWB, especially with high density packaging, is the mechanically drilled plated-through-hole (PTH) which is a path for electrical signals to arrive from the semiconductor chips and enter into the printed circuitry of the composite.

Because of the trend toward the multilayer boards often with different materials (metals, organic and inorganic composites), along with the emphasis on package density, the technical problem associated with the PWB analysis, design and manufacturing covers both

---

* AdTech Systems Research Inc., 1342 North Fairfield Road, Dayton, Ohio 45432

dimensional stability (especially with high density boards) and thermal stability (especially when using leadless ceramic chip carriers on non-ceramic substrates). Consequently, printed board technology encompasses most of the basic material sciences, with interdependence between chemical and metallurgical processes of manufacturing and the mechanical aspects such as material property evaluation and influence and stress/strain analysis [2]. This calls for a wide variety of analytical and experimental tools to monitor materials, processes and behavior of PWB's.

THERMAL COMPATIBILITY

The leadless chip carrier is attached to PWB by solders. During a thermal cyclic loading, the solder joints are subjected to strains. It has been established [3] that the thermal strains in the solder joints are directly proportional to the temperature excursion as well as the difference in the coefficients of thermal expansion (CTE) between the LCC and PWB. From the design point of view, optimizing the thermal fatigue life of solder joint connections is directly linked to a reduction in strains in the solder joints due to thermal cycling. Thus the difference in CTE between the connection materials has become a factor of major importance in the design and assembly techniques practiced in the industry. The importance of CTE mismatch between the chip carrier and the substrate in a PWB can be realized by considering the CTE values for different materials commonly used in these applications. Figure 1 graphically illustrates this situation where the in-plane CTE's are shown. The substrate materials have generally larger CTE values in x-y directions than the ceramic chip carriers. This forces the solder connections to absorb the differential thermal strains.

From the manufacturing point of view, the CTE mismatch can be remedied by one of the following methods [3]:

(i)   Compliant designs where an organic or inorganic sustrate has an inherently matching CTE with that of LCC. One of the common organic systems is Kevlar/epoxy with Kevlar fibers possessing negative CTE along their axis thus reducing the CTE of the composite system. Inorganic application includes porcelain-metal substrate or the use of a thick film of low expansion ceramic.

(ii)   Constrained designs in which low expansion materials such as copper-clad-invar (CIC for short) are incorporated in the substrate construction. Also included in this category is the technique of bonding the substrate to a core of low expansion material which can also act as a heat sink. A thick core, acting as a thermal mounting plate, provides extra stiffness to the assembly (thus providing mechanical integrity in resisting vibration and shock). However, the weight penalty imposed by thicker cores has also to be kept in mind during the final design stages.

(iii) Unconstrained design in which the CTE differences are corrected by a proper design of solder joints (such as thicker or deeper solders) and/or by controlling the solder composition and microstructure (such as fine or large grain size) during the LCC attachment.

As can be seen from figure 1, the in-plane CTE of ceramic is about 6.5 ppm/°C. Many attempts have been made both analytically and experimentally towards the development of low expansion PWB's whose CTE is close to that of the ceramic. Indeed, attempts have also been made to develop structural ceramic with nearly zero CTE by suitable chemical modification [4].

The other thermal expansion mismatch is between the laminate and the PTH barrel.

Usually, the most severe thermal excursion for the PTH occurs during the soldering operation [5]. With the more common glass/epoxy layers of the laminate construction, the board attains a temperature which is larger than the glass transition temperature for the layers of the laminate. Since the copper coating on the barrel has much lower CTE in the thickness direction (i.e., CTE-z) than that of glass/epoxy, a large thermal strain is induced in the barrel with a possible copper failure. Thus, an investigation into the variation of CTE-z with various composite material layers of PWB becomes significant.

MATERIALS

There is a variety of material combinations that can be considered for PWB package. The early efforts of minimizing CTE mismatch were to attach glass/epoxy boards to thick metal cores. The heavier packages and the unreliability of PTH passing through the metal core led to the exploration of other cores as well as different fiber systems for composites [6]. Although glass fiber in epoxy resin has been commonly used in many applications, other low CTE fibers such as Kevlar and quartz and resins such as polyimide have also been considered by military and industry establishments. Kevlar fibers are suited for high speed circuits because of their lower dissipation factor and dielectric constant. The negative CTE in the axial direction contributes to bringing down the CTE of the board to about 3 to 7 ppm/°C while at the same time being 20% lighter compared to glass fiber boards [7]. The main disadvantages of Kevlar boards are a high value of CTE-z and a high moisture absorption. These result in resin microcracking leading to failure in the barrel of PTH in Kevlar boards in comparison with the glass fiber boards. Also, Kevlar/epoxy materials are 2 to 3 times more expensive than glass/epoxy materials [8].

The lower CTE in the x and y directions of quartz fibers or fabrics has reduced the possibility of solder cracking in PWB packages. Also, these fabrics give rise to a lower value of CTE-z in composites compared to that of Kevlar boards. Polyimide/quartz composites, for example, have a CTE-z of about 30 to 35 ppm/°C in addition to possessing low water absorption, lower density and good dimensional stability and electrical characteristics [9]. They, however, tend to pose difficulty in the drilling process for holes because of higher hardness of quartz [10].

The ideal substrate for thermal compatibility with LCC is one made out of alumina. These are usually limited in size and are expensive for consumer type applications with low cost. This has led to the development of high temperature porcelain coated metal substrates with low expansion metal cores (such as copper-clad-inavar or CIC). Experimental work has been reported [11] where the CTE of porcelain coated CIC was observed to vary from 5.8 to 6.8 ppm/°C for temperatures below 200 °C.

OBJECTIVES OF THE PRESENT WORK

The previous discussion on the characteristics of the PWB under thermal loading has prompted some analytical and experimental investigations. The present paper reports one such study. The aspects of solder joints and chip size and the component interactions between LCC, solder joints and the PWB are not included here. Even so, a multitude of parameters has to be considered in the PWB analysis and thermal control [5]. The analytical investigations here pertain to the evaluation of both in-plane and out-of-plane thermal expansion coefficients (CTE-x y z) for a variety of combinations of metal and composite layers of practical interest. Several core materials are also considered for evaluation of some constrained designs of PWB. The comparison of CTE's and the composite material properties with those obtained by the industry sources (finite element analysis, for example) is also made to establish the simplicity and effectiveness of the present approach.

Some experimental results are also reported in this paper with regard to the parameters already described. Appropriate comparisons are drawn between the analytical and experimental results for verification purposes. Also, the reported experimental results from industry sources are considered wherever appropriate.

The analytical procedure of the present work is based on a simple composite model with the consideration of out-of-plane or z-direction properties also. Closed form solutions are obtained for the material properties of interest even when a large number of layers is present in the laminate. It is hoped that this investigation would, in the long run, help define relevant thermal design criteria and establish a thermal data base for rational and efficient solution.

ANALYSIS

The analysis is essentially based on the classical laminate theory with an extension to include calculations for transverse normal strain in each layer. In this particular instance, nonmechanical load, such as that due to temperature, is considered, although the effect of mechanical loads can be superposed. Each layer in the composite is considered to be thin compared to the other dimensions. The basic assumptions underlying thin plate theory are given in many sources of literature [12]. A summary of these assumptions is given below:

(i) Each layer of the PWB is orthotropic and perfectly bonded to the adjacent layers. The axes of material symmetry are at an angle θ with respect to x-y axes of the plate with the z-axis being considered in the thickness direction as shown in figure 2.

(ii) The displacements u, v and w are small compared to the plate thickness and the in-plane strains ε_x, ε_y and ε_{xy} are small compared to unity.

(iii) Each layer is in a state of plane stress. This implies that, in addition to transverse normal stress σ_z, the transverse shear strains ε_{xz} and ε_{yz} are also neglected.

In addition, the laminate is considered to be of symmetric lay-up. This is indeed the case in practice because an unsymmetric construction would result in warping of the cross-section of the PWB. The Duhamel-Neumann stress-strain relationships for the laminate can be written as [13]

$$\sigma_i = Q_{ij} (\varepsilon_j{}^m - \varepsilon_j{}^n) \tag{1}$$

where $\varepsilon_x{}^m$ and $\varepsilon_x{}^n$ refer to the mechanical and nonmechanical strains. The subscripts i and j refer to x and y directions of the laminate. The stress resultants (values of stresses integrated over the thickness) are then derived from eq.(1) as

$$N_i = A_{ij} \varepsilon_j{}^o + B_{ij} \kappa_j - N_i{}^n \tag{2}$$

where $\varepsilon_j{}^o$ and κ_j are mid-plane strain and curvature and $N_i{}^n$ correspond to the thermal stress resultants (nonmechanical loads). They are given by integrating the product $Q_{ij} \alpha_j \Delta T$ with respect to z over the thickness of the laminate (ΔT being the temperature difference). Since the laminate is symmetric about the mid-plane (i.e., $B_{ij} = 0$) and considering only nonmechanical loads (i.e., $N_i = 0$), eq.(2) can be rewritten as

$$N_i{}^n = A_{ij} \, \varepsilon_j{}^o \tag{3}$$

Since $N_i{}^n$ is known, the mid-plane strains $\varepsilon_j{}^o$ can be obtained by inverting eq.(2). In other words, the components $\varepsilon_x{}^o$ and $\varepsilon_y{}^o$ are known from evaluating A_{ij} of the laminate and the components $N_x{}^n$ and $N_y{}^n$. The effective in-plane CTE's for the laminate are then written as

$$\text{CTE-x} = \varepsilon_x{}^o / \Delta T$$
$$\text{CTE-y} = \varepsilon_y{}^o / \Delta T \tag{4}$$

In these calculations, ΔT can be chosen to be unity, so that the effective CTE's for the laminate can be calculated without specifying the actual temperature difference. However, it needs to be specified for evaluating the out-of-plane strain ε_z. By considering each layer to be in plane stress, the strain ε_z can be written for a layer as

$$\varepsilon_z = S_{13} \, \sigma_x + S_{23} \, \sigma_y + S_{36} \, \tau_{xy} - \alpha_z \Delta T \tag{5}$$

where S_{13}, S_{23} etc. are the compliance terms for the layer and σ_x, σ_y and τ_{xy} are the possible mechanical stresses and α_z is the CTE-z for the layer under consideration. This process is repeated for all the layers ($i = 1, N$) and the laminate strain is obtained as

$$\varepsilon_z{}^L = (1/H) \sum_1^N \varepsilon_z{}^i / h_i \tag{6}$$

where $\varepsilon_z{}^i$ and h_i are the strain and thickness, respectively, of layer i and H is the total thickness of the laminate. The summation here is carried out for $i = 1$ to N. Once the strain and the temperature difference are known, CTE-z is easily evaluated.

NUMERICAL RESULTS

The analytical descriptions given in the previous section are applied to solve some numerical problems. The calculations are based on numerical data appropriately drawn from different sources. As a simple check, the present modeling is used to check the established values of CTE for CIC material. Three ratios of copper to invar, namely 12.5:75:12.5, 16:68:16, 20:60:20 are considered for the purpose of tailoring to a selected CTE. The material properties chosen are

Copper: E = 17 msi ; G = 6.7 msi ; ν = 0.33 ; CTE = 16.5 ppm/°C
Invar: E = 21 msi ; G = 8.1 msi ; ν = 0.29 ; CTE = 1.4 ppm/°C

The CTE values obtained by the present model and those obtained elsewhere (references 7 and 14) are given in Table 1a. These values pertain to in-plane CTE's only and the correlation is seen to be quite good. The slight differences can be attributed to the differences in the individual material properties chosen from different sources. Table 1b shows the CTE-z obtained by the present model for the different cases of CIC. The actually measured value [15] for the 20:60:20 is also given in the same table. The values seem to correlate excellently.

A similar comparison, not only for CTE values, but for other material properties also is drawn in the next example. Here the PWB is constructed out of several "B" stage and "C" stage glass/epoxy (fabric) lay-up with intermediate copper signal layers. The lay-up geometry is given in reference 16 in which two sets of results are shown. One set corresponds to the predictions of in-plane properties by laminate plate theory. Another set is by using a simplified solid finite element model which predicts CTE-z also. Briefly, in the finite element model, each layer of the PWB is modelled by means of orthotropic plane strain elements. The material properties chosen for the analysis are given in reference 16. Again, two cases are considered here. One in which the simulated power and ground planes are of copper and the other in which CIC is used as those planes (with the hope of reducing the overall CTE for the PWB). Table 2a shows the results for the case with copper planes as obtained from different numerical approaches including the present one. Table 2b corresponds to the case of CIC planes. It can be observed from these two tables that the prediction from the present model is excellent in comparison with the other results. The CTE-z values are also very close to those obtained by the more elaborate finite element model. As expected, the use of CIC has reduced the CTE values. The CTE-z, for example is reduced by about 16%. Also, the use of CIC makes the board stiffer as evidenced by the higher values of the moduli.

The use of a core in the PWB substrate to reduce the CTE will be demonstrated next. In practice, layers of composite and copper are attached on either side of the core to form a symmetric construction. As such, only one half of the entire laminate need to be considered. The lay-up pattern, geometry for the layers are taken from reference 17. A thickness of 0.05 inch is considered for PWB on either side of the core. The properties of copper are as given before. For the purposes of comparison regarding the efficiency of different materials, three core materials are considered, namely, aluminum, graphite/epoxy and CIC. In each of the cases, the thickness of the core material is varied to study the corresponding variations of the in-plane and out-of-plane CTE's. The material properties chosen are listed below:

Substrate: Glass/epoxy: $E_1 = 2.78$ msi ; $E_2 = 2.08$ msi ; $E_3 = 1.16$ msi ; $G_{12} = 0.45$ msi
$\nu_{12} = 0.18$; $\nu_{23} = 0.32$; $\nu_{13} = 0.27$
CTE-x = 20.9 ppm/°C ; CTE-y = 16.0 ppm/°C ; CTE-z = 47.2 ppm/°C

Core: Aluminum: $E = 10.0$ msi ; $G = 3.76$ msi ; $\nu = 0.33$; CTE = 23.6 ppm/°C
CIC: $E = 19.5$ msi ; $G = 7.56$ msi ; $\nu = 0.29$; CTE = 6.4 ppm/°C
Graphite/epoxy: $E_1 = 9.34$ msi ; $E_2 = E_3 = 9.0$ msi ; $G_{12} = 0.8$ msi ; $\nu_{12} = 0.06$
$\nu_{13} = 0.3$; CTE-x = 1.87 ppm/°C ; CTE-y = 36.0 ppm/°C

Figures 3, 4 and 5 show the variations of in-plane CTE and CTE-z for the above cases with aluminum, CIC and graphite/epoxy cores respectively. The single values of in-plane CTE's shown in these figures correspond to the average values of CTE-x and CTE-y. Figure 3 indicates that while the CTE-z decreases with increasing aluminum thickness, the in-plane CTE increases correspondingly. Hence the choice of aluminum core is unsuitable for achieving a low in-plane CTE for the PWB. From figures 4 and 5, it can be observed that the in-plane CTE with CIC and graphite/epoxy cores reduces with increasing core thickness. In all the above three cases, CTE-z shows a decreasing trend with increase in core thickness. Of the three materials, CIC seems to be the best from the point of view of reducing CTE-z. This is because the CTE-z curve shows a maximum decreasing slope and also for the same thickness of core, CIC combination results in a much lower CTE-z value than the other two cases. The CTE-z in this case is about 25.4 ppm /°C for a CIC total core thickness of 0.12 inch whereas the corresponding value with graphite/epoxy core is about 40.6 ppm /°C. Of the CIC and graphite/epoxy core materials, the choice of graphite/epoxy is better from the point of view of reducing in-plane CTE, because for the same thickness, in-plane CTE is smaller with graphite/epoxy than with CIC. The in-plane CTE of the PWB with CIC core is about 9 ppm /°C when the total core thickness is about 0.12 inch, while for the same thickness the value with graphite/epoxy is about 7.2 ppm /°C.

The next two sets of results pertain to the calculation of out-of-plane strain (ε_z) in the PWB. In both the cases, the results from the present model are compared with those from the experiments. A thermal excursion experiment has been reported [18] in which strains ε_z are measured over a range of temperature. The strains were measured in the PTH barrel as well as in the laminate between PTH's. The PWB is 0.18 inch thick and the properties of the glass/epoxy and the copper layers are given below [18]:

Glass/epoxy: $E_1 = E_2 = 3.25$ msi ; $E_3 = 0.23$ msi ; $G_{12} = 0.09$ msi ; $\nu_{12} = 0.02$
CTE-x = CTE-y = 15 ppm /°C ; CTE-z = 90 ppm /°C
Copper: $E = 17.3$ msi ; $\nu = 0.35$; CTE = 17 ppm /°C

The details regarding the arrangement of copper and composite layers are taken from reference 18. The variations of ε_z (%) with respect to temperature as obtained from the present model and the experiments are shown in figure 6. The results from the present model pertain to the laminate strain only. It can be observed that there is an excellent agreement between the two sets of results till the glass transition temperature T_g of epoxy (~125 °C) around which the board expands drastically in the z-direction. The present model does not account for the flow effects (beyond T_g). It should be mentioned here that room temperature is assumed to be the stress-free temperature which corresponds to the curing temperature state in the present model. It can also be inferred from figure 6 that the PTH strains are quite close to the laminate strains within T_g for this particular case and the laminate strains are good, conservative estimates of the barrel strains.

The second set of experimental results has been reported [17] where the CTE-z were experimentally measured by a thermomechanical analyzer. Figure 7 shows the variation of the dimensional change (in z-direction) against temperature. It also shows the corresponding values obtained from the present model. The data for the lay-up and properties are the same as mentioned before for figures 3 through 5 [17]. It can be seen that the prediction from the present approach agrees fairly well with experimental one for this particular case. The discrepancy can be partly attributed to the CTE-z value assumed for the glass/epoxy. It could possibly be much higher (as high as 90 ppm /°C as in the previous example). A higher value tends to make the straight line from the present model move closer to the experimental curve. The slope of the experimetal curve gives a CTE-z equal to 54.7 ppm /°C whereas the value from the present model is 43 ppm /°C.

It is worth mentioning that, in the present research effort, experiments were conducted extensively to determine the material properties such as moduli and poisson ratios and CTE's for various composite systems at different temperatures. Specifically, glass/epoxy, Kevlar/epoxy and glass/polyimide composites have been considered in the -55°C to 125°C range. An extensive data base has thus been created for further use. Also, attempts were made to obtain approximate values of PTH strains by simple anisotropic elasticity solutions as well as estimates of transverse normal stresses in the PTH boundary. The results of these efforts have been reported in detail in reference 17.

CONCLUSIONS

The simple analytical model presented here to evaluate CTE's for multilayer laminates of PWB yield good results for the purposes of design and optimization of PWB packages. The method is validated by comparisons with results from other industry sources and from some experiments. It appears that the consideration of various combinations of materials, their geometry and lay-up to arrive at optimum design features is more important than establishing elaborate analysis techniques. The present model can be used for such optimization studies with regard to thermal compatibility in PWB packages.

ACKNOWLEDGEMENTS

The authors acknowledge the support of Dr. Ran Y. Kim of University of Dayton Research Institute in conducting the experimental work reported here and of Dr. N.J. Pagano and Mr. Donald H. Knapke of Air Force Materials Laboratory for helpful suggestions during this study. The work was sponsored by the Materials Laboratory of Air Force Wright Aeronautical Laboratories of Wright Patterson Air Force Base, Dayton, Ohio under the contract No. F33615-82-C-5071 through a subcontract from Texas Instruments Inc., Dallas, Texas.

REFERENCES

1. Lamoureux, R.T., "Leadless Chip Carrier Compatible PWB's", *28th National SAMPE Symposium*, 1983, pp. 1389-1398.
2. Seraphim, D.P., Lee, L.C., Appelt, B.K. and Marsh, L.L., "An Overview of Materials Science in Printed Circuit Packaging", *Proceedings of Material Research Society Symposium*, Vol. 40, 1985, pp. 21-48.
3. Lake, J.K. and Wild, R.N., "Some Factors Affecting Leadless Chip Carrier Solder Joint Fatigue Life", *28th National SAMPE Symposium*, 1983, pp. 1406-1414.
4. Roy, R. and Agrawal, D.K., "Successful Design of New Very Low Thermal Expansion Ceramics", *Proceedings of Material Research Society Symposium*, Vol. 40, 1985, pp. 83-88.
5. Nakayama, W., "Thermal Management of Electronic Equipment: A Review of Technology and Research Topics", *Applied Mechanics Reviews,* Vol. 39, No. 12, 1986, pp. 1847-1868.
6. Hanson, J.R. and Hauser, J.L., "New Board Overcomes TCE Problem", *Electronic Packaging and Production*, November 1986, pp. 48-51.
7. Markstein, H.W., "Low TCE Metals and Fibers Prove Viable for SMT Substrates", *Electronic Packaging and Production*, January 1985, pp. 52-59.
8. Korb, R.W. and Ross, D.O., "Direct Attachment of Leadless Chip Carriers to Organic Matrix Printed Wiring Boards", *27th National SAMPE Symposium*, 1982, pp. 260-270.
9. Mahler, B., "Polyimide-Quartz Laminate Materials for Leadless Chip Carrier Applications", *28th National SAMPE Symposium*, 1983, pp. 1236-1239.
10. Sanjana, Z.N., Valentich, J. and Marchetti, J.R., "Thermal Expansion of Circuit Board Materials for Leadless Chip Carriers", *28th National SAMPE Symposium*, 1983, pp. 1240-1250.
11. Hang, K.W., Prabhu, A.N., Andrus, J., Boardman, S.M. and Onyshkevych, L.S., "Low Expansion Porcelain-Coated Copper-Clad Invar Substrates", *RCA Review*, Vol. 45, 1984, pp. 33-48.
12. Whitney, J.M., *Structural Analysis of Laminated Anisotropic Plates*, Technomic Publishing Co., Pennsylvania, 1987.
13. Soni, S.R., *A Digital Algorithm for Composite Laminate Analysis - FORTRAN*, AFWAL-TR-81-4073 (Revised), Wright Patterson Air Force Base, Ohio, 1983.
14. *MANTECH Program Review Presentations*, Wright-Patterson Air Force Base, Ohio, May 1983.
15. *MANTECH Program Review Presentations*, Wright-Patterson Air Force Base, Ohio, October 1987.
16. *Hermetic Chip Carrier Compatible Printed Wiring Board*, AFWAL-TR-85-4082, Wright-Patterson Air Force Base, Ohio 1985.
17. Chandrashekara, S., Soni, S.R. and Kim, R.Y., *PWB and PTH Modeling and Analysis* ASR-8-TR-1988, Texas Instruments, Inc., Dallas, Texas, 1988.
18. Lee, L.C., Darekar, V.S. and Lim, C.K., "Micromechanics of Multilayer Printed Circuit Boards", *IBM Journal of Research and Development*, Vol. 28, No. 6, 1984, pp. 711-718.

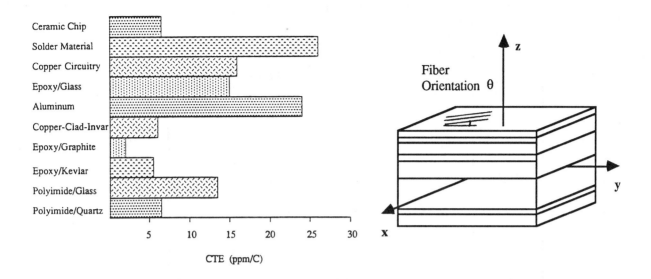

Figure 1. Printed Wire Board Construction With CTE Values

Figure 2. Geometry of PWB Laminate For Analysis

Copper: Invar: Copper Proportion	In-Plane CTE (ppm/C)		
	Ref. 7	Ref. 14	Present
12.5 : 75 : 12.5	2.8 - 3.5	--	4.76
16 : 68 : 16	4.0 - 4.5	5.0	5.74
20 : 60 : 20	5.0 - 6.0	6.3	6.90

1a. In-Plane CTE (CTE-x,y)

Copper: Invar: Copper Proportion	Out-of-Plane CTE (ppm/C)	
	Ref. 15	Present
12.5 : 75 : 12.5	--	6.01
16 : 68 : 16	--	7.21
20 : 60 : 20	8.90	8.53

1b. Out-of-Plane CTE (CTE-z)

Table 1. CTE Comparisons For Copper-Clad-Invar

Material Properties	LPT*	FEM**	Present	Material Properties	LPT*	FEM**	Present
E1 (msi)	6.50	6.50	6.66	E1 (msi)	10.2	10.2	10.14
E2 (msi)	6.49	6.49	6.61	E2 (msi)	10.2	10.2	10.11
G (msi)	1.88	--	1.94	G (msi)	3.42	--	3.45
$\upsilon 12$	.244	.243	.250	$\upsilon 12$	.267	.267	.270
$\alpha 1$ (ppm/C)	16.63	16.68	16.02	$\alpha 1$ (ppm/C)	10.37	10.39	10.20
$\alpha 2$ (ppm/C)	16.66	16.71	15.90	$\alpha 2$ (ppm/C)	10.37	10.40	10.10
$\alpha 3$ (ppm/C)	--	39.71	39.40	$\alpha 3$ (ppm/C)	--	33.05	32.94

* Laminated Plate Theory   ** Finite Element Model

2a. All Copper Layers   2b. With CIC Layers

Table 2. Comparisons Between LPT, FEM and Present Model

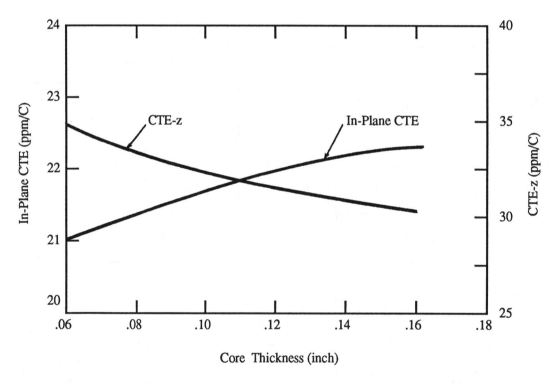

Figure 3. Variation of CTE with Core Thickness - Aluminum Core

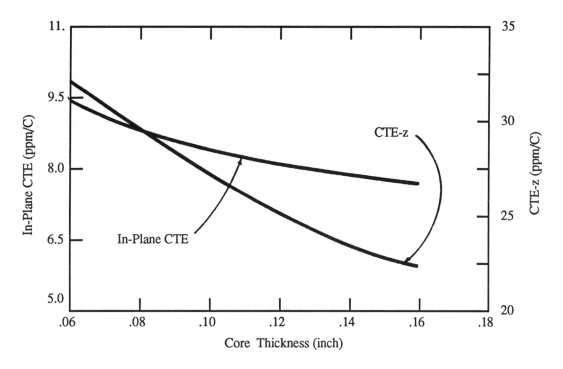

Figure 4. Variation of CTE with Core Thickness - CIC Core

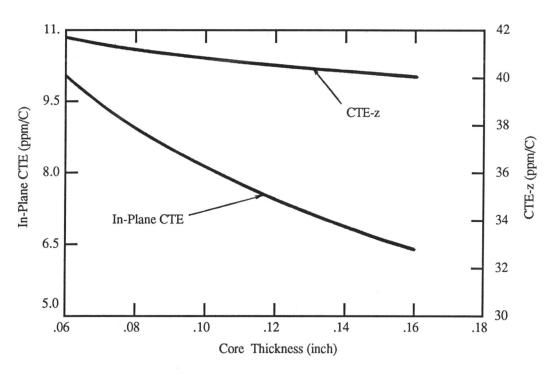

Figure 5. Variation of CTE with Core Thickness - Epoxy/Graphite Core

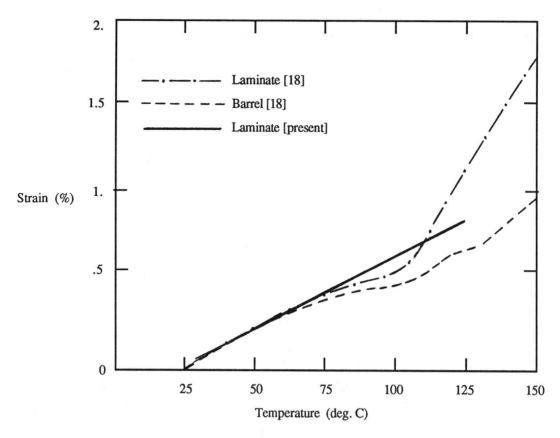

Figure 6. Transverse Normal Strain vs. Temperature in PWB

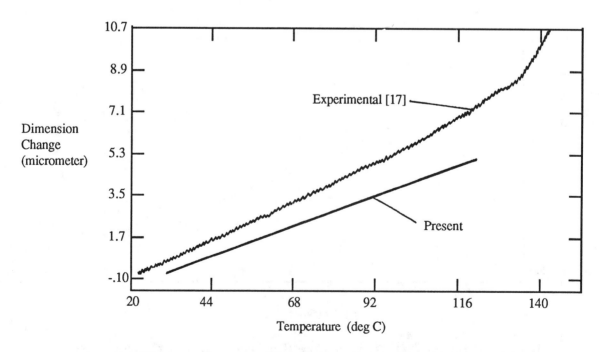

Figure 7. Transverse Dimension Change vs. Temperature for PWB

Measurement of Fiber Misalignment in Aligned Fiber Composites and the Influence of Lamination on Misalignment Distributions

S. W. YURGARTIS

Abstract

Fibers in nominally aligned-fiber composites are not absolutely straight, but in fact wander about the mean direction, resulting in local angular misalignment of the fibers. Not only is fiber misalignment suspected of having a significant affect on composite mechanical properties, but it is also suspected of affecting properties which are important in the processing of composite laminates. Fiber misalignment is likely to influence the deformation properties of the fiber-bed. It also appears to be related to fiber-bed permeability. These two effects implicate fiber misalignment, and fiber misalignment evolution, as having a significant role in influencing processing outcomes such as fiber volume fraction and void volume fraction. This paper presents a stereological method for measuring local fiber misalignment in typical high performance composites. The method has an angular resolution on the order of 0.25 degrees, and can determine both in-plane and out-of-plane misalignment distributions. Example distributions are shown for APC-2 composites and polycarbonate/carbon fiber composites. Standard deviations of the distributions range from 0.69 to 1.94 degrees. The influence of lamination on the distributions is discussed. In general, lamination broadens the distributions compared to the prepreg. The in-plane and out-of-plane distributions also become non-symmetric. Distribution changes are affected by stacking sequence. All-zero laminates show the most broadening. Initial fiber misalignment in prepreg is raised as an issue in material tailoring and prepreg quality control.

Introduction

Continuous, non-woven fibers in high performance composites wander about the mean direction, resulting in local fiber misalignment. The origins of fiber misalignment—sometimes referred to as fiber waviness—are still a matter of speculation. Swift [1] lists several possible sources. Many possible effects of fiber misalignment have been proposed. These include a significant influence on mechanical properties such as longitudinal compression strength, tensile modulus, transverse mechanical and transport properties, coefficient of thermal expansion, and delamination fracture toughness, among others [2, for references]. Fiber misalignment is also believed to have direct and indirect consequences on composite processing.

Fiber misalignment can be expected to strongly affect fiber-bed deformation behavior, both because of fiber cross-overs and because of neighboring fiber interactions. Fiber-bed deformation behavior plays a substantial role in lamination models by Gutowski, Morigaki, and Cai [3] and Dave, Kardos, and Dudukovic [4, 5]. For example, volume fraction distribution through the laminate thickness is believed to be controlled by fiber-bed behavior, as is the limiting fiber volume fraction. Gutowski, Cai, Bauer, Boucher, Kingery, and Wineman [6] have observed differences in fiber-bed stiffness among samples made

Steven W. Yurgartis, Assistant Professor, Mechanical and Industrial Engineering Dept., Clarkson University, Potsdam, NY 13676 (Telephone 315-268-6575)

from ostensibly identical material, and have speculated that these differences are due to differences in fiber misalignment in the prepreg. Since load transfer of the lamination pressure to the fiber-bed results in decreased resin pressure, it is reasonable to suspect that fiber misalignment will indirectly affect void fraction.

Fiber bed permeability is also affected by fiber misalignment. The so-called tortuosity of the porous medium is related to fiber shape and hence the misalignment distribution. furthermore, there is data that suggests that the transverse permeability of the fiber-bed, as reflected in the Kozeny 'constant', changes during laminate consolidation [6]. These results are attributed to fiber rearrangement during processing. Data presented below show that fibers do move substantially during lamination. Measurement of the evolution of fiber misalignment distribution can help reveal how permeability will be affected by such factors as initial fiber misalignment, stacking sequence, and processing path.

Fiber misalignment may also have a vital role in resin impregnation of fiber tows. In their model of thermoplastic resin impregnation into fiber tows, Lee and Springer [7] cite the length between fiber contacts as an important variable. This characteristic will be closely related to the fiber misalignment distribution, and in fact Lee and Springer make an estimate of the average fiber misalignment angle to get to an estimate of the average length between contacts.

Fiber misalignment is also of interest to post-lamination processing of thermoplastic matrix composites. Cogswell and Leach [8] identify four basic flow processes which may occur during processing. They speculate, based on preliminary experiments, that at least one process—axial intraply shear—is significantly affected by fiber twist, i.e. fiber misalignment. Experiments by Soll and Gutowski [9] in right angle forming revealed a correlation between fiber misalignment and a significant decrease in fracture strength as well as deviation from the desired part geometry.

Despite widespread speculation that fiber misalignment has a significant role in composite processing and properties, it has been extremely difficult to test these ideas due to an inability to accurately measure fiber misalignment distributions. This paper describes a method for making these measurements, and presents data on two thermoplastic/carbon fiber composites. It also includes data showing how processing can change the misalignment distributions.

Measuring Fiber Misalignment

Previous Work

Previous work to measure fiber misalignment has primarily been confined to direct optical methods, such as using an angular reticle in a microscope. Direct techniques do not have sufficient resolution to be applied to aligned fiber composites. McGee and McCullough [10] have developed an elegant diffraction technique to obtain a single parameter measure of short fiber misalignment. Colored tracer fibers have been used to measure gross misalignment, particularly as occurs in forming operations.

A useful but limited technique of serial transverse sectioning has been used to follow the path of a limited number of fibers. The technique is valuable in providing information not only about fiber misalignment, but also about fiber shape. Mansfield and Purslow [11] and Gutowski, et. al. [6] have contributed illuminating data from this technique. However, the technique is limited in resolution, is very time consuming, has limited sample size, and does not produce, as yet, a useful parametric measure of microstructure.

The Two-Dimensional Case

For ease of illustration the new method will be described for the two-dimensional case. Details of the method can be found in [2]. The technique is based on the simple observation that a plane-section of a circular cylinder is an ellipse. The major axis of the ellipse, ℓ, is related to the angle between the sectioning plane and the cylinder axis, ω, by,

$$\sin \omega = \frac{d}{\ell} \qquad (1)$$

where d is the cylinder diameter. If an array of cylinders were cut by the same plane, as indicated in Figure 1, then measuring the distribution of ℓ_i might give the distribution of ω_i, where the subscript indicates the ith cylinder.

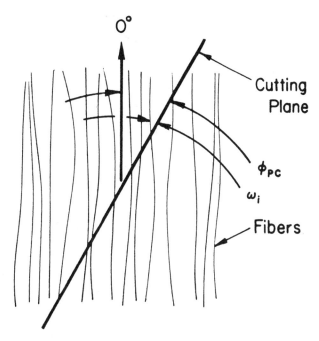

Figure 1: Plane Section of an Array of Fibers.

To apply these observations to the determination of fiber angle distribution in a composite, it must be assumed: (1) short sections of fibers are approximately straight, and (2) all fibers have the same circular diameter. These assumptions are examined and justified in [2].

In practice the method requires several steps. The composite sample must be sectioned and polished, the distribution of ℓ_i measured, and the distribution of ω_i calculated. A transformation is also needed from the angles ω_i to the angles ϕ_i, where ω_i is the angle that the ith fiber segment makes with respect to the sectioning plane, and ϕ_i is the angle of this same fiber segment, but with respect to the mean fiber direction. The mean fiber direction is defined a the zero degree direction.

A proper choice of ϕ_{pc} is a key factor in the technique described here. Four considerations suggest a limited choice of ϕ_{pc}. First, the sensitivity of ℓ to ϕ is greatest when ϕ is small, which leads to better angle resolution. Second, as ϕ approaches zero, ℓ rapidly becomes large and thereby weakens the assumption that measured fiber segments are straight. Third, the measured distribution folds back upon itself when $\omega < \phi_{pc}$. Finally, the calculated ω is less sensitive to variations in fiber diameter when ω is small.

Given these considerations a balanced choice for ϕ_{pc} must be made. For example, the data given below were measured at $\phi_{pc} \approx 5$ deg. Therefore most of the fibers—the fibers near the zero degree direction—are at an angle of about 5 deg to the sectioning plane. The ellipse axis lengths ℓ_i are then about 12 fiber diameters, and a 1 deg difference in ω will result in a 20% difference in ℓ. These dimensions have proven experimentally feasible to measure with good accuracy.

Correcting For Counting Bias

Transformation from the measured distribution $f(\ell)$ to the distribution $f(\omega)$ is not as straightforward as it might appear. The objective is to obtain a measure of the fraction of fiber volume in a given volume of fibers that is at a local misalignment angle near ω. Consider a cubic volume element of composite material with a single fiber traversing the element. A random plane section taken through this cube, parallel to the top face (ω_i=constant), is more likely to intersect a fiber with large ω_i than with small ω_i. A count of the number of fibers with ellipse axis length ℓ_i will be biased toward counting fibers at large ω_i (small ℓ_i) more frequently than fibers with small ω_i. Fortunately this is a long-standing stereological problem, and it is possible to use the generally developed theory to derive a specific correction for the circumstances at hand. Starting from the established stereological relationship [12],

$$N_{A_i} = N_{V_i} D_{V_i}(\omega) \tag{2}$$

where N_{A_i} is the average number of class i object counted per unit area, N_{V_i} is the average number of class i objects per unit volume, and D_{V_i} is the distance between tangent planes of the ith class object (restricted to two dimensions). In this case simple geometry gives,

$$D_{V_i} = L \tan \omega_i \tag{3}$$

where L is the side length of the cubic volume element. Neglecting the small difference in fiber-segment volume when the fiber is at different angles within volume L^3, the above relationships allow formation of the discrete distribution of fiber segments which lie in n^{th} interval centered on angle ω_i,

$$f_v(\omega) = \frac{\frac{N_i}{\tan \omega_i}}{\sum_{i=1}^{n}\left(\frac{N_i}{\tan \omega_i}\right)} \tag{4}$$

where $f_v(\omega_i)$ is the volume fraction of the total volume of fiber that is at an angle near ω_i, and N_i is the number of fiber segments counted in this interval. Equation 4 is in terms of measurable quantities. (For convenience in notation the subscripts v and i will be dropped.)

Transformation of the Distribution

It is necessary to transform from the angle ω, referenced to the sectioning plane, to the angle ϕ, referenced to the mean (zero degrees) fiber direction. The transformation is simply,

$$\phi = \omega - \phi_{pc} \tag{5}$$

Because ϕ_{pc} can not be known with good accuracy *a priori*, it is determined from $f(\omega)$. Since the zero degree direction is the mean of the fiber misalignment distribution, the mean of f(ω) equals ϕ_{pc}. Thus,

$$f(\phi) = f(\omega - \overline{\omega}). \tag{6}$$

Two factors may distort the estimated distribution $f(\phi)$. First, fiber segments with $\omega \approx 0$ will be unlikely to be properly measured because ℓ will be large and thus these fibers may not be counted at all if ℓ exceeds the measurement field boundaries, or more likely, because of fiber curvature, these fibers will be counted as having angles somewhat larger than they actually have. Second, fiber segments at an angle $\omega < 0$ will be measured as fibers with a positive ω, probably in the low angle part of the $f(\omega)$ distribution; in other words, the distribution at $\omega < 0$ folds back upon itself. It is possible to estimate the magnitude of these distortions and to compensate for them, as given in [2]. In the data sets presented here these distortions are not found to be a problem. Careful choice of ϕ_{pc} minimizes errors.

Generalization to Three Dimensions

Composites will of course have fibers misaligned in three dimensions. Using the natural geometry of typically planar composite structures, it is convenient to think of fiber misalignment occurring both (a) in the plane of the laminate, denoted by ϕ and (b) out of the plane of the laminate, denoted by θ. Geometric analysis shows that the major axis diameter ℓ, at any given angle ω, is independent of the angle θ. Using the same considerations it is also apparent that θ could be measured independently of ϕ.

The independence between ϕ and θ allows the distributions $f(\phi)$ and $f(\theta)$ to be determined independently. These distributions are so-called marginal distributions of the total bivariate, three dimensional distribution $f(\phi, \theta)$. If it is assumed that fiber segment misalignments are statistically independent—there is no correlation between θ and ϕ—then,

$$f(\phi, \theta) = f(\phi)f(\theta). \tag{7}$$

Experimental Results and Discussion

Measurements of fiber angle distribution were made on polyetheretherketone/XAS carbon fiber material known by the trade name APC-2 which is produced by ICI, Ltd. The material was chosen because it is available in the form of high quality, commercially marketed prepreg. It was felt that this material may have a narrow fiber angle distribution and thus be a good test of the measurement technique. In addition, a polycarbonate/AS-4 carbon fiber material was studied. This is an experimental 'model' material that was produced at NASA Langley Research Center using a drum-winding technique.

Outline of the Experimental Procedure

The fundamental measured quantity is ℓ_i. In this work ℓ_i was measured from micrographs using a digitizing table interfaced to a microcomputer. At the magnification on the micrograph (245:1) it is not possible to measure the fiber diameter of each fiber with sufficient resolution. A separate measurement of fiber diameter was made on a high magnification section taken transverse to the fiber axes. The XAS fibers have a fairly narrow diameter distribution with mean, d=6.95 μm and standard deviation of 0.31 μm. The AS-4 fibers have a mean diameter of 6.85 μm with a standard deviation of 0.40 μm. In addition, the assumption of circular cross-section, as checked with a circle template, appears to be very good in both cases. For computation it is assumed that all fibers are the average fiber diameter.

The program collects the ℓ_i data, calculates ω_i, and sorts datum into bins of width half a degree. About 800 ellipse lengths are measured for each distribution calculation. An uncertainty in fiber diameter of ± 1 standard deviation results in an angular uncertainty of about ± 0.25 deg. Previous error analysis [13] shows that this is the major source of uncertainty in the measurements. The program calculates a histogram corrected for the sample counting bias described earlier. Finally, the transformation between ω and ϕ (or θ) is made, in this case using equation 6 and assuming uniform distribution across each bin width.

Misalignment in APC-2 Prepreg

Figure 2 plots both in-plane and out-of-plane distributions of the APC-2 prepreg. The distributions are quite narrow—83% of the fiber volume is within ± 1 deg of being perfectly aligned. In-plane and out-of-plane distributions are, perhaps unexpectedly, axially symmetric. This is a clue about the original source of fiber misalignment in the prepreg. It suggests that prepregging does not bias the distribution, and that the initial misalignment is within the fiber tows. Misalignment data is unfortunately not available for the PC/AS-4 prepreg.

Misalignment Change Upon Lamination

When eight plys of APC-2 prepreg are laminated together (100 psi, 10 min, 380 deg C) the distributions are found to change. It should be noted that all of the APC-2 samples were from a contiguous length of prepreg less than 2 m long. Samples of about 2 cm by 1 cm were taken from the center of 10 cm square laminates. In Figure 3, the in-plane distributions are shown for a) the prepreg, b) the third ply of a $(0/90)_{2s}$ laminate, and c) an all-zero laminate. The same sequence, but for out-of-plane misalignment, is shown in Figure 4. A normal distribution fits the measured distributions moderately well. For example, using the 0/90 APC-2 data leads to the bivariate normal distribution shown in the surface plot, Figure 5.

In-plane misalignment deteriorates in processing, becoming quite broad for the all-zero laminate. The all-zero sample showed a weak in-plane grouping among fibers of about the same misorientation. The cause of this grouping appears to be due to a cooperative waviness among bundles of fibers; the distribution in Figure 3 reflects this behavior. Interestingly, it is not found in the out-of-plane distribution. Since no cooperative waviness was found in the prepreg, it is apparent that fibers can not only move substantially, but that their motion is influenced by their neighbors. It may be that a group of similarly shaped fiber segments acts as a nucleus which propagates similar fiber shapes. Some processing circumstances permit sufficient fiber mobility for this to occur, but clearly not all.

Out-of-plane alignment may slightly improve for the 0/90 laminate. Presumably the perpendicular fibers in the lamina above and below the third ply help straighten the fibers. Out-of-plane alignment in

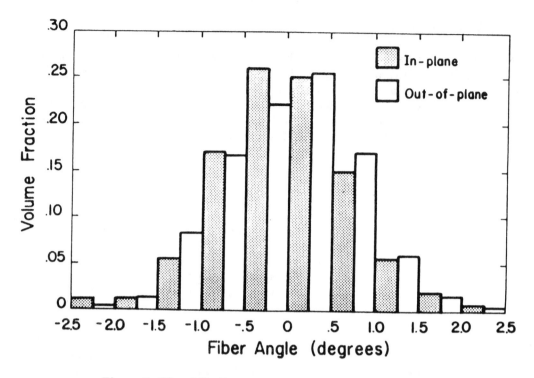

Figure 2: Fiber Misalignment Distributions in APC-2 Prepreg.

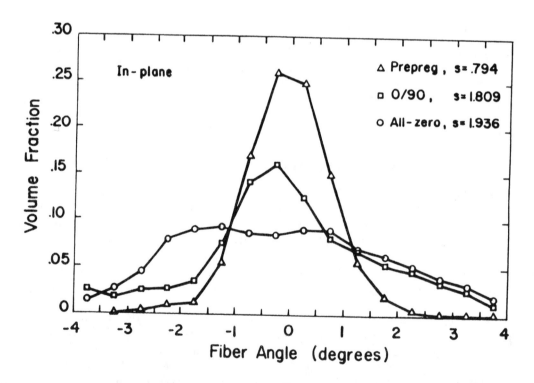

Figure 3: In-Plane Fiber Misalignment for APC-2 Prepreg, (0/90) Laminate, and All-Zero Laminate. Sample Sizes are 781, 1036, 2143, Respectively. Standard Deviations Shown in the Figure.

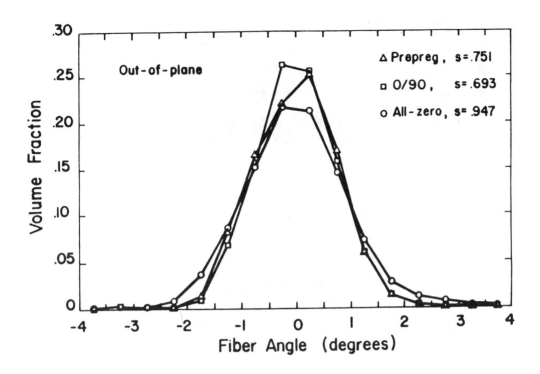

Figure 4: Out-of-Plane Fiber Misalignment for APC-2 Prepreg, (0/90) Laminate, and All-Zero Laminate. Sample Sizes are 1111,1133,1044, Respectively. Standard Deviations Shown in the Figure.

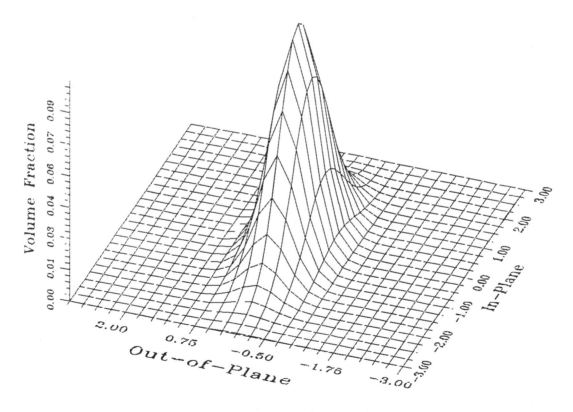

Figure 5: Fit of Normal Distribution Model to APC-2, (0/90) Laminate Fiber Misalignment.

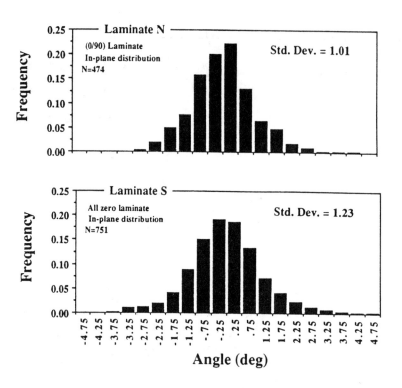

Figure 6: In-Plane Fiber Misalignment Distributions for PC/AS-4 Laminates. Prepreg Batch A.

the all-zero laminate worsens, but not as much as the in-plane alignment.

Similar results are found in the PC/AS-4 material. Figure 6 shows the in-plane distributions of an eight ply 0/90 laminate compared to a twelve ply all-zero laminate made from the same batch of prepreg. Figure 7 gives similar data from samples made from a different batch of prepreg. All PC/AS-4 laminates were processed identically. Again, stacking sequence is seen to affect the fiber mobility. Cooperative waviness is absent in these samples. The cause of the difference between batches of prepreg is not known, although it is known that different spools of fiber were used.

Closing Remarks

The technique presented above can accurately provide bivariate fiber misalignment distributions in high performance composites. The data shows the magnitude of misalignment that can be expected in these materials. It also clearly shows that that fibers move substantially during processing, and that the movement depends on the stacking sequence of the laminates.

A great deal of work remains to be done to correlate fiber misalignment to the properties discussed in the introduction. The method presented here for measuring fiber misalignment makes many experimental studies possible. Our laboratory is in the process of automating the measurements on image analysis equipment, making them much more accessible.

Further study of the evolution of fiber misalignment during processing is needed. These results will give insight useful for modelling changes in permeability and fiber-bed behavior. They are also likely to be relevant to the final mechanical properties of the laminate.

Determining the origin of initial fiber misalignment in prepreg would be a valuable study. Initial misalignment in prepreg may be an unappreciated issue in prepreg quality control. As mentioned previously, Gutowski, et. al. [6] have observed differences in fiber-bed behavior when samples were made from different rolls of ostensibly identical prepreg. Inherent inconsistency of prepreg may help explain scatter in mechanical properties and processing results.

It may also be possible to use initial misalignment in prepreg as another way to taylor these materials. For example, by increasing fiber misalignment the fiber-bed will become 'stiffer', and thereby it would

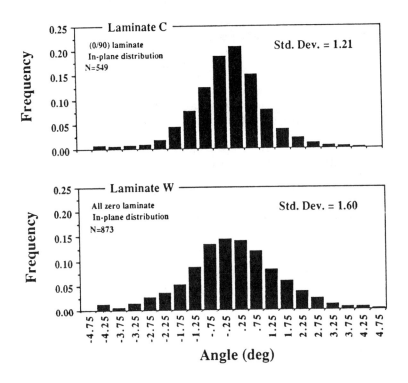

Figure 7: In-Plane Fiber Misalignment Distributions for PC/AS-4 Laminates. Prepreg Batch B.

be possible to laminate composites with a lower uniform fiber volume fraction and perhaps greater coupling between plys—both potentially beneficial to a property such as damage tolerance. Alternatively, better aligned prepreg would result in higher fiber volume fraction composites. It may be that increased misalignment will provide laminates with better 'conformability' for post-lamination forming; Chou's work on flexible composites is relevant here [14, for example]. Better understanding and control of local composite microstructure will become increasingly important as the technology of high performance composites matures.

References

[1] Swift, D. G., "Elastic Moduli of Fibrous Composites Containing Misaligned Fibers," *J. Phys. D: Appl. Phys.*, Vol. 8, 1975, pp. 223–240.

[2] Yurgartis, S. W., "Measurement of Small Angle Fiber Misalignments in Continuous Fiber Composites," *Composites Science and Technology*, Vol. 30, 1987, pp. 279–293.

[3] Gutowski, T. G., Morigaki, T., Cai, Z., "The Consolidation of Laminate Composites," *J. Composite Materials*, Vol. 21, 1987, pp. 172–188.

[4] Dave, R., Kardos, J. L., Dudukovic, M. P., "A Model for Resin Flow During Composite Processing: Part 1—General Mathematical Development," *Polymer Composites*, Vol. 8, 1987, pp. 29–38.

[5] Dave, R., Kardos, J. L., Dudukovic, M. P., "A Model for Resin Flow During Composite Processing: Part 2: Numerical Analysis for Unidirectional Graphite/Epoxy Laminates," *Polymer Composites*, Vol. 8, 1987, pp. 123–132.

[6] Gutowksi, T. G., Cai, Z., Bauer, S., Boucher, D., Kingery, J., Wineman, S., "Consolidation Experiments for Laminate Composites," *J. Composite Materials*, Vol. 21, 1987, pp. 650–669.

[7] Lee, W. I., Springer, G. S., "A Model of the Manufacturing Process of Thermoplastic Matrix Composites," *J. Composite Materials*, Vol. 21, 1987, pp. 1017–1055.

[8] Cogswell, F. N., Leach, D. C., "Processing Science of Continuous Fibre Reinforced Thermoplastic Composites," *SAMPE J.*, May/June 1988, pp. 11–14.

[9] Soll, W., Gutowski, T. G., "Forming Thermoplastic Composite Parts," *SAMPE J.*, May/June 1988, pp. 15–19.

[10] McGee, S. H., McCullough, R. L., "Characterization of Fiber Orientation in Short-Fiber Composites," *J. Appl. Phys.*, Vol. 55, 1984, pp. 1394–1403.

[11] Mansfield, E. H., Purslow, D. "The Influence of Fibre Waviness on the Moduli of Unidirectional Fibre Reinforced Composites," *RAE Tech. Report ARC-CP-1339*, 1974.

[12] DeHoff, R. T., Rhines, E. N., *Quantitative Microscopy*, McGraw Hill Co., NY, 1968.

[13] Yurgartis, S. W., "The Influence of Matrix Properties and Composite Microstructure on the Longitudinal Compression Strength of High Performance Composites," *Ph.D. Thesis, Rensselaer Polytechnic Institute*, Materials Engineering Dept., 1987.

[14] Chou, T.-W., Takahashi, K., "Non-Linear Elastic Behavior of Flexible Fibre Composites," *Composites*, January 1987, pp. 25–33.

SYMPOSIUM XIII

Composite Structure Design and Analysis

Global-Local Finite Element Method for Analysis of Composites

TEN-LU HSIAO AND SOM R. SONI

ABSTRACT

A global-local approach based on a mixed variational functional has been followed in this study. For global region of the laminate, the potential energy has been utilized, while the Reissner functional has been used for the local region. The field equations are based on an assumed thickness distibution of stress component within each layer of the local domain. The displacement field equations are assumed in the global region and are extended into local region for continuity. A set of equilibrium equations has been derived for the entire structure. The analysis transforms the 3D elasticity problems into 2D problems in a self-consistent approach that features realistic satisfaction of boundary and interface or interlayers condition. A set of problems (bending, uniform tensile and stretch) have been studied by global-local and local approaches. An important aspect of this research work is to use the frontal method to reduce the size of the structure matrix so that more plies and more refined mesh can be studied. Currently, the program can solve 3500 degree of freedom or more on IBM RT PC without having any memory problem.

INTRODUCTION

The study of delamination phenomenon in structural composite laminates began with analytical and experimental observations of the response of such bodies in the vicinity of a free edge. In 1967, Hayashi [1] presented the first analytical model treating interlaminar stresses in what has come to be known as the "free edge problem". Characteristically, this work focused on the computation of interlaminar shear stress, as in the early stages of composite research interlaminar and delamination effects were viewed as being synonymous with interlaminar shear. The presence of interlaminar normal stress, being of a more subtle origin and also seemingly defying common intuition, was not appreciated until many years after the pioneering work of Hayashi.

The development of the Hayashi model was based upon the implicit assumption that the in-plane stresses within a given layer did not depend upon the thickness coordinate. The magnitude of the maximum interlaminar shear stress was calculated to be a relatively large value in a glass epoxy [0/90] laminate. Unfortunately, however, owing to the omission of the interlaminar normal stress, the computed stress field within each layer does not satisfy moment equilibrium.

AdTech Systems Research Inc., 1342 N. Fairfield Road, Dayton, OH 45432.

The first reported experimental observations involving free edge delamination were made by Foye and Baker[2]. In that work, tremendous differences in fatigue life of boron-epoxy composite laminates as a function of layer stacking sequence were reported. Severe delaminations were witnessed in that work and were identified as the primary source of strength degradation in fatigue.

From the early work to the present time, the free edge laminate problem has been the most prominent device utilized in the study of composite delamination. Our emphasis shall be placed on the work which led to the development of the global-local mixed finite element model, a model which attempts to circumvent the overwhelming difficulties and complexities associated with stress analysis of multi-layered composite structures. In this model, 3D elasticity problems are transformed into 2D problems in a self-consistent approach that features realistic satisfaction of boundary and interface conditions. The present illustrative study includes; (1) The analysis of bending effects in laminates under prescribed uniform transverse load. (2) The analysis of edge effects in laminates under applied uni-axial uniform tensile stress or strain.

LITERATURE REVIEW

With regard to free edge problems, a finite width laminate under uniform extensional strain was first analyzed by Pipes and Pagano[3]. A finite difference scheme was used based on a formulation assuming linear elasticity. Subsequently, various authors have conducted investigations on this problem by using various techniques such as finite difference, finite element and series solutions.

References [4-7] lead to an understanding of various failure modes that have shown to occur in composite laminate. The bending and stretching of laminate deformation and Reissner-Mindlin plate theory have been incorporated in Reference [8,9], higher order [10,11] of CLT was applied to evaluate the interlaminar normal stress distribution, but only at a plane of symmetry.

In practical applications, a large number of layers may be present. Contemporary models are incapable of providing precise solution of the local stress fields in the vicinity of free edge. Reference [12] used an effective moduli global model, however only the extensional response of the regions was conducted, i.e. the flexural and flexural-extension coupling characteristics of laminated bodies were ignored. Reference[13] used a global representation upon a three dimensional laminate model which is a generalization and improvement of the material model given in reference[12]. However, in order to circumvent the complexity of an exact three dimensional elastic analysis [13], while at the same time being a reasonable precise method of studying the stress fields in laminate with moderately large number of layers the local representation [14] has been introduced. But the local model, in which each layer is represented as a homogeneous, anisotropic continuum, become intractable as the number of layers becomes large. The reference [14] has been recently extended into a formulation of global-local variational model [15]. In this model, a predetermined area, termed local region, of interest is represented by Reissner's functional that involves both stresses and displacements and the rest of the region, termed global region, is represented by potential energy in the variational functional. In this global-local model, the interface continuity, including three displacement components and the normal and shear stress components associated with the thickness direction are satisfied. For a large number of layers in reference [14,15] exceeding 16, the exponential parameter becomes very large and consequently cause computer overflow.

In the literature of the finite element solutions[12,16], the traction free edge conditions are satisfied in weighted integrated sense or lead to oscillating solutions near the edges. In reference [17], the authors have conducted the static and dynamic analysis of composite laminates using hybrid stress finite elements. Reference[18] has the free edge stresses in layered plates using eight nodes isoparametric elements. References[17,18], the laminate idealization for a reasonably accurate finite element analysis had to be very fine, i.e. a quarter of the laminate was divided into about 600 elements. No more than six layers would be considered for numerical calculations. For moderately large number of plies, these approaches will lead to computer storage/economic difficulties. In other studies a singular hybrid finite element model[19] is employed and a perturbation technique[20] as well as a series solution[21] has been considered to solve the free edge problem.

FORMULATION

Both local and global -local models were formulated by using the weighted displacement (model 1) or actual displacement (model 2) and interlaminar stress components. The laminate considered in the present investigation is shown in Figure 1. The laminate thickness comprizing N layers is divided into two parts viz; (i) local regioon (l) and global region (g). In this work, we shall assume that the interface between g and l is a plane z=const. The energy equation neglecting body forces for the global regions and local layers can be expressed as

$$
\pi = \int_V \left\{ W \, dV_g + \sum_{k=1}^{N} [\sigma^T \varepsilon - W] \, dV_k \right\} - \int_S \sigma^T U \, dS \tag{1}
$$

Where $\qquad W = W(U_i, e_{ij})$, $\qquad W = W(\sigma_{ij}, e_{ij})$

The first term represents the potential energy for the global region in terms of dispplacement U_i and e_{ij}, the expansional strain components, and the second term represents the Reissner variational functional for layers in local region in terms of stresses σ_{ij} and expansional strains e_{ij}. The third term expresses the work done by the external load. For each layer in the local region, the theory developed[14] has been used for the local regions. For the global region, theory developed[10] has been applied. The details of the derivation of equlibrium equation and continuity and boundary coonditions for the global and local domain are given in reference[15]. The interlaminar stress components σ_z, τ_{zy} and τ_{zx} at the top of the layer are denoted by p_2, s_2, t_2 respectively, while the corresponding stress components at the bottom of the layer are designated as p_1, s_1, t_1. Other parameters are, seven weighted dispacements in local domain and a like number of weighted displacements in the global region.

LOCAL DOMAIN :

Based on the reference[14], the simplest assumption consistent with a realistic stress analysis for the in-plane stress component are assumed to vary linearly through the thickness of each ply. The substitution of these stress components into the differential equations of equlibrium yields the interlaminar stress components in terms of tractions p_i, s_i, t_i (i=1,2), stress and moment resultants. Finally the weighted displacements are expressed in terms of stress resultants through constitutive relations.

GLOBAL MODEL :

Following theroretical development[10,15], in the global domain, an assumed continuous thickness distribution of displacement field is used. The substitution of the displacement functions into the strain-displacement relations of elasticity leads to the final stress-strain equations. Substituting the stress-strain equations from global fomulation as well as stress and stress resultant equations from local formulation into equation 1 and minimizing the energy equation, i.e. $\delta\pi = 0$, and following the standard finite element procedure[25,26,27], the final algebraic equation can be expressed as

$$[K] \ \{ U \} \ = \ [F]$$

Where the K is the stiffness matrix obtained from the volume integrals of the energy equations for the entire region. U denotes total number of discretizing weighted displacement as well as normal and shear stress at the top and bottom of layers. F is the load vector obtained from the area integral of the external tractions of whole boundary region.

EXAMPLE PROBLEMS AND NUMERICAL RESULT

The theoretical formulation is made for the analysis of composite structures with any geometric shape and size. Three cases of boundary value problems have been investigated. Case a deals with the bending of isotropic thick and thin rectangular plates with simply supported edges due to a uniformly distributed transverse load. Case b deals with the stress analysis of composite laminate under the influence of an applied in-plane unidirectional stress at the edges (x=constant) of the laminate. Case c deals with the stress analysis of a composite laminate under the influence of an applied in-plane strain at x=constant. The results are presented in graphical form. A comparison is made with existing results and show a good agreement.

Case a: Bending case:

For tesing the copmputer program written for the model 2, the solution to the problem of a plate transversely loaded by a uniformly distributed load with simply supported edges was obtained. The laminate geometry is shown in figure 1. Because of the isotropic laminate construction and the chosen symmetric geometry (see figure 2A), the following conditions were applied:

$$u = o \text{ at } x = o \qquad\qquad v = o \text{ at } y = o$$

The boundary conditions were chosen to be simply supported on the edges where x = a, y = b. This implies that at the edge of the laminate the transverse deflection W is set to zero for all nodes. Also, the shear stress components τ_{zx} and τ_{zy} are set to zero on edge x = a and y = b respectively. All the above mentioned conditions were applied to the assembled stiffness matrix representing the actual structure plate. The above solution of square plate as given in standard text book is

$$W_{max} = 0.00406 \, q \, a^4/D$$

Where q is the intensity of the applied loads, a is the side of the plate and D is the flexture rigidity. Material properties in the plane of elastic symmetry of each layer are given by

$$E_L = E_T = E_Z = 30 \times 10^6 \, psi, \qquad G_{LT} = G_{LZ} = G_{TZ} = 1.154 \times 10^6 \, psi$$

$$v_{LT} = v_{LZ} = v_{TZ} = 0.21,$$

where L, T and Z refer to fiber, transverse, and thickness directions respectively. The exact maximum deflection for square plate is given as 0.02365 q and 0.002956 q for plate thicknesses equal to .01 and .02 inch, respectively. The exact maximum deflection for rectangular plate with a/b = 2 is given as 0.00369 q and 0.000461 q for the plate thicknesses equal to 0.01 and 0.02 inch respectively.

The triangular elements and nodes are shown in figure 2B. Figure 3 shows the grid patterns used in the model. With grid patterns 1,2, 3 and 4 in the quarter laminate, the corresponding values for maximum deflection listed in these tables. It can be seen that the results from the global local finite element model converge very well to the exact value. In order to obtain a graphical idea about this aspect, the displacement parameter W calculated by the finite element method and given in tables 1 and 2 are plotted in figures 4 and 5 against the number of grids of the square and rectangular plates for both plate thicknesses.

By comparing the figures 4 and 5 in numerical sense, the error percentage change from 7.4% to 3.6% as the total plate thickness changes from .01 inch to .02 inch for the rectangular plate using grid 4. Similarly for square plate, the error percentage changes from 12.6% to 7.26%. The results in figures 4 and 5 show that the converging speed and error percentage of rectangular plate are better than those of square plate.

Case b: Applied Tensile Stress

This section shows the solution for the problem of a finite plate subjected to an in-plane tensile load $\sigma_x = 1$ at the edge, x = a by using models 1 and 2. The laminate geometry is shown in figure 1. Two cases of geometric configurations, the ratio of a/b, are studied using the first and second model. Also, the solutions for the thin and thick finite plate are compared. A $[0/90]_s$ symmetric laminate with the stacking of layers as 0°, 90°, 90°, 0° will be examined in this study. Comprehensive results based on the reference [14] will be employed to compare specific results given by the present theory.

The layers are of equal thickness h, the laminate width is 2b = 80h or 2b = 16h and material properties in the plane of elastic symmetry of each layer are given by

$$E_{11} = 20 \times 10^6 \text{ psi} ; \qquad E_{22} = E_{33} = 2.1 \times 10^6 \text{ psi}$$

$$G_{12} = G_{13} = G_{23} = .85 \times 10^6 \text{ psi} \qquad \upsilon_{12} = \upsilon_{13} = \upsilon_{23} = .21$$

where 1,2 and 3 refer to the fiber, transverse, and thickness directions respectively, and υ_{12} for example, is the Poisson ratio measuring strain in the transverse direction due to uniaxial tension in the fiber direction.

Because of the chosen symmetry of geometry in Figure 2A, the following conditions were applied

For model 1:   u, u* = o   at x = o   v, v* = o   at y = o

For model 2: u^t, u^b = o   at x = o v^t, v^b = o   at y = o

where parameters in model 1 are weighted displacements in x and y direction and are defined in references[14,15], and the superscripts t and b denote the respective variable at top and bottom of the layer, respectively. For applied in-plane tensile load case, the boundary conditions of the finite plate for both formulations are as follows:

The shear stress components τ_{zx} and τ_{zy} are set to zero at x = a and y = b respectively. At mid-surface, the transverse deflection W and shear stress components τ_{zy} and τ_{zx} are set to zero everywhere. All the above mentioned conditions were applied to the assembled stiffness matrix representing the actual structure plate.

The triangular elements and nodes are shown in Fig. 2B and different grid patterns are shown in figure 3. Figures 6 and 8 show the convergence study of both interlaminar stress components and weighted displacement for thick plates of aspect ratio a/b=5. The applied load is constant in-plane tensile stress at the edge of x=constant. The results obtained from model 1 have the same trend compared with reference[14], but magnitudes are different. That is because the applied

boundary condition in the present formulation is $\sigma_x = 1$ at $x = a$ while in reference[14] is $\varepsilon_x = \varepsilon$. The results of the stresses from the model 2 are almost equal to zero at free edge even though the actual displacements are reasonable.

Case c: Uniform Applied Strain

In this section the response of the laminate under the influence of the unidirectional constant strain applied at the edge of the plate has been studied. This case is simulated by applying the equivalently biaxial stress obtained by lamination theory. The convergence study of both interlaminal stress and weighted displacement of the thin and thick plate is similar to the one with uniform tensile load applied at the edge. Figure 7 shows the convergence pattern of interlaminar stresses for thick plate of aspect ratio a/b=5. The results of interlaminar stresses obtained from model 1 for constant stain case, have the same trend but different magnitude as reference[14]. Figure 11 shows the comparison of interlaminar stress distribution between the existing and present results. The differences in magnitude between the current model and those of reference[14] may be explained by the following points:

1. The present is 3D model and solution is obtained for finite length plate.

2. The meshes are not fine enough near the edge and/or more sublayers should be used in order to get better accuracy in results.

Figures 9 - 10 show the comparison of interlaminar stress of thick and thin plate for both constant tensile and strain load with respect to different aspect ratio (a/b).

SUMMARY AND CONCLUDING REMARKS

The convergence of current model is good as shown in the pictures. The results of interlaminar stress components and weighted displacement for both constant tensile and strain load from the Model 1 have the same trend but different magnitude compared with those of references[14]. The differences in magnitude between the current model and existing results may be explained as follows:

1. The present is 3D model and solution is obtained for finite length plate.

2. The meshes are not fine enough near the edge or more sublayers should be used in order to get better accurate results.

3. Applied boundary condition are different.

4. Geometric parameters are different.

The results of interlaminar stresses obtained from Model 2 are almost zero, even though the displacement trend makes sense. The reason may be explained as follows:

1. The meshes are not fine enough near the edge. For more accurate results, more refined mesh and/or more sub-layers should be used.

2. The transformation used to convert the weighted displacement into actual displacement may introduce additional applied loads, even though the displacement make sense.

3. The global region may need to include the interlaminar stresses as well as displacement.

For the thin plate, the results show that the shear stresses are more significant than the normal stress as compared to thick plate solution. It should be pointed out that because the mesh near the edge is not very fine, the models do not give prominent peek stresses in the presented thin plates solutions. This is because of the limited current hard disk space. A more refined mesh will be studied in future investigations. Convergence study, as shown in pictures, indicates that the results for constant strain applied load case are more stable than those for the contant tensile load case. The results by using the constant tensile load show slight oscillation. These oscillations reduce very significantly as the mesh size increases near the edge. The aspect ratio (a/b) may also have effect on the oscillation for the tensile load case.

The current code uses frontal solver and help reduce memory space requirements. For example the program can handle 3500 degree of freedom on IBM RT PC only using about 400 maximum frontal size, but the execution time is increased. The benefit of using frontal solution will increase with the solution of laminates with large number of layers.

The future studies include, modification of the computational procedure and the application of this model for investigating laminates with flaws and through the thickness holes. Also the curing stress study will be included.

ACKNOWLEDGEMENT :

The authors wish to acknowledge Dr. N. J. Pagano of the Air Force Material Laboratory for helpful discussion and useful suggestions on this study. The work was sponsored by the Nonmetallic Materials Division, Material Laboratory, Air Force Wright Aeronautical Laboratories, under Contract No. F33615-85-C-5034.

REFERENCES:

1.	T. Hayashi, "Analytical Study of interlaminar shear stresses in a Laminate Composite Plate" Trans. Japan Society for Aeronautics and Space Sciences Vol 10, No. 17 (1967), pp. 43-48.

2.	R. L. Foye and D. J. Baker, "Design of Orthotropic Laminates", 11th Annual AIAA structural Dynamics and Materials Conference, Denver, Colorao, April 1970.

3.	R.B. Pipes and N.J. Pagano, "Interlaminar Stress in Composite Laminates under Uniform Axial Extension", J. Comp. Mat. Vol.4, p. 538, 1970.

4.	N.J. Pagano and R.B. Pipes, "Some Observations on the Interlaminar Strength of Composite Laminates", Int. J. Mech. Sci., Vol.15, p. 679, 1973.

5.	L.B. Greszczuk, "Failure Mechanics of Composites Subjected to Compressive Loading", Air Force Material Laboratory Report V, AFML-TR-72-107, 1973.

6.	S.V. Kulkarni, J.S. Rice, and B.W. Rosen, "An Investigation of the Compressive Strength of Kevlar 49/Epoxy Composites" Vol.6, p. 127, 1975.

7.	F.H. Chang, D.E. Gordon, B.T. Rodini, R. H. McDaniel,"Real Time Characterization of Damage Growth in Graphite/Epoxy Laminates", J. Comp. Mat., Vol.10, p. 182, 1976.

8.	E. Reissner and Y. Stavsky, "Bending and Stretching of Certain Types of Heterogeneous Aelotropic Elastic Plates". J. Appl., Mech., Vol.28, p. 402, 1961.

9.   P.C. Yang, C.H. Norris, and Y. Stavsky, "Elastic Wave Propagation in Heterogeneous Plates", Int. J. Sol. Struct., Vol.2, p. 665, 1966.

10.   J.M. Whitney and C.T. Sun, "A Higher Order Theory for Extensional Motion of Laminated Composites", J. Sound Vib., Vol.30, p. 85, 1973.

11.   N.J. Pagano, "On the Calculation of Interlaminar Normal Stress in Composite Laminates", J. Comp. Mat., Vol.8, p. 65, 1974.

12.   A.S.D.Wang and F.W. Crossman, "Some New Results on Edge Effect in Symmetric Composite Laminate", J. Comp., Mat. Vol.11, p. 92, 1977.

13.   E.L. Stanton, L.M. Crain and T.L. Neu, "A Parametric Cubic Modeling System for General Solids of Composite Material", Int. J. Num Meth Engg. Vol.11, p. 653, 1977.

14.   N. J. Pagano, "Stress Fields in Composite Laminates", Technical Report, AFML-TR-77-114, Wright Patterson AFB, 1977. Also Int. J. Solids Struct., Vol.14, p. 385, 1978.

15.   N.J. Pagano and S.R. Soni, "Global-local Laminate Variational Model", Int. J. Solids Struct., Vol.19, p. 207, 1983.

16.   A. Harris, O. Orringer and E.A. Whitmer, "A Multilayer, Traction-Free-Edge, Quadrilateral Warping Element for the Stress Analysis of Composite Plates and Shell", MIT-ASRL-TR-193-1, 1979.

17.   R.L. Spilker and T.C.T. Ting, "Stress Analysis of Composite", Army Materials and Mechanics Research Center. Watertown, Mass. Tech. Rep. AMMRC-TR-81-5, 1981.

18.   I.S. Raju, J.D. Whitcomb and J.G. Goree, " A New Look at Numerical Analyses of Free Edge Stress in Composite Laminates", NASA Tech. Paper 1751, 1981.

19.   S.S. Wang and F.K. Yuan, "A Singular Hybrid Finite Element in Composite Laminates", Int. J. Sol. Struct. Vol.19, p. 825, 1983.

20.   P.W. Hsu and C.T. Herakovich, " Edge Effects in Angle-ply Composite Laminates", J. Comp. Mat. Vol.11, p. 422, 1977.

21.   J.T.S. Wang and J.N. Dickson, "Interlaminar Stress in Symmetric Composite Laminates", J. Comp, Mat. Vol.12, p. 390, 1978.

22.   B.M. Irons, "Roundoff Criteria in Direct Stiffness Solution", AIAA J., Vol.6,p. 1308-1312, 1968.

23.   B.M Irons, "A Frontal Solution Program",Int. J. Num. Meth. Eng. Vol.2, p. 532,1970.

24.   B.M. Irons, and A. Fawkes, "Students' Front Solution for Varying Degrees of Fereedom Per Node", Internal Report, Dept of Civil Engg., University College of Swansen, 1973

25.   R.H. Gallagher, "Finite Element Analysis-Fundamentals", Prentice Hall Inc., USA, 1975.

26.   O.C. Zienkiewicz, "The Finite Element Method", McGraw-Hill Book Co., UK, 1979.

27.   J. T. Oden and Graham F. Carey, "Finite Element Series", Prentice-Hall, Englewood Cliffs, New Jersey, USA, 1979.

TABLE 1   Comparison of Transverse Displacement, W, between existing and present results, Square Plate.

1a Thickness = .01 INCH

grid number	W FEM RESULTS	EXACT SOLUTION
1	.0185 q	.02365 q
2	.0189 q	
3	.0196 q	
4	.0206 q	

1b. Thickness = .02 INCH

grid number	W FEM RESULTS	EXACT SOLUTION
1	.00232 q	.002956 q
2	.00240 q	
3	.00258 q	
4	.00274 q	

TABLE 2   Comparison of Transverse Displacement, W, between existing and present results, Rectangular Plate.

2a Thickness = .01 INCH

grid number	W FEM RESULTS	EXACT SOLUTION
1	.00297 q	.00369 q
2	.00306 q	
3	.00324 q	
4	.00342 q	

2b. Thickness = .02 INCH

grid number	W FEM RESULTS	EXACT SOLUTION
1	.000373 q	.0004609 q
2	.000396 q	
3	.000427 q	
4	.000444 q	

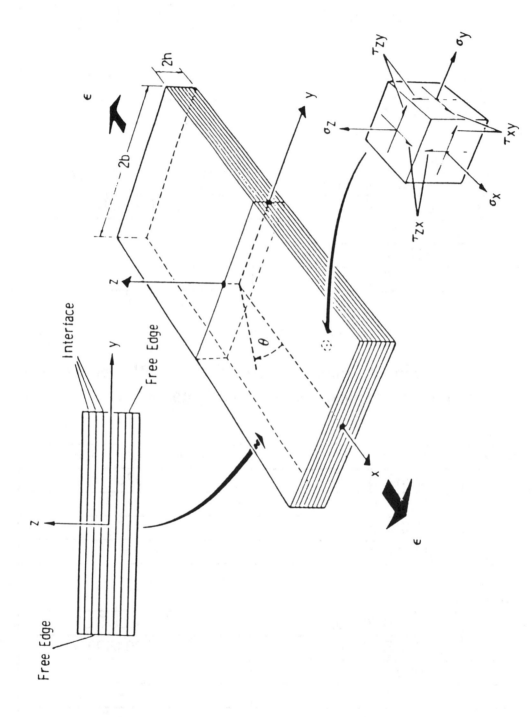

Fig. 1 Laminate Geometry

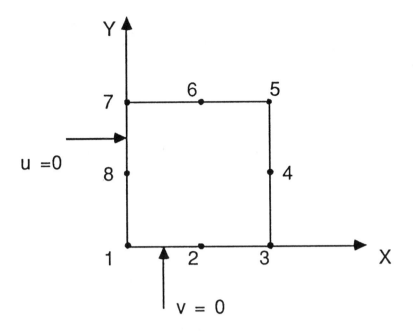

Fig. 2A Quarter plate laminate

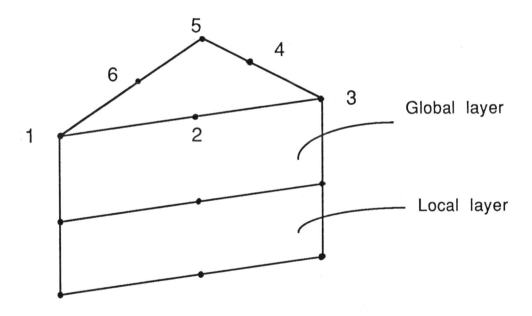

Fig 2B Triangle elements and nodes

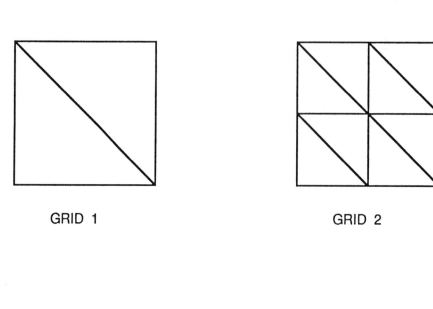

GRID 1   GRID 2

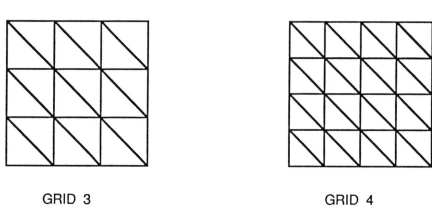

GRID 3   GRID 4

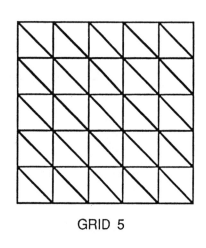

GRID 5

Fig. 3   Grid   Patterns

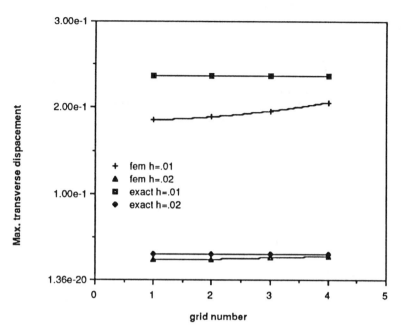

Fig. 4 Convergence study of transverse displacement for rectangular plate

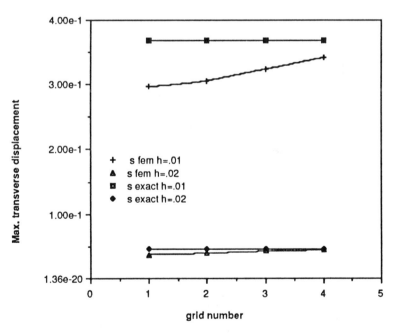

Fig. 5 Convergence study of transverse displacement for square plate

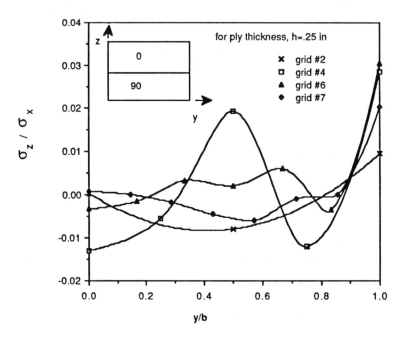

Fig. 6 Convergence study of normal stress at mid-surface for tensile load

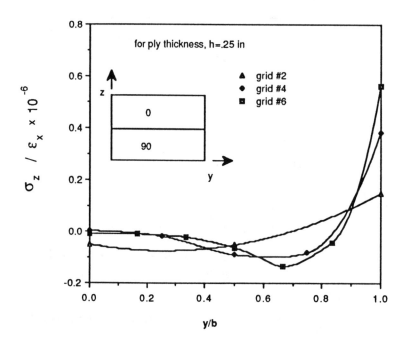

Fig. 7 Convergence study of normal stress at mid-surface for stretch load

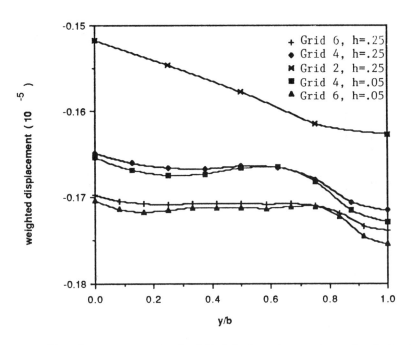

Fig. 8 Convergence study of weighted displacemt across top surface for tensile load

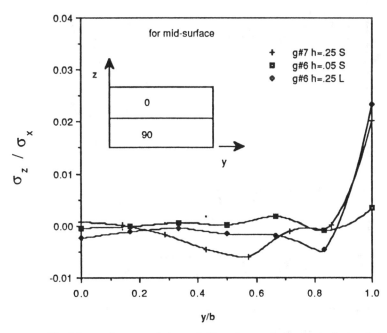

Fig. 9 Comparing mormal stress of different geometry parameter for tensile load

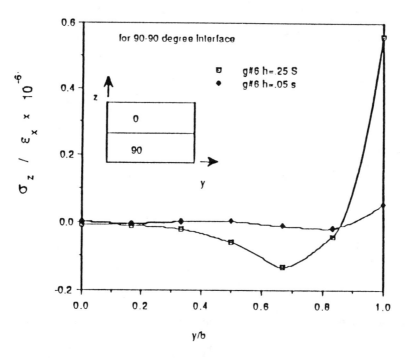

Fig. 10 Comparing normal stress of different geometry parameter for stretch load

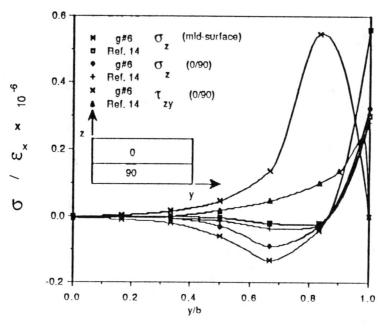

Fig. 11 Comparison of the Interlaminar Stress Distribution

between existing and current method

An Approach for Predicting the Crippling Strength of Thin-Walled Composite Airframe Structures Under Compression*

LAWRENCE W. REHFIELD AND **K. LEVEND PARNAS**

ABSTRACT

The objective of this paper is to present a criterion for one mode of postbuckled failure (crippling) of thin-walled composite stiffeners subjected to uniform compression.   Two general types of crippling mode have been observed in our experiments.   The first is a mode that is initiated by free edge delamination.   The second is local material failure which initiates in corners or junctures between contiguous elements and extends across webs. This is the mode for which a crippling criterion has been found.

The new crippling criterion is based upon a simple elastic postbuckling solution for compressively loaded thin composite plates and the use of a maximum strain failure criterion. The criterion yields a linear relationship between the buckling load and the crippling load. Experimental crippling results show excellent agreement with predictions based upon the new criterion.

NOMENCLATURE

A_{ij}, D_{ij}	Elements in laminate stiffness matrix
F	Airy stress function
K_{11}	Prebuckling stiffnesses of sections
N	Average inplane stress resultants at the loaded edges
N_{cc}	Crippling stress resultant
N_{cr}	Buckling stress resultant
N_{ult}	Ultimate stress resultant for the material
P_{cr}	Total buckling load of the I-section

* Research sponsored by ARO Contract DAAG 29-82-K-0094, Sikorsky Aircraft and the University of California.

Lawrence W. Rehfield, Professor, Department of Mechanical Engineering, University of California, Davis, CA 95616.

K. Levend Parnas, Aerospace Engineering Georgia Institute of Technology, Atlanta, Ga. 30332 (404) 894-8203.

W_{11} Amplitude of the normal displacement mode

a   Length of the I-beam

a_{ij} Elements of inverse axial stiffness matrix

b   Width of flange or web

m   Number of half sine waves in x-direction

u,v   Inplane displacements

w   Normal displacement

α Parameter in Equation (7)

$\bar{\alpha}$ Parameter defined in Equation (9)

μ a/mb

INTRODUCTION

The use of composite materials in secondary and primary airframe structures has resulted in weight savings, which directly result in performance enhancement, and improved durability due to relative freedom from fatigue and corrosion.   The extensive use of composites in rotorcraft structures has been accomplished in the Advanced Composite Airframe Program (ACAP) and the Bell-Boeing V-22 Osprey.   Projected composites usage in all military aircraft is 50-60 percent by the middle of the next decade.

The distinguishing features of the use of composites are substantial reductions in the number of fasteners used due to both reduced part count and widespread use of co-curing and secondary bonding to attach the different structural elements.   Semi-monocoque or stiffened construction continues to predominate in composite designs.   In relatively lightly loaded portions of an airframe, weight savings can be achieved in some cases by operating with a postbuckled skin and stable stiffeners. Sikorsky's ACAP utilized such a concept in the design.   Additional weight benefits are possible by permitting the stiffening members to operate in the postbuckled regime below the limit load.   The increased weight benefits achieved in this manner are particularly important to overall system performance in V/STOL aircraft.   Weight savings may be directly translated into increased payload, extended range or downsizing of the vehicle.

An obstacle in the way of achieving the promise of weight savings due to postbuckled design is the inability to predict failure in the postbuckling range (crippling) well enough.   This is the primary subject of this paper.

PREVIOUS WORK

A survey of previous work has recently been published in Ref. [1] which covers the local buckling, postbuckling and crippling of compressed composite elements. The works cited appear in Refs. [2-12]. Two important references [13,14] are missing from this survey. The authors' comments [1] are valuable, and no attempt to duplicate their efforts will be made here.

Renieri and Garrett [13] present test data on 128 specimens and attempt to correlate crippling strength using a semi-empirical approach borrowed from an existing method for metallic structures. We judge their approach inadequate. Log-log plots make the correlation appear better, but the discrepancies are often quite large.

The early work of Spier [2-5] is also devoted to the development of semi-empirical crippling curves. Spier's methodology is adopted and recommended in Ref. [14]. It is a modification of a semi-empirical approach, originally presented by Schuette in 1949 for metals [15], which accounts for the orthotropicity of the laminate.

The form of equation chosen in Refs. [13] and [15] to represent crippling in composites is of the same "power law" type used for ductile metals for which there is widespread yielding accompanying crippling. Since most of the composites tested to date are brittle thermosetting systems, it is not surprising that the empirical curves provide an unacceptable fit.

The prediction of crippling is complicated by the fact that there are two types of modes. Experiments in our laboratory on I-section specimens made of graphite-epoxy woven cloth [8,9] indicate two general types of crippling mode. The first is a mode that is initiated by free edge delamination. This mode was successfully captured on high speed 300 frame per second film [9]. The second is local material failure which initiates in corners or junctions between contiguous elements and extends across webs. This is the mode for which a crippling law can be found and which is studied in Refs. [2-5, 13, 14].

A thorough discussion of the failure modes observed in the recent tests conducted on specimens made of AS4/3502 graphite-epoxy unidirectional tape is given in Ref. 1. While the two classes of crippling mode are the same in these experiments as in Refs. [8,9], the picture is somewhat clouded by more variety of post-failure configurations.

The observed differences may be attributed to a variety of effects. Firstly, the woven cloth graphite-epoxy specimens [8,9] were all of quasi-isotropic layups. Consequently, residual curing stresses are at a minimum in these specimens. Secondly, some of the unidirectional tape specimens [1] had flanges which were machined after fabrication. It is likely that machining will greatly influence free edge delamination. Thirdly, the I-section configuration is a preferred one because this geometric shape creates a stable cross section. The symmetry reduces warp due to fabrication and minimizes or eliminates overall twisting deformations which occur in unsymmetrical configurations. This factor may help explain the different behavior found for I- and J-section specimens as compared to channel and zee specimens in Ref. [1]. Finally, as is common in stability critical structures, an examination of post-failure configurations often does not provide evidence of the initiation of the process. For this reason, together with the symmetry of the I-section configuration, we will utilize data from Refs. [8,9] for correlation in this paper.

OBJECTIVE AND APPROACH

In order to fully exploit composite structures in postbuckled applications, it is essential to have a good understanding of the crippling mechanisms and a reliable crippling load prediction methodology to guide the design. The objective of this work is to develop and present a crippling law for the local material strength failure mode. This has been facilitated by experiments on thin-walled I-section composite members made of woven cloth [8,9].

As mentioned earlier, the power law type of crippling criterion has proven useful for metallic structures which undergo widespread yielding with crippling. The data base for composites is composed mostly of results obtained on brittle thermosetting systems. Consequently, we abolish the semi-empirical power law approach as unlikely to be satisfactory.

A simple, straightforward approach has been adopted. An elastic postbuckling analysis is made of the no-edge-free configuration which permits the postbuckled strain field to be mapped. The maximum strain level, which occurs at the edges (junctures or corners of a stiffener), is then used to predict failure. This approach yields a linear relationship between the buckling load and the crippling load.

POSTBUCKLING ANALYSIS

Consider a thin rectangular composite plate element with all edges simply supported of length a and width b. Let x be the coordinate in the lengthwise direction and y correspond to the widthwise direction. The governing equations are taken to be of the nonlinear Von Karman type [16]. The two equations consist of the lateral equilibrium equation

$$D_{11}W,_{xxxx}+2(D_{12}+2D_{66})W,_{xxyy}+D_{22}W,_{yyyy}$$

$$-(F,_{yy}W,_{xx}-2F,_{xy}W,_{xy}+F,_{xx}W,_{yy}) = 0 \qquad (1)$$

and the compatibility equation

$$a_{11}F,_{yyyy}+(2a_{12}+a_{66})F,_{xxyy}+a_{22}F,_{xxxx} = W^2,_{xy}-W,_{xx}-W,_{xx}W,_{yy} \qquad (2)$$

For simple supports, the bending and twisting related boundary conditions are

$$W(o,y) = 0 \text{ and } W,_{xx}(o,y) = 0 \qquad (3)$$

$$W(a,y) = 0 \text{ and } W,_{xx}(a,y) = 0 \qquad (4)$$

$$W(x,o) = 0 \text{ and } W,_{yy}(x,o) = 0 \qquad (5)$$

$$W(x,b) = 0 \text{ and } W,_{yy}(x,b) = 0 \qquad (6)$$

The approximate solution methodology is a classical one. An admissible lateral displacement function is chosen which, in this case, is the exact buckling mode from classical linear buckling theory. It is

$$W(x,y) = W_{11}\sin(\frac{m\pi x}{a})\sin(\frac{\pi y}{b}) \tag{7}$$

Equations (3) - (6) are satisfied by this choice. This deflection is introduced into the compatibility equation (2) and a stress function can be found.

The stress function solution utilized herein leads to the following expressions for the stress resultants:

$$N_{xx} = F_{,yy} = -N - \frac{1}{8a_{11}\mu^2} (\frac{\pi}{b})^2 W_{11}^2 \cos\frac{2\pi y}{b} \tag{8}$$

$$N_{yy} = F_{,xy} = - \frac{1}{8a_{22}} (\frac{\pi}{b})^2 W_{11}^2 \cos\frac{2\pi x}{b} \tag{9}$$

$$N_{xy} = F_{,xy} = 0 \tag{10}$$

The inplane boundary conditions that are satisfied by this solution are

$$N_{xy}(x,o) = 0 \text{ and } N_{xy}(x,b) = 0 \tag{11}$$

$$N_{xy}(o,y) = 0 \text{ and } N_{xy}(a,y) = 0 \tag{12}$$

In addition, the unloaded edges y = o,b remain straight but are free to expand in Poisson manner. Conditions regarding the axial stress resultant are not satisfied at the ends x = o,a. Instead an average condition in the sense of St. Venant is applied which is equivalent to N being the average stress resultant across the plate width.

The approximate solution is obtained by substituting Eqs. (7) - (10) into the lateral equilibrium equation (1) and applying Galerkin's method. The result may be written in the following manner:

$$\frac{N_{cc}}{N_{ult}} = \frac{\alpha}{1+\alpha} + \frac{1}{1+\alpha} (\frac{N_{cr}}{N_{ult}}) \tag{13}$$

where
$$\alpha = \frac{1}{2} + \frac{1}{2} \frac{a_{11}}{a_{22}} \mu^4 \tag{14}$$

The buckling stress resultant N_{cr} is

$$N_{cr} = \frac{\pi^2}{b^2}[\frac{D_{11}}{\mu^2} +2(D_{12}+2D_{66})+\mu^2 D_{22}] \tag{15}$$

Equation (13) indicates that a linear relationship exists between buckling stress resultant and crippling stress resultant. This is a marked departure from power law crippling equations.

The predictions of Eq. (13) are compared with the experimental results from Refs. [8,9] in Fig. 1. Based upon the fact that scatter in crippling experiments of composite stiffener elements is normally very large, this degree of correlation is considered quite good.

A similar equation to Eq. (13) for strain levels may be obtained by using the stress-strain relations. The result is

$$\frac{\bar{\varepsilon}_{cc}}{\varepsilon_{ult}} = \frac{\bar{\alpha}}{1+\bar{\alpha}} + \frac{1}{1+\bar{\alpha}} \left(\frac{\varepsilon_{cr}}{\varepsilon_{ult}}\right) \tag{16}$$

where

$$\varepsilon_{cr} = a_{11} N_{cr} \tag{17}$$

$$\bar{\alpha} = \frac{1}{2} \left(1 + \frac{a_{11}}{a_{22}} \mu^4\right) + \frac{a_{12}}{a_{22}} \mu^2 \tag{18}$$

and ε_{ult} is the maximum axial strains at the unloaded edges of the plate. The average crippling strain is $\bar{\varepsilon}_{cc}$.

CONCLUDING REMARKS

A new crippling law for the local material strength failure mode has been developed and compared to experimental data obtained on composite I-sections made from graphite-epoxy woven cloth. The agreement is considered quite good. The conceptual basis for the development is clearly stated, and the brittle nature of the commonly used material systems is central.

The crippling mode for elements with a free edge that are initiated by free edge delamination will require a different type of analysis. While there is much left to do before an adequate crippling law is obtained for this mode, the basis has been established in Ref. [17].

ACKNOWLEDGEMENT

This research has been sponsored by ARO Contract DAAG 29-82-K-0094, Sikorsky Aircraft and the University of California.

REFERENCES

1. Bonanni, D. L., Johnson, E. R. and Starnes, J. H., Jr., "Local Crippling of Thin-Walled Graphite-Epoxy Stiffeners," AIAA Paper No. 88-2251, Proceedings of the 29th Structures, Structural Dynamics and Materials Conference, pp. 313-323.

2. Spier, E. E., "Crippling/Column Buckling Analysis and Test of Graphite/Epoxy Stiffened Panels," AIAA Paper No. 75-753, presented at the 16th Structures, Structural Dynamics, and Materials Conference, Denver, CO, May 27-29, 1975.

3. Spier, E. E. and Klouman, F. L., "Empirical Crippling Analysis of Graphite/Epoxy Laminated Plates," Composite Materials: Testing and Design (Fourth Conference), ASTM STP 617, 1977, pp. 255-271.

4. Spier, E. E., "Stability of Graphite/Epoxy Structures with Arbitrary Symmetrical Laminates," Experimental Mechanics, Vol. 18, No. 11, Nov. 1978, pp. 401-408.

5. Spier, E. E., "On Experimental Versus Theoretical Incipient Buckling of Narrow Graphite/Epoxy Plates in Compression," AIAA Paper No. 80-0686, in Proceedings of the 21st Structures, Structural Dynamics, and Materials Conference, May 12-14, 1980, pp. 187-193.

6. Spier, E. E., "Postbuckling Fatigue Behavior of Graphite-Epoxy Stiffeners," AIAA Paper No. 82-779, in Proceedings of the 23rd Structures, Structural Dynamics, and Materials Conference, Part 1, May 10-12, 1982, New Orleans, LA, pp. 511-527.

7. Tyahla, S. T. and Johnson, E. R., "Failure and Crippling of Graphite-Epoxy Stiffeners Loaded in Compression," Report No. CCMS-84-07 and VPI-E-84-19, Virginia Polytechnic Institute and State University, Blacksburg, VA, June 1984.

8. Reddy, A. D., Rehfield, L. W., Bruttomesso, R. I., and Krebs, N. E., "Local Buckling and Crippling of Thin-Walled Composite Structures Under Axial Compression," AIAA Paper No. 85-0672, in Proceedings of the 26th Structures, Structural Dynamics, and Materials Conference, Part 1, April 15-17, 1985, Orlando, FL, pp. 804-810.

9. Rehfield, L. W. and Reddy, A. D., "Observations on Compressive Local Buckling, Postbuckling, and Crippling of Graphite/Epoxy Airframe Structure," AIAA Paper No. 86-0923, in Proceedings of AIAA/ASME/ASCE/AHS 27th Structures, Structural Dynamics, and Materials Conference, Part 1, May 19-21, 1986, San Antonio, TX pp. 301-306.

10. Causbie, S. M. and Lagace, P. A., "Buckling and Final Failure of Graphite/PEEK Stiffener Sections," AIAA Paper No. 86-0921, in Proceedings of AIAA/ASME/ASCE/AHS 27th Structures, Structural Dynamics, and Materials Conference, Part 1, May 19-21, 1986, San Antonio, TX, pp. 280-287.

11. Wang, C., Pian, T.H.H., Dugundji, J., and Lagace, P. A., "Analytical and Experimental Studies on the Buckling of Laminated Thin-Walled Structures," AIAA Paper No. 87-0727, in Proceedings of AIAA/ASME/ASCE/AHS 28th Structures, Structural Dynamics, and Materials Conference, Part 1, April 6-8, 1987, Monterey, CA, pp. 135-140.

12. Bonanni, D. L., "Local Buckling and Crippling of Composite Stiffener Sections," Master of Science Thesis, Department of Aerospace and Ocean Engineering, Virginia Polytechnic Institute and State University, Blacksburg, VA, February 1988.

13. Renieri, M. P. and Garrett, R. A., "Investigation of the Local Buckling, Postbuckling and Crippling Behavior of Graphite/Epoxy Short Thin-Walled Compression Members, Report MDC A7091, 31 July 1981.

14. Deo, R. B. and Agarwal, B. L., "Design Methodology and Life Analysis of Postbuckled Metal and Composite Panels: Design Guide," Report AFWAL-TR-85-3096, Vol. III, December 1985.
15. Gerard, G., Handbook of Structural Stability, Part IV - Failure of Plates and Composite Elements, NACA TN 3784, 1957.

16. Stein, M., "Postbuckling of Orthotropic Composite Plates Loaded in Compression," <u>AIAA Journal</u>, Vol. 21, No. 12, December 1983, pp. 1729-1735.

17. Armanios, E. A. and Rehfield, L. W., "Interlaminar Analysis of Laminated Composites Using a Sublaminate Approach," Proceedings of the 27th Structures, Structural Dynamics, and Materials Conference, San Antonio, Texas, 19-21 May 1986. AIAA Paper No. 86-0969CP, Part 1, pp. 442-452.

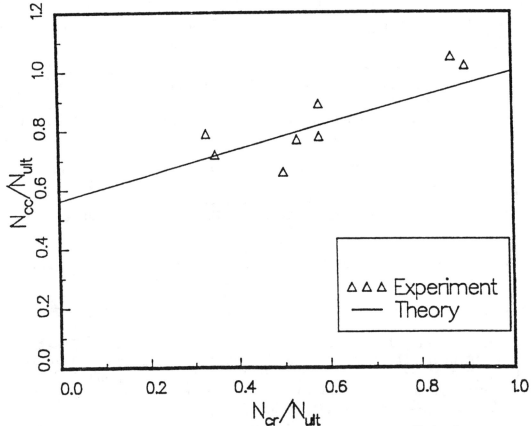

Figure 1. Comparison of Linear Crippling Law with Experimental Data [8,9].

Analysis of Interference Fit Pins in Composite Plates

T. S. RAMAMURTHY

ABSTRACT

Composites are finding increasing application in many advanced engineering fields like aerospace engineering due to their high specific strength and specific stiffness values. Inspite of these advantages the designers are not using them in primary structures due to lack of complete understanding of their behaviour in some crucial situations. Joining is one such area. Pin joints are one of the most commonly used method of joining becauses of its ability to provide dismantlable joints. In regions of high stress concentrations like wing root joints, interference pin joints are used to increase the fatigue life of the component. The behaviour of interference fit pin joint in a composite plate modelled as orthotropic lamina is the subject matter of this paper. The analysis technique incorporates proper boundary conditions at the pin plate interface. An inverse technique is successfuly applied to study the separation behaviour of the plate with a rigid pin subjected to push/pull type of loads.

INTRODUCTION

Composites are finding increasing application in many advanced engineering fields like aerospace engineering due to their high specific strength/stiffness. Their use in primary structures are limited due to lack of complete understanding in their behaviour in some situations like joints. Interference fit pin joints are used in highly stressed regions like wing root joints in aircraft to increase the fatigue life of the component. The present study is a contribution in understanding the behaviour of such joints. The composite plate is modelled as an orthotropic lamina and the pin is treated as a smooth rigid circular disc inserted in the plate hole. A two dimensional finite element analysis, with the ability to properly incorporate the pin/plate interface conditions is developed and used to obtain the results. An 'inverse' technique, first proposed by Rao [1], and later developed and extensively applied by the present author and his colleagues [2,3,4] is used.

T.S.RAMAMURTHY
(On leave from Department of Aerospace Engineering, I.I.Sc., Bangalore 560012. INDIA)
Resident Research Associate (NRC), AFWAL/ MLBM
WPAFB, OH 45433

PROBLEM DESCRIPTION

Consider a square lug ABCD with a hole of radius **a** Fig.1 located on the middle line A'D'. The elastic properties of the sheet are E_1, E_2, G_{12}, ν_{12} with 1 and 2 directions parallel to co-ordinate axes x-y. The hole is filled with a smooth rigid circular pin of radius **a_p** given by,

$$a_p = (1+\lambda)\, a \qquad\qquad (1)$$

The origin of the co-ordinate system is located at the centre of the pin. And λ is the nondimensional mismatch parameter, $\lambda > 0$ results in interference fit and $\lambda < 0$ results in clearance fit. The pin is subjected to a load P_x along the diameter EF and is reacted on the edge AB of the plate by a uniform direct stress s given by,

$$s = P_x / (2wt) \qquad\qquad (2)$$

In the general case of misfit pins, since the pin and the plate are not bonded together the pin/plate interface exhibits partial contact/separation behaviour and the contact region variousa nonlinearly with the load magnitude. The actual region of separtion for any given load is not known 'a priori'. As a limiting case of λ --- 0, one can get the configuration for a push or neat fit pin[1,3] and this configuration does not vary with the load magnitude but it is different for different types of load. Earlier researchers [5,6,7] have assumed the contact/separation area as a semi circular region. Tsujimoto and Wilson[6] further replaced the effect of pin contact by a cosinusoidal distribution of the radial pressure. This is based on the early work of Bickley [5] who first gave the approximate expression radial pressure distribution for the pin bearing effect Soni [7] has represented the effect of the pin bearing in terms of displacement conditions on the assumed 180° contact region for the case of rigid pin. Since for a specified load the configuration is not known apriori, the direct method of solution has to be necessarily iterative in nature [8,9,10].

In cases where the geometric and loading symmetries exist and physics of the problem enables one to specify the regions of separation/contact and the nature of their growth with the increase in the load magnitude one can take recourse to the 'Inverse Technique' [1,4,11]. In this method one starts with a specified regions of contact or separation and estimates the magnitude of the causative load/s. In the present paper the later technique is used to estimate the separation behaviour in an interference fit pin in orthotropic sheet with the loading on the pin.

METHOD OF SOLUTION

The problem is analysed using finite element technique and 'inverse method' of solution is adopted to get the load contact behaviour. Most of the results obtained so far are for the case when the pin load P_x acting in the direction F to E (Fig.1) and is reacted along the edge AB [4,9,10]. Because of the inherent symmetry in the problem it was sufficient to analyse one half of the domain say AA'D'D with appropriate conditions on the edge A'D'. The region of the plate is devided into isoparametric quadrilaterals and triangular elements. Based on the convergence studies conducted by the author [11], the hole region is devided into Nc (=65 in examples) equidistant nodes. To reduce the size of the problem and economise the computational effort the number of divisions on some successive arcs are halved to result in smaller number of elements in the outer regions where the stress gradients are not large. A typical finite element mesh is shown in the Fig.2.

To enable easier specification of the boundary conditions on the hole edge the degrees of freedom of the nodes on the hole edge are changed to radial and tangential directions . The corresponding element stiffness matricies are appropriately transformed before assembling them into global matrix. For convenience the nodes at the hole boundary are numbered as 1 at F to Nc at E.

The pin load P_x could act in the direction EF or FE. The two cases are identified as P_{xts} and P_{xas} symbolising as the load towards support (push) and load away from support (pull) respectively (Fig.3). It is seen from the literature for the case of P_{xas} loading separation is initiated at F and spreads symmetrically about F. In the inverse technique one starts with a specified configuration i.e. the region of separation/contact. The transition points between separation and contact regions are dentified as T_i's . In the finite element analysis the transition points T's are always located at the nodes on the hole boundary. For any general configuration the separation angle is θ_s and the corresponding pin load is P_{xas} .

BOUNDARY CONDITIONS

The nodal displacements which are unknowns of the problem are evaluated after proper boundary conditions are satisfied. These are specified below;

(i) For the nodes on the axis of symmetry (i.e. X -axis) the forces $F_x = 0$ and the normal displacements $U_y = 0$;

(ii) For nodes on the circular boundary
   a) $F_{\theta i} = 0$ for $1 \leq i \leq Nc$

   b) $F_{ri} = 0$ with $U_{ri} > a\lambda$ for $1 \leq i < T$

and $U_{ri} = a \lambda$ with $F_{ri} \geq 0$ for $T \leq i \leq Nc$
(iii) $F_{xj} = f_x$, $F_{yj} = f_y$ j's are nodes where external loads present (3)

Since T is known all the definitive displacement boundary conditions are incorporated in the matrix equations following the method of Zienkiewicz [12] and the resulting equations may be written a s

$$[K] \{\Delta\} = \{F\} \qquad (4)$$

where [K] is the final stiffness matrix
$\{\Delta\}$ is the unknown displacement vector,
$\{F\}$ is the known external force vector.

once the set of equation 4 is solved for any specified load vector {F}, the displacement vector {Δ} is determined.   Following the conventional steps all other quantities like stresses and strains are evaluated.

SOLUTION PROCEDURE IN INVERSE TECHNIQUE

As already pointed out, in the inverse technique one begins with the selection of the transition node T so that the configuration is specified andthe boundary conditions in eqn 3 are known and seek to evaluate the causative load.   At the transition node T only displacement condition is satisfied.   As explained in ref. [4,13] the proper load P_x must be such that the radial force F_r at the node T must be zero.   This load P_x is evaluated as follows:

(i) Solve the Finite Element Model for a given mis fit parameter λ and specified T with two load parameters $P_x = 0$ and $P_x = 1$.
(ii) the resulting displacement vectors may be denoted as {Δ_0} and {Δ_1} and the corresponding radial forces at node T be F_{rTo} and F_{rT1} .

Since the geometry and configuration are not changed, the radial force at the transition node must be a linear function of the pin load, resulting in

$$F_{rT} = A \lambda + B P_x \tag{5}$$

where A & B are constants for a given configuration.   Since two values of F_r are known the constants can be uniquely determined.   Then the true value of P_x is then determined fromn the condition of F_{rT} to be zero.   This yields the value of P_x as ,

$$P_x = -A\lambda/ B = F_{rTo} /(F_{rTo} - F_{rT1}) \tag{6}$$

The corresponding displacement vector is then evaluated as

$$\Delta_{true} = (1-P_x) \Delta_0 + P_x \Delta_1 \tag{7}$$

Then all the inequality constraints in eqn 3 are checked.   For all feasible configurations   the load from equation 6 always sastisfies the inequality constraints and a unique P_x is obtained.   The load P_x is expressed in terms of a nondimentional parameter as

$$F_1 = P_x / 2E_1 \lambda at \tag{8}$$

By sequentially varying the value of T, one obtains the variation of P_x with θ_s, and the information about all the stress fields.

EXAMPLE

For the examples the material properties are taken as $E_1 = 276.097$ GPa, $E_2 = 55.317$ GPa, $G_{12} = 32.888$ GPa, $\nu_{12} = .2278$ to enable comparison with resuits in the literature [2,13].   The geometric parameters are the plate size 2wx2w, hole radius **a** .   To determine the proper finite element model a sample problem of a square orthotropic plate with central rigid interference fit pin subjected to the load on the plate only is analysed.   Table I shows the comparision of load parameter for initiation of separation.   The values compare very well with those in reference [13]. The problem of square plate with subjected to pin load P_{xts} and P_{xas} is then analysed with $N_c = 65$. The numerical data for pin load towards the support is given in Table II. The variation of load parameter with separation angle is also shown in Fig. 4. It is seen from Fig.4 that the curve corresponding to P_{xas} shows a monotonically increasing trend where as the one corresponding to P_{xts} show a decreasing and increasing trend in the lower ranges of the separation angle resulting in non-unique angle of separation for some ranges of load parameter. When these results were closely observed it was found that in some regions of the hole boundary the inequality constraints are violated for the load esatimated from Eqn. 6.

The implication of this is that the initial configuration separation initiating at E and spreading symmetrically about it is erroneous. Another feasible configuration is for separation to initiate simultaneously at two symmetrical locations G and G' Fig. 5 and spread about them with increasing load magnitude. This concept is checked first by studying the radial stress field around the pin with full contact due to interference only and then studying its variation with the superposition of pin load towards the support. Fig. 6 shows the variation of the radial stress around the pin for some axial load parameters. It is seen that the lowest compressive stress occurs not at location E but at some interior location identified as G. As indicated in Fig. 6 at some load level, the compressive radial stress reaches a value zero at G. With further increase in load parameter the separation spreads about G. The extent of separation is indicated by $G_1 G_2$. For satisfying the symmetry conditions there is another location G' symmtrical to G and a region $G_1' G_2'$ where the plate has separated.

PROCEDURE FOR EVALUATION OF LOAD PARAMETER/S

As explained in the previous section the feasible configuration is to have separation from G_1 to G_2 and remaining region (E to G_1 and G_2 to F) in contact. Thus there are two transition points namely G_1 and G_2 in the domain of analysis. The pin load is initially presumed to be reacted by direct stresses on the two vertical sides DD' and AA'. Let the direct stress parameters for these two sides be r and s respectively. Thus the components of load on the pin plate combination could be split into three components viz., (i) $r=1, a\lambda=1$. (ii) $s=1, a\lambda = 1$ and (iii) only $a\lambda=1$ acting.

So for a specified configuration i.e. fixed G_1 and G_2 locations one can evaluate the the radial forces F_{ri} due to a unit magnitude of the force systems (i) (ii) and (iii). These forces could be identified by F_{ij}, i= 1,2 and j= 1,2,3 (i refers to the transition point and j referes to the load component). Then the net radial force at the transition locations for a general loading is given by

$$F_{rTi} = (F_{i1} - F_{i3})r + (F_{i2} - F_{i3})s + F_{i3} a\lambda = 0, \quad \text{for} \quad i= 1,2 \qquad (9)$$

For any specified transition locations G1 and G2 and given 'aλ' the radial forces at the two transition locations are to be zero. Thus solution of eqn. 9

enables one to evaluate the specific combination of r and s . From these sets of (r,s), the the solution for r = 0 is obtained as shown below:
  i) From the full contact solution with the pin load P_{xts} identify the location G and the corresponding pin load parameter for initiation of separation,
  ii) then select a location G_1 and sequentially vary G_2 and obtain the values of r and s from eqn 9,
  iii) at some location of G_2 there willbe a change of sign of r,
  iv) at this stage interpolate a location G_2 to correspond to r = 0 and the corresponding value of s.
  v) Then the load parameter s produces a separation of extent θ_s from G_1 to G_2.
  vi) Change the value of G_1 and repeat the steps from i to v. At some stage G_1 will coincide with E. Beyond this load parameter the earlier configuration of a single separation region symmetric about E will be the correct configuration.

NUMERICAL RESULTS

  The above procedure was applied to the problem of w/**a** = 3 and load of P_{xts}. The results are presented in Table III. The table shows the load parameter values for initiation of separation and its location and the growth of separation region and the corresponding load parameter values. The Fig.7 presents the data in graphical form as the variation of load parameter with the magnitude of the separation . The figure also presents the variation of load parameter evaluated using the equation 6. The figure shows that beyond some level of load parameter the two curves merge into single one implying a single continuous separation zone is feasible. As could be seen from the table there is a monotonic increase of the load parameter with monotonic increase in the separation zone and the ambiguity of two separation angles for a load parameter value (Fig 6) is now removed. Thus the configuration proposed in Fig. 6 satisfies all physical requirements and the boundary conditions. It is seen that both the separation bubble size and the causative load parameter increase monotonically. Figures 8 and 9 show the compari-son of maximum radial stress and maximum hoop stress with the load parameter for pull and push type of loads.

CONCLUSIONS

  An interference fit pin in a square orthotropic lug behaves differently when subjected to push type of load as compared to the pull type of load. For the example considered even the nature of behaviour is different, as bubble separation configuration occurs. This results in the significant variation in the maximum hoop and radial stresses for a given load parameter. The influence is more on the maximum hoop stress.

ACKNOWLEDGEMENTS

  The author acknowledges the support and encouragement from Dr.S.W.Tsai of AFWAL/ MLBM. The author would also like to thank NRC for awarding the Resident Research Associateship tenable at AFWAL, WPAFB, OH 45433 which enabled him to conduct these studies .

Table I : Comparison of results of Square orthotropic plate with a smooth rigid pin (Interference fit) : plate load only, $w/a = 3.0$

	$S_{xis}/E_1\lambda$
Present	.17769
Ref [2,13]	.17880

Table II : Separation transition and extent for square plate with central pin : Variation with load parameter F_1 (Fig. 3,4) $\mathbf{F_1} = \mathbf{P_{xts}}/(2E_1 at\ \lambda)$

N_T	θ_s deg	F_1
65	0.0	.84394
61	11.25	.84126
57	22.5	.82892
53	33.75	.81828
49	45.0	.81382
45	56.25	.82285
41	67.5	.86437
37	78.75	.99740
33	90.00	1.49537
29	101.25	30.69820

Table III Separation location and extent for square plate with central pin : Variation of load parameter with θ_s, Bubble configuration .

(Fig. 5,7) $\mathbf{F_{1ts2}} = \mathbf{P_{xts}}/(2E_1 at\ \lambda)$

G_2,G_1	θ_s deg	F_{1ts2}
49,49	0.0	.754502
49, 57.7	24.45	.798369
46.1, 61	41.90	.814210
45,63.4	51.75	.821048
44.6,65	57.40	.826421
41, 41'	67.5	.864310
37, 37'	78.75	.997400
33,33'	90.00	1.495367
29,29'	101.25	30.698199

REFERENCES

1. Rao,A.K. : "Elastic analysis of pin joints", J. Comp. & Struct., Vol. (9), p.125 -144, 1978.

2. Naidu, A.C.B.,Dattaguru,B.,Mangalgiri,P.D. & Ramamurthy, T.S.:"Analysis of finite composite plate with smooth rigid pin", J Comp. Struct., Vol(4), pp 197-216,1985.

3. Gosh,S.P., Dattaguru,B. & Rao,A.K.: "Load transfer from a smooth elastic pin to a large sheet", AIAA J, Vol.(19), pp 127-134, 1986.

4. Dattaguru, B., Naidu, A.C.B., Krishnamurthy, T. & Ramamurthy, T.S.:"Devlopment of special fastener elements", J.Comp. & Struct., Vol(24), pp127-134,1986.

5. Bickley, W.G. : "The distribution of stresses round a circular hole in a plate", Phil. Trans. R. Soc.(Lond) series A-227, pp 383-415, 1928.

6. Tsujimoto, Y. &Wilson, D.:"Elasto plastic failure analysis of composite bolted joints", J Comp. Matls., Vol(20), pp 236-252, 1986.

7.Soni, S.R.:"failure analysis of composite laminates with a fastener hole" ASTM STP 749, pp 145-164, 1981.

8. Rahman, M.U., Rowlands, R.E., Cook, R.D. & Wilkinson, T.L. : "An iterative procedure for finite element stress analysis of frictional contact problems", J Com. & Struct, Vol(18) pp 947-954, 1984.

9. Oplinger, D.W. & Gandhi, K.R.:"Stresses in mechanically fastened orthotropic laminates", Proc. 2nd Conf. in Fibrous compts. in Flight Veh. Design., Williamsburg, Va. pp 813-841, 1974.

10. Callinen, R.J. : " Aeronautical Research Laboratory - Structures Note 439, AR - 000-842, Melbourne, Australia, 1977.

11. Ramamurthy, T.S. , : " New studies on the effect of bearing loads in lugs with clearance fit pins" (To appear in Composite Structures an International Journal)

12. Zienkiewicz, O.C. : "Finite element method", 3rd Edn., McGraw -Hill Book Co., 1977.

13. Mangalgiri,P.D., Dattaguru,B. & Rao,A.K. : " Finite element analysis moving contact in mechanically fastened joints", Nucl. Engng & Desgn., Vol(78) pp 303-311, 1984.

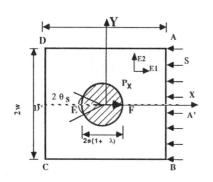

Fig.1 Problem Definition

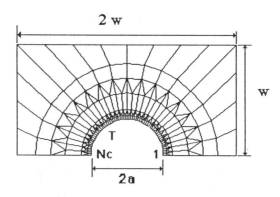

Fig.2 A Typical Finite Element Idealisation

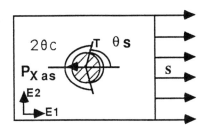 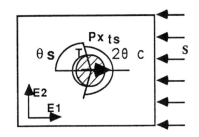

a) Pin load away from support   b) Pin load towards support

Fig.3 Schematic Behaviour Under Pull and
Push Loads

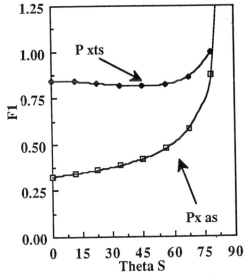

Fig.4 Variation of Load Parameter with
Separation Angles

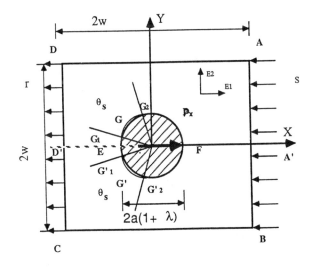

Fig.5 Bubble Separation Configuration

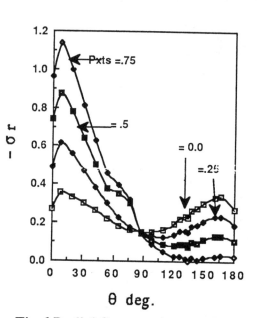

Fig.6 Radial Stresses Around Pin Under Push Loads

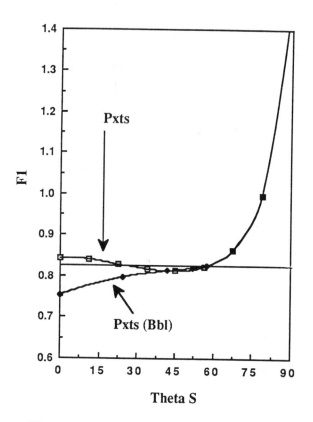

Fig.7 Load Parameter for Bubble Separation

F1 = Px/(2E1atλ)

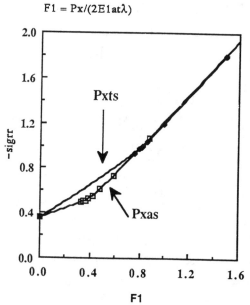

Fig.8 Maximum Radial Stress Variation with Load Parameter for Push and Pull Loads.

F1 = Px/(2E1atλ)

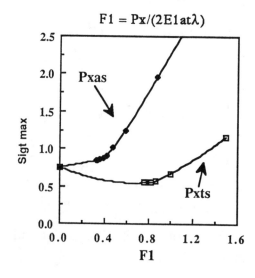

Fig.9. Maximum Hoop stress variation with Load parameter for Push and Pull Loads.

Tapered Laminates: A Study on Delamination Characterization

O. O. OCHOA AND W. S. CHAN

ABSTRACT

An important parameter in designing with composites is the ability to build components with contours. The contouring can be accomplished by terminating plies at appropriate intervals. The present work addresses the effect of ply termination on laminates subjected to multiaxial loads. Interlaminar stresses are evaluated with a special finite element developed for multiaxial response. The strain energy release rates are presented to guide the material and stacking sequence choices.

INTRODUCTION

The termination of a ply at a critical interface, a small distance from the free edge has been shown by the authors to prevent or delay delamination at a free edge [1]. Even though a localized region of stress concentration exits at the taper due to the formation of resin rich pocket, the interlaminar stresses are much smaller than those encountered at the free edge of laminates without ply termination. At present this observation is limited to laminates under uniaxial tensile loads in the absence of environmental concerns.

Recently, tension, torsion and bending loads coupled with thermal and hygroscopic effects are presented by Chan and Ochoa [2,3] for laminates without ply terminations. This effort studies the response of tapered laminates under multiaxial loads.

ANALYSIS

The approach developed by Chan and Ochoa [2] takes advantage of a conventional, simple isoparametric element that builds on the bending and torsion components. The element has eight nodes with three degrees of freedom per node, namely, the displacements $u(y,z)$, $v(y,z)$ and $w(y,z)$. This displacement field is modified to include the effects of bending and torsion as described below. The final form of the displacement field

O. O. Ochoa, Associate Professor, Mechanical Engineering Department, Texas A&M University, College Station, TX 77843

W. S. Chan, Associate Professor, Mechanical Engineering Department, University of Texas at Arlington, Arlington, Texas 76019

becomes

$$U(x,y,z) = (\varepsilon_0 + kxz) + \underline{u(y,z)}$$

$$V(x,y,z) = (Cxz) + \underline{v(y,z)}$$

$$W(x,y,z) = (-1/2kx^2 - Cxy) + \underline{w(y,z)}$$

where ε_0 is the uniform extension, k is the bending curvature, C is one half the twisting curvature. The terms that are underlined are the customary displacement field of an eight noded isoparametric element. The terms enclosed in the parenthesis are the modifications that take into account the bending and twisting curvatures.

The strain field corresponding to the displacement field shown above is stated as follows:

$$\{\varepsilon\} = \{\varepsilon\}_0 + \{\varepsilon\}_L$$

where $\{\varepsilon\}$ is the element strain and

$$\{\varepsilon\}_L = \left\{ \begin{array}{c} Kz + \varepsilon_0 \\ 0 \\ 0 \\ 0 \\ -Cy \\ Cz \end{array} \right\}$$

The matrix $\{\varepsilon\}_L$, with the terms ε_0, K, and C, represents the loading vector in the model. From lamination theory, the induced loading terms can be expressed in terms of the known load. The details of the element formulation are presented in reference 2.

COMPUTATIONAL MODELS

As shown in Figure 1, the cross section of $[45, -45, 0, 90]_s$, AS4/3501-6 laminate is modelled for tension, $[45, -45, 45, -45, 0_2, 90_2]_s$ for both bending and torsion. Untapered laminates will be referred to as baseline laminates. Figure 2 displays the tapered geometry for the tension case where the 90 ply is terminated at a distance of 28H from the free edge. Recall that H represents a single ply thickness. Immediately adjacent to the tapered ply, a triangular resin pocket of length of 4H and a height of H is formed. The thicknesses of adjacent layers are adjusted as shown in the figure. The adjustments are based on the experimental results reported in reference 1 and a micrograph of a typical cross section of a tapered laminate is presented in Fig. 3. The change in ply thickness between the termination and the free edge was caused by the nonbleeding vacuum bag procedure used to cure the coupons. Table I displays the material properties evaluated for different layer thicknesses used in the calculations. The original lamina thickness is designated by h and the modified on by h. The first (-45) ply from the outer surface is terminated for the bending and torsion cases.

RESULTS

Interlaminar Stresses

Tension:

The variation of interlaminar normal and shear stresses through the thickness, z, direction are presented for tapered and baseline laminates in Fig. 4. The stress values are obtained at a distance of half ply from the free edge and they are normalized with the applied uniform strain. It is readily observed that at the midplane, the interlaminar normal stress of the tapered laminate is only 15% of the baseline laminate. Thus a stress reduction of 85% is achieved by simply tapering the 90 ply. The stress reduction obtained for interlaminar shear stress is 43%.

The stress variation along the y-axis of the tapered laminate is depicted in Fig. 5. The values are obtained at 0.125H from the midplane. The highest normal stress occurs immediately outside of the resin pocket in the 90 region, whereas the highest shear stress is observed in the resin pocket.

Bending:

The variation of interlaminar normal stress for tapered and baseline laminates is depicted in Figure 6. The stress values are calculated at a single ply away from the free edge and are plotted through the thickness of the laminate. A slight reduction in magnitudes is observed. However, note that the -45 ply termination causes a switch from compressive to normal tensile stress at the outer angle plies. Interlaminar shear stresses, σ_{xz} and σ_{yz} are displayed in Figure 7 and Figure 8 respectively. The interlaminar shear stress σ_{xz} of the tapered laminate increases in magnitude with a sign reversal in the angle plies and shows a decrease in magnitude for the 0 and 90 interior plies. On the other hand, the interlaminar shear stress σ_{yz} goes through a sign reversal throughout the thickness direction with the exception of second layer of angle plies. The magnitude is reduced at the outer plies, but an increase is observed at all other interfaces.

Torsion:

The interlaminar normal and shear stresses of [45, -45, 45, -45, 0_2, $90_2]_s$ laminate subjected to a constant twisting curvature are presented in Figures 9-11. Note that the outermost -45 ply is terminated for this study. An increase in the magnitude of interlaminar normal stress in a tapered laminate is observed from Figure 9. It is noteworthy to point out the 30% decrease in the interlaminar shear stress σ_{xz} for the tapered geometry in comparison to the baseline laminate of Figure 10. However an increase in magnitude of shear stress σ_{yz} is observed for the tapered laminate as shown in Figure 11. Also, the sign of the shear stress changes for the inner angle plies.

<u>Strain Energy Release Rates</u>

Tension:

Fig. 12 displays the effect of delamination length on the total strain energy release rate. The delamination length is normalized by ply thickness, H, and G_T is normalize by the applied uniform strain. The delamination is at the midplane. The increase in G_T in the resin region, from 12H to 16H, is readily observed in Fig. 12. Note the decrease in the strain energy release rate outside the resin pocket.

CONCLUSION

Interlaminar normal and shear stress variations are presented for laminates with tapers. Three loading conditions, namely; tension, bending and torsion, are considered for understanding the laminate behavior. The stress reductions at the free edge are most pronounced for tensile loads. Also, the reduction in magnitude of the most dominant shear stress, σxz, for torsion is encouraging. However, bending results indicate that there is not an advantage to tapering the -45 ply as a tool to control the free edge for laminates subjected to constant bending loads. Further study of the strain energy release rates are underway for future discussions.

REFERENCES

1.   Chan, W.S., and Ochoa, O.O., "Suppression of Edge Delamination by terminating a Critical Ply Near Edges in Composite Laminates," AIAA Paper 88-2257, Proceedings of 29th SDM Conference, Williamsburg, Va., April 18-20, 1988, pp. 359-364.

2.   Chan, W.S., and Ochoa, O.O., "An Integrated Finite Element Model of Edge Delamination Analysis for Laminates due to Tension, Bending Torsion Loads," AIAA Paper 88-0704, Proceedings of 28th SDM Conference, Monterey, Ca., April 608, 1987, pp. 27-35.

3.   Ochoa, O.O., and Chan, W.S., "Hygrothermal Effects on Free Edge Characterization," Proceedings of 4th U.S.-Japan Conference on Composite Materials (to be published), Washington, D.C., June 27-29, 1988.

Table I. Material Properties

| | AS4/3501-6 | | | Matrix | |
	h/h = 1.00	h/h = 1.25	h/h = 1.50		
E1, Msi	19.3	15.6	13.1	E1,Msi	0.65
E2, Msi	1.62	1.24	1.08		
E12,Msi	1.02	0.63	0.49	G1,Msi	0.239
v	0.288	0.302	0.312	v	0.36

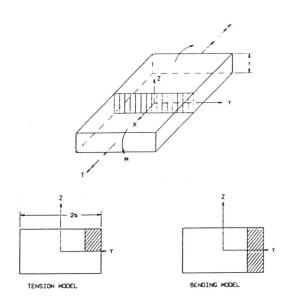

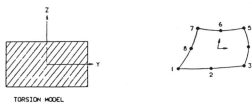

Figure 1. Schematic of a quasi three-dimensional finite-element model.

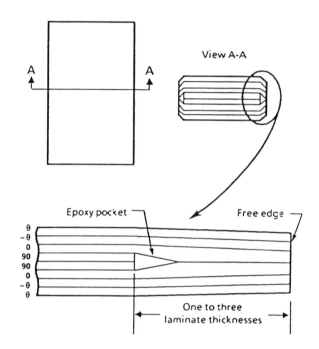

Fig. 2 Laminate configuration.

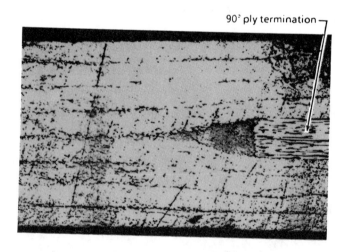

Fig. 3 Photomicrograph of laminate cross section
showing 90° ply termination.

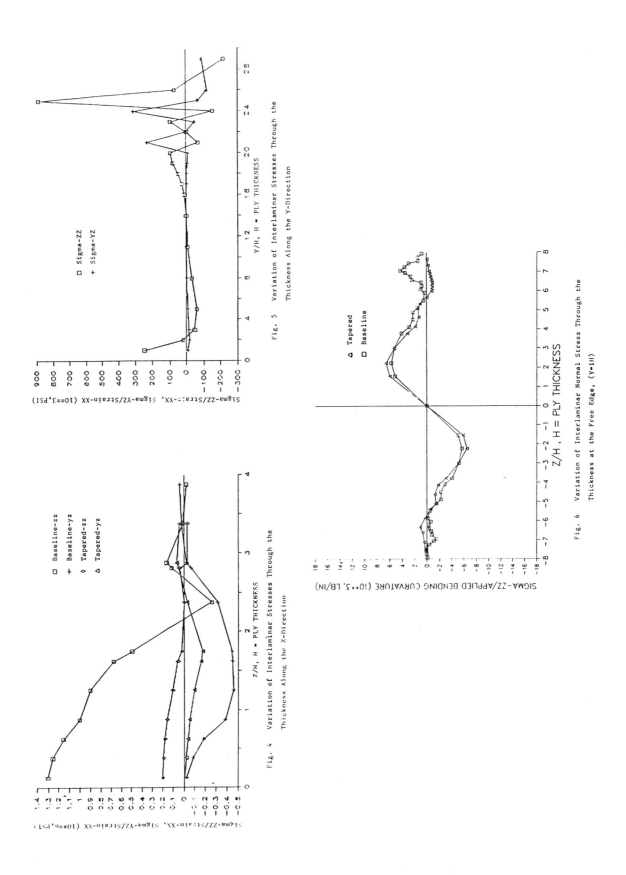

Fig. 5 Variation of Interlaminar Stresses Through the
Thickness Along the Y-Direction

Fig. 4 Variation of Interlaminar Stresses Through the
Thickness Along the Z-Direction

Fig. 6 Variation of Interlaminar Normal Stress Through the
Thickness at the Free Edge, (Y=1H)

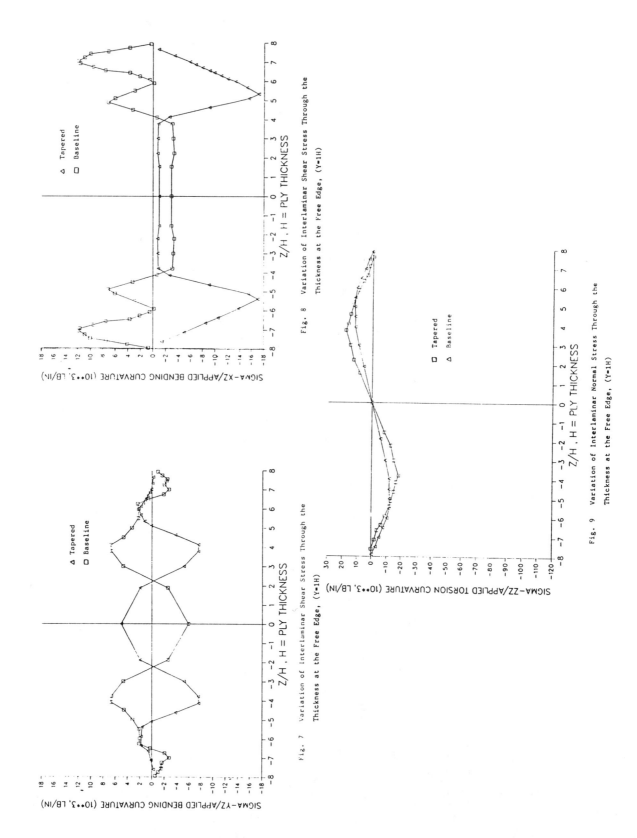

Fig. 7 Variation of Interlaminar Shear Stress Through the
Thickness at the Free Edge, (Y=1H)

Fig. 8 Variation of Interlaminar Shear Stress Through the
Thickness at the Free Edge, (Y=1H)

Fig. 9 Variation of Interlaminar Normal Stress Through the
Thickness at the Free Edge, (Y=1H)

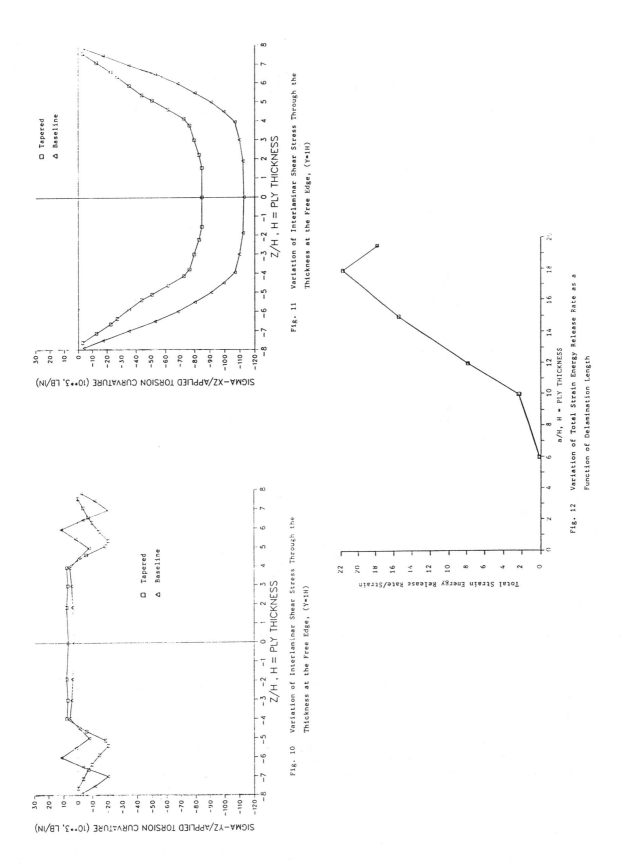

Fig. 10 Variation of Interlaminar Shear Stress Through the
Thickness at the Free Edge, (Y=1H)

Fig. 11 Variation of Interlaminar Shear Stress Through the
Thickness at the Free Edge, (Y=1H)

Fig. 12 Variation of Total Strain Energy Release Rate as a
Function of Delamination Length

641

Strain Energy Release Rate Analysis of Delamination in a Tapered Laminate Subjected to Tension Load

S. A. SALPEKAR*, I. S. RAJU** AND T. K. O'BRIEN

ABSTRACT

A tapered composite laminate subjected to tension load was analyzed using the finite element method. The $\{[0_7/(\pm45)]/_\uparrow[(\pm45)_3]/[0/(\pm45)/0]\}_s$ glass/epoxy laminate has a $(\pm45)_3$ group of plies dropped in three distinct steps, each 20 ply-thicknesses apart, thus forming a taper angle of 5.71 degrees. Steep gradients of interlaminar normal and shear stress on a potential delamination interface suggest the existence of stress singularities at the points of material and geometric discontinuities created by the internal plydrops. The delamination of the tapered laminate was assumed to initiate at the bottom of the taper on the -45/+45 interface indicated by the arrow in the laminate layup, and the delamination growth was simulated along the taper and into the thin region. The total strain-energy-release rate, G, and the Mode I and Mode II components of G, were computed at the delamination tip using the Virtual Crack Closure Technique. In addition, G was calculated from a global energy balance method. The strain-energy-release rate for a delamination growing a short distance into the thin laminate consisted predominantly of a Mode I (opening) component. For a delamination growing along the tapered region, the strain-energy-release rate was initially all Mode I but decreased with increasing delamination size until eventually it was all Mode II. These results indicated that a delamination initiating at the end of the taper will grow unstably along the taper and the thin laminate simultaneously.

INTRODUCTION

Composite rotor hubs are currently being designed and manufactured that are hingeless and bearingless to reduce weight, drag, and the number of parts in the hub. Such a design would involve tapering the laminate by dropping some plies in the flexure region of the hub. The plydrop in the laminate creates geometric and material discontinuities that create large interlaminar stresses and initiate delaminations. Therefore, there is a need to analyze tapered laminates with ply drops to understand their failure mechanisms. However, only a limited amount of literature is available on tapered laminates.

1) Research Scientist and 2) Senior Scientist, Analytical Services and Materials, Inc., Hampton, VA. 23666, and 3) Senior Scientist, Aerostructures Directorate, U.S. Army Aviation Research and Technology Activity (AVSCOM), NASA Langley Research Center, Hampton, VA. 23665-5225.

Adams et. al. [1] analyzed a $[0_{16}/(\pm 45)_5/90_4]$ graphite/epoxy laminate in which two zero degree plies were dropped. The effect of compressive load, moisture, and temperature due to the presence of the plydrop was studied using a 3-D finite element analyses with nonlinear orthotropic response. They concluded that all the interlaminar stresses induced by a 0 degree ply drop-off anywhere in the laminate were negligible compared to the in-plane stresses. However, they did not account for the low interlaminar strength of the composite compared to the in-plane strength.

Cannon [2] conducted experiments on graphite/epoxy tapered laminates from the $[\pm 45/0]_s$ and $[\pm 15/0]_s$ families, subjected to tension load. For most laminates the failure mode and the failure stress were similar to that of the untapered specimen at the thin (dropped) end of the laminate. An analysis based on the minimization of total potential energy which accounted for the effect of eccentricity due to the plydrop was used to predict the in-plane failure stresses in unsymmetric laminates. The tests on $[\pm 45/0/(\pm 45/0)_D]_s$, where D denotes dropped plies, showed that dropping a number of plies lumped together can change the initial damage from in-plane failure to delamination.

Kemp and Johnson [3] analyzed a tapered beam having a single plydrop using the finite element method. Symmetric and unsymmetric laminates were modelled as a generalized plane deformation problem subjected to a uniform strain in the longitudinal direction. The layups considered were $(\pm 45/0/90/0_{nD}/90/0/\pm 45)_T$ and $(0/90/\pm 45/0_{nD}/\pm 45/90/0)_T$ where n, the number of dropped zero degree plies, was chosen to be 1, 2, or 3. Failure strains were calculated corresponding to resin failure at the dropped plies, based on a maximum principal stress criterion, and intralamina failure in tension and compression, using the Tsai-Wu criterion. The first failure event in tension or compression was predicted to occur in the resin.

Although the stress distributions in the laminate help to identify the highly stressed critical areas, maximum stress or strain criteria cannot be used to predict delamination onset and growth if the stresses are singular. However, interlaminar fracture toughness, which is generic to a given composite material, is useful in determining the loads corresponding to the onset and propagation of delamination [4,5,6]. The delamination growth in a laminated composite structure may be predicted from the Mode I, and Mode II components of the strain-energy-release rate under static loading and from the the total strain-energy-release rate for fatigue loading [5,7].

Therefore, the purpose of this paper is to study the interlaminar stress distributions in a tapered beam subjected to tension loads and to determine the strain-energy-release rate for delamination growth that may occur due to the presence of plydrops. A typical stacking sequence used in a helicopter hub is $\{[0_9]/[(\pm 45)_3]/[(\pm 45)_2]\}_s$. The laminate used in this analysis has the same configuration but a somewhat different stacking sequence. A $\{[0_7/(\pm 45)]/_{\uparrow}[(\pm 45)_3]/[0/(\pm 45)/0]\}_s$ tapered laminate is analyzed using a two-dimensional finite-element analysis. The $(\pm 45)_3$ plies are dropped in three steps, 20 plies apart. The dropped plies result in a taper angle of 5.71°. The interlaminar normal and shear stress distributions along the taper interface, indicated by an arrow in the above layup are presented. Delaminations are assumed to initiate at a point of high interlaminar stress along this interface. The Mode I, Mode II and total strain-energy-release rate for various delamination lengths are presented. These results may be used to hypothesize the stability of delamination growth under static and fatigue loading.

NOMENCLATURE

a	delamination length along taper
b	delamination length in the thin region
E_{11}, E_{22}, E_{33}	Young's moduli
G	total strain-energy-release rate
G_I, G_{II}, G_{III}	Mode I, Mode II, and Mode III components of strain-energy-release rate, respectively
G_{12}, G_{13}, G_{23}	shear moduli
h	ply thickness
N_x	total load per unit width on symmetric half laminate
X, Y, Z	Cartesian Coordinates
σ_o	uniform tension load per unit area
σ_n	interlaminar normal stress
τ_{nt}	interlaminar shear stress
ν_{12}, ν_{13}, ν_{23}	Poisson's ratio

ANALYSIS

Specimen Configuration and Loading

Figure 1 shows the tapered laminate that was analyzed. The stacking sequence was assumed to be $\{[0_7/(\pm45)]/_\uparrow[(\pm45)_3]/[0/(\pm45)/0]\}_s$. The $(0_7/\pm45)$ ply group in the laminate of Fig.1 forms the belt area, and the $(0/\pm45/0)_s$ laminate in the center forms the core. The transition from the thick region at the left to the thin region at the right is achieved by dropping the group of $(\pm45)_3$ plies in three distinct steps, each 20-ply thicknesses apart. The shaded regions shown in Fig.1 are the resin pockets formed at the ends of the ±45 degree plies that are terminated. In similar laminates, delaminations have been observed at the interface indicated by the arrow in the layup above. Therefore, the delaminations are assumed to grow along the interface ABCD in Figure (1a). A typical delamination is shown in Fig. 1(b). The delamination forms at point C, and grows into the tapered region from tip "I" and into the thin region from point "H".

The tapered laminate was assumed to be made of S2/SP250 glass/epoxy and to be subjected to a uniform load at the thick end (X=0). Examination of the results indicates that the displacements are uniform in the neighborhood of x=60h. Thus, the uniform load condition at x=0 is equivalent to a uniform displacement condition. A fixed grip condition was assumed at the thin end. The material properties used in the analysis are given in Table 1. The in-plane properties for a unidirectional ply (e.g; E_{11}, E_{22}, G_{12}, ν_{12}) are similar to those used in reference 7. The out-of-plane properties (G_{13}, ν_{13}, G_{23}, ν_{23}) were assumed to be identical to the in-plane properties, and E_{33} was assumed equal to E_{22}.

Finite Element Model

A 3-D finite element analysis of the laminate is desirable, but such analyses are complex. Simple 2-D models, which do not account for the free edges, usually provide insight that can be used in 3-D analyses. Thus, as a first step, 2-D plane-strain analyses were performed in this study. Furthermore, the stacking sequence considered here contains only 0 degree and ±45 degree plies. With the absence of the 90 degree plies, the interlaminar Poisson mismatch between plies that causes edge delaminations was not considered significant [5]. Therefore, a two dimensional finite-element analysis should be reasonably accurate for this laminate.

A two dimensional finite element model was developed utilizing the symmetry of the laminate about the X-axis. The model had 7610 nodes and 2382 eight-noded, isoparametric, parabolic elements as shown in Fig.(2a). A refined mesh was used near plydrop points (B, E and F and C in Figure 1b) to capture the local influence of these geometric discontinuities and the corresponding stresses. The smallest element size used in the model was equal to one-quarter of the ply thickness. These small elements were provided near the plydrops on line BC, at the transition point B from the thick region to the tapered region, and the transition point C from the tapered region to the thin region. The element size immediately below line BC varied in the Z-direction due to the change in the resin thickness from two to zero ply thicknesses in the three resin pockets. Collapsed eight-noded elements were used at locations E, F, and C in the resin pockets. Figure (2b) shows local mesh detail at location E. A similar pattern was used at points F and C.

The nodes at the end of the thin region (at X = 180h in Figure 1a) of the laminate were constrained in both X- and Z- directions. A uniform tension per unit area, σ_o, (assumig unit width in the Y-direction) was applied along the X=0 line of the model. Plane strain conditions were used in the analysis.

To facilitate modeling delaminations along ABCD, duplicate nodes were created in the model all along lines AB, BC, and CD. Multi-point constraints were imposed for the corresponding duplicate nodes. Different size delaminations were simulated by relaxing the multi-point constraints for the appropriate nodes along lines BC and CD.

The material directions of plies in the laminate are oriented at an angle relative to the global coordinate system of the analysis. The material stress-strain relations for these plies were transformed to obtain the stress-strain relations in the global system.

Computation of Strain-Energy-Release Rate

The virtual crack-closure technique (VCCT) was used to obtain the strain-energy-release rate components, Mode I, and Mode II, based on the local forces ahead of, and the relative displacements behind, the delamination tip. These two components were calculated using the following equations.

$$G_I = - \frac{1}{2\,\Delta} \left[F_{ni} \left(v_k - v_{k'} \right) + F_{nj} \left(v_m - v_{m'} \right) \right] \qquad (1a)$$

$$G_{II} = - \frac{1}{2\,\Delta} \left[F_{ti} \left(u_k - u_{k'} \right) + F_{tj} \left(u_m - u_{m'} \right) \right] \qquad (1b)$$

where Δ is the element size, F_{ni} and F_{ti} are the normal (n) and tangential (t), forces, respectively, at node i, and $(v_k - v_{k'})$ and $(u_k - u_{k'})$ are the relative opening and sliding displacements, respectively, at node k (see Fig.3). Forces at node j and relative displacements at nodes m and m' are defined similarly. Equations 1 are similar to those given in references 8 and 9. The total strain-energy-release rate, G, was calculated as

$$G = G_I + G_{II} \tag{2}$$

The Mode III component of G was identically zero because plane strain conditions were assumed in the analyses.

Alternatively, the global energy change of the laminate due to delamination growth can also be used to calculate the total strain-energy-release rate, G. The strain energy of the laminate, U, can be conveniently computed as $U = 1/2(\sum f_i u_i)$ where f_i and u_i are the nodal forces and corresponding nodal displacements, respectively, for all nodes i on the line X=0 in Figure 1a. The strain-energy-release rate for successive delamination growth was calculated as

$$G = \frac{dW}{dA} - \frac{dU}{dA} \tag{3a}$$

where dW/dA and dU/dA are the rate of change of work and strain energy, respectively, with change in delamination area. In the finite-element analysis, Equation (3a) can be computed as

$$G = (U_{a+da} - U_a)/ da \tag{3b}$$

where U_{a+da} and U_a are the strain energies for delamination lengths a+da and da, respectively. The value of G thus calculated is considered to be the strain-energy-release rate at (a+ da/2), which is located at the center of the interval.

RESULTS AND DISCUSSION

First, the interlaminar stress distributions along the interface ABCD are presented. Next, the strain-energy-release rate variations for various size delaminations assumed along the interface line ABCD are shown. Finally, the peak values of the total strain-energy-release rate and the mode I component values are presented and their significance discussed.

Interlaminar Stresses

Figure 4 shows the normalized interlaminar normal stress, (σ_n/σ_0), along lines AB, BC, and CD in the laminate. Stresses were calculated in the local coordinate system, normal to the interface ABCD. The interlaminar normal stress shows peaks near the points of geometric and material discontinuity i.e. at points B, E, F, and C. The largest tensile value of the σ_n distribution occurred at the transition point C. At the plydrops, points B, E, F, the stresses changed from a high compressive value immediately to the left of the plydrop to a high tensile value immediately to the right of plydrop. The variation of normalized interlaminar shear stress, (τ_{nt}/σ_0), along the same interfaces AB, BC, and CD is shown in Figure 5. The shear stress also shows peaks at points B, E, F and C.

These sudden changes in the normal and shear stress distributions at points B, E, F, and C are not unexpected. At these points, the material stiffness is different in different directions (see Fig. 6). Therefore, at points B, E, F, and C, stress singularities probably exist [10].

In order to investigate if this is true, a two-dimensional finite-element analysis of a homogeneous tapered laminate was performed with the same model as in Fig. 2. The tapered laminate was assumed to be of an isotropic material. The normalized interlaminar normal stress (σ_n/σ_0)

distribution along the line ABCD is presented in Fig. 7. At points E and F, the stiffness is same in different directions irrespective of how these points are approached. No sudden changes in stress distribution are expected at these points. This behavior is confirmed in Figure 7. The normals to lines AB and BC at point B have different directions. Similarly, normals to lines BC and CD are different at point C. Thus, except for very small discontinuities at these points, the stress distribution all along ABCD is smooth. This confirms that the sharp changes in stresses observed in Figures 4 and 5 are solely due to material discontinuities at the points B, E, F, and C.

Strain-Energy-Release Rate Analysis

Delamination growth in a laminated composite structure may be predicted from the Mode I, and Mode II components of the strain-energy release rate under static loading and from the total strain-energy-release rate for fatigue loading [5,7]. The computation and the use of the strain-energy-release rate in delamination prediction for the tapered laminate are discussed below.

As seen in Figure 4, point C has the highest value of interlaminar normal stress, σ_n, compared to any other location on the interface line ABCD. Therefore, the delamination was assumed to initiate at this point. Delamination lengths a and b (see Fig. 1b) were assumed within the tapered region along CB and in the thin laminate along CD, respectively. The strain-energy-release rate values G, G_I, and G_{II} were computed using the finite element analysis and Equations 1-3 for various values of a and b.

The total strain-energy-release rates were calculated using two different methods; VCCT (equation 2) and from global energy change (equation 3). These G values normalized by N_x^2/h, (where N_x is defined as the product of uniform tension stress σ_o and half the laminate depth at X = 0, and h is the ply thickness), are plotted for comparison in Figure 8. For this case, no delamination was assumed along the taper, CB, (i.e. a=0) and the values of G were obtained for various values of delamination lengths, b, along CD in the thin region of the laminate. Excellent agreement between the G values computed by the two methods was obtained. Similar agreement was found for all the cases studied. The G values obtained by using equation 3 are presented in this paper because more data points were available for this computation and values of the individual modes, G_I and G_{II}, were taken from the VCCT calculation.

Figure 9 presents a composite of G distributions for delamination growth in the thin and thick regions. In the right hand portion of the figure, the G distributions were plotted for a fixed value of normalized delamination length, a/h, along the taper and for various values of normalized delamination length, b/h, in the thin region. Similarly, the left hand portion of figure 9 presents the G distributions for a fixed value of normalized delamination length b/h in the thin region and for various values of normalized delamination length, a/h in the tapered region.

Referring to the right side of figure 9 (where a/h is held constant and b/h varies), the G at the delamination tip H (Fig. 1(b)) in the thin laminate is higher for larger initial values of delamination length a/h along the tapered region. The total strain-energy-release rate initially increased rapidly as the delamination grows into the thin laminate along line CD. This trend is also seen in figure 8 for the case of a/h=0. The G attains a peak value and drops slightly with further delamination growth. This drop is more gradual for the larger values of initial delamination length a/h considered and eventually does not occur for the largest b/h considered.

In a complementary situation shown on the left side of Figure 9, the delamination distance, b/h, in the thin region is held constant and the G values at the delamination tip I in the tapered region (Fig. 1(b)) were evaluated for various values of a/h. The total strain-energy-release rate increases initially, and then is relatively constant, or drops slightly, before approaching the plydrop. In the proximity of the dropped plies, however, G values increase rapidly and attain peak values at the plydrops (a=20h and a=40h).

The results of Figure 9 suggest that a delamination initiating at point C will grow in an unstable manner simultaneously along the tapered interface CB as well as in the thin laminate along CD. This can be explained as follows. Consider a small delamination initiating at point C in Fig. 9. If the G distributions on the left and right sides control growth along the thin and tapered regions, respectively, then delaminations would arrest after they had grown to the peak values. However, delamination growth in one direction will increase the G, causing growth in the other direction. Hence, as soon as a stable situation occurs on one side, it will increase the G on the other side causing further growth. Hence, as the peaks in the G distributions on the left and right side of Figure 9 increase monotonically with increasing a/h and b/h, a delamination initiating at point C will grow unstably in both directions simultaneously.

As discussed above, the values of G shown in Fig. 9 reached peak values for delamination growth along the thin side CD for a given value of (a/h) or along tapered side CB for a given value of (b/h). For delamination growth along the thin side CD, the plots of G vs. b/h are similar to those obtained for edge delamination, where G is initially zero at b/h=0 and raises to a plateau at some distance, usually b/h≈ 2 to 3 [5-7]. The distance at which G reaches a plateau for the edge case may vary with the interface analyzed [5]. Similarly, the distance at which G vs. b/h reaches a peak will vary with delamination size, a/h, in the tapered region. The b/h distance required to reach this peak, however, is of little consequence since it is assumed that the plateau, or peak value of G in the distribution governs the delamination onset at the edge or, in this case, at the initial point of the taper [5]. The peak values of G on the right side in Figure 9 are plotted on the right side of Figure 10 as a function of the corresponding a/h, in the tapered region. Similarly, the peak G values between C and F along the taper, not considering the region near the plydrop F, are plotted on the left side of Figure 10 as a function of b/h, in the thin region. The peak G value at a/h=0 (for b/h=4.5) and at b/h=0 (for a/h=5.75) have non zero values and are nearly equal. These values may be hypothesized as the total strain-energy-release required for the onset of the delamination in a tapered laminate at point C under fatigue loading. This value may be compared to the threshold for delamination onset under fatigue loading in order to predict delamination onset in these tapered laminates [6].

If the delamination initiates at point C under static tension loading, its growth will be governed by a mixed-mode criterion [5] because both the Mode I and Mode II components of G are present due to the tapered configuration. Figure 11 shows the percentages of Mode I and Mode II at delamination tip H (Fig. 1(b)) in the thin laminate, corresponding to a value of a=24h. The Mode I component is predominant for all values of b/h≤18.

 In contrast, as shown in Figure 12, a delamination along the tapered region CB (point I in Fig. 1(b)) initially consists of a large Mode I component at delamination tip "I", but at a/h=18 the Mode I component is only 50 percent of the total and continues to decrease with increasing a/h.

Near the plydrops, G_I drops suddenly, but then recovers. Overall G_I decreases with increasing delamination distance, b/h, and G_{II} increases with increasing delamination length, a/h. The value of G_{II} equals the value of G (i.e. G becomes 100 percent Mode II) at a/h=54. This distance depends on the initial delamination length, b/h, in the thin laminate.

The Mode I component of G is predominant for small values of delamination length, initiating from point C (Fig.1) and growing either along CD (Fig.11) or along CB (Fig.12). The corresponding peak G_I values for various a/h and b/h ratios are shown in Fig. 13. This figure is constructed in similar manner to Fig.10. The peak G_I values at a/h=0 for growth into the thin region or at b/h=0 for growth into the tapered region may be compared to G_{IC} for the composite to predict delamination onset under static tension loading[5].

CONCLUDING REMARKS

A tapered composite laminate subjected to tension load was analyzed using the finite element method. The stacking sequence of the laminate was assumed to be $\{[0_7/(\pm45)]/_\uparrow[(\pm45)_3]/[0/(\pm45)/0]\}_s$. The group of $(\pm45)_3$ plies was dropped in three distinct steps, each 20 ply-thicknesses apart, thus forming a taper angle of 5.71 degrees. Neat resin pockets are assumed at the ends of ±45 degree plies that were terminated. The material of the laminate was assumed to be S2/SP250 glass/epoxy.

A two-dimensional plane strain analysis was performed to determine stress distributions in the laminate without a delamination. The interlaminar normal stress and interlaminar shear stress distribution along the tapered interface, indicated by an arrow in the above stacking sequence, were calculated. Then delaminations were assumed to initiate at the point of intersection of the tapered interface and the thin region of the laminate. Delamination growth in the finite element model was simulated along the taper and into the thin region. The total strain-energy-release rate, G, and the Mode I, and Mode II components, G_I and G_{II}, were computed at the delamination tip using the Virtual Crack Closure Technique (VCCT). Alternatively, G was obtained from a global energy balance. Based on the analysis performed here, the following conclusions were reached:

1) Steep gradients of interlaminar normal and shear stress exist at the points of material and geometric discontinuities created by the internal plydrops. The largest value of interlaminar normal stress appears to occur at the intersection of the tapered interface and the thin region of the laminate. This is probably the site where the onset of delamination occurs.

2) The strain-energy-release rate, G, was calculated for a delamination initiating at a point, located at the intersection of the taper and the thin laminate, and lying on the interface indicated by the arrow in the layup. The G values increase continually as the delamination grows into the thin laminate portion or along the taper. This indicates that a delamination initiating at the end of the taper will grow unstably along the taper and the thin laminate simultaneously.

3) The strain-energy-release rate for a delamination growing a short distance into the thin laminate consists predominantly of Mode I (opening) component throughout its growth.

4) For a delamination growing along the tapered region, the strain-energy-release rate was initially all Mode I but decreased with increasing delamination size until eventually it was all Mode II.

These results may help understand the delamination behavior in the tapered laminates and may be useful in predicting the onset and growth of the delamination under static and fatigue loading.

Acknowledgement: This work was performed under NASA contracts NAS1-18235 and NAS1-18599.

REFERENCES

[1] Adams, D. F., Ramkumar, R. L., Walrath, D.E., "Analysis of Porous Laminates in the Presence of Ply Drop-offs and Fastener holes," Northrop Technical Report 84-113, May 1984, One Northrop Avenue, Hawthorne, CA 90250 and University of Wyoming, Mechanical Engineering Department, Laramie, WY 82071.

[2] Cannon, R. K., "The Effect of Ply Dropoffs on the Tensile Behavior of Graphite/epoxy Laminates," TELAC Report 87-12, May 1987, M.I.T., Cambridge, MA. 02139.

[3] Kemp, B. L., Johnson, E. R., "Response and Failure Analysis of a Graphite-Epoxy Laminate containing Terminating Internal Plies," Paper No. AIAA-85-0608, Proceedings of the AIAA, ASME, ASCE, AHS, 26th Structures, Structural Dynamics and Materials Conference, Orlando, Florida, 1985, pp 13-24.

[4] O'Brien, T.K., Murri, G.B., and Salpekar, S.A., "Interlaminar Shear Fracture Toughness and Fatigue Thresholds for Composite Materials," NASA TM 89157, August, 1987, Presented at the 2nd ASTM symposium on Composite Materials: Fatigue and Fracture, Cincinnati, OH, April, 1987.

[5] O'Brien, T. K., "Mixed-Mode Strain-Energy-Release Rate Effects on Edge Delamination of Composites," ASTM STP 836, Effects of Defects in Composite Materials, American Society for Testing and Materials, Philadelphia, PA, 1982, p.125.

[6] O'Brien, T. K., "Towards a Damage Tolerance Philosophy for Composite Materials and Structures," Presented at the 9th ASTM Symposium on Composite Materials: Testing and Design, Reno, Nevada, April 27-29, 1988, (NASA TM 100548, May, 1988).

[7] Chan, W. S., Rogers, C., Aker, S., "Improvement of Edge Delamination Strength of Composite Laminates Using Adhesive Layers," Composite Materials :Testing and Design, ASTM STP 893, 1986, p.266.

[8] Rybicki, E. F., and Kanninen, M. F., "A Finite Element Calculation of Stress-Intensity Factors by a Modified Crack-Closure Integral," Engineering Fracture Mechanics, Vol. 9, 1977, pp. 931-938.

[9] Raju, I. S., "Calculation of strain-energy-release rates with Higher Order and Singular Finite Elements." Engineering Fracture Mechanics, vol.28, 1987, pp. 251-274.

[10] Erdogan, F., Stress distribution in bonded dissimilar materials with cracks, J. appl. Mech. 32, Series E, 403(1965).

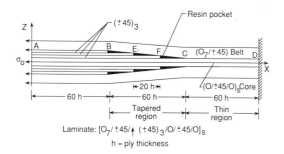

Laminate: $[O_7/\pm45/\uparrow(\pm45)_3/O/\pm45/O]_S$
h = ply thickness

Fig. 1(a) - Tapered laminate configuration
and loading.

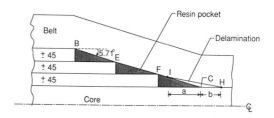

Fig. 1(b) - Typical delamination.

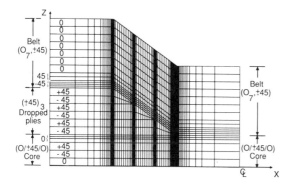

Fig. 2(a) - Finite-element model of
the tapered laminate.

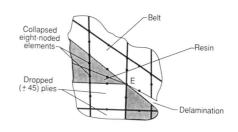

Fig. 2(b) - Mesh detail showing
delamination tip at point E.

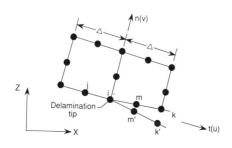

Fig. 3 - Finite-element
idealization near the
delamination tip for
calculating G using VCCT.

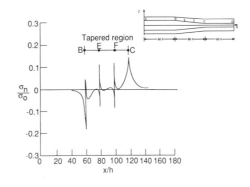

Fig. 4 - Normalized interlaminar normal
stress distribution along
the tapered interface BEFC.

651

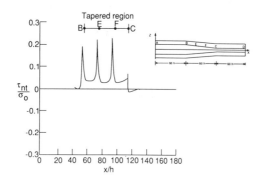

Fig. 5 - Normalized interlaminar shear stress distribution along the tapered interface BEFC.

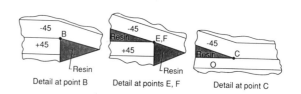

Fig. 6 - Details of geometric and material discontinuities along interface ABCD.

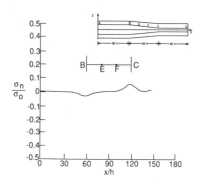

Fig. 7 - Normalized interlaminar normal stress distribution along interface ABCD in the tapered isotropic (resin) beam.

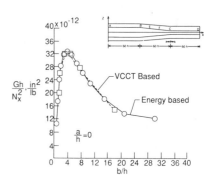

Fig. 8 - Normalized total strain-energy-release rate at delamination tip H along interface CD.

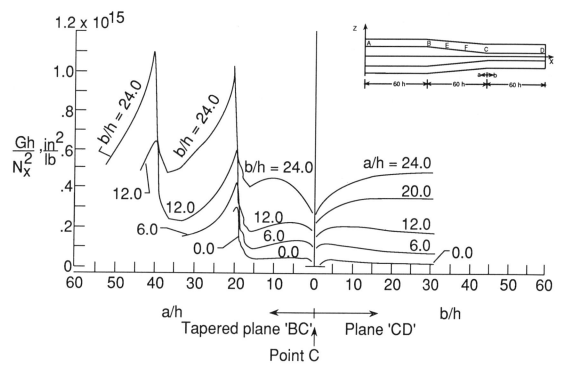

Fig. 9 - Normalized total strain-energy-release rate at delamination
tip I along interface CB and tip H along interface CD.

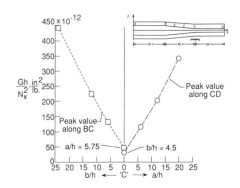

Fig. 10 - Peak values of total
strain-energy-release rate along
either side of C.

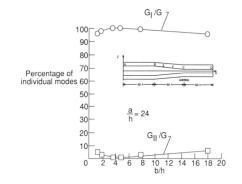

Fig. 11 - G_I/G and G_{II}/G at
delamination tip H along
interface CD.

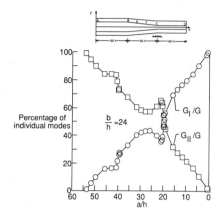

Fig. 12 - G_I/G and G_{II}/G at delamination tip I along interface BC.

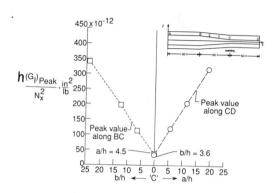

Fig. 13 - Peak values of G_I on either side of C.

Scaling Effects in the Static Large Deflection Response of Graphite-Epoxy Beam-Columns

KAREN E. JACKSON

ABSTRACT

Scale models of graphite-epoxy composite beams ranging from 1/6 to full scale were statically tested under an eccentric axial compressive load to investigate scaling effects in their large deflection response and failure. Beams having unidirectional, angle ply, cross ply, and quasi-isotropic laminate stacking sequences were included in the study. Comparisons between the experimental results, a one dimensional large rotation beam analysis, and a finite element beam model are presented. Static test results indicate that the beam response in the small deflection region scales as predicted by the model law. Failure mechanisms between the model and prototype were similar for all laminate types. However, a significant scale effect was observed in failure behavior. It is important that this phenomenon be understood before strength testing of scale model composite structures can be utilized.

INTRODUCTION

Scale model technology represents one method of investigating the behavior of advanced, weight efficient composite structures under a variety of loading conditions. Testing of scale models of composite structures can provide a cost effective alternative to destructive testing of expensive composite prototypes. The results obtained from these tests can be used to verify predictions obtained through non-linear finite element analyses. It is important, however, to understand the limitations involved in testing scale model structures. Scaling effects in the response and failure of composite structures must be characterized before the technique can be used to full advantage. Previous research [1] in which scale model composite beams were impacted transversely has demonstrated the validity of the technique. It was shown that classical scaling laws apply for elastic behavior, but a significant scale effect was observed in failure behavior.

The objective of the current research is to characterize scaling effects in the large deflection response of composite beams. The scaled beams are loaded with an eccentric axial compressive load designed to produce large bending deflections of the beam. This testing configuration was chosen since it promotes global failure of the beam in bending away from the supported ends. A dimensional analysis was performed on the beam-column system using methods outlined in Baker [2] to determine the

K. E. Jackson, Aerospace Engineer, U.S. Army Aerostructures Directorate, NASA Langley Research Center, M.S. 495, Hampton, VA 23665-5225

nondimensional parameters or Pi terms which govern the scaled response. An experimental program designed to validate the scaling laws was performed and initial results are reported in this paper. Also, a one dimensional large rotation analysis and a DYnamic Crash Analysis of STructures (DYCAST) [3] finite element model of the composite beam were developed for comparison with experimental results.

EXPERIMENTAL PROGRAM

Beams having unidirectional, angle ply, cross ply, and quasi-isotropic laminate stacking sequences were constructed of a high modulus graphite fiber and a Hercules[*] epoxy matrix system designated as AS4/3502 for the static tests. The full scale beam had dimensions of 3.0" by 30.0" by 48 plies thick with a nominal ply thickness of 0.0054". The scale model beams were constructed by applying seven different geometric scale factors including 1/6, 1/4, 1/3, 1/2, 2/3, 3/4, and 5/6, to the full scale beam dimensions. The thickness dimension was scaled by reducing the number of layers in each angular ply group of the full scale laminate stacking sequence. Using this approach, it was not possible to fabricate a 1/4 or 3/4 scale quasi-isotropic beam. A set of scaled beams are illustrated in Figure 1 and the dimensions and lay-ups of the beams are listed in Table 1. Static tests were conducted on three full scale and three scale model beams of each laminate stacking sequence. The beams were machined from panels which were hand constructed from pre-preg tape and cured according to manufacturer's specifications. Slight variations were observed in the thickness dimensions of the cured beam specimens. The maximum deviation in normalized thickness was approximately five per cent between the 1/6 scale and full scale beam specimens.

During the tests each beam specimen was gripped in a set of hinges which were designed to offset the axial load with a moderate eccentricity, as shown in Figure 2. Eight sets of scaled hinges were constructed to ensure that the end condition was properly scaled for each test. The hinges were pinned to the platens of a standard load test machine which applied the compressive vertical load. The hinged-pin connection allowed the beam to undergo large rotations during deformation. Beam specimens were loaded until catastrophic failure, defined as loss of load carrying capability.

Each beam was instrumented with back-to-back strain gages located at distances one-quarter and two-thirds along the length and with strain gage rosettes at the midpoint. Vertical load, end displacement, and strain data were recorded using a personal computer based data acquisition system. The analog signals were amplified and filtered prior to being digitized and converted to engineering units. Only the load versus end displacement data will be presented in this report.

ANALYSIS

A one dimensional large rotation "elastica" type solution was developed to predict the response of the composite beam-column under eccentric axial load. The governing equation for the beam was derived from equilibrium of the forces and moments on a beam element. The exact

---

[*]Identification of commercial products and companies in this paper is used to describe adequately the test materials. The identification of these commercial products does not constitute endorsement, expressed or implied, of such products by the U.S. Army, the National Aeronautics and Space Administration, or the publishers of these conference proceedings.

expression relating moment and curvature was incorporated in the analysis, thus allowing the solution to predict large rotation response. The solution of the governing equation is outlined in Timoshenko and Gere [4] for the "elastica" problem and was adapted for this problem by applying the end moment boundary conditions produced by the eccentric vertical load. The solution is given in terms of elliptic integrals and predicts the end displacement, transverse displacement of the midpoint of the beam, and end rotation for increasing load.

The beam bending stiffness was derived based on the method described by Whitney [5]. The bending stiffness, EI, from classical beam theory is replaced by an equivalent stiffness for the composite beam by summing the modulus of each ply multiplied by its moment of inertia about the midplane of the laminate. The equivalent modulus incorporates the effect of shear coupling which is important for angle ply and quasi-isotropic laminates.

In addition to the beam analysis, the nonlinear finite element structural analysis computer program DYCAST [3] was used to model the composite beam-column. The model consisted of 60 beam elements which were constrained to permit only planar deformations, as shown in Figure 2. The hinges at the top and bottom of the beam were modeled by two rigid beam elements. The model assumed pinned conditions between the load machine and the hinge, and clamped conditions between the hinge and beam. The bending stiffness used in the DYCAST model was the same as used in the beam analysis outlined previously.

RESULTS

Normalized load versus end displacement plots and corresponding photographs of a complete (1/6 through full scale) set of failed beam specimens for the unidirectional, angle ply, cross ply, and quasi-isotropic laminates are shown in Figures 3-6. Vertical load was normalized by the Euler column buckling load for the beam and end displacement was normalized by the gage length. Since three repeat tests were performed for each laminate type and size of beam, the results from one representative test are presented here. Repeatability between the three tests was good.

Normalized Load Versus End Displacement Results

In general, the load versus displacement curves show that the response scales in the small deflection, elastic range. Deviation from scaled response is observed for all laminate types as the beams undergo large deflections and the response becomes nonlinear. The angle ply beams show the most pronounced deviation from scaled response, as seen in Figure 4(a). In the nonlinear response region the small scale beams exhibit a higher normalized load for any given value of normalized end displacement than the full scale beam. They also fail at higher normalized load and end displacement levels. This observed scale effect in failure behavior is significant. The 1/6 scale beams fail at an end displacement to length ratio from 2 to 10 times the value for the full scale beam depending on the laminate type.

Failure Mechanisms

The photographs shown in Figures 3(b) through 6(b) indicate that while the failure modes for the laminate types considered in this study are different from each other, they are similar between scaled beams within the laminate family. Failure modes appear to be independent of

specimen size. The unidirectional beams, shown in Figure 3(b), failed by
fiber fracture near the midpoint of the beam. This failure mode is
typical of all the unidirectional beams 1/6 through full scale. Failure
of the angle ply beams occurred by a transverse matrix crack along the 45
degree fiber line. There was no evidence of fiber breakage, as shown in
Figure 4(b). The cross ply laminates exhibited combined failure
mechanisms of transverse matrix cracking and fiber fracture. As the cross
ply beam underwent large rotations, the 90 degree plies which were
sandwiched between outer 0 degree plies developed transverse matrix
cracks. The cracks were evenly spaced and resulted in uniform pieces of
debris, some of which are shown in Figure 5(b) for the 5/6 scale beam.
The ultimate failure of the cross ply beam was caused by fiber fractures
in the 0 degree plies. The quasi-isotropic beams failed through a
combination of matrix cracking, delamination, and some fiber failure.
Although the photograph in Figure 6(b) does not give a good indication,
the damaged quasi-isotropic beams are highly curved. The sequence of
failure events occurred such that the remaining intact section of the beam
consisted of an unsymmetric laminate, resulting in the observed curvature.

Analytical Results

Comparison of the experimental data for the 1/6 and full scale
specimens with the large rotation beam analysis and the DYCAST finite
element analysis is plotted in Figures 7(a) through 7(d) for each of the
laminate types. Good correlation is obtained in the small deflection,
elastic region with both the beam analysis and DYCAST for all laminate
types. The DYCAST solution typically underpredicts the experimental beam
response in the large deflection region, except for the angle ply laminate
in which the DYCAST prediction falls between the 1/6 scale and full scale
experimental data, as seen in Figure 7(b). The slope of the response
curve in this region as predicted by DYCAST is in good agreement with
experiment.

The one dimensional large rotation beam analysis compares well with
the experimental data in the large deflection region up to an end
displacement to length ratio of approximately 0.4. At this point the
analysis predicts a stiffening effect which is not observed in the
experimental data. This effect may be due in part to certain assumptions
made in the analysis including inextensible beam and constant stiffness
assumptions. The inextensible beam does not allow in-plane deformations
due to membrane loads, even though these loads are introduced in the beam
through the testing configuration. Also, the stiffness of the beam is
reduced due to damage events such as transverse matrix cracking which are
not modeled by the analysis.

DISCUSSION

The results presented here indicate that a significant scale effect
exists in the failure behavior since the smaller scale beams fail at a
much higher normalized load and end displacement value than the full scale
beam. Stress and strain based failure criterion such as maximum stress,
maximum strain, Tsai-Hill, or Tsai-Wu, would not be able to predict the
observed scale effect. According to classical scaling laws, stress and
strain should scale as unity. Consequently, under perfectly scaled
experimental test conditions the stress and strain in a model beam will be
the same as for the prototype. Any stress analysis of the scaled test
will predict one value of end displacement to length ratio at which
failure should occur. Morton [1] discusses a linear elastic fracture

mechanics approach to the strength scaling of transversely impacted composite beams and shows that a theory for a notch-sensitive or brittle material can predict scaling effects in a cracked plate. Application of these theories to a stress analysis of the beam-column problem is planned as a continuation of the experimental and analytical results presented here.

CONCLUDING REMARKS

Scaling effects in the large deflection response and failure behavior of graphite-epoxy composite beams was investigated. A series of static tests on scale model composite beams having unidirectional, angle ply, cross ply, and quasi-isotropic laminate stacking sequences was conducted. The beams were loaded under an eccentric axial compressive load to promote large bending deformations and global failure. Plots of normalized load versus end displacement were generated to compare with a one dimensional large rotation composite beam analysis and a DYCAST finite element model.

Results from the experiments show that beam response scales in the small deflection, elastic region; however, deviations from scaled response appear as the beams undergo large deflections and rotations. In general, the small scale beam response is stiffer than the full scale beam response for all laminate types tested. A significant scale effect in strength behavior was observed even though failure modes were consistent between scale model beams and the prototype within the same laminate family. The one dimensional large rotation beam analysis and DYCAST finite element model gave good agreement with the experimental data in the small deflection region. The DYCAST model response was less stiff in the large deflection region than the experiment, but predicted the shape of the response curve well. The beam analysis gave good agreement in the large deflection response region up to a limiting value of end displacement to length ratio of 0.4. At this point a stiffening effect was observed in the analysis which was not observed experimentally.

The results of this study indicate that an important scale effect exists in the modeling of failure behavior of composite structures. Further work is required to identify the micromechanical mechanisms involved in this effect and to understand how they interact on a macroscopic level to produce the observed scale effect in ultimate failure of the structure.

REFERENCES

1. Morton, John., "Scaling of Impact Loaded Carbon Fiber Composites," Proceedings of the 28th SDM Conference, Part 1., Monterey, CA., April 1987, pp. 819-826.

2. Baker, W.E., Westine, P.S., and Dodge, F.T., <u>Similarity Methods in Engineering Dynamics</u>, Hayden Book Co., Rochelle Park, N.J., 1973.

3. Pifko, A. B., Winter, R., and Ogilvie, P., "DYCAST - A Finite Element Program for the Crash Analysis of Structures," NASA Contractor Report 4040, January 1987.

4. Timoshenko, S., and Gere, J., <u>Theory of Elastic Stability</u>, 2nd edition, McGraw-Hill, New York, 1961.

5. Whitney, J., <u>Structural Analysis of Laminated Anisotropic Plates</u>, Technomic Publishing Co., Lancaster, 1987.

Table 1. Scale model beam test specimen dimensions and lay-ups.

SCALE	BEAM DIMENSION	UNIDIRECTIONAL	ANGLE PLY	CROSS PLY	QUASI-ISOTROPIC
1/6	0.5" X 5.0"	$[0]_{8T}$	$[45_2/-45_2]_S$	$[0_2/90_2]_S$	$[-45/0/45/90]_S$
1/4	0.75" X 7.5"	$[0]_{12T}$	$[45_3/-45_3]_S$	$[0_3/90_3]_S$	————
1/3	1.0" X 10.0"	$[0]_{16T}$	$[45_4/-45_4]_S$	$[0_4/90_4]_S$	$[-45_2/0_2/45_2/90_2]_S$
1/2	1.5" X 15.0"	$[0]_{24T}$	$[45_6/-45_6]_S$	$[0_6/90_6]_S$	$[-45_3/0_3/45_3/90_3]_S$
2/3	2.0" X 20.0"	$[0]_{32T}$	$[45_8/-45_8]_S$	$[0_8/90_8]_S$	$[-45_4/0_4/45_4/90_4]_S$
3/4	2.25" X 22.5"	$[0]_{36T}$	$[45_9/-45_9]_S$	$[0_9/90_9]_S$	————
5/6	2.5" X 25.0"	$[0]_{40T}$	$[45_{10}/-45_{10}]_S$	$[0_{10}/90_{10}]_S$	$[-45_5/0_5/-45_5/90_5]_S$
6/6	3.0" X 30.0"	$[0]_{48T}$	$[45_{12}/-45_{12}]_S$	$[0_{12}/90_{10}]_S$	$[-45_6/0_6/-45_6/90_6]_S$

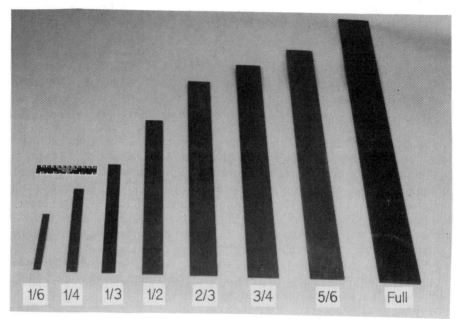

1/6 1/4 1/3 1/2 2/3 3/4 5/6 Full

Figure 1. Photograph of scaled composite beam specimens.

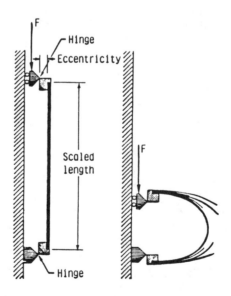

Figure 2. Schematic drawing of the test configuration.

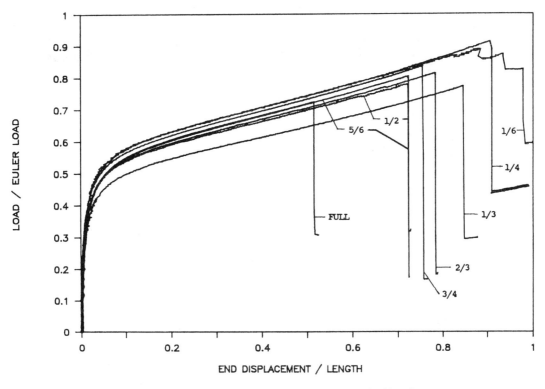

(a) Normalized load versus end displacement.

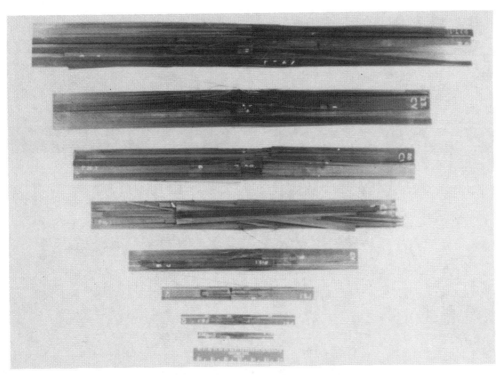

(b) Failed beam specimens.

Figure 3. Unidirectional graphite-epoxy composite beam results.

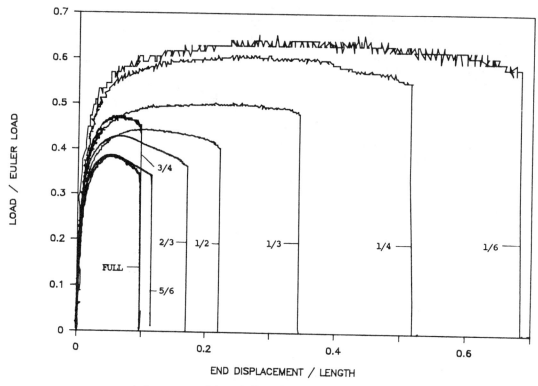

(a) Normalized load versus end displacement.

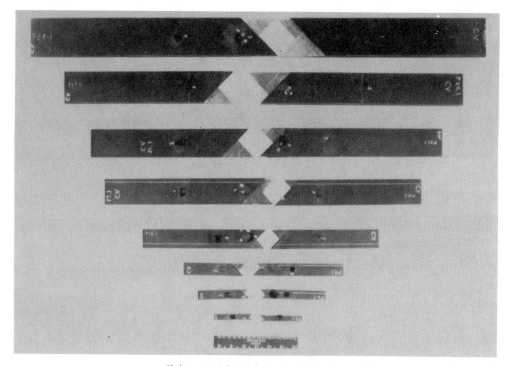

(b) Failed beam specimens.

Figure 4. Angle ply graphite-epoxy composite beam results.

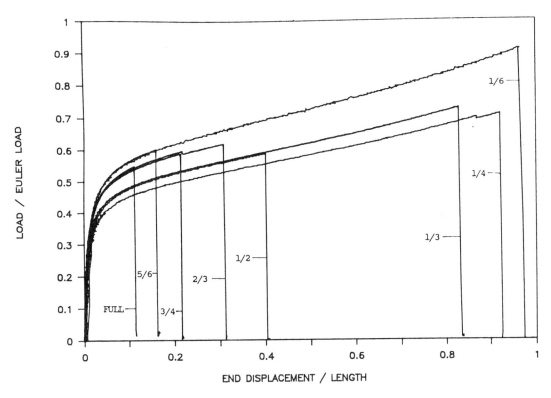

(a) Normalized load versus end displacement.

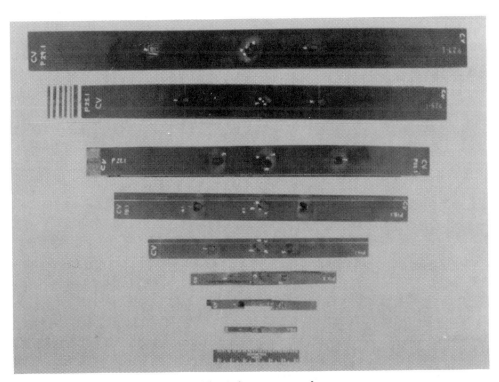

(b) Failed beam specimens.

Figure 5. Cross ply graphite-epoxy composite beam results.

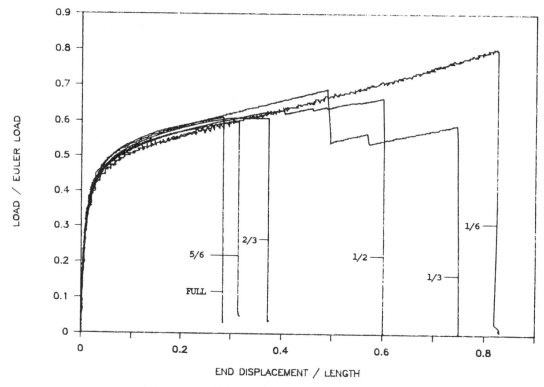

(a) Normalized load versus end displacement.

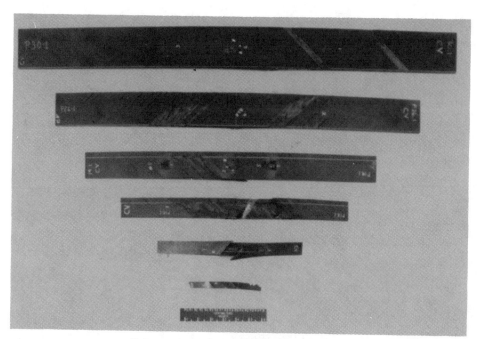

(b) Failed beam specimens.

Figure 6. Quasi-isotropic graphite-epoxy composite beam results.

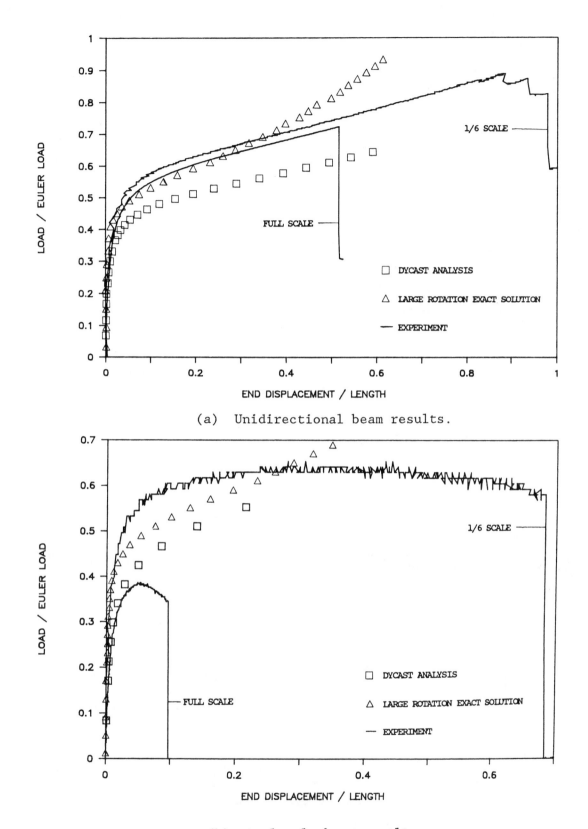

(a) Unidirectional beam results.

(b) Angle ply beam results.

Figure 7. Comparison of normalized load versus end displacement
experimental data for 1/6 and full scale beams with DYCAST finite element
analysis and large rotation exact solution.

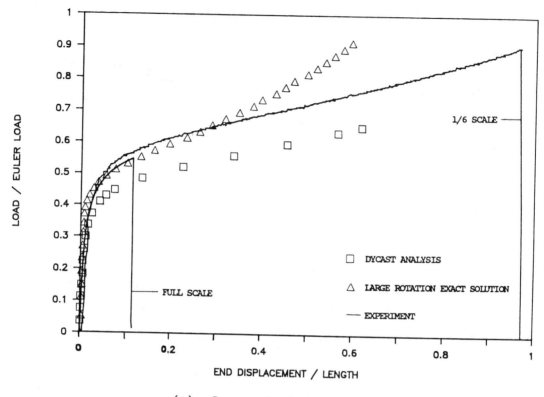

(c) Cross ply beam results.

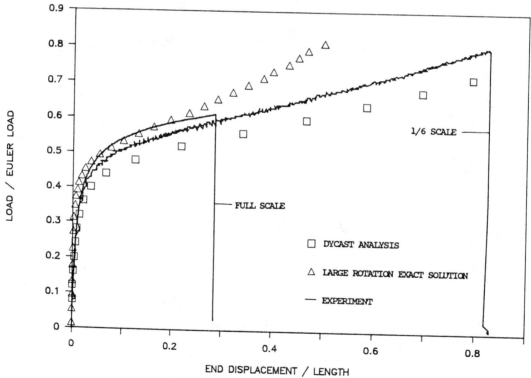

(d) Quasi-isotropic beam results.

Figure 7. Comparison of normalized load versus end displacement
experimental data for 1/6 and full scale beams with DYCAST finite element
analysis and large rotation exact solution.

SYMPOSIUM XIV

Fatigue and Fracture

Dependence of Incipient Edge Delamination Strength on Neat Resin Fracture Toughness

ALAN R. WEDGEWOOD AND **MARILYN W. WARDLE**

ABSTRACT

The incipient edge delamination strength of laminates based on Avimid*K polyimide prepregs was shown to increase with the fracture toughness of the K-polymer matrix resin. The investigation covered a wide range of resin fracture toughness, G_{IC}, with values ranging from 0.3-1.7 KJ/m^2. Both visual and acoustic emission were used to detect the onset of edge delamination. A description of the acoustic emission detection procedure, which gave improved data reproducibility, is provided.

INTRODUCTION

When examined on a specific weight basis, continuous fiber reinforced composite laminates offer desirable in-plane properties. In contrast, these composites usually perform poorly when subjected to out-of-plane loads, under which interlaminar cracks easily initiate and may cause delamination failure of the structure. To address these concerns, high performance thermoplastic matrix composites with good inherent toughness have been developed. Some of these material systems are: Avimid*K and Avimid*N [1-3], a polyimide composite and composites resin; a polyetherketoneketone (PEKK) polymer [4] and polyamide J1 (semi-crystalline) and J2 (amorphous) polymers [5,6].

Alan R. Wedgewood, Senior Research Engineer, and Marilyn W. Wardle, Research Associate, E.I. du Pont de Nemours & Co., Inc., Fibers and Composites Development Center, Wilmington, Delaware 19898
* Du Pont Registered Trademark

Previous investigations have shown that the good toughness characteristics of these thermoplastic resins, represented by their high mode I plane-strain fracture toughness, is directly translated into their respective composites [7]. In this study, the relationship between the plane-strain fracture toughness and the incipient edge delamination strength was established by modifying the chemistry of the polyimide to change the G_{IC} This investigation covered a wide range of resin fracture toughnesses, G_{IC}, with values ranging from 0.3-1.7 KJ/m^2. In addition, an acoustic emission method was developed to provide a more reliable procedure for detecting the onset of delamination than the more common visual observation method.

POLYMER CHEMISTRY

The polymer base for Avimid®K is an amorphous linear condensation polyimide produced from monomeric solutions by reacting an aromatic diethylester diacid with a proprietary aromatic diamine in N-methyl pyrrolidone (NMP) solvent. The reaction proceeds with the loss of water and ethanol to form the imide ring. These resins offer an excellent balance in mechanical, hot-wet service, environmental resistance and damage tolerance properties [1,2]. A series of experimental resins based on this chemistry with a wide range of toughness values were produced for this study. This was achieved by adjusting the molecular weight distribution of the polymer. Except for properties that are affected by molecular weight distribution, all of these resins can be viewed as structurally similar polymers derived from identical polyimide chemistry.

NEAT RESIN FRACTURE TOUGHNESS

Three point bending of single-edge-notched (SEN) specimens was used to determine the mode I fracture toughness, G_{IC}. Care was taken in measuring the neat resin fracture toughness to ensure that the plane-strain test conditions were satisfied. The importance of using plane-strain fracture toughness values when making correlations with composite properties has been illustrated in an earlier study. Details of this study and SEN fracture toughness test procedure are provided elsewhere [7].

LAMINATE FABRICATION AND QUALITY ASSURANCE

The test laminates were prepared from tacky, drapeable, wet prepregs of Avimid® K, which are suitable for processing with currently available tape lay-down and vacuum bag autoclave molding technologies. The prepreg is a combination of a monomeric (polyimide precursor) solution and a continuous reinforcing fiber. The laminates referred to in this paper were prepared from 12 inch wide Magnamite® AS4 or IM6 carbon fiber based prepregs, using an autoclave molding process [1,8]. The volatile content of the prepreg, which includes both the condensation polymerization by-products and solvent was 15-17 weight percent. The final laminates contained 57 +/-2 volume percent reinforcing fiber.

The quality of the laminates were assessed by microscopy and ultrasonic C-scan to determine void content and distribution. The total void volume was also estimated by comparing Archimedes density measurements with the calculated theoretical laminate density [1]. All the laminates used in this work were found to have finely dispersed voids, equal to or less than 0.5% the total volume.

EDGE DELAMINATION CHARACTERIZATION

Incipient edge delamination strength testing provides a technique for qualitatively assessing a material's interlaminar toughness. This results because the unbalanced $[(+/-25)_2,90]_s$ laminates tested are designed to yield high interlaminar normal stresses when tested in tension [9]. The fabricated test laminates, which were approximately 0.147 cm (0.058") thick, were cut into 2.54 cm x 30.48 cm (1" x 12") test coupons with the 0° fiber direction defined parallel to the specimen's long axis. The edge machining was done using a grinder/ slicer (Harig 618) and a diamond wheel saw blade (Norton 690478). The edges were kept parallel within 0.0127 cm (0.005 inches) and visually examined to check that the surface was free of machining defects. All specimens were tested as machined.

The specimen was tested in tension in the long axis direction and the load at which a crack first forms along either of the specimens' edges was recorded. The sample was loaded at a cross-head speed of 0.127 cm/min (0.05 inch/min). The edge delamination strength (EDS) was defined as this onset load divided by the specimens crossectional area. Both visual observation and acoustic emission (AE) were used to detect the onset of edge delamination (crack formation). A minimum of ten specimens were tested from each laminate.

To aide visual detection the specimen edge was painted with white typist's correction fluid. During loading two operators are required to observe both edges of the sample simultaneously. Although the cracks usually first appear in the center region of the sample, the operators had to continuously scan the sample edge to ensure detection of the first crack.

Detection of the onset was also done using acoustic emission. A single channel AE detection system (Denegon-Endevco Model 702) was used with a Dunegan Model D-750 B transducer attached to the face of the specimen with rubber bands. Although this equipment is obsolete, compared with today's state of the art in AE, it proved to be quite effective for the purpose, giving a reliable indication of crack initiation.

The analogue output of the AE system, which is proportional to the total AE counts, was feed to the computer data acquisition system (Measurements Technology, Inc. 16/64) which was being used to record the load and displacement. A schematic plot of load versus total acoustic counts in real time is shown as Figure 1. At the onset of delamination a significant increase in acoustic counts occurs. Extrapolation back to the vertical axis defines the load at which cracking initiates. As compared with the visual observation, the AE method gave essentially the same results with less data scatter. In addition, the use of AE eliminates the need for a second operator to observe for crack initiation.

CORRELATION OF RESIN AND COMPOSITE TOUGHNESS

Figure 2 shows the edge delamination strength (EDS) of the tested composite laminates plotted against the mode I plane-strain fracture toughness of the neat matrix resins. As shown, the typical EDS value for composites based on standard Avimid®K prepreg was 290 +/- 23 MPa (42.0 +/- 3.4 Ksi). The measured EDS, which gives a qualitative measure of the composites' interlaminar toughness, was observed to increase as the resin toughness was increased. It is apparent from this result that the mode I plane-strain fracture toughness of these resins is directly translated into the interlaminar toughness of their corresponding composites.

A similar observation was made when the mode I interlaminar fracture toughness of a series of thermoplastic composite laminates was compared against the mode I plane-strain fracture toughness of their matrix resins [7]. These results showed that as the resin toughness increased, the interlaminar fracture toughness approached the plane-strain fracture toughness of the neat resin. However, for lower toughness resins (on the order of less than 1 KJ/m^2), the interlaminar fracture toughness was in many cases significantly greater than the matrix resin toughness. This difference was attributed to the more complex fracture surface topography in the composite fracture specimens and the availability of other energy mechanisms in the composite, such as fiber bridging and pullout, not available in the neat resin specimens. The above observations indicate that the contributions of these other energy mechanisms can be significant for low toughness composite systems. Although the extent to which these other energy mechanisms contribute to the measured EDS cannot be easily determined, their contribution in the case of low toughness composite systems may also be significant.

CONCLUSIONS

High performance thermoplastic composites derived from Avimid®K prepreg have been shown to have excellent toughness properties. The interlaminar fracture toughness of these composites, as reflected by their incipient edge delamination strength, is directly related to the mode I plane-strain fracture toughness of the matrix resin. An acoustic emission procedure for detecting the onset of the edge delamination was shown to give improved data reproducibility when compared to the more common visual observation method.

ACKNOWLEDGEMENTS

The following Du Pont employees made notable contributions to this work by either preparing samples or assisting in their testing. Special gratitude is owed D. C. Grant, M. D. Neal and A. E. Wilkins.

REFERENCES

1. R. J. Boyce, T. P. Gannett, H. H. Gibbs and A. R. Wedgewood, "Processing, Properties and Applications of K-Polymer Composite Materials Based on Avimid®K-III Prepregs", 31st Annual SAMPE Symposium and Exhibition, Anaheim, California, April 6-9, 1987, pp. 169-184.

2. H. H. Gibbs, "The Processing and Properties of Damage Tolerant Composites Based on K-Polymer Materials", Fifth Annual Conference on Composite Materials (ICCM-V), San Deigo, California, July 29-August 1, 1985, pp. 971-993.

3. D. A. Scola, "Thermo-oxidative Stability of Recent Graphite Fiber/HT Resin Composites", HI Temple Workshop VII, Sacramento, California, January 26-30, 1987.

4. I. Y. Chang, "PEKK as a New Thermoplastic Matrix for High Performance Composites", 33rd International SAMPE Symposium, Anaheim, California, March 7-10, 1988, pp. 194-205.

5. I. Y. Chang, " Thermoplastic Matrix Continuous Filament Composites of Kevlar® Aramid or Graphite Fiber", Composite Science and Technology, Vol. 24, 1985, pp.61-79.

6. I. Y. Chang, " Thermoplastic Matrix Composites", Composites '86: Recent Advances in Japan and the United States, Proceedings of the Third Japan-U.S. Conference on Composite Materials, June 1986, Tokyo, Japan.

7. A. R. Wedgewood, K. B. Su and J. A. Nairn, " Toughness Properties and Service Performance of High Temperature Thermoplastics and Their Composites", 19th International SAMPE Technical Conference, Crystal City, Virginia, October 13-15, 1987, pp. 454-467.

8. A.R. Wedgewood, "Autoclave Processing of Condensation Polyimide Composites Based on Prepregs of Avimid®K-III", 19th International SAMPE Technical Conference, Crystal City, Virginia, October 13-15, 1987, pp. 420-434.

9. R. Byron Pipes, "Boundary Layer Effects in Composite Laminates", Fiber Science and Technology, Vol. 13, 1980, pp. 49-71.

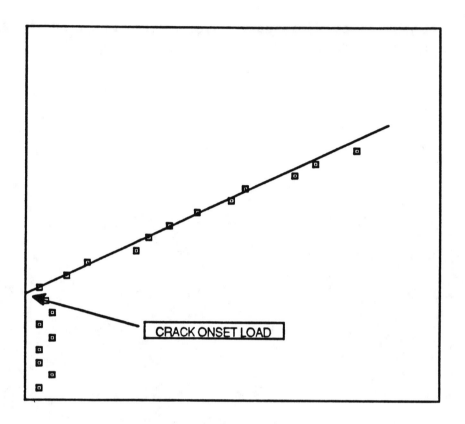

Figure 1. Schematic of load versus total acoustic counts in real
time. Extrapolation to determine onset load is
illustrated by line.

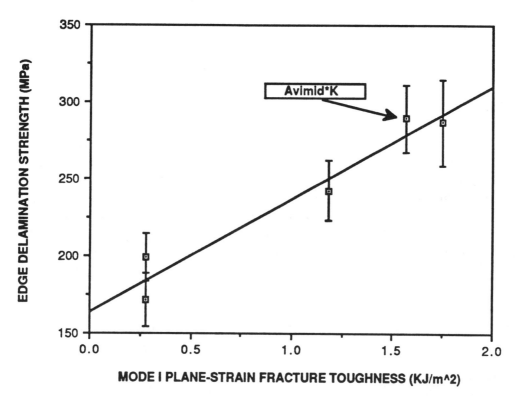

Figure 2. Comparison of Incipient Edge Delamination Strength and Neat
          Resin Fracture Toughness for K-Polymer Based Composites
 * Du Pont Registered Trademark

Predicted and Observed Effects of Stacking Sequence and Delamination Size on Instability-Related Delamination Growth

JOHN D. WHITCOMB*

ABSTRACT

Compressive loads can cause local buckling in composite laminates that have a near-surface delamination. This local buckling causes load redistribution, which in turn causes interlaminar stresses and delamination growth. In this study, a three-dimensional, geometrically nonlinear finite element analysis (NONLIN3D) was used to calculate strain-energy release rate distributions for several laminates which exhibited this instability-related delamination growth. Two stacking sequences ($(0/90/90/0)_6$ and $(90/0/0/90)_6$) were considered. The specimens were fabricated with a double layer of .013mm Kapton film located between the fourth and fifth plies, thereby simulating a delamination. Initial delamination sizes were 30, 40, or 60mm in diameter. The static strain corresponding to delamination growth was determined experimentally. Comparison of the analytical and experimental results indicated that delamination growth was governed by the magnitude of G_I.

INTRODUCTION

Delaminations reduce the strength of composites loaded in compression. If the loads are sufficiently large, a delaminated group of plies will buckle. The postbuckling deformation causes load redistribution, which can lead to delamination growth (Herein, referred to as instability-related delamination growth). Because this growth may be rapid and very extensive, it is valuable to be able to predict when a delamination will begin to grow.

This paper will present the results of a combined experimental and analytical investigation of the effects of stacking sequence and initial delamination size on instability-related delamination growth. The configuration consisted of a 24-ply laminate with a single delamination located between the fourth and fifth plies. A schematic of this configuration is shown in fig. 1. The material system was IM7/8551-7, which is a toughened graphite/epoxy composite system manufactured by Hercules. Two stacking sequences and three delamination sizes were considered.

Strain-energy release rate analysis was performed for all six stacking sequence and size combinations. A geometrically nonlinear, three-dimensional finite element program,

---

* J. D. Whitcomb, Research Engineer, Fatigue and Fracture Branch, NASA Langley Research Center, Hampton, Va. 23665

NONLIN3D (ref. 1), was used to calculate the distributions of the components of strain-energy release rate.

Before discussing the results of this study, the analysis and experimental procedure will be presented.

ANALYSIS

This section will give a brief description of the finite element program used to obtain the strain-energy release rates. The finite element program, which is named NONLIN3D, was developed at NASA Langley Research Center. Further details may be found in ref. 1 Also, a typical finite element model and the material properties will be described.

NONLIN3D is a three-dimensional, geometrically nonlinear finite element program. The primary type of geometric nonlinearity considered is that due to significant rotations. This is accounted for by using the Lagrangian nonlinear strain-displacement relations (ref. 2). The governing equations were derived by minimization of the total potential energy. A Newton-Raphson iterative procedure was used to solve the nonlinear governing equations.

Figure 2 shows a typical finite element model. There are 3549 nodes and 648 elements. The elements are 20-node isoparametric hexahedrons. Because of symmetry only one-fourth of the specimen was modeled and the constraints $u = 0$ on $x = 0$ and $v = 0$ on $y = 0$ were imposed. There was also a constraint $w = 0$ on $z = 0$. This constraint was imposed to prevent global bending. Along the boundary $x = W$, all u displacements are specified to equal $x\varepsilon_o$, where ε_o is the specified axial strain.

Strain-energy release rates were the primary output from the analysis. The strain-energy release rates were calculated using a three-dimensional version of the virtual crack closure technique in ref. 3. Details of the strain-energy release rate calculation are given in ref. 1. The strain-energy release rates were calculated relative to a local coordinate system. The local coordinate system is shown in fig. 3. One axis is normal to the delamination front and one axis is tangent. For all the cases considered, the local z-axis was parallel to the global z-axis. Fig. 3 also defines the perimeter coordinate "S", which is the distance along the delamination front measured from the y-axis.

The properties for the IM7/8551-7 were based on unpublished data generated by the University of Wyoming under NASA grant NAG1-674, which began in July, 1986. This grant determined the inplane properties, E_{11}, E_{22}, G_{12}, and ν_{12}. Because of a lack of other data, the remaining 3-D properties were assumed. The unidirectional material properties used are

$E_{11} = 162$ GPa ; $\quad E_{22} = E_{33} = 8.14$ GPa
$G_{12} = G_{13} = G_{23} = 6.48$ GPa
$\nu_{12} = \nu_{13} = \nu_{23} = .22$

EXPERIMENTAL PROCEDURE

Twenty-four ply specimens were fabricated from IM7/8551-7, which is a toughened graphite/epoxy material system manufactured by Hercules. Two stacking sequences were used: $(0/90/90/0)_6$ and $(90/0/0/90)_6$. The specimens were fabricated with a double layer of .013mm Kapton film located between the fourth and fifth plies, thereby simulating a delamination. Three implant diameters were used: 30, 40, and 60mm. The width of the specimens was 102mm.

The specimens were loaded in uniaxial compression until delamination extension occurred, which caused a change in the boundaries of the buckled region. This change was monitored with the naked eye. To prevent global buckling, steel guide plates lined with a layer of Teflon were used. These are shown in fig. 4. The plates are 19mm thick and window is 82mm in diameter.

RESULTS and DISCUSSION

This section will discuss typical strain-energy release rate distributions, location of delamination growth, strain and strain-energy release rates corresponding to delamination growth, and damage other than delamination which accompanied the delamination growth.

Figs. 5 and 6 show typical strain energy release rate distributions along the delamination front for the two stacking sequences. These particular results are for a delamination diameter of 30mm. The trends were the same for the other delamination sizes. The figures do not show G_{III} because it was always much smaller than G_I and G_{II}. There is a large variation of G_I, G_{II}, and G_T along the delamination front. For the same strain, the G's are larger for the $(90/0/0/90)_6$ laminate. Also, the gradient in G_I is much larger for the $(90/0/0/90)_6$ laminate. In figure 6, part of the curves are dashed for G_I. The dashed region indicates the portion of the delamination front where the faces overlapped (analytically). To prevent the overlapping would have required application of nonlinear contact constraints. However, including constraints to prevent overlapping further complicates an already complicated stress analysis problem. Based on results in ref. 4, the maximum magnitudes of G_I and G_{II} are not affected much by not including the contact constraints. Of course, in the overlap region the G_I is zero.

Also shown in figs. 5 and 6 is the variation of maximum G_I, G_{II}, and G_T versus strain. The strain-energy release rates are zero until bifurcation buckling occurs. For example, in fig. 5, the bifurcation buckling strain is a little less than -.003. The vertical line labeled "critical strain" indicates the strain level at which delamination growth initiated. These results show that delamination growth did not occur immediately after bifurcation buckling.

Based on analytical results like those in figs. 5 and 6, the delamination growth was expected to occur preferentially near $S = 0$, i.e., perpendicular to the load direction. Also, delamination growth should occur over only a small portion of the delamination front. Radiography was used to measure the extent of delamination growth. The sketch in fig. 7 of a typically observed growth pattern shows that the analysis predicted the correct trend.

Fig. 8 shows the strain at which delamination growth began for the six configurations. Two specimens were tested for each configuration. The critical strain decreases with increased delamination diameter. The critical strains are significantly larger for the $(0/90/90/0)_6$ laminate than they are for the $(90/0/0/90)_6$ laminate.

Fig. 9 shows calculated G_I and G_{II} corresponding to the strains at which delamination occurred. The magnitude of G_I corresponding to delamination growth is about the same for five of the six cases. For the 30mm delamination with a $(0/90/90/0)_6$ stacking sequence, the G_I was quite low. This was probably because a very high compressive strain would have been required to cause a large G_I. When the compressive strains are high, other failure mechanisms (such as fiber microbuckling and matrix cracking) can augment the effect of the interlaminar stresses caused by the postbuckling. The critical magnitude of G_I based on the data in fig. 9 is not much different than the double-cantilevered beam measurements of 392-513 J/m^2 reported by Hercules (ref. 5). The magnitude of G_{II} varies widely. Based on the results in fig. 9, it appears that delamination growth is controlled by G_I.

Microscopic and X-ray examination of several specimens revealed that the delamination growth was accompanied by other damage mechanisms. There was significant ply cracking. Also, at the higher strains, there was a tendency for microbuckling to occur on the inner surface of the postbuckled region. Fig. 10 shows a typical micrograph. Also, when delamination growth occurred in the $(0/90/90/0)_6$ laminate, the delamination tended to switch interfaces from the original location between the fourth and fifth plies to the interface between the third and fourth plies. These results indicate that even when a configuration starts out fairly simple (in this case a single delamination locally separating the laminate into two balanced and symmetric laminates), the damage development tends to be complicated.

CONCLUSIONS

A combined analytical and experimental study of instability-related delamination growth was performed. The analysis was performed using a three-dimensional, geometrically nonlinear, finite element program (NONLIN3D). The specimens were fabricated using IM7/8551-7, which is a toughened graphite/epoxy material system. Kapton implants were used to simulate initial delaminations. Two stacking sequences ($(0/90/90/0)_6$ and $(90/0/0/90)_6$)) and three delamination sizes (30, 40, and 60mm diameter) were examined. The main results from this study are

1. Delamination growth did not occur until the bifurcation buckling was significantly exceeded.

2. The compressive strains for buckling and delamination growth were smaller for the $(90/0/0/90)_6$ laminate, for the same delamination size.

3. For each stacking sequence, the compressive strain for delamination growth decreased with increased delamination size.

4. Delamination growth was transverse to the load direction and occurred over only a small portion of the delamination front. This behavior was predicted by the analysis.

5. Delamination growth appeared to be controlled by the magnitude of G_I.

6. Ply cracking occurred in the postbuckled region. Sometimes there was also fiber microbuckling.

REFERENCES

1. Whitcomb, J. D. : Three-Dimensional Analysis of a Postbuckled Embedded Delamination. NASA TP 2823, 1988.

2. Frederick, D. and Chang, T. S.: Continuum Mechanics. Scientific Publishers, Inc., Cambridge, 1972, pp. 79-82.

3. Rybicki, E. F. and Kanninen, M. F.: A Finite Element Calculation of Stress Intensity Factors by a Modified Crack Closure Integral. Engineering Fracture Mechanics, 1977, Vol. 9, pp. 931-938.

4. Whitcomb, J. D. : Instability-Related Delamination Growth of Embedded and Edge Delaminations. Ph.D. Thesis, Virginia Polytechnic Institute and State University, May, 1988.

5. U. S./European Aircraft IM7/8551-7 Qualification Test Results. Hercules Aerospace Products Group publication, 1987.

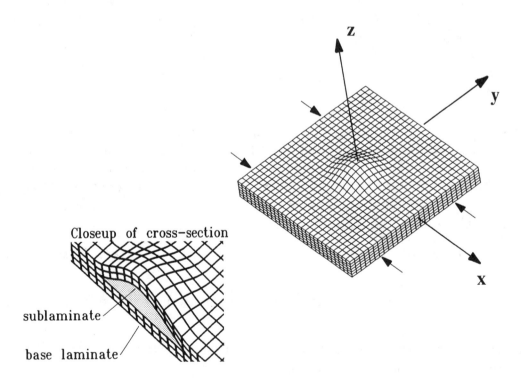

Fig. 1 Schematic of specimen configuration.

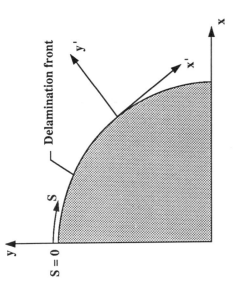

Fig. 3 Global and local coordinate systems and perimeter coordinate S.

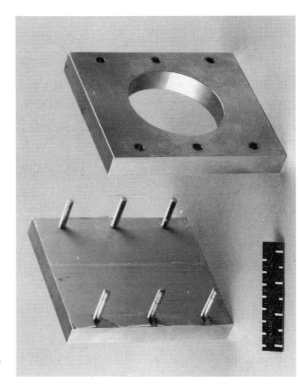

Fig. 4 Guide plates.

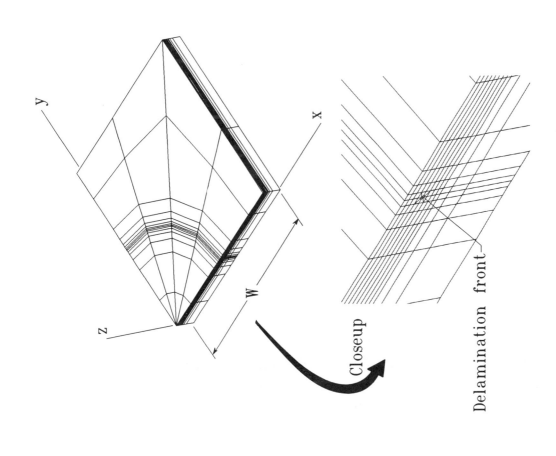

Fig. 2 Typical finite element model.

683

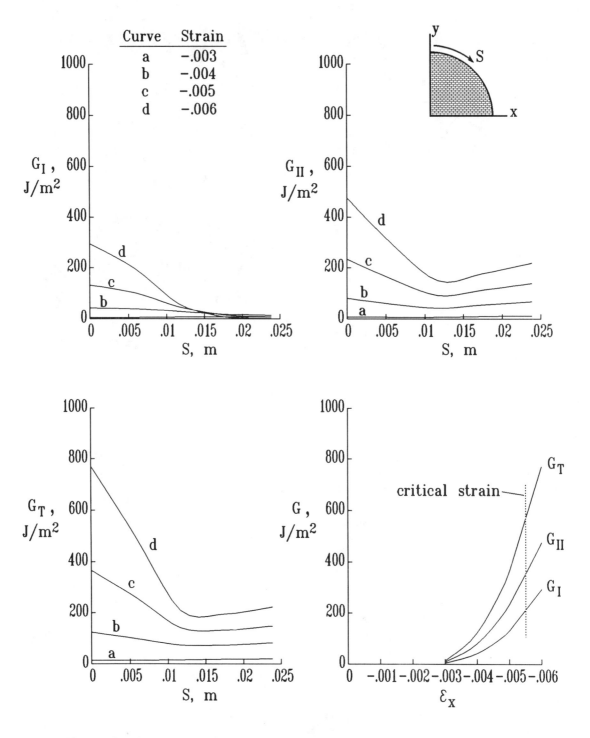

Fig. 5 Strain-energy release rate distributions and maximum strain-energy release rate vs. strain. The sublaminate is 30mm in diameter and has a stacking sequence of (0/90/90/0).

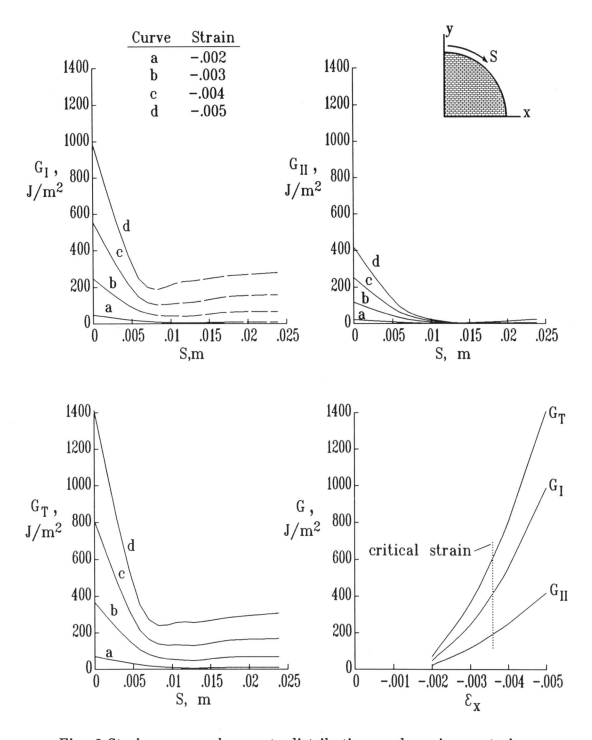

Fig. 6 Strain-energy release rate distributions and maximum strain-energy release rate vs. strain. The sublaminate is 30mm in diameter and has a stacking sequence of (90/0/0/90).

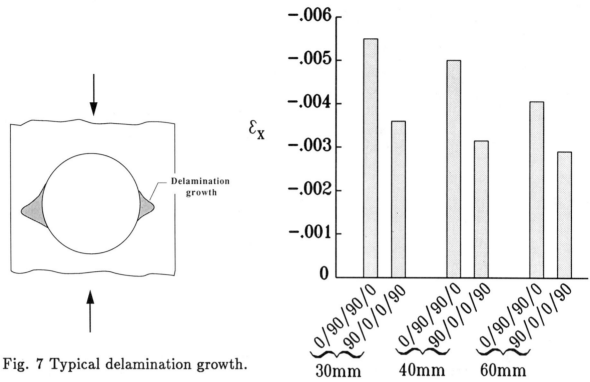

Fig. 7 Typical delamination growth.

Fig. 8 Strain at which delamination growth began for six configurations.

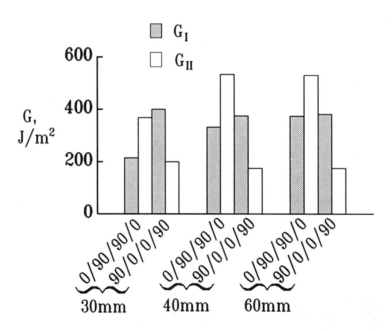

Fig. 9 Mode I and Mode II strain-energy release rates for delamination growth. Results are shown for two stacking sequences and three delamination sizes.

Sublaminate

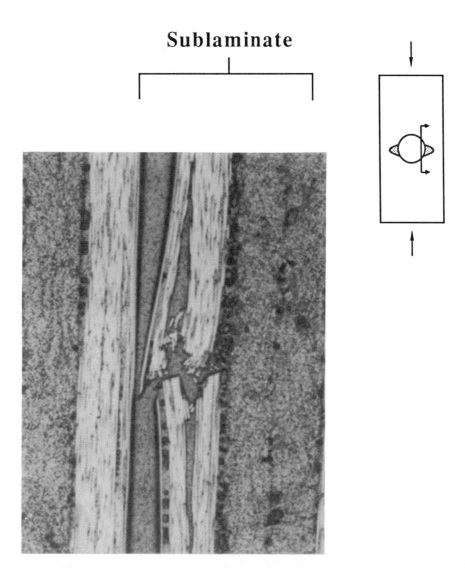

Fig. 10 Cross-section of laminate with a (0/90/90/0) sublaminate and an initial delamination diameter of 40 mm.

Effect of Initial Delamination on G_{Ic} and G_{Ith} Values from Glass/Epoxy Double Cantilever Beam Tests

RODERICK H. MARTIN

ABSTRACT

The effect of insert thickness and method of pre-cracking on mode I interlaminar fracture toughness, G_{Ic}, and delamination fatigue threshold, G_{Ith}, values obtained from a glass/epoxy double cantilever beam specimen were determined. The results of the static tests showed that pre-cracking in tension would cause fiber bridging and thus may yield unconservative values of G_{Ic} and G_{Ith}. Pre-cracking in shear yields suitable values of G_{Ic} but overly conservative values of G_{Ith}. For the glass/epoxy composite used, an insert thickness of 0.5 mil was most suitable for determining G_{Ic} and G_{Ith} values although an insert thickness up to 3 mil was acceptable. Inserts thicker than 3 mil were not acceptable for determining G_{Ic} and G_{Ith} values.

INTRODUCTION

A common failure mechanism in laminated composite structures under static and cyclic loads is delamination. The double cantilever beam specimen, DCB, fig.1, has been used to characterize mode I delamination caused by tensile loads perpendicular to the delamination surface [1-4]. The DCB specimen is a unidirectional laminate with a midplane insert at one end to simulate a delamination. The tensile loads are applied through hinge tabs bonded to the end of the specimen. The critical load at delamination onset from the insert is used to determine the fracture toughness, G_{Ic}, in the DCB test. Although several values of G_{Ic} may be obtained for longer delamination lengths, the DCB specimen experiences "fiber bridging", where fibers join the delaminated surfaces, thus increasing the apparent strain energy release rate, G_I, required to extend the delamination down the length of the beam [5,6]. However, in structures a delamination invariably occurs between plies of different orientation [7] and fiber bridging may not occur. At the first point of delamination growth from the simulated delamination in a DCB specimen, there is no fiber bridging, and hence this

---

R. H. Martin, National Research Council Research Associate, NASA Langley Research Center, Hampton VA 23665-5225

point may yield a valid characterization of delamination in a structure. However, a resin pocket may form in front of a thick insert. The size of this resin pocket is dependent on the thickness of the insert. Other factors that may effect the size of the resin pocket are fiber stiffness and viscosity of the resin in its liquid state at cure temperature. This resin pocket increases the static fracture toughness measured using an end-notched flexure, ENF, specimen [8] and an edge delamination test, EDT, specimen [9] when measured from the end of the insert. This increase in fracture toughness is analogous to the increase in fracture toughness with bondline thickness in an adhesively bonded DCB test specimen [10]. Therefore, the maximum acceptable insert thickness which will yield conservative values of mode I fracture toughness from a DCB specimen must be found.

Under cyclic loads the rate of delamination growth per cycle has been related to the cyclic strain energy release rate using a power law of the form $da/dN = A \, G^B$ [11-14]. However, for composite materials the exponent B in these power laws is often large. Thus, small uncertainties in the anticipated design loads may lead to large uncertainties in the predicted delamination growth rate. Hence, for composite materials emphasis must be placed on the threshold strain energy release rate, G_{th} [7,14,15,16]. Structural components may be designed for loads below this threshold level to prevent any delaminations from growing, or starting to grow, from a discontinuity.

A value of mode I threshold strain energy release rate, G_{Ith}, may be obtained from cyclic delamination growth rate tests [14,17,18] where the applied cyclic strain energy release rate G_{Imax} is reduced so that delamination arrest occurs and the delamination ceases to grow. However, this method may yield very unconservative values of G_{Ith} because fiber bridging may occur as the delamination extends along the beam. Also, there will be load history effects, such as crack closure and surface roughness, helping to delay the opening of the delamination during the fatigue cycle [14]. A more appropriate value of G_{Ith} for no delamination growth may be obtained from plots of the maximum cyclic strain energy release rate, G_{Imax}, versus the number of cycles to delamination growth onset [14]. The applied G_{Imax} value at which delamination growth onset occurred after 10^6 cycles may be considered as G_{Ith} [14,16,19]. However, obtaining G_{Ith} by this method may be largely dependent on the size and extent of the damage zone ahead of the delamination front caused by forming the simulated delamination. This damage zone can be very large if, for example, the delamination was formed in shear [20]. In many situations a fracture mechanics approach may be used to predict delamination growth onset in structures that do not have a current delamination, but have a discontinuity which acts as a delamination initiator. These discontinuities can take the form of an edge [7,9], a ply drop in a tapered laminate [21] or more commonly a simple crack in the matrix of an off-axis ply [17,22]. Therefore, an overly conservative value of strain energy release rate threshold may be obtained from a DCB specimen that had been statically pre-cracked in shear and had a damage zone ahead of the delamination front. If the specimens were pre-cracked in tension then fiber bridging would occur also yielding unconservative values of G_{Ith}. However, testing to delamination growth onset from the end of the insert, which has no damage zone in the undelaminated part of the specimen, would yield an accurate value of G_{Ith} provided the effect of the resin pocket is minimal.

This paper documents the influence of the insert thickness on G_{Ic} and G_{Ith} for a glass/epoxy composite. This was achieved by testing to delamination growth onset from the end of the insert in DCB specimens using four different thicknesses of insert. A comparison was made of the G_{Ic} and G_{Ith} values obtained from specimens with different insert thicknesses and also G_{Ic} and G_{Ith} values obtained from specimens that had been statically pre-cracked in shear. Based on the results, a range of suitable insert thicknesses was suggested for accurately determining G_{Ic} and G_{Ith} for this material.

MATERIALS

Unidirectional, 24 ply, glass/epoxy (S2/SP250) panels with a midplane insert at one end were manufactured at NASA Langley Research Center according to the manufacturer's instructions. The average volume fraction of the specimens was 64.6 percent determined following ASTM procedure D-3171. The inserts used in the panels were different thicknesses of Kapton film. The Kapton films used were 0.5 mil (0.0127mm), 1 mil (0.0254mm), 3 mil (0.0763mm) and 5 mil (0.127mm) thick, where 1 mil =0.001 in. Single layers of Kapton were used in each panel except one in which two layers of 5 mil Kapton were used in an attempt to create a total insert thickness of 10 mil. To prevent adhesion of the Kapton to the epoxy, the Kapton was sprayed with a release agent and heated for fifteen minutes at 70°C prior to lay-up of the prepreg. After the panels had been cured they were cut into 152 by 25mm DCB specimens. Aluminum alloy hinges were bonded to the test coupon using a two part room temperature cure adhesive. The specimens were dried according to a General Dynamics drying procedure (specification FPS-2003A). The procedure consisted of heating the specimens for one hour at 95°C, one hour at 110°C, sixteen hours at 120°C and one hour at 150°C. This drying cycle also served as a post cure for the room temperature adhesive used to bond the hinges. The specimens were then allowed to cool and stored in a desiccator prior to testing.

TEST PROCEDURE

Static Testing

Static tests were conducted on three specimens of each different insert thickness. The tests were conducted in displacement control at a cross-head rate of 0.5mm/min. A load-displacement plot was obtained during the test. The specimens were loaded until the delamination had extended approximately 12mm from the end of the insert and were then unloaded. The onset of static delamination growth was clearly indicated by a deflection in the load-displacement plot. The values of load and displacement at this deflection were used to determine the static fracture toughness, G_{Ic} from the end of the insert.

Pre-cracking in Shear

Static shear pre-cracks were initiated in the DCB specimens with a 0.5 mil insert thickness by positioning them in a three point bend loading fixture, described in reference [14], so that the end of the insert was approximately 12mm away from the center loading roller. Then, the specimens were loaded under displacement control at a cross-head rate of 2.54 mm/min, until the delamination extended to a point under the loading roller. These specimens were used to determine a value of G_{Ic} and G_{Ith} from a static shear pre-crack.

Fatigue Delamination Growth Onset Determination

The fatigue tests were run in displacement control at a displacement ratio of R=0.5 and a frequency of 10Hz. Delamination growth onset was observed using an optical microscope of magnification 60X and the number of cycles was recorded. Several specimens were tested at different maximum cyclic displacements to obtain a G_{Imax} versus N curve for each insert thickness and with a shear pre-crack, where G_{Imax} is the maximum cyclic strain energy release rate and N is the number of cycles to delamination growth onset. The value of G_{Imax} where no delamination growth had occurred up to 10^6 cycles was considered as the threshold strain energy release rate, G_{Ith}

RESULTS AND DISCUSSION

Micrographs of the Resin Pocket

Figures 2 through 5 show the resin pockets for the different thicknesses of insert used. A resin pocket is not obvious with an insert thickness of 0.5 mil but becomes apparent at insert thicknesses of 1 mil and thicker. An attempt was made to fold a 5 mil insert to form a 10 mil insert. Fig.6 shows that the fold was not flat and the actual thickness was greater than 10 mil. Films less than 5 mil thick may fold flatter than the 5 mil film shown in fig. 6. Also, other techniques such as creasing the film with a warm soldering iron may produce a flatter insert. However, if a folded insert is used then careful monitoring of the insert thickness at the fold is necessary. No static or fatigue tests were run on the specimens with folded inserts.

Static Testing

Figure 7 shows the plot of fracture toughness, G_{Ic}, versus delamination length for a specimen with a 3 mil thick insert. The large increase in G_{Ic} was typical for any insert thickness used and was caused by fibers bridging the delaminated surfaces of the DCB. Figure 8 shows a plot of G_{Ic} versus insert thickness where G_{Ic} is the initial value when the delamination starts to grow from the end of the insert. The mean of three specimens is shown along with the scatter band. In all the tests, except with a 5 mil insert, the delamination grew stably from initiation through the resin pocket. With a 5 mil insert the delamination grew unstably through the resin pocket, fig. 9, at a G_{Ic} approximately 30 percent higher than that of the thinner inserts. The delamination then grew stably through the remainder of the beam's length. Fracture toughnesses measured using specimens with shear pre-cracks and with insert thicknesses of 0.5 mil, 1 mil and 3 mil were very similar. The results shown in this work are consistent with those of reference 10 on bond thickness effects of an epoxy resin on mode I fracture toughness, where an increase in fracture toughness was noted at bond thicknesses above 2.5 mil (0.06mm).

Fatigue Testing

Figure 10 shows a plot of G_{Imax} versus the number of cycles to delamination growth onset for a 3 mil thick insert. The plot is similar to those obtained at the different insert thicknesses and with a shear pre-crack. Included in figure 10 is G_{Ic} obtained from the end of an insert 3 mil thick. The curve drawn through the fatigue data is a best fit. The value of G_{Ith} is equated to the value of G_{Imax} at which delamination growth onset occurred at 10^6 cycles.

Figure 11 shows a plot of G_{Ith} versus the insert thickness. Included in fig.11 is the G_{Ith} value determined from a shear pre-crack. Testing with a shear pre-crack, unlike the static test results, gave a lower value of G_{Ith} than with using an insert. This result is consistent with the findings of reference 14 where a graphite/thermoplastic DCB specimen with and without a shear pre-crack was tested. When the specimen is pre-cracked in shear, cracks form in the resin rich layer ahead of the delamination front perpendicular to the direction of the principal tensile stress [19]. In fatigue, these matrix cracks easily coalesce giving a lower value of G_{Ith} than in previously undamaged material.

The threshold values obtained with insert thicknesses of 0.5 mil, 1 mil and 3 mil show little variation in G_{Ith} and give a value approximately 25 percent of the static fracture toughness. At an insert thickness of 5 mil an approximate 40 percent increase in G_{Ith} from those of the thinner inserts is noticeable; this result is consistent with the 30 percent increase of fracture toughness using a 5 mil insert.

CONCLUDING REMARKS

Artificially high values of G_{Ic} and G_{Ith} may be obtained from the DCB specimen because of fiber bridging and load history effects. Therefore, it is important to obtain G_{Ic} and G_{Ith} values from the end of the insert where there is no fiber bridging. For an S2/SP250 composite, no noticeable resin pocket was observed at the end of an insert 0.5 mil thick. However, a resin pocket was observed at an insert thickness of 1 mil and above. Despite the presence of this resin pocket there was no noticeable increase in G_{Ic} or G_{Ith} values with insert thicknesses of 1 mil and 3 mil. However, at an insert thickness of 5 mil there was a 30 percent increase in G_{Ic} and a 40 percent increase in G_{Ith}. Pre-cracking the specimens in shear gave similar values of G_{Ic} to those of the thinner inserts, however, a 50 percent decrease in G_{Ith} was observed.

Therefore, to obtain accurate values of G_{Ic} and G_{Ith} from a glass/epoxy DCB specimen a 0.5 mil insert is most suitable because of the lack of a resin pocket. However, an insert thickness up to 3 mil is acceptable. Inserts thicker than 3 mil should not be used. Also pre-cracking in shear should not be conducted because of the damage caused by the pre-cracking process. Pre-cracking in tension should not be conducted because fiber bridging will occur.

ACKNOWLEDGEMENTS

This work was done while the author held an NRC-NASA Research Associateship. The author wishes to acknowledge the support of the U.S. Army Aerostructures Directorate at NASA Langley Research Center.

REFERENCES

1. Wilkins, D.J., Eisenmann, J.R., Camin, R.A., Margolis, W.S. and Benson, R.A., "Characterizing Delamination Growth in Graphite-Epoxy," Damage in Composite Materials, ASTM STP 775, K.L.Reifsnider Ed., 1982, pp. 168-183.

2. Whitney, J.M., Browning, C.E. and Hoogsteden, W., "A Double Cantilever Beam Test for Characterizing Mode I Delamination of Composite Materials," Journal of Reinforced Plastics and Composites, Vol. 1, October 1982, pp.297-313.

3. Carlile, D.R. and Leach, D.C., "Damage and Notched Sensitivity of Graphite/PEEK Composites," Proceedings of the 15th National SAMPE Technical Conference, October 1983, pp.82-93.

4. Keary, P.E., Ilcewicz, L.B., Shaar, C. and Trostle, J., "Mode I Interlaminar Fracture Toughness of Composite Materials Using Slender Double Cantilever Beam Specimens," Journal of Composite Materials, Vol.19, March 1985, pp.154-177.

5. Russell, A.J., "Factors Affecting the Opening Mode Delamination of Graphite/Epoxy Laminates," Defence Research Establishment Pacific (DREP), Canada, Materials Report 82-Q, December 1982.

6. Johnson, W.S. and Mangalgiri, P.D., "Investigation of Fiber Bridging in Double Cantilever Beam Specimens," Journal of Composites Technology and Research, Vol.9, No.1, Spring 1987, pp. 10-13.

7. O'Brien, T.K., "Generic Aspects of Delamination in Fatigue of Composite Materials," Journal of the American Helicopter Society, Vol. 32, No. 1, January 1987, pp.13-18.

8. Murri, G.B. and O'Brien, T.K., "Interlaminar G_{IIc} Evaluation of Toughened Resin Matrix Composites Using the End-Notched Flexure Test," AIAA-85-0647, Proceedings of the 26th AIAA/ASME/ASCE/AHS Conference on Structures, Structural Dynamics and Materials, Orlando, Florida, April, 1985, p197.

9. O'Brien, T.K., Johnston, N.J., Raju, I.S., Morris, D.H. and Simonds, R.A., "Comparisons of Various Configurations of the Edge Delamination Test for Interlaminar Fracture Toughness," Toughened Composites, ASTM STP 937, N.J.Johnston Ed., 1987, pp.275-294.

10 Chai, H., "Bond Thickness Effect in Adhesive Joints and its Significance for Mode I Interlaminar Fracture of Composites," Composite Materials: Testing and Design (7th Conference), ASTM STP 893, J.M. Whitney Ed., 1986, pp.209-231.

11.   Mall, S., Yun, K.T., and Kochhar, N.K., "Characterization of Matrix Toughness Effect on Cyclic Delamination Growth in Graphite Fiber Composites," Presented at the 2nd ASTM Symposium on <u>Composite Materials: Fatigue and Fracture</u>, Cincinnati, Ohio, April 1987.

12.   Prel, Y.J., Davies, P., Benzeggagh, M.L., and de Charentenay, F.X., "Mode I and Mode II Delamination of Thermosetting and Thermoplastic Composites," Presented at the 2nd ASTM Symposium on <u>Composite Materials: Fatigue and Fracture</u>, Cincinnati, Ohio, April 1987.

13.   de Charentenay, F.X., Harry, J.M., Prel, Y.J. and Benzeggagh, M.L., "Characterizing the Effect of Delamination Defect by Mode I Delamination Test," <u>Effect of Defects in Composite Materials</u>, ASTM STP 836, 1984, pp.84-103.

14.   Martin, R.H. and Murri, G.B., " Characterization of Mode I and Mode II Delamination Growth and Thresholds in Graphite/PEEK," NASA TM 100577, April 1988, Presented at the 9th ASTM Symposium on <u>Composite Materials: Testing and Design</u>, Reno, Nevada, April 1988.

15.   Johnson, W.S. and Mall, S., "A Fracture Mechanics Approach for Designing Adhesively Bonded Joints," <u>Delamination and Debonding of Materials</u>, ASTM STP 876, W.S. Johnson Ed., 1985, pp. 189-199.

16.   O'Brien, T.K., Murri, G.B., and Salpekar, S.A., "Interlaminar Shear Fracture Toughness and Fatigue Thresholds for Composite Materials," NASA TM 89157, August 1987, Presented at the 2nd ASTM Symposium on <u>Composite Materials: Fatigue and Fracture</u>, Cincinnati, Ohio, April 1987.

17.   O'Brien, T.K., "Towards a Damage Tolerance Philosophy for Composite Materials and Structures," NASA TM100548, April 1988, Presented at the 9th ASTM Symposium on <u>Composite Materials: Testing and Design</u>, Reno, Nevada, April 1988.

18.   Gustafson, C.G., Jilken, L. and Gradin, P.A., "Fatigue Thresholds of Delamination Crack Growth in Orthotropic Graphite/Epoxy Laminates," <u>Delamination and Debonding of Materials</u>, ASTM STP 876, W.S.Johnson Ed., 1985, pp. 200-216.

19.   Bathias, C. and Laksimi, A., "Delamination Threshold and Loading Effect in Fiber Glass Epoxy Composites," <u>Delamination and Debonding of Composite Materials</u>, ASTM STP 876, W.S.Johnson Ed., 1985, pp. 217-237.

20.   Corleto, C., Bradley, W. and Henriksen, M., "Correspondence Between Stress Fields and Damage Zones Ahead of the Crack Tip of Composites Under Mode I and Mode II Delaminations," Proceedings of the 6th International Conference on Composite Materials, ICCM VI, Vol. 3, pp. 378-387.

21.   Salpekar, S.A., Raju, I.S. and O'Brien, T.K., "Strain Energy Release Rate Analysis of Delamination in a Tapered Laminate Subjected to Tension Loads," To be presented at the American Society of Composites Third Technical Conference on Composite Materials, September 26-29, 1988, Seattle, Washington.

22.   O'Brien, T.K., "Analysis of Local Delaminations and Their Influence on Composite Laminate Behavior," Delamination and Debonding of Materials, ASTM STP 876, W.S.Johnson Ed., 1985, pp. 282-297.

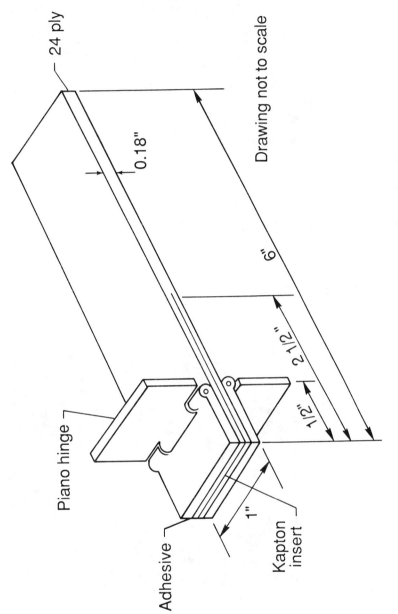

24 ply

0.18"

Drawing not to scale

6"

2 1/2"

1/2"

Piano hinge

Adhesive

1"

Kapton insert

Fig. 1 Double Cartilever Beam Specimen.

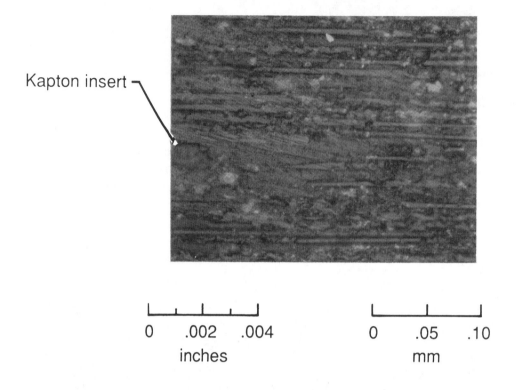

Kapton insert

0   .002   .004
inches

0   .05   .10
mm

Fig. 2   0.5 mil (0.0127 mm) Kapton Insert.

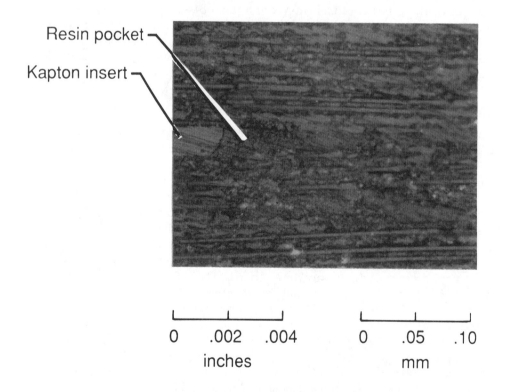

Resin pocket

Kapton insert

0   .002   .004
inches

0   .05   .10
mm

Fig. 3   1 mil (0.0254 mm) Kapton Insert.

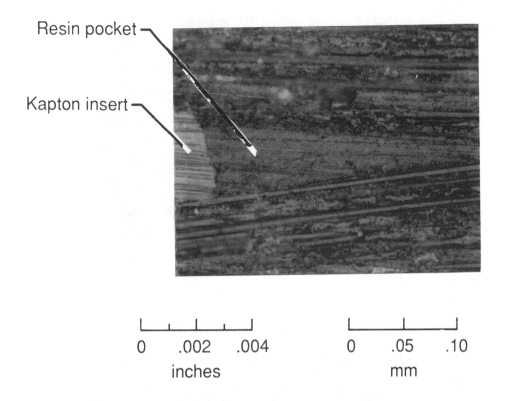

Resin pocket

Kapton insert

| 0 | .002 | .004 |
inches

| 0 | .05 | .10 |
mm

Fig. 4   3 mil (0.0762 mm) Kapton Insert.

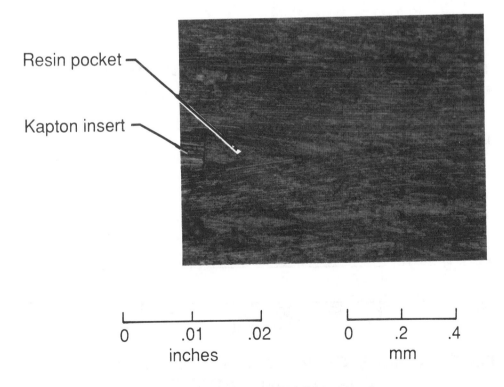

Resin pocket

Kapton insert

| 0 | .01 | .02 |
inches

| 0 | .2 | .4 |
mm

Fig. 5   5 mil (0.127 mm) Kapton Insert.

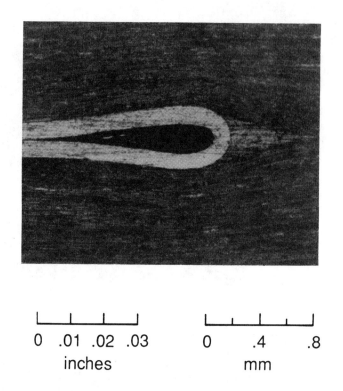

Fig. 6   Folded 5 mil (0.127 mm) Kapton Insert.

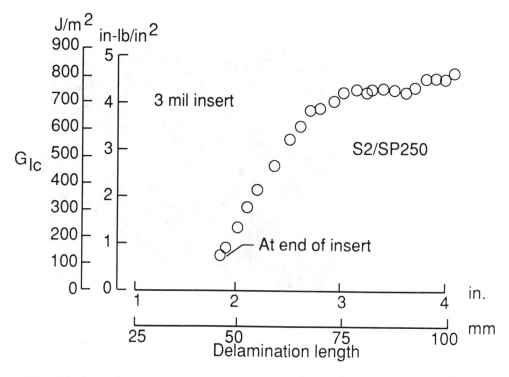

Fig. 7   Increase in G_{Ic} with Delamination Length Caused by Fiber Bridging in a DCB Test.

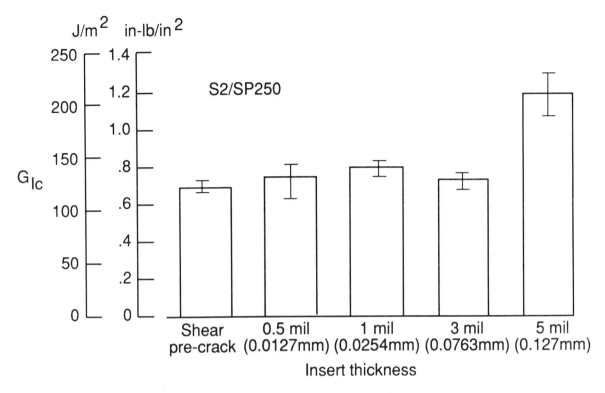

Fig. 8 Effect of Insert Thickness on G_{Ic}

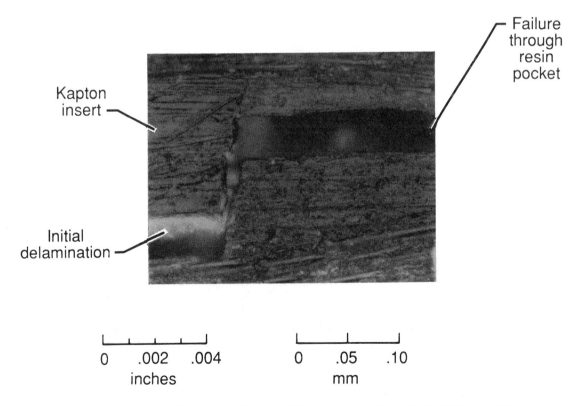

Fig. 9 Failure Path Through Resin Pocket of a 5 mil (0.127 mm) Insert.

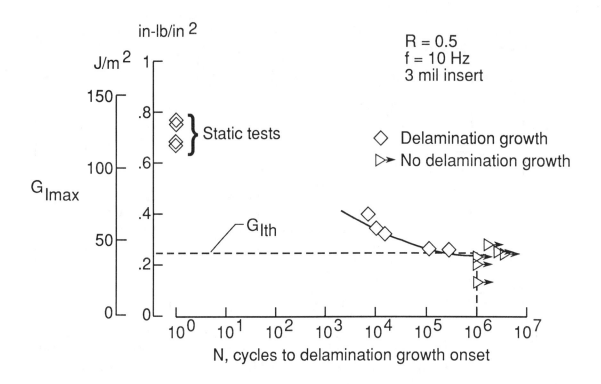

Fig. 10 G_{Imax} as a Function of Cycles to Delamination Growth Onset.

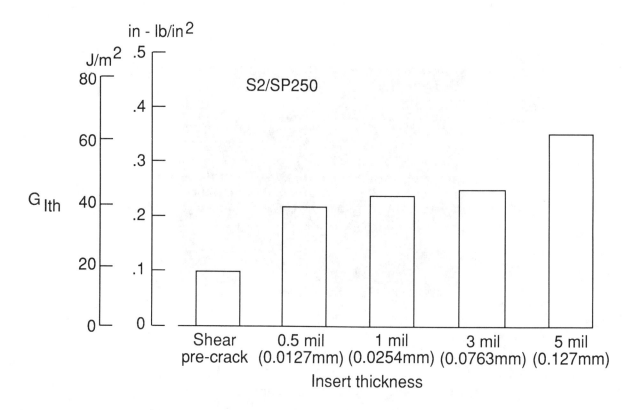

Fig. 11 Effect of Insert Thickness on G_{Ith}

The Effect of Special Orientation on the Fracture Behavior of Graphite/Epoxy Laminates

J. CHAUDHURI AND Q. JANG

ABSTRACT

An improved fracture behavior of laminates with special configuration, as compared with the fracture behavior of balanced laminates is reported in this paper. The difference in fracture behavior of these laminates are believed to be due to the presence of in-plane negative Poisson's ratio in the laminates.

INTRODUCTION

For composite materials, in-plane Poisson's ratio can be zero as well as negative, which is not true for conventional materials [1]. Figure 1 shows the longitudinal Poisson's ratio versus the bisector of angles between two laminae. It can be seen that there is steep descent of the Poisson's ratio near the thirty-degree bisector. Also, a small change in the angle can result in a large change in Poisson's ratio. Unfortunately, no attention has been given so far to the properties of laminates with a negative Poisson's ratio.

In the present study, the fracture behavior of laminates with negative Poisson's ratio was compared to the fracture behavior of balanced laminates of similar mechanical properties. The fracture toughness values are predicted following the theory of general fracture toughness parameter developed by C.C. Poe, Jr. [2,3,4]. However, Poe has derived the singular stresses ahead of the crack tip for orthotropic materials. In this paper, the stresses near the crack tip were analyzed using the theory of elasticity for anisotropic materials.

J. Chaudhuri, Assistant Professor, Mechanical Engineering Department and Associate, Institute for Aviation Research, The Wichita State University, Wichita, Kansas 67208
Q. Jang, Graduate Student, Mechanical Engineering Department, The Wichita State University, Wichita, Kansas 67208

EXPERIMENTAL TECHNIQUES

The orientations of the unbalanced and balanced laminates were [0/15/75/15]s and [0/90/60/-60]s, respectively. These orientations were carefully chosen such that some of the mechanical properties of these two laminates are similar. Table 1 shows the estimated properties of the laminates. The material used were AS4/3501-6 Graphite/Epoxy. The specimens were 12" long and 1.5" wide. Notched geometry used was center crack with five different crack lengths, .125, .25, .375, .5 and .625 inches, respectively. The cracks were sharpened at the edges using jewelers saw of width .012 in. Tensile tests were conducted in a MTS machine. The plots obtained during the test were load versus remote axial strain and load versus crack opening displacement, respectively.

RESULTS AND DISCUSSION

The fracture stress S_c, remote axial strain ϵ_c and maximum crack opening displacement $CODu$ are listed in Tables 2 and 3 for [0/15/75/15]s and [0/90/60/-60]s laminates, respectively. Figure 2 shows the fracture stress plotted against the crack length. The gross area fracture stress of [0/90/60/-60]s laminates drops faster than the gross area fracture stress of [0/15/75/15]s laminates. Furthermore, the net area fracture stress of [0/15/75/15]s laminates remain constant while the net area fracture stress of [0/90/60/-60]s laminates decreases with the increasing crack length. This indicates that the notch sensitivity of the unbalanced laminates [0/15/75/15]s is quite small.

FRACTURE DAMAGE

Photographs of failed laminates are shown in Figure 3. Although the fracture paths were self-similar in the balanced laminates [0/90/60/-60]s, the unbalanced laminates [0/15/75/15]s were broken along the direction making an angle of 75 to the original crack. It should be clearly observable that there is no 0 splitting in either laminate.
Figure 4 shows the optical photomicrograph of the failed regions in the laminates. The failure of [0/15/75/15]s can be characterized by only one failure mode; the breakage of the fiber and the matrix. In contrast, the failure of the balanced laminates [0/90/60/-60]s consisted of several failure modes; the breakage of the fiber and the matrix, delamination between the plies and splitting across the plies [5]. This clearly indicated that the unbalanced laminates had a clean fracture surface. On the other hand, considerable amount of damage existed near the crack tip in the balanced laminate prior to fracture. The scanning electron photomicrographs as shown in Figure 5 also support this view. The average damage zone size near the

crack tip in the unbalanced and balanced laminates were
.13 and .30 inches, respectively [2].

FRACTURE TOUGHNESS

Experimental data was used to calculate the
fracture toughness (i.e. critical stress intensity
factor, KQ) of the laminates from the method proposed by
Waddoups, Eisenmann and Kaminski [6] [Table 3]. It can
be seen from Table 3 that the fracture toughness of
unbalanced laminates were quite high as compared to the
fracture toughness of balanced laminates.

Fracture toughness values were also predicted
theoretically. For an anisotropic homogeneous material
with mode I loading, the stress distribution near the
crack tip is defined

$$\sigma_x = \frac{K}{\sqrt{2\pi r}} Re \left(\frac{1}{\mu_1 - \mu_2} \left(\frac{\mu_1}{(Cos\theta + \mu_2 Sin\theta)^{\frac{1}{2}}} - \frac{\mu_2}{(Cos\theta + \mu_1 Sin\theta)^{\frac{1}{2}}} \right) \right)$$

$$\sigma_y = \frac{K}{\sqrt{2\pi r}} Re \left(\frac{\mu_1 \mu_2}{\mu_1 - \mu_2} \left(\frac{\mu_2}{(Cos\theta + \mu_2 Sin\theta)^{\frac{1}{2}}} - \frac{\mu_1}{(Cos\theta + \mu_1 Sin\theta)^{\frac{1}{2}}} \right) \right) \quad (1)$$

$$\tau_{xy} = \frac{K}{\sqrt{2\pi r}} Re \left(\frac{\mu_1 \mu_2}{\mu_1 - \mu_2} \left(\frac{1}{(Cos\theta + \mu_1 Sin\theta)^{\frac{1}{2}}} - \frac{1}{(Cos\theta + \mu_2 Sin\theta)^{\frac{1}{2}}} \right) \right)$$

where K is the stress intensity factor, 0° is the
direction of crack-propagation, r is the distance from
the crack tip, μ_1 and μ_2 are dimensionless elastic
constants and are the roots of the following equation

$$a_{22}\mu^4 - 2a_{26}\mu^3 + (2a_{12} + a_{66})\mu^2 - 2a_{16}\mu + a_{11} = 0 \quad (2)$$

where a_{ij} are the components of the compliance matrix.
The strains in the principal direction of the principal
load carrying ply (i.e. i th ply) can be obtained as

$$\begin{pmatrix} \epsilon_1 \\ \epsilon_2 \\ \gamma_{12} \end{pmatrix}_i = \frac{K}{\sqrt{2\pi r}} \begin{pmatrix} \xi_1 \\ \xi_2 \\ \xi_3 \end{pmatrix}_i \quad (3)$$

$$
\begin{pmatrix} \xi_1 \\ \xi_2 \\ \xi_3 \end{pmatrix}_i = \begin{pmatrix} T \end{pmatrix}_i \begin{pmatrix} a_{ij} \end{pmatrix} Re \begin{pmatrix} \frac{1}{\mu_1-\mu_2}\left(\frac{\mu_1}{(Cos\theta+\mu_2 Sin\theta)^{\frac{1}{2}}} - \frac{\mu_2}{(Cos\theta+\mu_1 Sin\theta)^{\frac{1}{2}}} \right) \\ \frac{\mu_1\mu_2}{\mu_1-\mu_2}\left(\frac{\mu_2}{(Cos\theta+\mu_2 Sin\theta)^{\frac{1}{2}}} - \frac{\mu_1}{(Cos\theta+\mu_1 Sin\theta)^{\frac{1}{2}}} \right) \\ \frac{\mu_1\mu_2}{\mu_1-\mu_2}\left(\frac{1}{(Cos\theta+\mu_1 Sin\theta)^{\frac{1}{2}}} - \frac{1}{(Cos\theta+\mu_2 Sin\theta)^{\frac{1}{2}}} \right) \end{pmatrix} \quad (4)
$$

where [T]i is the transformation matrix of the i th ply. The values of ξ_1, ξ_2 and ξ_3 along the crack propagation direction are listed in Table 4 [7].

Poe's criterion assumed that a laminate fails when the fiber strain of the principal load carrying laminae reach a critical level [2]. He introduced a general fracture toughness parameter Qc which depends only on the fiber ultimate tensile strain ϵ_{tuf}. The fracture toughness K_Q was defined

$$
K_Q = \frac{Q_c}{(\xi_1)_i} \quad (5)
$$

where

$$
Q_c = .3\epsilon_{tuf} \; in^{\frac{1}{2}} \quad (6)
$$

Since the damage zone size in the unbalanced laminates was about a factor of 2 lower than the damage zone size in the balanced laminates, fracture toughness was calculated considering the distance from the crack tip r as .014 in for balanced laminates [2] and .007 in for unbalanced laminates. Theoretical fracture toughness values were listed in Table 3. It can be seen from Table 3 that there was an excellent agreement between the theoretical prediction and experimental results for the balanced laminates (i.e. the difference is 6%). On the other hand, the theoretical value was about 24% lower than the experimental ones for the unbalanced laminates. This could possibly be due to the non-zero shear strain present in the [0/15/75/15]s laminates, which was not considered in Poe's fracture model.

CONCLUSION

Fracture toughness of laminates with negative in-plane Poisson's ratio were found to be considerably higher as compared to the fracture toughness of conventional laminates with similar mechanical properties. Further investigation is under way to determine the fracture toughness of this type of laminates with different negative Poisson's ratio or even with zero Poisson's ratio.

REFERENCES

1. Tsai, S. W., and Hahn, H. T., Introduction to Composite Materials, Technomic Publ. Co., Inc., Lancaster, PA, 1980.

2. Poe, C. C., Jr., and Sova, J. A., "Fracture Toughness of Boron/Aluminum Laminates With Various Proportions of 0o and +45o Plies," NASA Technical Paper 1707, 1980.

3. Poe, C. C., Jr., "Fracture Toughness of Fibrous Composite Materials," NASA Technical Paper 2370, 1984.

4. Poe, C. C., Jr., "A Unifying Strain Criterion for Fracture of Fibrous Composite Laminates," Eng. Frac. Mech., Vol. 17, No. 2, 1983, pp 153-171.

5. Agarwal, B. D., and Broutman, L. J., Analysis and Performance of Fiber Composites, John Wiley and Sons, New York, 1980.

6. Waddoups, M., Eisenmann, J., and Kaminski, B., "Macroscopic Fracture Mechanics of Advanced Composite Materials," J. Composite Materials, Vol. 5, 1971.

7. Jang, Q., "The Effect of Special Orientation on The Fracture Behavior of Gr/Epoxy Laminates," MS Thesis, Mech. Eng. Dept, The Wichita State University, 1988.

Table 1. Estimated Material Properties of the Two Laminates

Configuration	Poisson's Ratio	Stiff. x (GPA)	Ult. Str. x (MPA)	Stiff. Y (GPA)	Ult. Str. Y (MPA)
[0/15/75/15]s	-0.125	81.96	375	32.22	325
[0/90/60/-60]s	0.166	46.23	397	78.61	750

Table 2 Measured Fracture Properties of Balanced and
Unbalanced Laminates

2a(in)	[0/15/75/15]s			[0/90/60/-60]s		
	Sc (ksi)	ϵ_c (x10-2)	CODu (x10-3 in)	Sc (ksi)	ϵ_c (x10-2)	CODu (x10-3 in)
0	54.4684	0.7536	---	57.5729	1.1690	---
0.125	52.7536	0.4921	8.7410	47.2679	1.2730	0.1281
0.25	44.2967	0.4310	11.4200	36.8528	0.8897	0.0887
0.375	40.7737	0.4085	11.1790	31.1136	0.6271	0.1121
0.50	36.4153	0.3695	11.9250	30.2727	0.5758	0.1091
0.625	32.8377	0.3301	14.7390	23.6569	0.4600	0.0993

Table 3 Fracture Toughness of Balanced and

Unbalanced Laminates

2a(in)	[0/15/75/15]s		[0/90/60/-60]s	
	$K_Q(Ksi\sqrt{in})$ Measured	$K_Q(Ksi\sqrt{in})$ Predicted	$K_Q(Ksi\sqrt{in})$ Measured	$K_Q(Ksi\sqrt{in})$ Predicted
0.125	100.5189		37.1644	
0.25	50.2824		30.9636	
0.375	49.8852	37.419	36.3156	30.9464
0.50	49.8364		34.9287	
0.625	49.5919		29.6515	
Avg	49.8990*		33.8008	

* Average of last four values

Table 4   Values of Balanced and Unbalanced Laminates

Configuration	ξ_1	ξ_2	ξ_3
[0/15/75/15]s	0.02066	0.01256	−0.03350
[0/90/60/−60]s	0.01696	0.01301	0

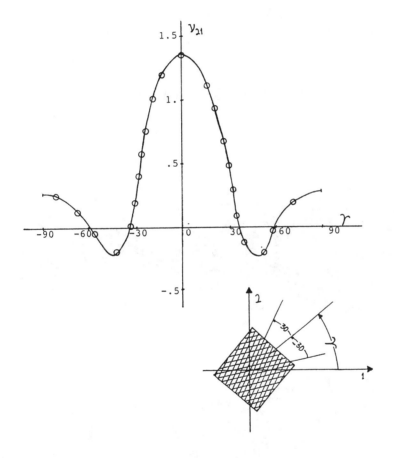

Figure 1. Poisson's Ratio VS Bisector of Angle For
Angle Ply Laminates (ref. 1)

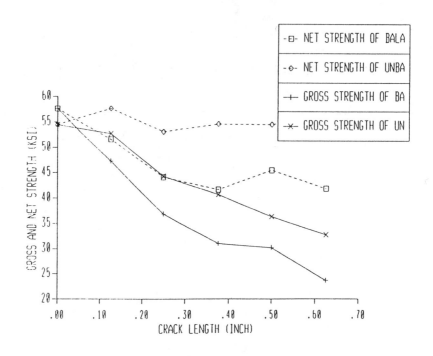

Figure 2.   Fracture Stress VS Crack Length

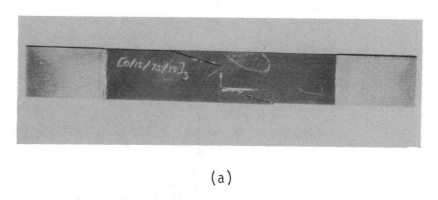

(a)

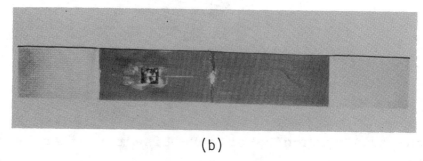

(b)

Figure 3.   Failed Specimens (a) [0/15/75/15]s

(b) [0/90/60/-60]s

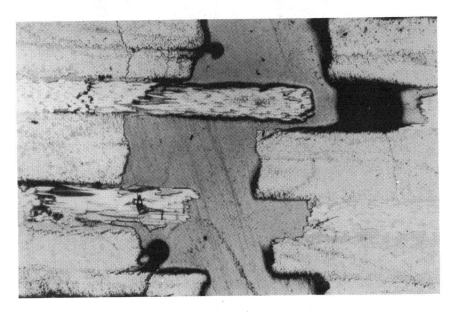

(a)

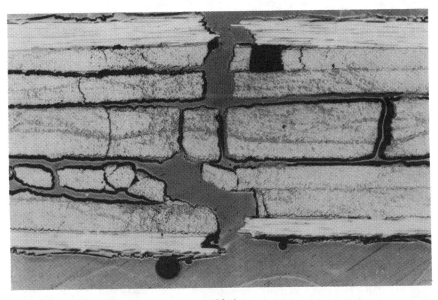

(b)

Figure 4. Optical Photomicrographs Illustrating

Microscopic Fracture Modes (a) [0/15/75/15]s
(b) [0/90/60/-60]s Magnification 45x

(a)

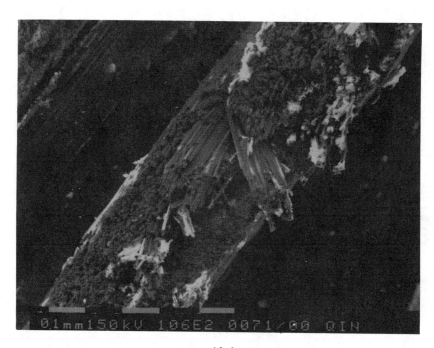

(b)

Figure 5. SEM Photomicrographs Illustrating Typical
Fracture Surfaces (a) [0/15/75/15]s
(b) [0/90/60/-60]s

Mechanical Characterization of a Thick Composite Laminate

RAN Y. KIM

University of Dayton Research Institute
300 College Park
Dayton, OH 45469-0001

FRANCES ABRAMS AND **MARVIN KNIGHT**

Materials Laboratory
Air Force Aeronautical Laboratories
Wright-Patterson Air Force Base, OH 45433-6533

INTRODUCTION

Many current and future applications of advanced composite materials involve thick laminates. Design, failure, and stress analysis of such thick laminate structures require elastic constants and strengths in the three principal directions (x, y, and z). Although some of these properties, mainly elastic properties, are sparsely reported in the literature [1], systematic and complete properties are not available to the author's best knowledge. It is well known that the mechanical behavior can vary widely, dependent upon the fabrication method used for the laminate. It is more critical for a thick laminate case because of difficulties in the cure process for such a thick laminate. It is the objective of this paper to present an experimental approach to obtain three-dimensional elastic and strength properties of thick laminates from direct measurement and to study the effect of process conditions on the mechanical properties of the thick laminate. To achieve the objectives, tension, compression, and shear tests were conducted. The results are presented and discussed in detail.

EXPERIMENTAL PROCEDURE

A thick laminate (256 layers) of unidirectional graphite/epoxy (AS4/3501-6) 12" x 12" x 1.5" was fabricated in an autoclave. The laminate was cured with an expert system developed in Reference 2. The material for expert system cure was specially requested to have a 34 percent net resin content for the prepreg. The laminate was cured using a closed-loop controller developed at the US Air Force Materials Laboratory. This controller uses dielectric data and temporal and spatial temperature gradients to develop the cure cycle in real time. It attempts to cure the part as rapidly as possible without allowing uncontrolled exotherms or large gradients of temperature from surface to middle of the part. Compaction, however, is still largely controlled by damming and bagging procedure since no sensor is available to determine progress towards that goal. The part is simply bagged to limit dimensional change and bleed. After the cure, it was discovered that the prepreg resin content was higher than expected. The resin content was determined to be about 55 percent for cured specimen. The panel was successfully cured without bleed. Because of the incorrect assumption of resin content, the laminate was resin rich.

The laminate was inspected by X-ray to ascertain the quality. X-ray radiographs were made of sections of the laminates. Two slices were taken from the ends of the laminate. One slice had the fibers parallel to the long axis of the slide and the other had the fibers to perpendicular. The radiograph showed no apparent damages such as delaminations through the thickness of the laminates. Also, fluorescent penetrant inspection was used on the two slices. Only two very small pore-like defects were found on each slice.

Laminates were cut into various sizes and shapes of specimens depending upon the desired property to be determined. Figure 1 shows the cutting diagram for specimens. A photomicrographic technique was used for determining fiber volume and varion at various points. The specimen for fiber volume determination was polished using polishing powder. An area method was employed for determination of fiber volume content from the photomicrographs taken. In-plane tensile and compressive properties were determined using flat specimens (.08 inches thick nominally) which were sliced from the laminate block. Considerable care was taken in cutting to get uniform thickness. The variation of thickness within each specimen was less than 0.005 inches. The specimen dimensions were 0.5 - 0.75 inches wide and 8 - 10 inches long for tension and 0.25 inches wide and 0.5 inches long for compression. However, the properties in the z-direction were determined by using short specimens of gage length 0.7 inches for tension and 0.5 inches for compression. All specimens were end-tabbed with low fiber volume glass cloth/epoxy material using a structural adhesive. The elastic properties in shear were obtained by torsion tests using 0.5-inch diameter cylinders. The cylinders with fiber along and perpendicular to the cylinder axis were machined from rectangular blocks. Double notch shear specimens, 0.5 inches thick, were used for shear strengths determination. Strains necessary for determination of elastic constants were measured with strain gages .125 - 0.25 inches long.

All tests were conducted on two closed-loop hydraulic testing machines with load control for tension and stroke control mode for compression. In any cases, the loading rate was equivalent to 0.005 - 0.01 inches/ minute. Load and strains were monitored continuously using an $x-y_1$ and y_2 recorder until final failure of the specimen.

RESULTS AND DISCUSSION

Figure 2 shows examples of the photomicrographs indicating variation of fiber distribution in microlevel. Fiber volume was determined from the microgrpahs using the area method. Figure 3 shows the fiber volume distribution in a thin slab (10 inches x 1.5 inches x 0.08 inches) that was taken from the middle of the laminate block. The resin rich areas exist throughout the specimen mainly due to difficulty of resin flow during the consolidation process. However, it should be noted that the photomicrographs technique gives more variation in the result because a very small area is involved in the measurement compared with direct measurements (6.8×10^{-5} vs. 0.25 inches^2). The average fiber volume determined from 20 photomicrographs is 55 percent for the composite block used in this test. Some small voids were present as shown in Figure 2.

Table 1 shows the summary of the elastic constants and strengths determined from the laminate block. The number of replicate and coefficient of variation are also included in the table. The longitudinal property data, both moduli and strengths in Table 1 are the test values converted to a fiber volume of 66 percent in order to compare with the published data for thinner laminates [3]. The modulus agrees very well with the data obtained from the thin laminate. However, the strength is considerably lower than the data for thin laminate. The main reason for this appears to be due to the flaw sensitivity of strength. The compressive strength is slightly lower than the tensile strength and buckling was the predominant failure mode. Although there is a slight difference for longitudinal modulus in tension and compression in Table 1, the difference is minimal and considered to be within the experimental scatter rather than a material properties variation.

The transverse properties agree very well with the published data for the thin laminate. Shear modulus, Gxy and Gxz is in the range of published data but transverse shear modulus, Gyz is unable to be compared because of difficulty in finding the value in the literature. The shear strengths, Sxy and Sxz are smaller than the result obtained from the ±45 laminate (11 ksi vs. 13.5 ksi). The main reason for this is the concentration of stress at the root of the notches in the double-notch shear specimen. It should be noted that the double-notched shear specimen is not suitable for generation of shear properties of composite laminates. Development of an improved shear test method for unidirectional laminate is eminent.

The experimental data support the relations for transverse isotropy assumption as shown in Table 2. The difference is less than six percent at most in both elastic constants and strengths. The test procedure employed in this work is reasonably reliable.

CONCLUSIONS

Based on the experimental results, the following conclusions can be drawn: 1) Elastic constants measured from thick composite agree very well with the published results for thin laminate and the self-consistence of the data checked. 2) Strength in the fiber direction is lower than the published data and its scatter is very high. 3) Tensile and compressive strengths in the y-direction and in the z-direction are practically equal and compared well with the data obtained from a thin laminate. 4) The lower shear strengths appear to be due to the shear test method rather than the quality of the specimens.

REFERENCES

1. Knight, M., "Three-Dimensional Elastic Moduli of Graphite/Epoxy Composites," Journal of Composite Materials, Vol. 16, (March 1982).

2. Abrams, F., et al., "Qualitative Process Automatic for Autoclave Curing of Composites," AFWAL-TR-87-4083, Wright-Patterson Air Force Base, OH, 1987.

3. Tsai, S. W., Composites Design, 4th Edition, Think Composites, Dayton, OH, 1988.

TABLE 1 – SUMMARY OF TEST RESULTS

$$\nu_f = 66\%$$

		Modulus			Poisson's Ratio	Strength		
		N	$\bar{X}$, msi	r, %		N	$\bar{X}$, ksi	r, %
Longitudinal Tension	X (y)	8	20.9	7.0	0.31	11	239.5	23.3
	X (z)	8	21.5	4.9	0.30	11	222.6	13.7
Compression	X' (y)	3	19.9	–	–	10	197.4	7.9
	X' (z)	3	20.4	–	–	10	189.0	7.9
Transverse Tension	Y (x)	10	1.4	8.5	–	10	7.6	14.8
	Y (z)	7	1.4	2.0	0.52	10	8.5	15.9
Compression	Y' (x)	2	1.5	–	–	7	28.9	6.8
	Y' (z)	2	1.4	–	–	7	30.7	9.3
Transverse Tension	Z (x)	5	1.4	4.5	–	11	7.4	8.7
	Z (y)	6	1.3	2.1	0.53	11	9.0	16.0
Shear	S_{xy}	7	0.76	12.2	–	4	11.1	12.5
	S_{xz}	3	0.72	2.4	–	6	10.9	14.5
	S_{yz}	4	0.44	6.0	–	3	15.0	13.4

N = Number of replicate

$\bar{X}$ = Mean

r = Coefficient of variation

TABLE 2 - TRANSVERSE ISOTROPY

Elastic Constants	Experiment
ν_{xy}/ν_{xy}	1.03
ν_{zy}/ν_{yz}	1.02
E_{yy}/E_{zz}	1.04
G_{xy}/G_{xz}	1.06
$1/G_{yz}/2(1+\nu_{yz})/E_{yy}$	1.02

Strengths	
Y(z)/Z(x)	1.03
Z(y)/Y(z)	1.06
S_{xy}/S_{xz}	1.02

*All the ratios must be unit for transverse isotropy.

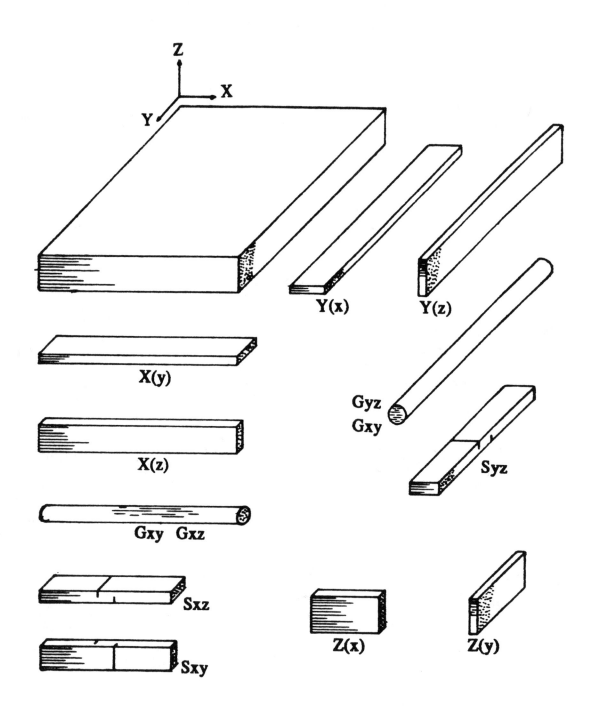

Fig. 1. Specimen cutting diagram.

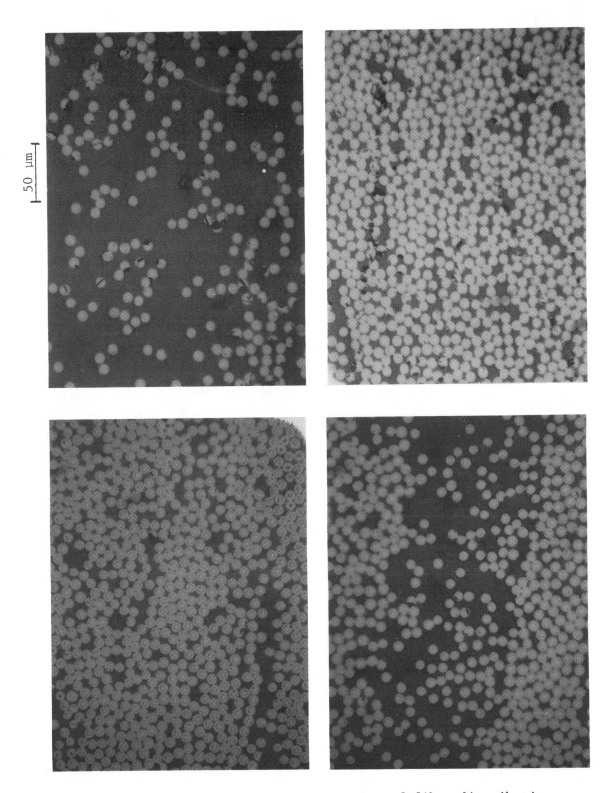

Fig. 2. Photomicrographs showing variation of fiber distribution.

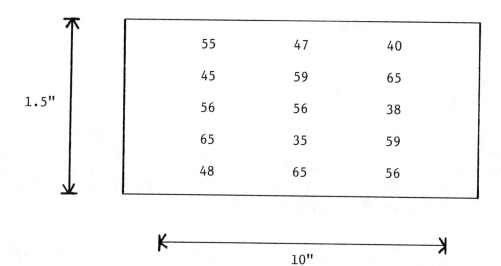

Average ν_f = 53%

Coefficient of Variation = 10%

Fig. 3. Fiber volume distribution in a slab, %.

Novel Design Concepts Utilizing
Shape Memory Alloy Reinforced Composites

CRAIG A. ROGERS

ABSTRACT

Shape Memory Alloy (SMA) reinforced composites are a new class of composite materials that has been recently developed and demonstrated at Virginia Polytechnic Institute and State University. A shape-memory-alloy (Nitinol) may be used as fiber reinforcement or actuation in a laminated, fiber-reinforced composite. Adaptive materials are a relatively new class of materials that have the capability of changing their physical geometry or of altering their physical properties. Possible applications for this class of adaptive materials are: in structures that are part of long-duration, unattended space missions (for which the material must be able to compensate for damage by redistributing the load around failed portions of the structure); in active vibration control of large flexible structures, in active acoustic control for aircraft to reduce interior sound levels, and in robotic manipulators.

The adaptive response of the SMA reinforced composites has been demonstrated experimentally by controlling the motion of a fiberglass cantilever beam with embedded Nitinol 'fibers'. The cantilever beam is a unique composite material containing several shape memory alloy fibers (or films) in such a way that the material can be stiffened or controlled by the addition of heat (i.e., apply a current across the fibers). Shape memory alloys and the mechanism by which they exhibit the characteristic shape memory effect (SME) will be explained in some detail. Several novel design concepts that utilize different aspects of material and structure interaction will be presented such as Active Modal Modification, Active Strain Energy Tuning, and Active Structural Modification.

INTRODUCTION

'Smart', 'Intelligent', 'Sense-able', and 'Adaptive' have all been used to described and/or classify materials and structures which contain their own sensors, actuators and computational/control capabilities and/or hardware. The definition of Smart Materials, which has become an item of discussion, may be materials that posses adaptive capabilities to external stimuli such as load or environment with inherent or dedicated intelligence. The control or intelligence of the material could perhaps be 'programmed' by material composition, processing, defect and microstructure, or conditioning to adapt in a controlled manner to various levels of stimulus or be the result of integrated and/or dedicated control hardware. Smart structures may simply

Craig A. Rogers, Director, Smart Materials & Structures Laboratory, Assistant Professor, Mechanical Engineering Department, Virginia Polytechnic Institute and State University, Blacksburg, Virginia 24061.

be constructed of Smart materials or may have dedicated or integrated actuators, sensors, and intelligence in a more discrete form. The early 'Smart Materials' contained for the most part embedded and/or distributed sensors for strain and temperature. However, the complexity and utility of smart materials has increased rapidly to the present time where major advancements seem to be occuring on a monthly basis in the areas of materials, actuators, sensors, and controls. Although smart materials and structure concepts may be applied to the design and implementation of buildings, dams, bridges, pipelines, ships, and ground-based vehicles, recent research efforts have been concentrated on potential aerospace applications in advanced aircraft, launch vehicles, and large space-based platforms.

Smart materials and structures, which contain distributed actuators, sensors, and microprocessor capabilities, can be used in many applications requiring a high degree of adaptability to changing external and internal conditions. External conditions may consist of environment, loads, or the desire to change the scope, purpose, or geometry of the structure after it has been built and is in service. Internal conditions may be damage or failure to isolated portions of the material or structure. The number of applications requiring or benefitting from such adaptability is increasing rapidly and more are sure to follow. One of the current needs is for long-duration unattended materials and structures that can be used in isolated environments (i.e., aircraft, marine vessels, defense vehicles, and the space station) or in biomedical applications. Using adaptive/intelligent materials may result in structures with self-inspection and self-identification capabilities which can direct the adaptive response based on the environment and/or damage to the structure. For example, a structural member made of an intelligent adaptive material can compensate for deterioration in absorptivity and thermal expansion properties that result in excessive change in length of that or other members as well as control the motion and vibration of the structure. The same material can be used to change load paths in a structure or within the material so that the component can be replaced or repaired before it causes catastrophic failure of the system or unacceptable degradation of performance.

The definition of 'Smart Materials and Structures' has been a topic of discussion and controversy since the late 1970's when a 'Smart Material' simply consisted of optical fiber sensors embedded in a composite material. Some definitions state that the material or structure simply have integral (perhaps embedded) sensors, actuators, and 'intelligence'. The intelligence is most often dedicated (or integral) computation/control hardware. However, other definitions state that all sensing, actuating and intelligence capabilities be inherent to the material or structure. The actuator capabilities must be tied to the sensing of external stimuli such as temperature, load, strain, light, radiation, etc. Examples of this later case may be photochromic glasses and many multi-viscosity lubricants.

Potential applications for smart materials and structures include:

- Active vibration control and acoustic suppression for submarines, robot manipulators, propeller aircraft, large flexible structures, etc.
- Failure detection/prevention of structures (i.e., bridges, walkways, phone and electrical cables, and mechanical components).
- Active control of helicopter rotor blades.
- Thermal expansion balancing.

- Robot manipulators (fingers).
- Thermally activated valves, ducts, and switches.
- Structural dimension adjustment and environment adaptation for antennas.

The development and subsequent production of this class of materials could have tremendous impact on several diverse technological fields, i.e., material science, vibrations and controls, ocean and aerospace structures, defence technologies, biotechnology, and may act as a catalyst for the development of many new devices and technologies.

Introduction to Shape Memory Alloy (SMA) Reinforced Composites

The development of the material referred to as SMA reinforced composites in this paper is simply a composite material that contains shape memory alloy fibers (or films) in such a way that the material can be stiffened or controled by the addition of heat (i.e., apply a current through the fibers) [1]. Shape memory alloys and the mechanism by which they exhibit the characteristic shape memory effect (SME) is explained in some detail in the background section below and in references [2-8]. One of the many possible configurations of the SMA reinforced composite material is one in which the shape memory alloy fibers are embedded in a material off of the neutral axis on both sides of the beam in agonist-antagonist pairs. Before embedding the fibers, the shape memory alloy fibers are plastically elongated and constrained from contracting to their 'normal' or 'memory' length upon curing the composite material with high-temperature. The fibers are therefore an integral part of the composite material and the structure. When the fibers are heated, generally by passing a current through the shape memory alloy, the fibers 'try' to contract to their 'normal' length and therefore generate a uniformly distributed shear load along the entire length of the fibers. The shear load offset from the neutral axis of the structure will then cause the structure to bend in a known and predictable manner.

The second configuration consists of creating 'sleeves' within the composite laminate which the plastically elongated shape-memory alloy can be inserted and then clamped to both ends. When the shape memory alloy is heated, the fibers try to contract in the same fashion as explained above. The fibers in a sleeve will exert a concentrated force on the ends of the structure in a direction that is always tangent to structure at the point where the fibers are clamped to the structure. The difference between the embedded fibers and the fibers in a sleeve is that in the first case the force of the shape memory alloy is distributed over the length of the fiber and in the later case the force is concentrated at the end of the structure.

Both of the design concepts described above have been incorporated into prototypes and their potential demonstrated on a limited scale. It appears that both designs have unique advantages and disadvantages. The two design concepts described above are in all likelihood only the beginning of the design possibilities for this remarkable material. Other design possibilities have been proposed and are currently being investigated. For instance, if the shape memory alloy fibers are placed on the neutral axis and activated, the result would be placing the structure in a state of 'residule' stress which in turns changes the modal response of the structure without causing any induced motion or vibrations. Other design strategies will be described below.

Introduction to Shape Memory Alloys

In 1965, Buehler and Wiley of the U.S. Naval Ordnance Laboratory received a United States Patent on a series of engineering alloys that possess a unique mechanical (shape) "memory" [9]. The generic name of the series of alloys is 55-Nitinol. Theses alloys have chemical compositions in the range of 53 to 57 weight percent nickel. A great deal of effort was expended over the next ten years in characterizing the material and developing new applications to exploit its remarkable shape memory effect (SME) and its unusual mechanical properties. The Naval Ordnance Laboratory (now known as the Naval Surface Weapons Center) was and still is the leader in characterizing Nitinol. Several other laboratories have made significant contributions to the understanding of the Nitinol, in particular is Battelle Memorial Institute and NASA.

The shape-memory effect (SME) can be described very basically as follows: an object in the low-temperature martensitic condition, when plastically deformed and the external stresses removed will regain its original (memory) shape when heated. The process, or phenomenon, is the result of a martensitic transformation taking place during heating. Although the exact mechanism by which the shape recovery takes place is a subject of controversy, a great deal has been learned about the unique properties of this class of materials in the past twenty years [10-12]. It appears clear however that the process of regaining the original shape is associated with a reverse transformation of the deformed martensitic phase to the higher temperature austenite phase.

Many materials are known to exhibit the shape memory effect. They include the copper alloy systems of Cu-Zn, Cu-Zn-Al, Cu-Zn-Ga, Cu-Zn-Sn, Cu-Zn-Si, Cu-Al-Ni, Cu-Au-Zn, Cu-Sn, and the alloys of Au-Cd, Ni-Al, Fe-Pt, and others. The most common of the shape memory alloys or transformation metals is a nickel-titanium alloy known as Nitinol.

Nickel-titanium alloys (Nitinol, NiTi) of proper composition exhibit unique mechanical "memory" or restoration force characteristics. The name is derived from Ni (Nickel) - Ti (Titanium) - NOL (Naval Ordinance Laboratory). The shape recovery performance of Nitinol is phenomenal. The material can be plastically deformed in its low-temperature martensite phase and then restored to the original configuration or shape by heating it above the characteristic transition temperature. This unusual behavior is limited to NiTi alloys having near-equiatomic composition. Plastic strains of typically six-to-eight percent may be completely recovered by heating the material so as to transform it to its austenite phase. Restraining the material from regaining its memory shape can yield stresses of 100,000 psi (the yield strength of Nitinol is approximately 12,000 psi) as shown in Fig. 1.

Substantial progress has been made in understanding the nature of the "shape memory effect" (SME). The characteristic transition temperature on heating, which corresponds approximately to the top of the temperature range through which the material must be heated to restore it to its memory configuration, varies from about -50° to +166° C. A great deal of literature has been published over the past twenty years presenting detailed thermal, electrical, magnetic, and mechanical characterizations of this unusual alloy. However, there is still much to be learned about the influence of residual stress and high temperatures on the extent, duration and repeatability on SME.

NOVEL DESIGN CONCEPTS

Shape Memory Alloy (SMA) reinforced composites have tremendous potential for creating new paridgms for material-structures interaction. The list of scientific areas that can be influenced by novel approaches possible with SMA reinforced composites is quite large. For example, vibration control can be accomplished by using the distributed force actuator capabilities similar to the common piezoelectric systems. However, two unique approaches to active control are possible with a material that can change its stiffness and physical properties; i) Active Strain Energy Tuning, and

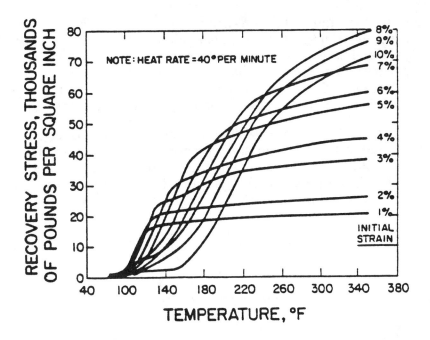

Figure 1.   Shape Recovery Stress vs. Temperature

ii) Active Modal Modification. Simulation results showing the potential for SMA reinforced composites to vary the modal response of a composite plate will be presented below.

Applications for SMA reinforced composites extend far beyond vibration control tasks. Active buckling control or more generically active structural modification schemes can be imagined in which SMA fibers are stiffened within a composite to alter the critical buckling load in a transient sense of the structure. SMA composites that are used for various vibration control tasks could also be used for motion or shape control, allowing a structure to maintain a given shape or orientation for an extended period of time. Motion and shape control will in all likelihood involve the simultaneous use of force actuators (SMA) and stiffness actuators (the technique in which the SMA is heated to change its modulus of elasticity) to create a structure that behaves much like a mechanical muscle.

Transient and steady-state vibration control can be accomplished with SMA reinforced composites using several techniques. Transient vibration control is defined here as the ability to suppress or damp structural vibration by applying forces (distributed and/or point) to the structure in such a way as to dissipate the energy within the structure. This is accomplished generally by applying point transverse loads to the structure or applying an 'actuator film' to the surface of the structure. The approach with SMA reinforced composites is to simply embed the actuators (shape memory alloys) in the structure such that, when actuated correctly, they exert agonist-antagonist forces off the neutral axis thereby reducing vibrations [1].

Active Modal Modification

Steady-state vibration control which may also be used for structural acoustic control can be accomplished with SMA reinforced composites using a novel technique termed "Active Modal Modification". The modal response of a structure or mechanical component (i.e., plate or beam) can be tuned or modified by simply heating the SMA fibers or lamina to change the stiffness of all or portions of the structure. When

Nitinol is heated to cause the material transformation from the martensitic phase to the austenite phase, the Young's modulus changes by a factor of approximately four as shown in Figs. 2 and 3. Not only is the stiffness increased by a factor of four but the yield strength also increases by a factor of ten. This change in the material properties occur because of a phase transformation and does not result in any appreciable force and does not need to be initiated by any plastic deformation.

Simulations of the effect of tuning SMA fibers within a composite structure will be presented by evaluating the free vibration response of a SMA reinforced quasi-isotropic plate. The following formulation and discussion will be limited to a midplane symmetric laminate ($B_{ij} = 0$) without inplane lateral loads ($q = N_x = N_y = N_{xy} = 0$) for which the governing differential equation for free vibration is

$$D_{11}\frac{\partial^4 w}{\partial x^4} + 4D_{16}\frac{\partial^4 w}{\partial x^3 \partial y} + 2(D_{12} + 2D_{66})\frac{\partial^4 w}{\partial x^2 \partial y^2}$$

$$+ 4D_{26}\frac{\partial^4 w}{\partial x \partial y^3} + D_{22}\frac{\partial^4 w}{\partial y^4} = q$$

The boundary conditions are:

at $x = 0$ and a $w = M_x = -D_{11}\frac{\partial^2 w}{\partial x^2} - 2D_{16}\frac{\partial^2 w}{\partial x \partial y} - D_{12}\frac{\partial^2 w}{\partial y^2} = 0$

at $y = 0$ and b $w = M_y = -D_{12}\frac{\partial^2 w}{\partial x^2} - 2D_{26}\frac{\partial^2 w}{\partial x \partial y} - D_{22}\frac{\partial^2 w}{\partial y^2} = 0$

Using the Ritz method to obtain an approximate solution to the governing equation yields the energy expression

$$\frac{1}{2}\int_0^b \int_0^a \left[D_{11}\left(\frac{\partial^2 w}{\partial x^2}\right)^2 + 2D_{12}\frac{\partial^2 w}{\partial x^2}\frac{\partial^2 w}{\partial y^2} + D_{22}\left(\frac{\partial^2 w}{\partial y^2}\right)^2 \right.$$

$$+ 4\left(D_{16}\frac{\partial^2 w}{\partial x^2} + D_{26}\frac{\partial^2 w}{\partial y^2}\right)\frac{\partial^2 w}{\partial x \partial y} + 4D_{66}\left(\frac{\partial^2 w}{\partial x \partial y}\right)^2$$

$$\left. - 2qw \right] dx\, dy = \text{stationary value}$$

The assumed solution for the energy expression using the separation of variables is

$$w = \sum_{m=1}^{M} \sum_{n=1}^{N} A_{mn} X_m(x)\, Y_n(y)$$

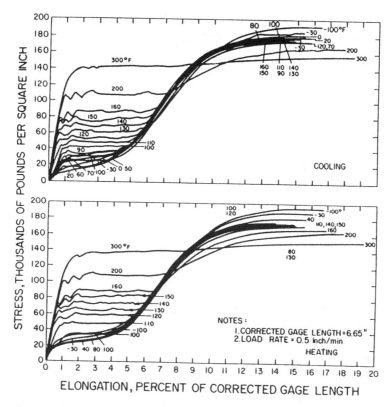

Figure 2.   Stress vs. Elongation at Various Temperatures

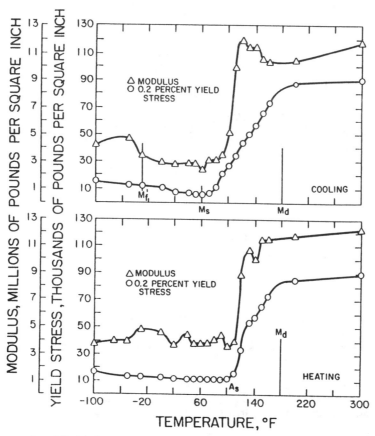

Figure 3.   Yield Stress and Elastic Modulus vs. Temperature

Substituting this series into the energy expression the following set of equations result.

$$\sum_{i=1}^{M} \sum_{j=1}^{N} \left\{ D_{11} \int_0^a \frac{d^2 X_i}{dx^2} \frac{d^2 X_m}{dx^2}\, dx \int_0^b Y_j\, Y_n\, dy \right.$$

$$+ D_{12} \left[\int_0^a X_m \frac{d^2 X_i}{dx^2}\, dx \int_0^b Y_j \frac{d^2 Y_n}{dy^2}\, dy \right.$$

$$+ \left. \int_0^a X_i \frac{d^2 X_m}{dx^2}\, dx \int_0^b Y_n \frac{d^2 Y_j}{dy^2}\, dy \right]$$

$$+ D_{22} \int_0^a X_i\, X_m\, dx \int_0^b \frac{d^2 Y_j}{dy^2} \frac{d^2 Y_n}{dy^2}\, dy$$

$$+ 4 D_{66} \int_0^a \frac{dX_i}{dx} \frac{dX_m}{dx}\, dx \int_0^b \frac{dY_j}{dy} \frac{dY_n}{dy}\, dy$$

$$+ 2 D_{16} \left[\int_0^a \frac{d^2 X_i}{dx^2} \frac{dX_m}{dx}\, dx \int_0^b Y_j \frac{dY_n}{dy}\, dy \right.$$

$$+ \left. \int_0^a \frac{dX_i}{dx} \frac{d^2 X_m}{dx^2}\, dx \int_0^b Y_n \frac{dY_j}{dy}\, dy \right]$$

$$+ 2 D_{26} \left[\int_0^a X_m \frac{dX_i}{dx}\, dx \int_0^b \frac{dY_j}{dy} \frac{d^2 Y_n}{dy^2}\, dy \right.$$

$$+ \left. \int_0^a X_i \frac{dX_m}{dx}\, dx \int_0^b \frac{d^2 Y_j}{dy^2} \frac{dY_n}{dy}\, dy \right] A_{mn}$$

$$= q_0 \int_0^a X_m\, dx \int_0^b Y_n\, dy \quad \begin{matrix} m = 1,2,\ldots,M \\ n = 1,2,\ldots,N \end{matrix}$$

The $M \times N$ linear simultaneous equations are then rewritten in matrix form as

$$[K]\{A_{mn}\} = \{q_{mn}\}$$

Using the Ritz method to solve the energy equation allows the assumed solution to only satisfy the displacement boundary conditions. For a simply-supported plate, the double sine-series is a viable solution

$$X_m(x) = \sin \frac{m\pi x}{a}$$

$$Y_n(y) = \sin \frac{n\pi y}{b}$$

Since the *MxN* linear simultaneous equations are homogeneous, a nontrivial solution can be obtained only if the determinant of the coefficient matrix, [K], is zero. Therefore the eigenvalues of [K] are determined which then reflect the natural frequencies of free vibration.

The formulation must be further expanded for SMA reinforced composites as the bending stiffnesses (D_{ij}) are functions of temperature and can be tuned by activating individual plys of a laminate resulting in a change of the fiber modulus (see Fig. 3) by as much as a factor of four. The change in the fiber modulus occurs over a relatively small temperature range (selectable from 10 to 20°C) and is a result of a solid phase material transformation between the martensite and austenite phases. Therefore, the superscripts 'M' and 'A' are used to denote the physical and mechanical properties in each phase. Unsuperscripted values are intended to represent instantaneous values which are 'tunable' between the martensitic and austenitic values.

Variation of Stiffness: The simulation results have been generated based upon the following assumptions:

- Square plate
- Midplane symmetric at all times (activation of a ply occurs on both sides of the midplane simultaneously)
- [+45, − 45,0,90]$_s$ laminate
- Material is entirely Nitinol/Epoxy
- Fiber (Nitinol) modulus increase from martensite to austenite is a factor of four
- Fiber volume fraction = 50 percent
- Macroscopic lamina properties determined from the rule-of-mixtures

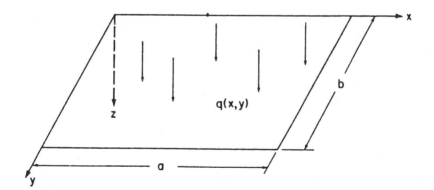

Figure 4. Simply-Supported Plate Geometry

The stiffness of a composite plate, extension or bending, can be taylored within reasonable bounds. However, SMA reinforced composites can be taylored to not only have a specified stiffness but can be taylored to have a range of material properties that can be controlled or tuned. For example, Fig. 5 shows the change in the flexural stiffness (D_{11}) of the square plate when one or all of the individual lamina are activated such that the fiber modulus increases. Simply activating the top and bottom + 45° ply can modify the plate flexural stiffness by over 40 percent. Activating all of the plys to increase the fiber modulus by a factor of four increases the flexural stiffness by approximately 90 percent. By utilizing the numerous permutations of activated laminae to unactivated laminae and using modulation schemes in which some lamina can be only partially activated can result in subtle and versatile control possibilites. Changing the stiffness of a composite structure has some important practical implications. One prime example is in active structural acoustic control where the radiated sound pressure levels can be reduced dramatically by reducing the amplitude of the structural acoustic vibrations. Obviously changing the stiffness of the plate also changes its dynamic response and has other significant implications on vibration and acoustic control which will be described below.

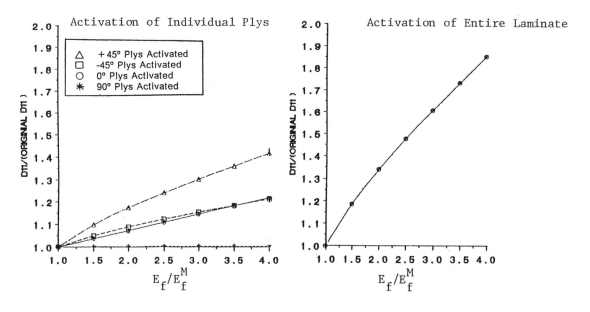

Figure 5.   Flexural Stiffness Tuning of Quasi-Isotropic Plate

Variation of Natural Frequencies and Mode Shapes: Naturally, changing the stiffness of a structure impacts on more than the maximum deflection but also modifies the modal response of the structure, hence the term 'Active Modal Modification'. One of the objectives of Active Modal Modification is to tune the structure based upon various performance criteria or external conditions such as periodic force or pressure inputs to the structure that may be near resonant frequencies or result in low transmission loss. Active or adaptive control of the stiffness of the structure will influence the nature of the modal response of the structure by changing the natural frequencies and the characteristic mode shapes. Utilizing classical composite technology in which structures are fabricated with taylored properties and various orientations of individual plys allow for tremendous flexibility in the structural design of these tunable structures for various applications.

Figure 6 illustrates the potential for changing the natural frequency of a square quasi-isotropic plate by activating one or all of the individual plys. Again, the greatest authority is achieved, for single lamina activation, by activating the +45° plys that are positioned on the top and bottom surfaces of the plate. However, by activating various permutations of lamina the control of the natural frequencies and stiffnesses can be accomplished in a more sophisticated manner perhaps allowing for dual-requirements associated with the orthotropy of the structure and the modal response in a coupled fashion.

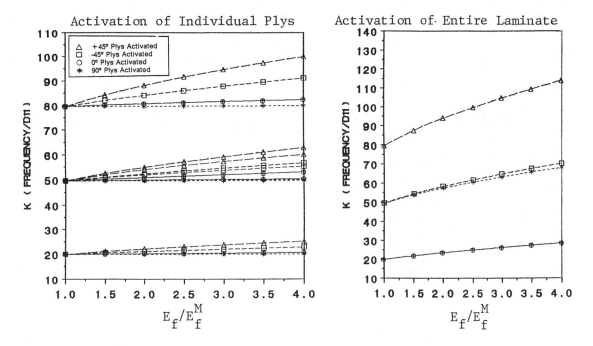

Figure 6.   Natural Frequency Tuning of Quasi-Isotropic Plate

The authority of SMA reinforced composites is quite dramatic for active modal modification. Activating the +45° plys result in approximately a 25 percent increase in the natural frequencies and activating all of the plys increase the natural frequencies by about 50 percent. The impact of tuning the stiffness of the fibers is also seen in the modification of the mode shapes which naturally occur because of the increased orthotropy introduced by changing the stiffness of a ply or plys of an initially quasi-isotropic structure. Again, the concept of tuning the mode shapes of a structure is another novel approach to composite design.

Modification of the mode shapes associated with the fourth natural frequency are shown in Figs. 7 and 8. Figure 7 shows the mode shape for the quasi-isotropic square plate without any activated fibers or plys. Note that quasi-isotropic plate does not have an anti-node line in the center of the plate as it is only quasi-isotropic in extension and is but only a close approximation to isotropic in bending. When the +45° plys are activated in the plate, the flexural stiffness increases, the natural frequency increases and the characteristic mode shapes also change as illustrated in Fig. 8. Comparing Figs. 7 and 8 show the dramatic change in the location of the nodes and anti-node lines which also indicate the possiblity of tuning the impedance and mobility of any point on the plate. Lastly, Fig. 9 show some of the variations in mode shapes and natural frequencies that can be accomplished by activating individual plys of the entire structure.

CONCLUSIONS

An introduction to a new class of composite materials has been given and results indicating some of the potential for new design concepts. The development and subsequent production of this class of materials could have tremendous impact on several diverse technological fields, i.e., material science, vibrations and controls, ocean and aerospace structures, biotechnology, and may act as a catalyst for the development of many new devices and technologies. The demonstrated ability of SMA reinforced composites to change stiffness, modal response and mechanical behavior presents numerous possibilities for further study and implementation.

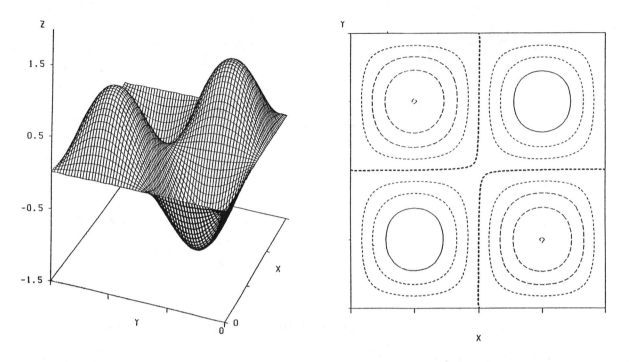

Figure 7.   Mode Shape of Unactivated [+45, − 45,0,90]$_s$ Plate

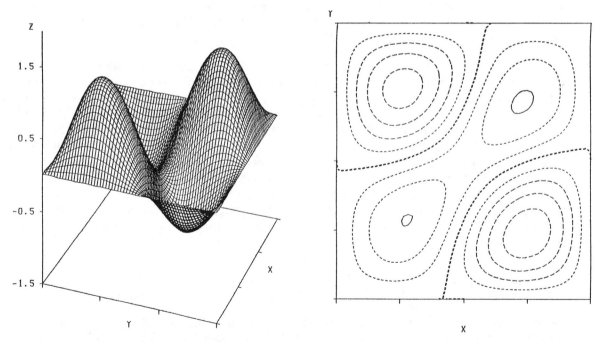

Figure 8.   Mode Shape of [+45, − 45,0,90]$_s$ Plate with Activated + 45° Plys

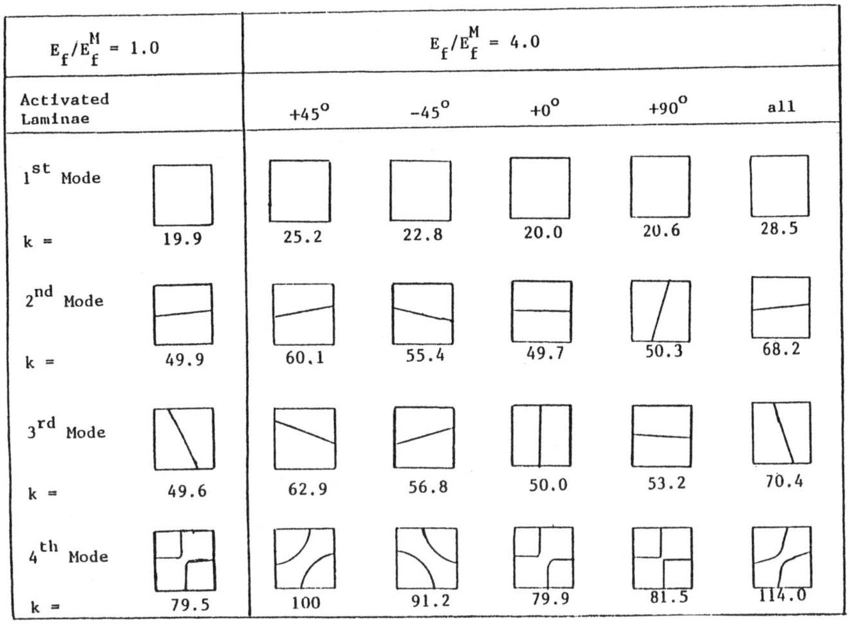

Figure 9. Comparison of Mode Shapes for [+45, − 45,0,90], SMA Reinforced Plate

ACKNOWLEDGEMENTS

The author respectfully acknowledges the support for this research effort by the Virginia Center for Innovative Technology, U. S. Nitinol, and the Office of Naval Research Young Investigator Program.

REFERENCES

1. Rogers, C. A., and H. H. Robertshaw, "Shape Memory Alloy Reinforced Composites," *Engineering Science Preprints 25*, ESP25.88027, Society of Engineering Sciences, Inc., June 20-22, 1988.

2. Wayman, C. M., and K. Shimizu, "The Shape Memory ('Marmem') Effect in Alloys," *Metal Science J.*, Vol. 6, 1972, p. 175.

3. Perkins, J., ed., *Shape Memory Effects in Alloys*, Plenum Press, New York, 1975.

4. Goldstein, D., "A Source Manual for Information on Nitinol and NiTi," Naval Surface Weapons Center, Silver Spring, Maryland, Report NSWC/WOL TR 78-26, 1978.

5. Schetky, L., "Shape Memory Alloys," *Scientific American*, Vol. 241, 1979, p.74.

6. Jackson, C. M., H. J. Wagner, and R. J. Wasilewski, "55-Nitinol - The Alloy with a Memory : Its Physical Metallurgy, Properties, and Applications," *NASA-SP-5110*, 1972, 91 p.

7. Buehler, W. J., and R. C. Wiley, "Nickel-Base Alloys," U. S. Patent 3,174,851, March 23, 1965.

8. Delaey, R. V., H. Tas Krishnan, and H. Warlimont, "Thermoelasticity, Pseudoelasticity and the Shape Memory Effects Associated with Martensitic Transformations," *Journal of Material Science*, Vol. 9, 1974, pp. 1521-1545.

9. Saburi, T., and C. M. Wayman, "Crystallographic Similarities in Shape Memory Martensites," *Acta Metallurgica*, Vol. 27, 1979, p. 979.

10. Cross, W. B., A. H. Kariotis, and F. J. Stimler, "Nitinol Characterization Study," *NASA CR-1433*, Sept. 1969.

Author Index